T0206094

Second Edition

Interfacial Forces in Aqueous Media

Carel J. van Oss

CRC Press
Taylor & Francis Group
Boca Raton London New York

CRC Press is an imprint of the
Taylor & Francis Group, an **informa** business

CRC Press
Taylor & Francis Group
6000 Broken Sound Parkway NW, Suite 300
Boca Raton, FL 33487-2742

First issued in paperback 2020

© 2006 by Taylor & Francis Group, LLC
CRC Press is an imprint of Taylor & Francis Group, an Informa business

No claim to original U.S. Government works

ISBN 13: 978-0-367-57786-5 (pbk)
ISBN 13: 978-1-57444-482-7 (hbk)

**Visit the Taylor & Francis Web site at
http://www.taylorandfrancis.com**

**and the CRC Press Web site at
http://www.crcpress.com**

Library of Congress Cataloging-in-Publication Data

van Oss, Carel J.
    Interfacial forces in aqueous media / Carel J. van Oss.-- 2nd ed.
        p. cm.
    Includes bibliographical references and index.
    ISBN-13: 978-1-57444-482-7 (acid-free paper)
    ISBN-10: 1-57444-482-4 (acid-free paper)
    1. Surface chemistry. 2. Solution (Chemistry) I. Title.

QD506.V36 2006
541'.33--dc22
                                                               2005035590

Preface

This is not only an updated edition of *Interfacial Forces in Aqueous Media* that includes new data and concepts which have emerged during the last dozen years, but it also contains a new Part II, treating the "Interfacial Properties and Structure of Liquid Water" (Chapters VIII–XI), plus a new, last chapter (XXV) entitled "Kinetics and Energetics of Protein Adsorption onto Metal Oxide Surfaces." Furthermore, as a consequence of an unusual development occurring shortly after the appearance in print of the first edition of this book, many changes had to be made in all those chapters with a heavy dependence on, and even in those with a more modest involvement in, the interfacial tension between condensed-phase compounds or materials (i) and water (w), i.e., γ_{iw}, or in the closely related interfacial free energy of interaction between molecules, particles or surfaces (i), immersed in water (w): ΔG_{iwi} (where $\Delta G_{iwi} = -2\gamma_{iw}$).

To explain these changes, a chapter (XIII) was added, devoted to "Interfacial Tension Determination." Apart from this (90% new) Chapter XIII, Chapters IV, XVII, XVIII and XIX have been significantly revised because of γ_{iw} or ΔG_{iwi} considerations, while Chapters III, IX–XII, XIV, XVI and XXI–XV also have undergone corrections and/or additions for the same reason. This is all a consequence of the fact that, soon after the first edition of this work appeared in 1994, Professor Robert J. Good and I proved that the methodology by which many of the γ_{iw} values had been determined in the first half of the 20th century, was fatally flawed (van Oss and Good, 1996). The result was that most existing data concerning interfacial tensions between polar organic liquids and water which had been determined by drop-shape or drop-weight methods had to be discarded as much too low. In a number of cases new, correct γ_{iw} values have since been determined, mainly from aqueous solubilities, and are listed in Chapter XIII.

With the exception of Chapters I and II of the 1994 edition of this book, all other chapter numbers have been changed in the new edition, or have been given numbers that were not used before (i.e., XXII to XV). Old Chapter XI has now become Chapter III. Old Chapters III to VI have shifted up, to become Chapters IV to VII. The next four chapters (VIII to XI) are the new ones comprising the new Part II, treating the "Interfacial Properties and Structure of Liquid Water." Former Chapters VII and VIII are now combined into one (Chapter XII). Chapter XIII is a new one, on "Interfacial Tension Determination." Old Chapters IX and X have become XIV and XV; Chapters XII to XV have become XVI to XIX; former Chapters XVI and XVII were combined into one, as Chapter XX. Former Chapters XVIII to XXI have become XXI to XXIV and Chapter XXV is new.

Author

Carel J. van Oss, Ph.D., is professor emeritus of microbiology and immunology, adjunct professor of chemical and biological engineering (since 1980) and of geology (since 1995), at the State University of New York at Buffalo. Dr. van Oss was associate professor of microbiology (1968–1972), professor of microbiology (1972–1998), and head, Immunochemistry Laboratory (1968–1998), Department of Microbiology, School of Medicine, State University of New York at Buffalo and a member of the Ernest Witebsky Center for Immunology (formerly The Center of Immunology) (1972–1998). From 1965–1968, he was director of the Serum, Plasma, and Immunochemistry departments, Milwaukee Blood Center; adjunct associate professor of biology, Marquette University, Milwaukee, Wisconsin; and a consultant on biological and biochemical applications of membranes to the Amicon Corporation, Lexington, Massachusetts, and to the General Electric Company, Vallecitos Nuclear Center, Pleasanton, California. From 1958–1965, he was assistant head of the Department of Microbiology and Immunology, Montefiore Hospital and Medical Center in New York (1963–1965); director of the Laboratory of Physical Biochemistry, National Veterinary College of Alfort (Paris, France) (1958–1963); and from 1962 maître de recherche at the French National Institute of Agronomical Research and a consultant on membrane separation and purification processes to Byla Laboratories, Paris, France.

From 1970–1990, Dr. van Oss was a member of the USRA-NASA Committee on Separation Methods in Space, and of the committee on Surface Phenomena under Microgravity Conditions (1970–1982); Immunology Panel, FDA, Medical Devices (1975–1983); the Advisory Committee to the American Red Cross on Reducing the Infectious Potential of Transfused Red Blood Cells (1991–1992); and a visiting professor at the University of Amsterdam/Central Laboratory of the Blood Transfusion Service of the Netherlands Red Cross (1979), the University of Bristol, Department of Physical Chemistry (1986), and the University of Paris XII/CNRS, Laboratory of Physical Chemistry of Biopolymers (1990).

Dr. van Oss has served as founding editor (1971–1998) and a member of the Editorial Advisory Board (since 1999) of *Preparative Biochemistry and Biotechnology,* a trimonthly journal; founding editor (1973–1976 and 1982 to the present) and a member of the Editorial Committee (since 1972) of *Immunological Investigations,* a bimonthly journal; and founding editor (1972–1988) and a member of the Editorial Advisory Board (since 1989) of *Separation and Purification Methods,* a semiannual journal. He is also a member of the Editorial Boards of *Colloids and Surfaces (B)* (1986–present) and the *Journal of Protein Chemistry* (1984–present). He has published over 300 journal articles and book chapters and has authored/co-authored or edited/co-edited 12 books.

Dr. van Oss was Chairman for 1984 of the Gordon Research Conference on *Separation and Purification;* fourteen times invited speaker at various Gordon Research Conferences 1964–1987. He also was the (Honorary) Consul of the

Netherlands in Western New York (1970–1990) and has been a member of the Board of Directors of the International Institute of Buffalo since 1971. He has been awarded the Netherlands' Commemorative Resistance Cross for his activities in the Dutch Resistance during the German occupation of the Netherlands in World War II, and has been awarded a Knighthood in the Order of Orange-Nassau by H.M. Queen Beatrix, in 1985.

He is a member of the American Association for the Advancement of Science and the American Chemical Society and its Division of Colloid and Surface Chemistry.

Acknowledgments

The second edition of this book has been made possible to an important degree by the continuing scientific collaboration with Professor Rossman F. Giese, Jr. (Department of Geology, State University of New York at Buffalo) and Dr. Wenju Wu (Johnson & Johnson Pharmaceutical R&D Division, Raritan, New Jersey). In addition, Chapter XXV ("Kinetics and Energetics of Protein Adsorption") is to a significant extent based on the doctoral thesis work of, and concomitant publications with, Dr. Aristides Docoslis (now Assistant Professor of Chemical Engineering and Canada Research Chair in Colloids and Nanoscale Engineering, Queen's University at Kingston, Ontario, Canada).

I am most obliged to Dr. John Hay (Grant T. Fisher Professor and Chairman of the Department of Microbiology and Immunology, State University of New York at Buffalo), who made it possible to work on this edition, while surrounded with all the indispensable facilities and help. Among the latter, I am especially grateful for the frequent secretarial, organizational, and electronic help of Vicki Weigle and Hillary A. Hurwitz. I also thank Scott Suckling of MetroVoice Publishing Services for his excellent, expert help in the production stage of this book.

I am also much indebted to my wife, Rosine, to my son, James, and to the graduate students at the Office of Medical Computing (School of Medicine and Biomedical Sciences, State University of New York at Buffalo) for frequent assistance and for teaching me how to cope with the word processing involved in the preparation of the manuscript for this edition.

Contents

PART II Interfacial Properties and Structure of Liquid Water

PART III Experimental Measurement Methods

PART IV   Associated Phenomena and Applications

Introduction

NON-COVALENT INTERACTIONS

In this work we treat non-covalent interactions between biological and non-biological macromolecules, surfaces and particles. For the sake of simplicity of nomenclature we include hydrogen-bonding and other electron-donor-electron-acceptor interactions among the non-covalent interactions, even though some authors (see, e.g., Drago *et al.*, 1971) classify these interactions as at least partly covalent. The non-covalent interactions under consideration are:

1. Electrodynamic, or Lifshitz–van der Waals (LW), interactions,
2. Polar, electron-donor–electron-acceptor, or Lewis acid-base (AB) interactions,
3. Electrostatic (EL) interactions, and
4. Brownian movement (BR).

VAN DER WAALS FORCES

The existence of a general attractive interaction between neutral atoms was first postulated by van der Waals in 1873, to account for certain anomalous phenomena occurring in non-ideal gases and liquids. These "van der Waals" forces were subsequently shown to comprise three different, but closely related phenomena: 1. randomly orienting dipole-dipole (or orientation) interactions, described by Keesom (1915, 1920, 1921); 2. randomly orienting dipole–induced dipole (or induction) interactions, described by Debye (1920, 1921); 3. fluctuating dipole–induced dipole (or dispersion) interactions, described by London (1930). Of these three, Keesom and Debye interactions are only found among molecules which have permanent dipole moments. The London interaction, however, is universal and is present in atom-atom interactions as well. All three interaction energies between atoms or small molecules decay very steeply with distance (ℓ) as ℓ^{-6}. Of these three only van der Waals–London (dispersion) interactions are of preponderant importance between macroscopic bodies, *in condensed systems* (Overbeek, 1952; Fowkes 1983; Chaudhury, 1984); especially

1

in aqueous media which contain electrolytes, the already small Keesom interaction is virtually completely screened out (Israelachvili, 1991, p. 256). Hamaker (1937a) developed a theory of van der Waals–London interactions between macroscopic bodies and showed that the (presumptive) additivity of these interactions renders them considerably more long-range, i.e., the dispersion energy between two semi-infinite parallel flat slabs decays with distance as ℓ^{-2}. However this proved to be accurate only for relatively short distances ($\ell < 100$ Å); due to retardation it decays as ℓ^{-3} at greater distances (Casimir and Polder, 1948; Overbeek, 1952).

The first theories of the stability of hydrophobic colloids were based on these relatively long-range attractive forces, as a balance between the van der Waals attraction and an electrical double layer repulsion, by Hamaker (1936, 1937b and c) and Derjaguin and Landau (1941) and, independently, by Verwey and Overbeek (1946, 1948). This general theory of colloidal stability, based on these considerations, has become known as the DLVO theory, by combining the initials of the latter four authors. In these older treatments, however, it was not always thoroughly realized how much the van der Waals–London attraction between two macromolecules, particles or surfaces becomes attenuated when the interaction takes place in a liquid, even though Hamaker (1937a) had already developed the equations describing that general situation. Indeed, Hamaker indicated (1937a) that it was actually possible for the van der Waals–London interaction between two different materials, immersed in a liquid, to be repulsive. This was reaffirmed by Derjaguin (1954), and Visser (1972, 1976) subsequently established the precise conditions necessary for the occurrence of repulsive van der Waals–London forces. Fowkes (1967) first indicted a few (theoretical) examples of such repulsions, and van Oss, Omenyi and Neumann (1979) demonstrated the actual existence of many repulsive systems, leading to phase separation and to particle exclusion phenomena (Neumann, Omenyi and van Oss, 1979; van Oss, Chaudhury and Good, 1989); see also Chapter XXI.

Thus, the van der Waals–London interaction energy between macromolecules, particles or surfaces, immersed in liquids, although fairly long-range in nature, usually is quantitatively relatively slight (see also Napper, 1983). It has been shown, using the Lifshitz approach (1955), that in condensed, macroscopic systems, the two other electrodynamic interactions, i.e., the van der Waals–Keesom and van der Waals–Debye interactions, can (and should) be treated in the same manner as van der Waals–London interactions (Chaudhury, 1984). When thus grouped together, these electrodynamic interactions will be alluded to as Lifshitz–van der Waals (LW) interactions. See Chapters II and III.

POLAR, OR LEWIS ACID-BASE (AB) INTERACTIONS

It has long been surmised that physical forces other than van der Waals attractions and electrostatic repulsions could play an important role in colloidal interactions. An important force among these is often alluded to as "hydrophobic interaction" (Franks, 1975), an effect that has resisted quantitative experimental determination as well as precise theoretical definition until relatively recently (van Oss *et al.*, Advan. Coll. Interf. Sci., 1987; Separ. Sci. Tech., 1987). Closely linked to "hydrophobic interaction," which is generally so called when attractive, is "hydration pressure" (LeNeveu *et al.*, 1977), which is its repulsive counterpart (van Oss, Chaudhury and Good, Separ. Sci. Tech., 1987; Chem. Rev., 1988; see also Israelachvili, 1985); see Chapters XVIII and XIX.

A certain looseness of definition as to what is polar and what is not polar has long pervaded the terminology of colloid and interface science, but whilst a case certainly can be made for the exclusive use of the adjective "polar" for dipole-dipole and dipole–induced dipole interactions (Fowkes, 1983), there remains a sufficiently widespread usage of the word "polar" for, e.g., the hydrogen-bonding type of interfacial interaction (see, e.g., Busscher *et al.*, 1986, and Dalal, 1987) to warrant the continued use of the adjective in this sense. Even though electron-donor–electron-acceptor interactions frequently are not "dipole" interactions *sensu stricto* (Fowkes, 1983; van Oss *et al.*, Advan. Coll. Interf. Sci., 1987), their strong electrostatic component (Kollman, 1977) and especially their pronounced asymmetry (van Oss *et al.*, Advan. Coll. Interf. Sci., 1987) are still most aptly described as "polar"; we shall designate these interactions as AB, for (Lewis) acid-base interactions. To avoid confusion however, we shall neither allude to electrodynamic (dipole-dipole, dipole-induced dipole, or fluctuating dipole–induced dipole interactions), nor simple electrostatic interactions (based on electrokinetic potentials), as polar.

AB interactions, which are of polar, but not of electrodynamic or electrostatic origin, in the attractive ("hydrophobic interaction") as well as in the repulsive mode ("hydration pressure") represent energies that may be up to two decimal orders of magnitude higher than the LW and electrostatic (EL) interactions which are commonly described as the components of the traditional DLVO energy balances (van Oss, Cell Biophys., 1989; in Glaser and Gingell, 1990; van Oss, Giese and Costanzo, 1990). These polar interactions, which are based on electron-acceptor–electron-donor (Lewis acid-base) interactions between polar moieties, in polar media (such as water) (van Oss *et al.*, Advan. Coll. Interf. Sci., 1987; van Oss *et al.*, Chem. Rev., 1988) are at the origin of virtually all the anomalies that have beset the interpretation of interfacial interactions in polar media for many years (Good and Girifalco, 1960; Girifalco and Good, 1957; Fowkes, 1962). See Chapter IV.

In aqueous media the exponential rate of decay of AB interactions is rather steep: in the repulsive mode (manifested as hydration pressure), as well as in the attractive mode (hydrophobic attraction), the interactions typically decay with a decay length of ≈ 10 Å. See Chapter VII.

ELECTROSTATIC (EL) INTERACTIONS

In polar, and especially in aqueous media, very few biological macromolecules or particles (or indeed any inorganic or organic surfaces) are completely devoid of electrical surface charge. When immersed in a polar liquid such as water, polymers or particles with the same sign of charge will repel each other. In most such systems the resulting energy of intermolecular or interparticulate repulsion may well be too strong, in comparison with the LW and AB interaction energies, to be neglected in determining the total energy balance. Thus, in all cases where the outcome of the total energy balance of a polar system (or even of an apolar system) is necessary for the prediction of particle stability, polymer solubility, polymer compatibility, particle or cell adhesion, or polymer adsorption, the electrostatic interaction energy (EL) component must also be measured.

The rate of decay of EL interactions is relatively steep; like that of AB interactions, the decay is exponential, but unlike AB interactions, it is strongly dependent on the ionic strength of the liquid medium. Under physiological conditions (i.e., in 0.15 M NaCl), EL interactions are noticeable to a distance of $\approx$80 to 100 Å. See Chapter V.

OTHER NON-COVALENT FORCES

There are a variety of secondary non-covalent forces, all closely linked to one or more of the primary non-covalent forces already mentioned above. See Chapter XX.

BROWNIAN MOVEMENT (BR) FORCES

Among particles suspended, or macromolecules, and other solutes dissolved in liquid media, Brownian movement always occurs at temperatures higher than 0°K. All particles (or molecules), big or small, are endowed with a Brownian (BR) free energy of +1½ kT, when they have three degrees of freedom. Thus especially for the smaller colloidal-sized particles and for macromolecules and small solutes, BR forces may have to be taken into account. See Chapter VI.

GENERAL BIBLIOGRAPHY

When this book is used in connection with graduate or post-graduate studies, the reader is advised also to consult a number of other works on various aspects of colloid and interface science and of physical biochemistry:

Kruyt, *Colloid Science* (1952); an older but still a useful text; the first volume was largely written by Overbeek; the second volume (1949) deals with Polymer Science, Coacervation and Complex Coacervation, Gels and Micelles.
Adamson, *Physical Chemistry of Surfaces* (1990), 5th edition; the major textbook on Surface Science.[*]
Hiemenz and Rajagopalan, *Principles of Colloid and Surface Science* (1997); a useful textbook on Colloid Science.
Israelachvili, *Intermolecualr and Surface Forces* (1985, 1991); a masterful modern work, thoroughly treating the theoretical background as well as the newer experimental findings, by the author who played the major role in developing the force balance in liquid media. This work, however, does not treat the use of the force balance as such; for that purpose the reader is referred, e.g., to a dissertation, by:
Claesson, *Forces between Surfaces Immersed in Aqueous Solutions* (1986); a dissertation discussing the use of the force balance, and some of the results obtained with it.

---

[*] There is a 6th edition (1997), by Adamson and Gast, but that edition lacks some of the Tables that are essential for calculating surface tensions of liquids, using the drop-shape method.

Mahanty and Ninham, *Dispersion Forces* (1976); a theoretical treatment of London-van der Waals intractions.

Hunter, *Zeta Potential in Colloid Science* (1981) and Kitahara and Watanabe, *Electrical Phenomena at Interfaces* (1984) deal with different theoretical aspects of electrokinetics and electrostatic interactions.

Hunter, *Foundation of Colloid Science*, 2 Vols. (1987, 1989); a textbook.

Righetti, van Oss and Vanderhoff, *Electrokinetic Separation Methods* (1979); treat some electrokinetic theory, but they mainly deal with practical methods for measuring and utilizing electrokinetic potentials.

Cantor and Schimmel, *Biophysical Chemistry*, 3 Vols. (1980); a textbook.

Derjaguin, Churaev and Muller, *Surface Forces* (1987); Derjaguin, *Theory of Stability of Colloids and Thin Films* (1989); up-to-date overviews of the contributions of the Russian school of Colloid Science, from the 1930s to the present.

Van de Ven, *Colloidal Hydrodynamics* (1989); one of the rare texts combining Rheology (especially in aqueous media) with Colloid Science.

Morra (Ed.), *Water in Biomaterials Surface Science* (2001); Lewis acid-base forces are treated in a number of the chapters.

Giese and van Oss, *Colloid and Surface Properties of Clays and Related Minerals* (2002); Contains the newest data on the surface thermodynamic properties of clays and other mineral particles, made possible by the use of thin layer wicking, with many applications of the new results.

Apart from a few indications by Israelachvili (1985, 1991) and some papers by Parsegian *et al.* (1985, 1987), little mention will be found in the above list of publications of the influence of polar (AB) forces, with the exception of Morra (2001) and Giese and van Oss (2002). Early treatments of AB forces will be found in a number of publications by Fowkes (e.g., 1983, 1987). More up-to-date treatments of various aspects of AB forces can be found in various papers by the present author and his colleagues (see References) and in this work.

Part I

Theory

II Lifshitz–van der Waals (LW) Interactions

VAN DER WAALS FORCES

J.D. van der Waals observed that deviations of the ideal gas law:

$$P\overline{V} = RT \tag*{[II-1]*}$$

(where P is the pressure, $\overline{V}$ the atomic or molar volume, R the gas constant and T the absolute temperature) occurred, apparently caused by an interatomic or intermolecular attraction between neutral gas atoms or molecules (van der Waals, 1873, 1899). These non-covalent, non-electrostatic inter-molecualr interactions were subsequently named "van der Waals forces." Their nature became clear during the early part of the 19th century. Keesom (1915, 1920, 1921) described the interaction between permanent dipoles (orientation forces). Debye (1920, 1921) described the interaction between a permanent dipole and a dipole induced by it (induction forces). Finally, London (1930) showed that even in neutral atoms rapidly fluctuating dipoles arise, which in turn induce dipole moments in other atoms, and thus attract them (dispersion forces). Keesom energies are proportional to the fourth power of the dipole moment (μ); Debye energies are proportional to the polarizability (α) and to the square of the dipole moment; London energies are proportional to the characteristic energy (hν) corresponding to the main dispersion frequency (ν), and to the square of the polarizability. All three van der Waals interaction energies are inversely proportional to the sixth power of the interatomic distance (ℓ) (Overbeek, 1952):

$$V_{Keesom} = -\mu^4/kT\ell^6 \tag{[II-2]}$$

$$V_{Debye} = -\alpha\mu^2/\ell^6 \tag{[II-3]}$$

$$V_{London} = -3/4\ \alpha^2 h\nu/\ell^6 \tag{[II-4]}$$

where k is Boltzmann's constant and T the absolute temperature.

---

* A list of symbols can be found on pages 399–406.

The total constant A_{ii}, governing the interaction between two bodies of material i, at short distances, *in vacuo*, in van der Waals–London (dispersion) interactions (which is alluded to as the Hamaker constant), is expressed as:

$$A_{ii} = \pi^2 q_i^2 \, \beta_{ii} \qquad [\text{II-5}]$$

where q_i is the number of atoms per unit volume and β the constant in London's equation for the interaction between two atoms i, and $\beta = 3/4 \, \alpha^2 h\nu$ (see eq. [II-4]):

$$V_{ii} = -\beta_{ii}/\ell^6 \qquad [\text{II-6}]$$

where ℓ is the distance between the atoms i (see eq. [II-4]). Then, as we may assume that, for two materials i and j:

$$\beta_{ij} = \sqrt{\beta_{ii}\beta_{jj}} \qquad [\text{II-7}]$$

it follows that:

$$A_{ij} = \sqrt{A_{ii}A_{jj}} \qquad [\text{II-8}]$$

MACROSCOPIC APPROXIMATION

Hamaker (1937a) first calculated the dispersion (van der Waals–London) interaction energy for larger bodies by a pair-wise summation of the properties of the individual molecules (assuming these properties to be additive, and non-retarded). Using this ("macroscopic") approximation, the total attractive dispersion energy for two semi-infinite flat parallel bodies (of material i), separated by a distance ℓ, in air or *in vacuo*, becomes (for ℓ greater than a few atomic diameters):

$$V_{\text{London}} = -A/12\pi\ell^2 \qquad [\text{II-9}]$$

For two materials 1 and 2, singly or together embedded or immersed in medium 3, the combining rules are respectively described by:

$$A_{131} = A_{11} + A_{33} - 2A_{13} \qquad [\text{II-10}]$$

and:

$$A_{132} = A_{12} + A_{33} - A_{13} - A_{23} \qquad [\text{II-11}]$$

Given the applicability of eq. [II-8], eqs. [II-10] and [II-11] can also be expressed as (Visser, 1972):

$$A_{131} = (\sqrt{A_{11}} - \sqrt{A_{33}})^2 \qquad [\text{II-10A}]$$

and:

$$A_{132} = (\sqrt{A_{11}} - \sqrt{A_{33}})(\sqrt{A_{22}} - \sqrt{A_{33}}) \qquad \text{[II-11A]}$$

It is clear that A_{131} (eq. [II-10]) always is positive (or zero), however, equally clearly, A_{132} can assume negative values, i.e., when:

$$A_{11} > A_{33} > A_{22} \qquad \text{[II-12A]}$$

and when:

$$A_{11} < A_{33} < A_{22} \qquad \text{[II-12B]}$$

(see Visser, 1972).

These conditions (eqs. [II-12A] and [II-12B]) result in *repulsive* Lifshitz–van der Waals forces (van Oss, Omenyi and Neumann, 1979; Neumann, Omenyi and van Oss, 1979). It should be clear that these conditions are by no means rare or exceptional. Hamaker already indicated the possibility of such repulsive (dispersion) forces (1937a), which possibility was reiterated by Derjaguin in 1954. The precise conditions under which a repulsion could be expected to occur were first given by Visser (1972, 1976). All of these considerations initially only applied to van der Waals–London interactions, utilizing Hamaker constant combining rules. However, as shown in Chapter III, the surface thermodynamic approach for obtaining eqs. [II-10] and [II-11] is essentially the same as Hamaker's approach (1937a), as long as eq. [II-9] is valid. This validity, however, remains strictly limited to Lifshitz–van der Waals (LW) interactions, see below.

It should be emphasized that there is nothing self-contradictory in repulsive van der Waals forces. The LW interaction between two molecules or particles (identical or different) *in vacuo* is always attractive. And the LW interaction between two *identical* molecules or particles immersed in a liquid is always attractive, although it can become zero, when $A_{11} = A_{33}$ (see eq. [II-10A]). But when two *different* materials 1 and 2 interact, immersed in liquid 3, and when $A_1 \neq A_2$ and the conditions in eqs. [II-12A] or [II-12B] prevail, a net repulsion occurs; see Chapter XXI. To quote Israelachvili (1982): "Repulsion is not confined to van der Waals attractive fields but has a close analogy, e.g., to Archimedes' principle as applied to gravitational fields. For example, even though a gravitational attraction occurs between the earth and both cork and iron in air, the net interaction is very different in water. Cork floats on water (i.e., there is a net repulsion) whereas iron sinks (i.e., in there is a net attraction)." In terms of the oscillator model, this amounts to a condition such as given in eqs. [II-12A] and [II-12B], where a situation prevails in which one material with a positive excess polarizability interacts with another, in a net negative polarizability state (see below). Consequently, when an instantaneous dipole of one material induces a dipole in another, the dipoles orient in the same spatial direction, which then leads to a repulsion. Curiously, it is possible to encounter repulsive LW forces at short distances (e.g., up to 150 Å), which, however, become attractive at greater distances, due to retardation (Chaudhury and Good, 1983); see Chapter VII. Thus, between a layer of glycerol and a surface of Teflon FEP, a stable layer of n-hexane can remain, on which the glycerol can permanently "levitate" above the Teflon.

Contrary to earlier hypotheses (Overbeek, 1952; Schenkel and Kitchener, 1960) the right-hand side of eqs. [II-10] and [II-11] do *not* require modification by a multiplication factor (van Oss and Good, 1984). For various approximations for the factor β (eq. [II-6]), see Visser (1972). By assuming additivity, by Hamaker's (1937a) procedure, using a process of summation it can be shown that for two identical spheres of radius R, which are a distance ℓ apart, as a first approximation, one may express the (dispersion) attractive energy as:

$$\Delta G = -AR/12\ell \qquad\qquad [II\text{-}13]$$

whilst the force between the same two spheres is:

$$F = AR/6\ell^2 \qquad\qquad [II\text{-}14]$$

For two parallel flat plates of thickness t:

$$\Delta G = -(A/12\pi)\,[1/\ell^2 - 2/(\ell+t)^2 + 1/(\ell+2t)^2] \qquad\qquad [II\text{-}15]$$

(see Nir, 1976). For two semi-infinite flat plates (see eq. [II-5]), this becomes:

$$\Delta G = -A/12\pi\ell^2 \qquad\qquad [II\text{-}16]$$

and:

$$F = A/6\pi\ell^3 \qquad\qquad [II\text{-}17]$$

(see Chapter VII). For more complex configurations, see Nir (1976) and for the connection between the Hamaker constant (A) of the given compound and the LW component of the surface tension (γ^{LW}), see Chapter III.

THE LIFSHITZ APPROACH

In contrast with Hamaker's approach (1937a), which started with single interatomic interactions, to arrive at the total interaction energy for larger bodies by a process of summation (the microscopic approach, Visser, 1972), the macroscopic approach followed by Lifshitz (1955) treats the bulk material *ab initio*, as well as the shielding effect.

Lifshitz's theory of condensed media interaction (1955) has its origin in Maxwell's equations, where the electric and magnetic fields are subjected to fast temporal fluctuations. In order to accommodate the temporal fluctuations of the fields, Lifshitz utilized the fluctuation theory developed by Rytov (1953). The integration of the Fourier transform of the normal component of Maxwell's stress tensor over all the allowed frequencies resulted in an expression for the van der Waals pressure of two semi-infinite parallel slabs separated by a finite distance in vacuum or in another dielectric phase. The van der Waals pressure according to Lifshitz's theory can be expressed in terms of the continuum properties, i.e., dielectric susceptibilities of the interacting phases under consideration. Later, Dzyaloshinskii *et al.* (1961) gave an alternate derivation of Lifshitz's formula by adopting a more sophisticated approach of quantum electrodynamics. Recently, Parsegian (1973, 1975) has offered a heuristic derivation of Lifshitz's formula for the free energy of interaction of two phases through a third dielectric by summing over the individual oscillator free energies over an allowed set of frequencies, where the latter was derived from the solution of Fourier transformed Maxwell's equations. Historical reviews are given by Mahanty and Ninham (1976) and by Nir (1976).

According to Lifshitz's theory (1955) (see also Dzyaloshinskii *et al.*, 1961), the approximate expression for the free energy of interaction, $G_{132}(\ell)$, between two semi-infinite parallel slabs of phases 1 and 2 and separated by a film of phase 3 of thickness, ℓ, is:

$$\Delta G_{132}(\ell) = \frac{kT}{\pi c^3} \int_{\ell}^{\infty} \sum_{p=1}^{\infty}{}' \sum_{n=0}^{\infty} \varepsilon_3^{3/2} \omega^2 n \int_{\ell}^{\infty} P^2 \left[\frac{\exp(2P\omega_n \ell \varepsilon_3^{1/2} e)}{\Delta_1 \Delta_2} \right]^{-1} dP d\ell \qquad [11\text{-}18]$$

where k is Boltzmann's constant, T the absolute temperature, c the velocity of light *in vacuo*, and P an integration parameter. Δ_1, Δ_2 and ω_n are given as follows:

$$\Delta_1 = \frac{\varepsilon_1(i\omega_n) - \varepsilon_3(i\omega_n)}{\varepsilon_1(i\omega_n) + \varepsilon_3(i\omega_n)} \qquad [\text{II-19}]$$

$$\Delta_2 = \frac{\varepsilon_2(i\omega_n) - \omega_3(i\omega_n)}{\varepsilon_2(i\omega_n) + \varepsilon_3(i\omega_n)} \qquad [\text{II-20}]$$

$$\omega_n = \frac{4\pi^2 nkT}{h} \qquad [\text{II-21}]$$

where h is Plank's constant, ω_n the frequency, n the quantum number of the relevant oscillation and $\varepsilon(i\omega_n)$ the dielectric susceptibility along the complex frequency axis $i\omega_n$ (where $i = \sqrt{-1}$). The prime in the summation of eq. [II-18] indicates that the zero frequency term is to be divided by two.

On performing the integration in eq. [II-18], it is found that:

$$\Delta G_{132}(\ell) = -\frac{kT}{8\pi\ell^2} \sum_{n=0}^{\infty}{}' \sum_{j=1}^{\infty} (\Delta_1 \Delta_2)^j \left(\frac{X_o}{j^2} + \frac{1}{j^3} \right) \exp(-jX_o) \qquad [\text{II-22}]$$

where:

$$X_o = (2\omega_n \ell \varepsilon_3^{1/2})/c \qquad [\text{II-23}]$$

and j = 1, 2, 3. . . . . when the separation distance, ℓ, is very small, i.e., when $\ell \to 0$, X_o also tends to zero. One then obtains an expression for the non-retarded van der Waals energy as:

$$\Delta G_{132}(\ell) = -\frac{kT}{8\pi\ell^2} \sum_{n=0}^{\infty'} \sum_{j=1}^{\infty} \frac{(\Delta_1\Delta_2)^j}{j^3} \qquad [\text{II-24}]$$

If the materials interacting with each other are of similar kinds and the phase 3 is vacuum, eq. [II-24] in view of eqs. [II-19] and [II-20] reduces to:

$$\Delta G_{11}(\ell) = \frac{kT}{8\pi\ell^2} \sum_{n=0}^{\infty'} \sum_{j=1}^{\infty} \left[\frac{\varepsilon_1(i\omega_n)-1}{\varepsilon_1(i\omega_n)+1} \right]^{2j} \cdot j^{-3} \qquad [\text{II-25}]$$

The free energy of interaction, when expressed in conventional form, involving Hamaker's constant A_{ii} (see above) appears as:

$$\Delta G_{ii}(\ell) = -\frac{A_{ii}}{12\pi\ell^2} \qquad [\text{II-16}]$$

where A_{ii} is the familiar Hamaker constant of interaction of material i *in vacuo*. Combining eqs. [II-16] and [II-25], it is found that:

$$A_{ii} = \frac{3kT}{2} \sum_{n=0}^{\infty'} \sum_{j=1}^{\infty} \left[\frac{\varepsilon_i(i\omega_n)-1}{\varepsilon_i(i\omega_n)+1} \right]^{2j} \cdot j^{-3} \qquad [\text{II-26}]$$

In the low temperature limit, the sum over n in eq. [II-26] can be replaced by an integral eq. [II-25]. Equation [II-26] can then be expressed as:

$$A_{ii}(\ell) = \frac{3h}{4\pi} \int_0^{\infty} d\omega \sum_{j=1}^{\infty} \left[\frac{\varepsilon_1(i\omega_n)-1}{\varepsilon_1(i\omega_n)+1} \right]^{2j} \cdot j^{-3} \qquad [\text{II-26A}]$$

If the dielectric permeability of a material is known along the imaginary frequency axis, $i\omega_n$, the Hamaker constant of the material and hence the free energy of interaction at small separation distances can be calculated from eqs. [II-16] and [II-26], see Chapter III and Table III-3.

The dielectric permeability of the material along the imaginary frequency axis $i\omega_n$ can be expressed in terms of the Kramers-Kronig relation (see Lifshitz, 1955) as:

$$\varepsilon(i\omega_n) = 1 + \frac{2}{\pi} \int_0^{\infty} \frac{\omega\varepsilon''(\omega)\,d\omega}{(\omega^2 + \omega_n^2)} \qquad [\text{II-27}]$$

where $\varepsilon''(\omega)$ is the loss component of the frequency dependent dielectric function $\varepsilon(\omega) = [\varepsilon'(\omega) - i\varepsilon''(\omega)]$. For dielectrics Ninham and Parsegian (1970) have considered the major contribution to $\varepsilon(i\omega_n)$ arising from microwave, infrared and ultraviolet relaxations and obtained a simplified expression for $\varepsilon(\omega_n)$ as:

$$\varepsilon(i\omega_n) = 1 + \frac{\varepsilon_\infty - \varepsilon_0}{1 + \dfrac{\omega_n}{\omega_{MW}}} + \frac{\varepsilon_0 - n_0^2}{1 + \left(\dfrac{\omega_n}{\omega_{IR}}\right)^2} + \frac{n_0^2 - 1}{1 + \left(\dfrac{\omega_n}{\omega_{UV}}\right)^2} \qquad \text{[II-28]}$$

where ε_∞ is the static dielectric constant, ε_0 the dielectric constant when microwave relaxation ends and the infrared relaxation begins, n_0 the refractive index in the visible range, and ω_{MW}, ω_{IR} and ω_{UV} are the characteristic microwave, infrared and ultraviolet absorption frequencies.

For general situations, the microwave component to $\varepsilon(i\omega_n)$ is better expressed as:

$$\frac{\varepsilon_\infty - \varepsilon_0}{1 + \left(\dfrac{\omega_n}{\omega_{MW}}\right)^{(1-\alpha)}} \qquad \text{[II-29]}$$

where α is a Cole-Cole parameter (Chaudhury, 1984).

Israelachvili (1974) assumed that the major part of the dispersion interaction originates from electronic excitation in the ultraviolet frequency range. In this case, eq. [II-28] simplifies to:

$$\varepsilon(i\omega_n) = 1 + \frac{n_0^2 - 1}{1 + \left(\dfrac{\omega_n}{\omega_{UV}}\right)^2} \qquad \text{[II-30]}$$

With the above simplification, and considering only the first terms of the j-summation, Israelachvili obtained by integrating eq. [II-26A] the following expression:

$$A_{ii} = \frac{3}{16\sqrt{2}} \frac{(n_0^2 - 1)^2}{(n_0^2 + 1)^{1.5}} n\omega_{UV} \qquad \text{[11-31]}$$

Assuming a characteristic value of ω (2.63×10^{16} rad/sec), Israelachvili (1974) calculated the Hamaker constants of different liquids from the data of their refractive indices. He then calculated the surface tensions of these liquids by combining eqs. [II-10] and [II-18] to obtain:

$$\gamma_i = \frac{A_{ii}}{24\pi\ell_0^2} \qquad \text{[II-32]}$$

where γ_i stands for the (apolar component of) the surface tension of substance i, according to:

$$\gamma_i \equiv -\frac{1}{2}\Delta G_{ii} \qquad \text{[II-33]}$$

where ΔG_{ii} is the free energy of cohesion of species i *in vacuo*, and ℓ_0 is the separation distance between two semi-infinite slabs when they are in van der Waals contact. Israelachvili (1974) ignored the Born repulsion (see van Oss and Good, 1984) in his treatment. This approximation is discussed below (see also Chapter III).

INTERFACIAL LIFSHITZ–VAN DER WAALS
INTERACTIONS

Surface tension (or surface free energy per unit area) of a liquid or solid is defined as minus one-half of the free energy change due to cohesion (see eq. [II-33]) of the material *in vacuo* (Good, 1967a), where ΔG_{ii} is the free energy of cohesion i *in vacuo*.

Since the free energy of cohesion is contributed by a number of more or less independent forces, Fowkes (1963) proposed that the surface tension can also be broken down into its separate components, i.e.:

$$\gamma_i = \sum_j \gamma_i^j \qquad [II\text{-}34]$$

where j stands for dispersion, dipolar, induction, H-bonding and metallic interactions and γ_i^j stands for the component of the surface tension arising from the j'th type of interaction.

For all exclusively Lifshitz–van der Waals interactions (which are referred to by the superscript symbol LW), i.e., interactions between two completely apolar compounds 1 and 2, the Good-Girifalco–Fowkes combining rule (Good and Girifalco, 1960; Fowkes, 1963) is applicable:

$$\gamma_{12}^{LW} = (\sqrt{\gamma_1^{LW}} - \sqrt{\gamma_2^{LW}})^2 \qquad [II\text{-}35]$$

which may also be written as:

$$\gamma_{12}^{LW} = \gamma_1^{LW} + \gamma_2^{LW} - 2\sqrt{\gamma_1^{LW}\gamma_2^{LW}} \qquad [II\text{-}35A]$$

The interaction energy between materials 1 and 2 *in vacuo* is, according to the Dupré equation:[*]

$$\Delta G_{12}^{LW} = \gamma_{12}^{LW} - \gamma_1^{LW} - \gamma_2^{LW} \qquad [II\text{-}36]$$

and the interaction energy between molecules or particles of material 1, immersed in a liquid 2, is:

$$\Delta G_{121}^{LW} = -2\gamma_{12}^{LW} \qquad [II\text{-}33A]$$

while the cohesive energy of material 1 is:

$$\Delta G_{11}^{LW} = -2\gamma_1^{LW} \qquad [II\text{-}33B]$$

Finally, the interaction energy between materials 1 and 2, immersed in a liquid 3, is:

---

[*] The Dupré equation (eq. [II-36]) is not precisely the equation that was used by Dupré (1869); Dupré gave this equation in a slightly different form, containing an (erroneous) factor 2 before the γ_{12} term. However, following universal usage, we shall continue to allude to the correct form given in eq. [II-36] as the Dupré equation.

$$\Delta G_{132}^{LW} = \gamma_{12}^{LW} - \gamma_{13}^{LW} - \gamma_{23}^{LW} \qquad [\text{II-37}]$$

ΔG_{11}^{LW}, ΔG_{12}^{LW}, ΔG_{121}^{LW} and ΔG_{132}^{LW}, thus defined, are linked to the respective Hamaker constants as:

$$\Delta G^{LW} = -A/12\pi\ell_o^2 \qquad [\text{II-16A}]$$

where, in each case, A takes the same subscripts as ΔG^{LW}.

Inserting the LW interfacial tensions γ_{ij}^{LW} (eq. [II-35A]) in the right-hand term of eq. [II-37]:

$$
\begin{aligned}
\Delta G_{132}^{LW} &= \gamma_1^{LW} + \gamma_2^{LW} - 2\sqrt{\gamma_1^{LW}\gamma_2^{LW}} - \gamma_1^{LW} - \gamma_3^{LW} \\
&\quad + 2\sqrt{\gamma_1^{LW}\gamma_3^{LW}} - \gamma_2^{LW} - \gamma_3^{LW} + 2\sqrt{\gamma_2^{LW}\gamma_3^{LW}} \\
&= -2\gamma_3^{LW} - 2\sqrt{\gamma_1^{LW}\gamma_2^{LW}} + 2\sqrt{\gamma_1^{LW}\gamma_3^{LW}} \\
&\quad + 2\sqrt{\gamma_2^{LW}\gamma_3^{LW}}
\end{aligned}
\qquad [\text{II-38}]
$$

As:

$$\Delta G_{12}^{LW} = -2\sqrt{\gamma_1^{LW}\gamma_2^{LW}} \qquad [\text{II-39}]$$

(from eqs. [II-35A] and [II-36]) we can rewrite eq. [II-38] as:

$$-\Delta G_{132}^{LW} = \Delta G_{33}^{LW} + \Delta G_{12}^{LW} - \Delta G_{13}^{LW} - \Delta G_{23}^{LW} \qquad [\text{II-38A}]$$

thus proving the correctness of Hamaker's combining rule (eq. [II-11]) (Neumann *et al.*, 1982) via a surface thermodynamic approach, provided that: (a) the geometric combining rule (eq. [II-9]) holds the LW interactions (which is the case), and (b) ℓ_o (eqs. [II-32] and [II-16]) has the same value for ΔG_{132}^{LW}, ΔG_{12}^{LW}, ΔG_{33}^{LW}, ΔG_{13}^{LW} and ΔG_{23}^{LW}, which also can be demonstrated (van Oss and Good, 1984); see also Chapter III. The correctness of Hamaker's combining rule for A_{131} (eq. [II-10]) can also be demonstrated by the surface thermodynamic approach, in the same manner.

ADDITIVITY OF THE THREE ELECTRODYNAMIC CONTRIBUTIONS TO THE SURFACE TENSION

Using the Lifshitz approach for van der Waals interactions in condensed media, Chaudhury (1984) showed that the dispersion (London), induction (Debye) and dipole (London) contributions to the Lifshitz–van der Waals (or apolar) surface component γ^{LW}, are simply additive:

$$\gamma^{LW} = \gamma^L + \gamma^D + \gamma^K \qquad [\text{II-40}]$$

This can be proven experimentally in a rather simple manner. Girifalco and Good (1957) introduced an interaction parameter, Φ, to reconcile the large deviations between the observed interfacial tensions between "polar" liquids (for the definition of "polar", see Chapter IV) and the interfacial tensions calculated by eq. [II-35A]. To that effect the factor Φ was inserted into eq. [II-35A]:

$$\gamma_{12} = \gamma_1 + \gamma_2 - 2\Phi\sqrt{\gamma_1\gamma_2} \qquad \text{[II-41]}$$

Obviously when only the apolar (LW) component of the surface tension is considered, and if eq. [II-40] is correct, Φ should be equal to unity, and eq. [II-41] should then revert to eq. [II-35A] (see also Krupp, 1967). The interaction parameter may be defined as:

$$\Phi_{12} = A_{12}/\sqrt{A_{11}A_{22}} \qquad \text{[II-42]}$$

We can express A_{ij} (see also eq. [II-26]) as:

$$A_{ij} = \frac{3}{2}kT\sum_{n=0}^{\infty}\sum_{j=1}^{\infty}\left\{\frac{[\varepsilon_i(i\omega_n)-1]\cdot[\varepsilon j(i\omega_n)-1]}{[\varepsilon_i(i\omega_n)+1]\cdot[\varepsilon j(i\omega_n)+1]}\right\}^j \cdot j^{-3} \qquad \text{[II-43]}$$

From numerical calculations of Φ^{LW} for a number of apolar interactions using eqs. [II-26], [II-42] and [II-43], Chaudhury (1984) found $0.98 < \Phi^{LW} < 1.00$ (see also van Oss et al., 1986b). It must be stressed again that eqs. [II-26] and [II-43] contain not only the London (dispersion) interactions, but also the Debye (induction) and the Keesom (dipolar) forces. The zero frequency component contains the induction and dipolar (orientation) terms, which in rarefied systems give rise to Debye and Keesom forces.

Thus, for strictly LW systems, eq. [II-40] is indeed correct so that in condensed systems, using the macroscopic approach (Lifshitz, 1955), *all three electrodynamic interactions obey the same general equations and must be treated in the same manner.* This leaves us with the clear-cut situation where we can develop the treatment of "polar" interactions of hydrogen-bonding origin in an entirely separate fashion, without needless concern for putative encroachments of, or overlaps between the electrodynamic LW forces treated above, and the "polar" interactions, which are the subject of Chapter IV.

CODA

On a macroscopic level all three van der Waals interfacial interactions (Keesom, Debye and London) decay at the same rate as a function of distance, ℓ (at $\ell \le 10\text{nm}$), and may be treated together as the total of apolar, or Lifshitz–van der Waals (LW) interactions. Thus the aggregate macroscopic Hamaker constant, A, is proportional to the apolar component of the surface tension, γ^{LW}, and, for condensed-phase material i, is readily determined form γ_i^{LW}, the value of which can be obtained *via* contact angle measurement with a high-energy apolar liquid (see also Chapters III and IV). The interfacial tension between condensed-phase materials, i and j, is the square of the difference between the square roots of γ_i^{LW} and γ_j^{LW}, and thus is always positive (or zero).

III Relation Between the Hamaker Constant and the Apolar Surface Tension Component

PROPORTIONALITY FACTOR A/γ^{LW}

From eqs. [II-26 to 29] the Hamaker constant A_{11} of a homogeneous material (1) in the condensed state (i.e., either liquid or solid) can be obtained via the Lifshitz (1955) approach; see Chapter II. However, the determination of the necessary spectroscopic data can be tedious and, in many cases, difficult. But if it can be shown that there is a reliable proportionality between A_{11} and the apolar surface tension component γ_i^{LW} of material (i), the determination of the A_{ii} value of various liquid or solid materials can be much simplified.* Using eq. [II-32]:

$$\gamma_i = \frac{A_{ii}}{24\pi\ell_o^2}$$

which can be rewritten as:

$$A_{ii} = 24\pi\ell_o^2\,\gamma_i^{LW} \qquad\qquad \text{[III-1]}^\dagger$$

that proportionality factor is $24\pi\ell_o^2$. The minimum equilibrium distance ℓ_o does not, in this case, take the Born repulsion (van Oss and Good, 1984; van Oss et al., Chem. Rev., 1988) into account, and thus is not an accurate measure of the "real" equilibrium distance, which may not be precisely measurable by this approach in any event (see below). For a fairly large number of cases, where A_{11} can be determined from permitivities and spectroscopic data (see Tables III-1 and 2), as well as from known values of γ_i^{LW}, a comparison can be made (see Table III-3). From this variegated collection of materials it can be seen that the value of ℓ_o thus found in all cases is close to the average value of $\ell_o = 1.57$ Å, with a standard deviation (S.D.) of ± 0.09 Å (van Oss et al., Chem. Rev., 1988). The proportionality factor $24\pi\ell_o^2$ then is 1.8585 (± 0.0065 S.D.) $\times 10^{-14}$ cm². Thus, for all practical purposes, for single, pure substances:

---

* It should be recalled that for liquids, γ_L^{LW} can be found by means of contact angle determination on apolar solids, and for solids, γ_S^{LW} may be determined through contact angle measurements with apolar liquids; see Chapter XII, and eq. [IV-28].
† A list of symbols can be found on pages 399–406.

$$A_{ii} = 1.8585 \ (\pm 0.0065 \ \text{S.D.}) \times 10^{-14} \ \gamma_i^{LW} \qquad\qquad [\text{III-2A}]$$

when γ_i^{LW} is expressed in ergs/cm², or in mJ/m², to obtain A_{ii} in ergs. Or one may use:

$$A_{ii} = 1.8585 (\pm 0.0065 \ \text{S.D.}) \times 10^{-21} \gamma_i^{LW} \qquad\qquad [\text{III-2B}]$$

(here also γ_i^{LW} is expressed in ergs/cm², or in mJ/m²), to obtain A_{ii} in J.

Thus, the Hamaker constants of single pure substances in the condensed state can be derived from their apolar surface tension component γ_i^{LW} *via* the above proportionality factor, e.g., eqs. [III-2A, 2B], or from the averaged ℓ_o value of 1.57 Å using eq. [III-1], plus or minus less than 0.7% (S.D.).

The averaged value for ℓ_o found from the materials listed in Table III-3 agrees well with the values found by Hough and White (1980, Table III-4) for alkanes with a fairly high carbon number ($n \geq 9$); for the lower alkanes ($n \geq 7$) these authors found a slightly higher value for ℓ_o. Israelachvili (1974) was the first to establish the relation between A and γ^{LW} as a fairly constant and potentially useful parameter; he found a value for ℓ_o of the order of 2.0 Å; in 1985 the same author arrived at $\ell_o \approx$ 1.65 Å (Israelachvili, 1985, Table XVII). Neumann *et al.* (1979) found values for $\ell_o \approx 1.8$ Å.

SIGNIFICANCE OF THE AVERAGED ℓ_o VALUE

The average ℓ_o value found above *via* eq. [III-1] (see Table III-3) is not necessarily the minimum equilibrium distance ℓ_{eq} (ℓ_o may perhaps best be alluded to as the "practical" equilibrium distance). To begin with, to find ℓ_{eq}, one must take the Born repulsion into account, and instead of eq. [II-32], use:

$$\gamma_i^{LW} = \frac{A_{ii}}{32\pi\ell_{eq}^2} \qquad\qquad [\text{III-3}]$$

TABLE III-1
Dielectric Dispersion Data of Some Materials in the Microwave Region

Material	ε_∞	ε_o	α (Cole-Cole)	ω_{MW} (rad/sec)	Ref.
Ethanol	25.07	4.2	0	6.97 × 109	a
Methanol	33.64	5.7	0	1.88 × 1010	a
Glycerol	42.5	4.16	0.3	8 × 108	b
Water	80.1	5.2	0	1.06 × 1011	b

[a] *Tables of Dielectric Dispersion Data of Pure Liquids and Dilute Solutions*, National Bureau of Standards, Washington, DC, circular, p. 589.
[b] Chaudhury, 1984.
Source: van Oss *et al.*, Chem. Rev., 1988.

TABLE III-2
Spectroscopic Constants of Several Materials

	(square of refr. index in the visible range) n_0^2	IR × 10–14 (rad/sec)	UV × 10–16 (rad/sec)	Ref.
Helium	1.048	—	3.73	a
Hydrogen	1.228	—	2.33	a
Nitrogen	1.4	—	2.36	a
Argon	1.53	—	2.39	a
Hexane	1.864	5.54	1.873	b
PTFE	1.846	2.27	1.793	b
Heptane	1.899	5.54	1.87	b
Octane	1.925	5.54	1.863	b
Water	1.755	5.66	1.793	b
Methanol	1.7349	3.52	1.87	a,c,d
Ethanol	1.831	2.588	1.924	a,c,d
Decane	1.965	5.54	1.873	b
Dodecane	1.991	5.54	1.877	b
Hexadecane	2.026	5.54	1.848	b
Benzene	2.179	2.165	1.348	a,c,d
Chlorine	2.1	—	1.93	c
Carbon disulfide	2.49	3.33	1.05	a,c,d
Glycerol	2.189	5.54	1.915	e

[a] Weast, 1972.
[b] Hough and White, 1980.
[c] Pouchert, 1975.
[d] *American Institute of Physics Handbook*, 3rd Ed., McGraw-Hill, New York.
[e] *Tables of Dielectric Dispersion Data of Pure Liquids and Dilute Solutions*, National Bureau of Standards, Washington, DC, circular, p. 589.
Source: van Oss *et al.*, Chem. Rev., 1988.

(see van Oss and Good, 1984). According to eq. [III-3], the average value for ℓ_{eq} obtained from the data of Table III-3 then becomes ≈1.36 Å (van Oss *et al.*, Chem. Rev., 1988). However, even that value does not necessarily represent the "real" ℓ_{eq} distance, because the Lifshitz approach is not accurate at these extremely small values of ℓ. Nevertheless, various indications (especially those based on immunochemical data) suggest that the "real" ℓ_{eq} distance is not likely to be shorter than 1 Å, nor longer than 2 Å.[*]

---

[*] The ℓ_{eq} distance is the minimum equilibrium distance between the Born-Kihara shells of atoms or molecules, i.e., between those surfaces which surround all atoms or molecules, at which any closer approach is prevented by the Born repulsion (van Oss and Good, 1984).

TABLE III-3
Table of Hamaker Constants A_{ii} and γ_i^{LW} Values (at 20°C) for Various Liquids, Showing the Minimum Equilibrium Distance ℓ_o from eq. [III-1]

	Temperature (°C)	Hamaker constant (× 1013 ergs; × 1020 J)	γ_i^{LW} (ergs/cm²) (mJ/m²)	ℓ_o (Å)
Helium	−271.5	0.0535	0.353	1.418
Hydrogen	−255	0.511	2.31	1.713
Nitrogen	−183	1.42	6.6	1.689
Argon	−188	2.33	13.2	1.530
Hexane	25	3.91	18.4	1.679
Polytetrafluoroethylene	25	3.8	19	1.629
Heptane	25	4.03	20.14	1.629
Octane	25	4.11	21.8	1.581
Water	25	4.62	21.8	1.677
Methanol	25	3.94	18.5	1.681
Ethanol	25	4.39	20.1	1.702
Decane	25	4.25	23.9	1.536
Dodecane	25	4.35	25.4	1.507
Tetradecane	25	4.38	26.6	1.478
Chloroform	25	5.34	27.14	1.615
Hexadecane	25	4.43	27.7	1.456
Benzene	25	4.66	28.9	1.462
Chlorine	50	5.4	29.2	1.566
Carbon disulfide	25	5.07	32.3	1.443
Glycerol	25	6.7	34	1.617
Polystyrene	25	6.58	42	1.441
Polymethylmethacrylate	25	7.11	40	1.535
Mercury	25	33.0a	200	1.479

$$\ell_o = 1.568 \pm 0.093 \text{ Å (S.D.)}$$

[a] For more recent calculations on A and γ^{LW} of mercury ($\gamma^{LW} = 211 \pm 5$ m] /m²), see Chaudhury (1987); see also Chaudhury (1984).
Source: van Oss *et al.*, Chem. Rev., 1988.

However, the average value of ℓ_o, as defined in eq. [III-1], as found from the data given in Table III-3, provides a perfectly useful practical link between the apolar surface tension component γ_i^{LW} and the Hamaker constant A_{ii} of any condensed material (i).

APPLICABILITY OF EQ. [III-1] TO A_{12}, A_{121}, A_{132}

For apolar materials (see eq. [II-16]):

$$\Delta G^{LW} = \frac{-A}{12\pi\ell_o^2} \qquad \text{[III-4]}$$

TABLE III-4
Table of Hamaker Constants A_{12} and Apolar Interfacial Free Energy ΔG_{12}^{LW} at 20°C, Between Alkanes (1) and Water (2), Giving the Practical Equilibrium Distance ℓ_o [a]

Alkane (carbon number)	$-\Delta G_{12}^{LW}$	=	γ_1^{LW} [b]	+	γ_2^{LW}	−	γ_{12}^{LW} [c]	A calc. [d] (1013 ergs; 1020 J)	ℓ_o (Å)
5	37.41	=	16.05	+	21.8	−	0.44	3.63	1.604
6	40.06	=	18.4	+	21.8	−	0.14	3.78	1.582
7	41.91	=	20.14	+	21.8	−	0.03	3.89	1.569
8	43.42	=	21.62	+	21.8	−	0	3.097	1.557
9	44.64	=	22.85	+	21.8	−	0.01	4.05	1.550
10	45.58	=	23.83	+	21.8	−	0.05	4.11	1.547
11	46.37	=	24.66	+	21.8	−	0.09	4.14	1.538
12	47.02	=	25.35	+	21.8	−	0.13	4.20	1.539
13	47.61	=	25.99	+	21.8	−	0.18	4.21	1.532
14	48.13	=	26.56	+	21.8	−	0.23	4.23	1.523
15	48.59	=	27.07	+	21.8	−	0.28	4.25	1.523
16	48.94	=	27.47	+	21.8	−	0.33	4.28	1.523

Average $\ell_o = 1.55 \pm 0.03$ Å, (S.D.)

[a] From $\Delta G_{12}^{LW} = -A_{12}/12\pi\ell_o^2$ (eq. [II-16A]).
[b] From Jasper, 1972.
[c] From $\gamma_{12}^{LW} = (\sqrt{\gamma_1^{LW}} = \sqrt{\gamma_2^{LW}})^2$ (eq. [II-35]).
[d] From Hough and White, 1980.

which proportionality holds for ΔG_{12}^{LW} and A_{12}, for ΔG_{121}^{LW} and A_{121}, and for ΔG_{132}^{LW} and A_{132}. We also know that according to the Dupré equation (1869):

$$\Delta G_{12}^{LW} = \gamma_{12}^{LW} - \gamma_1^{LW} - \gamma_2^{LW} \qquad [\text{II-36}]$$

$$\Delta G_{121}^{LW} = -2\gamma_{12}^{LW} \qquad [\text{II-33A}]$$

$$\Delta G_{132}^{LW} = \gamma_{12}^{LW} - \gamma_{13}^{LW} - \gamma_{23}^{LW} \qquad [\text{II-37}]$$

There is therefore a simple relation between A_{12}, A_{121}, A_{132} on the one hand, and γ_i^{LW} and γ_{ij}^{LW} on the other (where i and j represent materials 1, 2 or 3). The proportionality of A_{ii}/γ_i^{LW} to ℓ_{oi}^2 is shown in eq. [III-1] (see Table III-3); it has become clear that for all condensed materials (i) the value of ℓ_o does not deviate much from $\ell_o = 1.57$ Å (± 0.09 Å, S.D.). The question now is whether the value of ℓ_{oii} is the same as that of ℓ_{oij} or, in other words, whether the (practical) minimum equilibrium distance between molecules of the same substance is the same as the (practical) minimum equilibrium distance between molecules of two different substances (always, as related to eq. [III-1]). Unfortunately, whilst γ_{ij}^{LW} values are easily determined with a fair degree of accuracy, calculated A_{iji} values of, e.g., polymer-water and quartz-water systems fluctuate rather widely (see, e.g., Israelachvili, 1985, Table XVI; Hough and White, 1980, Table 5). However, the data of Hough and White (1980, Table 4) on hydrocarbon-water systems (of which the γ_i^{LW} and the γ_{ij}^{LW} values can be established with greater precision than those for polymer-water or quartz-water systems) show less fluctuation. These data (Hough and White, 1980, Table 4) yield a value for ℓ_0 (derived from ΔG_{ij}^{LW} values) of 1.55 Å ($\pm$ 0.03 Å, S.D.) (see Table III-4), which is close to our earlier determined value of $\ell_0 = 1.57$ Å obtained from γ_i^{LW} values (see Table III-3). It should, however, be noted that here also a slightly higher value is found for ℓ_0 with the lower alkanes. From alkane-water apolar interfacial tension components, γ_{ij}^{LW} and the A_{iji} values calculated by Hough and White (1980), the obtainable ℓ_0 values fluctuate too widely to arrive at a precise conclusion concerning the relation between γ_{13}^{LW} and A_{131} *via* this approach. However, as A_{131} can be written as $-12\pi\ell_0^2 \Delta G_{131} = -12\pi\ell_0^2 (\Delta G_{11}^{LW} + \Delta G_{33}^{LW} - 2\Delta G_{13}^{LW})$ (see eqs. [II-9] and [II-16A]) and as it can be shown that for both ΔG_{ii}^{LW} and for ΔG_{ij}^{LW}, $\ell_0 = 1.57$ ($\pm$ 0.9) Å (see Tables III-3 and III-4), and as further, $\Delta G_{131}^{LW} = -2\gamma_{13}^{LW}$, γ_{ij}^{LW} also must obey the relation shown in eq. [III-1], i.e.:

$$A_{iji} = 24\pi\ell_0^2 \, \gamma_{ij}^{LW} \qquad\qquad \text{[III-1A]}$$

where also $\ell_0 \approx 1.57$ Å. Van Oss and Good (1984) also arrived at the conclusion that ℓ_0 is invariable for A_{ii}, A_{ij}, A_{iji} and A_{ijk}, on more general grounds.

Thus, in general, the Hamaker constants of single substances, across a vacuum, as well as across a liquid, *and* the Hamaker constants pertaining to the interaction between two different substances, across a vacuum as well as across a liquid, can all be derived from the appropriate apolar surface tension components and/or from the appropriate apolar interfacial tension components, *via* a proportionality factor of 1.86×10^{-14} (eq. [III-2A]), or *via* a (practical) equilibrium distance $\ell_0 = 1.57$ Å, according to:

$$A = 24\pi\ell_0^2 \, \gamma^{LW} \qquad\qquad \text{[III-1B]}$$

CODA

The connection between γ^{LW} and the (macroscopic) Hamaker constant, A, is demonstrated. It has been shown that for all materials tested (from liquid He to liquid Hg), $\ell_0 = 1.57 \pm 0.09$ Å.

IV Polar or Lewis Acid-Base Interactions

INTERFACIAL LEWIS ACID-BASE INTERACTIONS

Now that it has become clear that the dipolar (Keesom, or orientation) component of the interfacial van der Waals interactions must be classified with, and treated as the other electrodynamic forces, a clear-cut division becomes possible between Lifshitz–van der Waals (LW), or apolar, interactions and other interactions in condensed media which (with the exclusion of metallic forces) all are held to be "polar." In aqueous media these polar interactions mainly comprise the interactions between hydrogen-donors and hydrogen-acceptors (or between Brønsted acids and Brønsted bases). It is, however, preferable to extend the concept of polar interactions more widely, and to define them to comprise all electron-acceptor–electron-donor, or Lewis acid-base interactions, designated by the superscript AB (Fowkes, 1987; van Oss et al., Advan. Colloid Interface Sci., 1987).[*]

As the apolar and polar components of the free energies of interfacial interaction are additive (Fowkes, 1983; van Oss et al., Advan. Colloid Interface Sci., 1987; Langmuir, 1988):

$$\Delta G = \Delta G^{LW} + \Delta G^{AB} \qquad [IV\text{-}1]^{\dagger}$$

and given that:

$$\gamma_i \equiv -\frac{1}{2}\Delta G_{ii} \qquad [II\text{-}33]$$

we may also state:

$$\gamma_i = \gamma_i^{LW} + \gamma_i^{AB} \qquad [IV\text{-}2]$$

---

[*] In a number of earlier publications (Chaudhury, 1984; van Oss, Moore et al., 1985; van Oss et al., 1986a; b; c; van Oss, Chaudhury and Good, 1987b; van Oss, Good and Chaudhury, 1987a; b), these polar interactions were indicated by the superscript SR, for short-range. However, the growing realization that these forces, in liquid media, often have a medium-range or even long-range effects (see Chapters VII and XXIII) made us prefer the designation "AB", as less equivocal and more precise.

[†] A list of symbols can be found on pages 399–406.

The equations governing γ_i^{LW} and γ_{ij}^{LW} have been treated above, in Chapter II. A crucial relation is:

$$\Delta G_{12}^{LW} = -2\sqrt{\gamma_1^{LW}\gamma_2^{LW}} \qquad\qquad [II-39]$$

It continues to be assumed by a number of authors (see Chapter XIV) that an equation of the type of eq. [II-39] is also valid for AB interactions. However, as Fowkes (1983) already pointed out a number of years ago, this is incorrect.

Electron-acceptor–electron-donor interactions are essentially asymmetrical (van Oss et al., Advan. Colloid Interface Sci., 1987) in the sense that of a given polar substance i the electron-acceptor and the electrondonor parameters are usually quite different. Also, one of the parameters is not manifested at all, unless the other parameter is either present in another part of the same molecule of substance i, or in another polar molecule j with which molecules of i can interact "reciprocally" (Jensen, 1980). Thus, given that electron-acceptor and electron-donor parameters of most substances are not equal to each other, and that electron-acceptors of substance i will react with electron-donors of substance j, and vice versa:

$$\Delta G_{ij}^{AB} \equiv -2\sqrt{\gamma_i^{\oplus}\gamma_j^{\ominus}} - 2\sqrt{\gamma_i^{\ominus}\gamma_j^{\oplus}} \qquad\qquad [IV-3]$$

where $\gamma^{\oplus}$ stands for the electron-acceptor parameter and $\gamma^{\ominus}$ for the electron-donor parameter of the surface tension of substance i or j. Kollman (1977) also expressed this interaction in a similar fashion. It should be noted that eq. [IV-3] is not to be confounded with a simple geometric combining rule (as in eq. [II-39]), but rather that it expresses a double, asymmetrical interaction, of an analogous nature to electrostatic interactions (Kollman, 1977), between two different substances.

The polar cohesive energy of substance i then is:

$$\Delta G_{ii}^{AB} \equiv -4\sqrt{\gamma_i^{\oplus}\gamma_i^{\ominus}} \qquad\qquad [IV-4]$$

As:

$$\gamma_i \equiv -\frac{1}{2}\Delta G_{ii} \qquad\qquad [II-33]$$

(which is valid irrespective of polarity or apolarity), we may define:

$$\gamma_i^{AB} \equiv 2\sqrt{\gamma_i^{\oplus}\gamma_i^{\ominus}} \ ^* \qquad\qquad [IV-5]$$

In eq. [IV-3], the negative sign is made imperative by the thermodynamic convention that a negative sign for ΔG signifies an attraction (given that ΔG_{ij}^{AB} is always attractive, or zero, whilst the two right-hand terms under the square root signs always are positive, or zero).

---

* The factor 2 (eqs. [IV-3, IV-5]) is necessitated by the concern to maintain values of the correct orders of magnitude for $\gamma_i^{\oplus}$, $\gamma_i^{\ominus}$ and γ_i^{AB} (van Oss et al., Advan. Colloid Interface Sci., 1987).

Using the Dupré equation (see eq. [II-36]), which is also valid irrespective of polarity or apolarity:

$$\Delta G_{12}^{AB} = \gamma_{12}^{AB} - \gamma_1^{AB} - \gamma_2^{AB} \qquad [IV-6]$$

one can express the interfacial tension γ_{12}^{AB} between substances 1 and 2 as follows:

$$\gamma_{12}^{AB} = \Delta G_{12}^{AB} + \gamma_1^{AB} + \gamma_2^{AB} \qquad [IV-6A]$$

Inserting the values for ΔG_{12}^{AB} defined in eq. [IV-3] and for γ_1^{AB} and γ_2^{AB} and defined in eq. [IV-5] into eq. [IV-6A], one obtains:

$$\gamma_{12}^{AB} = 2(\sqrt{\gamma_1^{\oplus}\gamma_1^{\ominus}} + \sqrt{\gamma_2^{\oplus}\gamma_2^{\ominus}} - \sqrt{\gamma_1^{\oplus}\gamma_2^{\ominus}} - \sqrt{\gamma_1^{\ominus}\gamma_2^{\oplus}})^* \qquad [IV-7]$$

which can also be written as:

$$\gamma_{12}^{AB} = 2(\sqrt{\gamma_1^{\oplus}} - \sqrt{\gamma_2^{\oplus}})\ (\sqrt{\gamma_1^{\ominus}} - \sqrt{\gamma_2^{\ominus}}) \qquad [IV-7A]$$

It is interesting to note that an equation of the type of eq. [IV-7] was proposed long ago, by Small (1953). Comparing the expression for the apolar (LW) interfacial tension component:

$$\gamma_{12}^{LW} = (\sqrt{\gamma_1^{LW}} - \sqrt{\gamma_2^{LW}})^2 \qquad [II-35]$$

with eqs. [IV-7] and [IV-7A], it is obvious that an important difference between γ_{12}^{LW} and γ_{12}^{AB} lies in the fact that whilst γ_{12}^{LW} clearly can only be positive, or zero, γ_{12}^{AB} can readily be negative (van Oss, Chaudhury and Good, 1987a; 1988a), i.e., when:

$$\gamma_1^{\oplus} > \gamma_2^{\oplus} \ \text{ and } \ \gamma_1^{\ominus} < \gamma_2^{\ominus} \qquad [IV-8A]$$

or when

$$\gamma_1^{\oplus} < \gamma_2^{\oplus} \ \text{ and } \ \gamma_1^{\ominus} > \gamma_2^{\ominus} \qquad [IV-8B]$$

Thus, contrary to earlier dogma (based on the exclusive applicability of eq. [II-35]), and applying eq. [IV-2], it is obvious that when $\gamma_{12}^{LW} < |\gamma_{12}^{AB}|$ and when the conditions of eq. [IV-8A] or eq. [IV-8B] prevail, the total interfacial tension between polar substances 1 and 2:

$$\gamma_{12} = (\sqrt{\gamma_1^{LW}} - \sqrt{\gamma_2^{LW}})^2 + 2(\sqrt{\gamma_1^{\oplus}\gamma_1^{\ominus}} + \sqrt{\gamma_2^{\oplus}\gamma_2^{\ominus}}$$
$$- \sqrt{\gamma_1^{\oplus}\gamma_2^{\ominus}} - \sqrt{\gamma_1^{\ominus}\gamma_2^{\oplus}}) \qquad [IV-9]$$

---

* A comparison between the approach given in eq. [IV-7] and the hypothesis that γ_{12}^{AB} "γ_{12}^{P}" $= \sqrt{\gamma_1^{P}\gamma_2^{P}}$ which still has some adherents, is given in Chapter XIV.

can be *negative* (van Oss *et al.*, Advan. Colloid Interface Sci., 1987). The implications of this finding are far-reaching; they are discussed later in this chapter; see also Chapters XIII and XIV.

THE YOUNG-DUPRÉ EQUATION

In a paper presented to the Royal Society near the end of 1804, Thomas Young described (without using equations) the connection between the work of adhesion between the surface tension of a solid (γ_S) and a liquid (γ_L), the interfacial tension between the solid and the liquid (γ_{SL}) and the contact angle θ made by a drop of liquid L (at the triple point: solid liquid-air, tangential to the drop and measured through the drop), deposited on a (flat) horizontal solid surface S. Young's equation is taken to be the quintessence of his verbal description (Young, 1805) and is usually expressed as:

$$\gamma_L \cos \theta = \gamma_S - \gamma_{SL} \qquad \text{[IV-10]}$$

Dupré (1869) (see footnote on page 16) expressed the relation between the work of adhesion between solid and liquid as:

$$\Delta G_{SL} = \gamma_{SL} - \gamma_S - \gamma_L \qquad \text{[IV-6B]}$$

Inserted into the Young equation (eq. [IV-9]) this yields the Young-Dupré equation:

$$-\Delta G_{SL} = \gamma_L (1 + \cos \theta) \qquad \text{[IV-11]}$$

Taking both ΔG^{LW} and ΔG^{AB} into account (e.g., eq. [IV-1]), the Young-Dupré equation can be more precisely expressed as:

$$-\Delta G_{SL}^{LW} - \Delta G_{SL}^{AB} = \gamma_L (1 + \cos \theta)^* \qquad \text{[IV-12]}$$

Upon combination with eqs. [II-39], [IV-3], [IV-5] and [IV-6] the complete Young-Dupré equation then becomes:

$$(1 + \cos \theta)\gamma_L = 2(\sqrt{\gamma_S^{LW}\gamma_L^{LW}} + \sqrt{\gamma_S^{\oplus}\gamma_L^{\ominus}} + \sqrt{\gamma_S^{\ominus}\gamma_L^{\oplus}})^\dagger \qquad \text{[IV-13]}$$

---

* In all cases of non-zero contact angles, θ, the spreading pressure term π_e in the Young-Dupré equation (see Adamson, 1990) will be neglected. In many such cases π_e is indeed negligible; in other instances, especially in some cases involving hydrophilic surfaces, π_e may well have some influence which, however, in the present state of the art, usually is not easy to assess with an appreciable degree of confidence. Fowkes *et al.* (1980) have described a useful method for measuring spreading pressures on hydrophobic surfaces.

† A comparison between this version of the Young-Dupré equation and an older, intuitive (but erroneous) version, in which our $\sqrt{\gamma_S^{\oplus}\gamma_L^{\ominus}} + \sqrt{\gamma_S^{\ominus}\gamma_L^{\oplus}}$ is replaced by a single term, $\sqrt{\gamma_S^P\gamma_L^P}$, can be found in Chapter XIV. The influence (if any) of the "spreading pressure" described by Bangham and Razouk (1937) is really a consequence of the deposition of vapor molecules from the contact angle liquids onto the solid flat surface. This influence is however exceedingly small; see van Oss, Giese and Wu (1998) and see also Chapter XII.

Thus, even if for one or several liquids L, γ_L, γ_L^{LW}, $\gamma_L^{\oplus}$ and $\gamma_L^{\ominus}$ are known, and θ can be simply measured, while of course:

$$\gamma_L = \gamma^{LW} + 2\sqrt{\gamma_L^{\oplus}\gamma_L^{\ominus}} \qquad [IV\text{-}14]$$

there still remain three unknowns in eq. [IV-13], i.e.: γ_S^{LW}, $\gamma_S^{\oplus}$ and $\gamma_S^{\ominus}$. It is therefore only possible to determine these three (independently variable) entities of γ_S by measuring contact angles θ with three different liquids L (each with completely determined γ_L, γ_L^{LW}, $\gamma_L^{\oplus}$, $\gamma_L^{\ominus}$), and using eq. [IV-13] three times (van Oss *et al.*, Advan. Colloid Interface Sci., 1987; Chem. Rev., 1988).

It should be noted that for solid/liquid systems, Young's equation in its simplest form, i.e.:

$$\gamma_{SL} = \gamma_S - \gamma_L \cos\theta \qquad [IV\text{-}10A]$$

is always true (cf. eqs. [IV-10, 13]). As the value for γ_L normally is readily obtainable, the ease of obtaining γ_{SL} directly *via* eq. [IV-9] thus hinges on the determination of γ_S:

$$\gamma_S = \gamma_S^{LW} + 2\sqrt{\gamma_S^{\oplus}\gamma_S^{\ominus}} \qquad [IV\text{-}2A]$$

(see eqs. [IV-2,5]). Thus, for determining γ_S, it remains necessary to do at least three contact angle determinations (with three different liquids), using eq. [IV-13]. However, once γ_S has been determined *via* eq. [IV-2A], there is some merit in obtaining γ_{SL} from eq. [IV-9], *via* contact angle determinations with the same liquid L in which the solubility of solid S is being studied, especially in very polar systems.

ATTRACTIVE AND REPULSIVE POLAR FORCES

Using the Dupré equation in the form of:

$$\Delta G_{132} = \gamma_{12} - \gamma_{13} - \gamma_{23} \qquad [IV\text{-}15]$$

(see eq. [II-37]) to describe the interaction between materials 1 and 2, immersed in liquid 3, both the LW and AB interactions can be taken into account, by inserting eq. [II-35] (for γ_{ij}^{LW}) and eq. [IV-7] (for γ_{ij}^{AB}), yielding:

$$\Delta G_{132} = 2[\sqrt{\gamma_1^{LW}\gamma_3^{LW}} + \sqrt{\gamma_2^{LW}\gamma_3^{LW}} - \sqrt{\gamma_1^{LW}\gamma_2^{LW}} - \gamma_3^{LW}$$

$$+ \sqrt{\gamma_3^{\oplus}}(\sqrt{\gamma_1^{\ominus}} + \sqrt{\gamma_2^{\ominus}} - \sqrt{\gamma_3^{\ominus}}) + \sqrt{\gamma_3^{\ominus}}(\sqrt{\gamma_1^{\oplus}} \qquad [IV\text{-}16]$$

$$+ \sqrt{\gamma_2^{\oplus}} - \sqrt{\gamma_3^{\oplus}}) - \sqrt{\gamma_1^{\oplus}\gamma_2^{\ominus}} - \sqrt{\gamma_1^{\ominus}\gamma_2^{\oplus}}]$$

Similarly, for describing the interaction between molecules or particles of substance 1, immersed in liquid 3:

$$\Delta G_{131} = -2\gamma_{13} = -2(\sqrt{\gamma_1^{LW}} - \sqrt{\gamma_3^{LW}})^2 - 4(\sqrt{\gamma_1^{\oplus}\gamma_1^{\ominus}}$$

$$+ \sqrt{\gamma_3^{\oplus}\gamma_3^{\ominus}} - \sqrt{\gamma_1^{\oplus}\gamma_3^{\ominus}} - \sqrt{\gamma_1^{\ominus}\gamma_3^{\oplus}}) \qquad\qquad [\text{IV-17}]$$

The interaction between substances 1 and 3 *in vacuo* is:

$$\Delta G_{13} = -2(\sqrt{\gamma_1^{LW}\gamma_3^{LW}} + \sqrt{\gamma_1^{\oplus}\gamma_3^{\ominus}} + \sqrt{\gamma_1^{\ominus}\gamma_3^{\oplus}}) \qquad\qquad [\text{IV-18}]$$

(see eqs. [II-39] and [IV-3]).

The interaction between two substances 1 and 3 *in vacuo* clearly always is attractive, as ΔG_{13} (eq. [IV-18]) must be negative in all cases. ΔG_{13} cannot be zero, as γ_1^{LW} of all materials in the condensed state has a finite, positive value. Thus, water is always *attracted* to even the most "hydrophobic" surfaces, with an energy that is rarely less than 40 mJ/m² (which represents approximately the attractive energy of water to Teflon). This is one of the reasons why the term "hydrophobic" interaction is misleading. The continued use of the term may well have contributed materially to the persisting misunderstanding of the actual mechanism of these interactions among most workers, especially in the Biological Sciences; see, however, Israelachvili (1985, footnote, page 105), on the word "hydrophobic substances"—"Originally coined to describe the water-hating properties of these substances. It is well to note that their interaction with water is actually attractive, due to the dispersion force, though the interaction of water with itself is much more attractive. Water simply loves itself too much to let some substances get in its way"—a statement which admirably establishes the perspective of the role of water in these interactions.

The actual "hydrophobic" interaction is more closely linked to eq. [IV-17]. When, in aqueous systems (water is then substance 3), low energy (or "hydrophobic") substances interact with each other, γ_{13} is positive and ΔG_{131} negative, giving rise to an attraction (see Chapter XIV). In most cases of "hydrophobic," or preferably *interfacial* attraction, between low-energy substances, 1, in water, 3, γ_{13}^{LW} is close to zero, because γ_3^{LW} for water (21.8 mJ/m²) is close to the γ_1^{LW} for "hydrophobic" materials, which varies from 18.5 mJ/m² (for Teflon) through 20 to 27 mJ/m² for most alkanes (Jasper, 1972), to 42 mJ/m² for polystyrene (Good and Kotsidas, 1979). Even in the latter case, (polystyrene vs. water), γ_{13}^{LW} is only 3.3 mJ/m². In all cases of negligible polarity in the "hydrophobic" material, the only other term of importance in eq. [IV-17] is $-4\sqrt{\gamma_3^{\oplus}\gamma_3^{\ominus}}$ (representing the polar contribution to the cohesive energy of water), amounting to $\Delta G_{131}^{AB} \approx -102$ mJ/m². This is the principal contribution to the interfacial attractions between low-energy substances immersed in water (van Oss and Good, J. Dispersion Sci. Tech., 1988) (see below, and Chapter XIV).

Repulsions occur when ΔG_{131} (eq. [IV-17]) is positive. This happens when the situation described in eq. [IV-8A] or eq. [IV-8B] prevails, and when $\gamma_{13}^{LW} < |\gamma_{13}^{AB}|$. In actual practice such negative interfacial tensions occur quite often when two condensed phases are brought into contact. Negative interfacial tensions usually (but not always) persist for only a short time, in which case they are the driving forces of: solubilization, microemulsion formation, or stabilization of particle suspensions (van Oss *et al.*, Advan. Colloid Interface Sci., 1987). In some cases, however, e.g., in the deposition of drops of water on materials of very low solubility, such as glass and other inorganic surfaces (Chapters XIX, XXI and XXII), negative interfacial tensions may persist for a long time, or even indefinitely. The mechanism of solubilization, microemulsion formation and stabilization of particle suspensions by negative interfacial tensions is most easily visualized as giving rise to a positive value for ΔG_{131} (eq. [IV-17]), which means that particles or molecules, 1, must repel each other when immersed in liquid 3, which then gives rise to the dispersion (or solubilziation) of particles 1 in liquid 3.

RELATIVE VALUES OF THE ELECTRON-ACCEPTOR AND ELECTRON-DONOR PARAMETERS OF γ

No absolute values of $\gamma^{\oplus}$ and/or $\gamma^{\ominus}$ are known at present for any compound (not even for water, see below). If one nevertheless wishes to estimate $\gamma^{\oplus}$ or $\gamma^{\ominus}$ values for a given compound, these will have to be based on an assumed or postulated value for a reference compound, for instance, water (van Oss *et al.*, Advan. Colloid Interface Sci., 1987; van Oss, 2000); see below. However, for the derivation of values for γ_{12}^{AB}, ΔG_{12}^{AB}, ΔG_{121}^{AB} and ΔG_{132}^{AB}, it is *not* necessary to know the absolute values of $\gamma_i^{\oplus}$ and $\gamma_i^{\ominus}$ of any substance i. It suffices to use *polarity ratios*, of $\gamma_i^{\oplus}$ and $\gamma_i^{\ominus}$ relative to the values $\gamma_R^{\oplus}$ and $\gamma_R^{\ominus}$, respectively, of a reference compound, such as, for instance, water. Using the subscript W for water, such polarity ratios, $\delta^{\oplus}$ and $\delta^{\ominus}$ are then defined as follows:

$$\delta_{iW}^{\oplus} = \sqrt{\gamma_i^{\oplus}/\gamma_W^{\oplus}} \qquad \text{[IV-19A]}$$

and

$$\delta_{iW}^{\ominus} = \sqrt{\gamma_i^{\ominus}/\gamma_W^{\ominus}} \qquad \text{[IV-19B]}$$

$\delta_{iW}^{\oplus}$ and $\delta_{iW}^{\ominus}$ being the relative Lewis acid and base polarities, respectively, of substance i, with respect to water. Using water as the contact angle liquid, the complete Young-Dupré equation (eq. [IV-13]) may be written as:

$$(1+\cos\theta_{iw})\gamma_W = 2(\sqrt{\gamma_i^{LW}\gamma_W^{LW}} + \sqrt{\gamma_i^{\oplus}\gamma_W^{\ominus}} + \sqrt{\gamma_i^{\ominus}\gamma_W^{\oplus}}) \qquad \text{[IV-13A]}$$

Because for water, $\gamma_W^{AB} = 51$ mJ/m² (Fowkes, 1965; Chaudhury, 1984), and assuming for the moment that $\gamma_W^{\oplus} = \gamma_W^{\ominus} = 25.5$ mJ/m², eq. [IV-13A] may also be written as:

$$(1+\cos\theta_{iw})\gamma_W = 2(\sqrt{\gamma_i^{LW}\gamma_W^{LW}} + 25.5\sqrt{\gamma_i^{\oplus}/\gamma_W^{\oplus}}$$
$$+ 25.5\sqrt{\gamma_i^{\ominus}/\gamma_W^{\ominus}}) \qquad \text{[IV-13B]}$$

so that (see eqs. [IV-19A] and [IV-19B]) the complete Young-Dupré equation (for L = water) may be expressed as:

$$(1+\cos\theta_{iw})\gamma_W = 2\sqrt{\gamma_i^{LW}\gamma_W^{LW}} + 51\,\delta_{iW}^{\oplus} + 51\,\delta_{iW}^{\ominus} \qquad \text{[IV-20]}$$

(It can easily be shown that eq. [IV-20] is true, even if a ratio for $\gamma_W^\oplus/\gamma_W^\ominus$ other than unity is adopted.) Thus eq. [IV-20] is a variant of the complete Young-Dupré equation [IV-13], of which all terms can be determined quantitatively without any need for knowing the ratio $\gamma_W^\oplus/\gamma_W^\ominus$. The polarity ratios $\delta_{iW}^\oplus$ and $\delta_{iW}^\ominus$ can be precisely established for any compound i (van Oss *et al.*, Advan. Colloid Interface Sci., 1987).

In the same way equations of the type of eq. [IV-20] can be developed for other test liquids. An especially useful high-energy polar liquid, next to water, is glycerol (G). It is known that $\gamma_G^{AB} = 30$ mJ/m^2 (Chaudhury, 1984). The complete Young-Dupré equation for the use of glycerol, instead of water, as the contact angle liquid then becomes, in analogy with eq. [IV-20]:

$$(1+\cos\theta)\gamma_G = 2\sqrt{\gamma_i^{LW}\gamma_G^{LW}} + 30\,\delta_{iG}^\oplus + 30\,\delta_{iG}^\ominus \qquad [IV-21]$$

Given that:

$$\delta_{iG}^\oplus = \delta_{iW}^\oplus/\delta_{GW}^\oplus \qquad [IV-19C]$$

and:

$$\delta_{iG}^\ominus = \delta_{iW}^\ominus/\delta_{GW}^\ominus \qquad [IV-19D]$$

eq. [IV-21] can also be expressed as:

$$(1+\cos\theta)\gamma_G = 2\sqrt{\gamma_i^{LW}\gamma_G^{LW}} + 30\,\delta_{iW}^\oplus/\delta_{GW}^\oplus$$
$$+ 30\,\delta_{iW}^\ominus/\delta_{GW}^\ominus \qquad [IV-21A]$$

By measuring contact angles on a number of different monopoalr surfaces (see the last section of this chapter for the definition of monopolarity) with both water and glycerol, it has been determined that $\delta_{GW}^\oplus = 0.392$ and $\delta_{GW}^\ominus = 1.5$ (van Oss, Good and Busscher, 1990). It is therefore not necessary to know the values for $\delta_G^\oplus$ and $\delta_G^\ominus$, or even of $\delta_{iG}^\oplus$ and $\delta_{iG}^\ominus$ in eq. [IV-21], as we may use, for glycerol as the contact angle liquid:

$$(1+\cos\theta)\gamma_G = 2\sqrt{\gamma_i^{LW}\gamma_G^{LW}} + 76.53\,\delta_{iW}^\oplus + 20.0\,\delta_{iW}^\ominus \qquad [IV-22]$$

In all cases where γ_i^{LW} is known, or can be determined with an apolar liquid (see Chapter XII), after contact angle measurements with both water and glycerol, equations [IV-20] and [IV-22] can be solved for $\delta_{iW}^\oplus$ and $\delta_{iW}^\ominus$. The AB component of substance i can then be obtained by means of:

$$\gamma_i^{AB} = 51\,\delta_{iW}^\oplus\delta_{iW}^\ominus \qquad [IV-23]$$

The polar free energy of adhesion between substances i and j can be expressed as:

$$\Delta G_{ij}^{AB} = -51(\delta_{iW}^\oplus\delta_{jW}^\ominus + \delta_{iW}^\ominus\delta_{jW}^\oplus) \qquad [IV-24]$$

The polar interfacial tension between substances i and j becomes:

$$\gamma_{ij}^{AB} = 51(\delta_{iw}^{\oplus}\delta_{iw}^{\ominus} + \delta_{jw}^{\oplus}\delta_{jw}^{\ominus} - \delta_{iw}^{\oplus}\delta_{jw}^{\ominus} - \delta_{iw}^{\ominus}\delta_{jw}^{\oplus})$$ [IV-25]

From γ_{ij}^{AB}, ΔG_{iji}^{AB} is obtained by using:

$$\Delta G_{iji}^{AB} = -2\gamma_{ij}^{AB}$$ [IV-26]

(see eq. [II-33A]) and, finally, the polar free energy of adhesion between two different substances 1 and 2, immersed in liquid 3 is:

$$\Delta G_{132}^{AB} = 102\,[\delta_{3w}^{\oplus}(\delta_{1w}^{\ominus} + \delta_{2w}^{\ominus} - \delta_{3w}^{\ominus}) + \delta_{3w}^{\ominus}(\delta_{1w}^{\oplus}$$
$$+ \delta_{2w}^{\oplus} - \delta_{3w}^{\oplus}) - \delta_{1w}^{\oplus}\delta_{2w}^{\ominus} - \delta_{1w}^{\ominus}\delta_{2w}^{\oplus}]$$ [IV-27]

Thus γ_1^{AB}, γ_{12}^{AB}, ΔG_{12}^{AB} and ΔG_{132}^{AB} can all be precisely determined without knowing the ratio $\gamma^{\oplus}/\gamma^{\ominus}$ for water (or for any other system). The only case where it may be desirable for $\gamma_i^{\oplus}$ and $\gamma_i^{\ominus}$ to be expressed separately is for single substances, i. For the time being, this is only possible when an assumption is made about the $\gamma^{\oplus}/\gamma^{\ominus}$ ratio for, e.g., water (see also van Oss, Ju et al., 1989, and below).

MONOPOLAR SURFACES AND SUBSTANCES

Referring back to eq. [IV-5], it clearly is at least theoretically possible that a given material i only manifests a $\gamma_i^{\oplus}$, or only a $\gamma_i^{\ominus}$ parameter. Its γ_i^{AB} is then equal to zero and its total surface tension γ_i then is simply equal to its γ_i^{LW} (eq. [IV-2]). However, such substances, which are designated as *monopolar* (van Oss et al., Advan. Colloid Interface Sci., 1987; Chem. Rev., 1988; Langmuir, 1988) can strongly interact with bipolar materials, and with monopolar materials of the opposite polarity, notwithstanding the seemingly apolar nature of their surface tension.

It was found that a great many substances are wholly, or mainly, monopolar (van Oss et al., Advan. Colloid Interface Sci., 1987; Langmuir, 1988) and that many of these, especially biopolymers, manifest largely, or exclusively, electron-donor reactivities, i.e., they all tend to have $\gamma^{\ominus}$ monopolar properties, except for DNA, which also has a pronounced electron-acceptor propensity (van Oss et al., J. Chromatog., 1987). With monopolar substances it becomes especially desirable to be able to express the strength of their monopolar parameters in quantitative terms that allow them to be compared with, e.g., the γ^{LW} and γ^{AB} components of the same as well as of other substances. It is of course possible to revert completely to the use of ratios, such as $\delta^{\oplus}$ and $\delta^{\ominus}$, defined above, but it is less complicated to deal directly with surface tension (γ) parameters and it is therefore preferable to give quantitative values to the terms $\gamma^{\oplus}$ and $\gamma^{\ominus}$. Only one reference point has to be determined (at a given temperature) to enable us to express an entire polar (or monopolar) system quantitatively; this requires the establishment of a $\gamma^{\oplus}/\gamma^{\ominus}$ ratio for a reference polar system. The best choice for such a system appears to be water at 20°C, and the most reasonable and practical choice for the value of that ratio ($\gamma_w^{\oplus}/\gamma_w^{\ominus}$) is unity (van Oss *et al.*, Advan. Colloid Interface Sci., 1987; van Oss, 2000).* As $2\sqrt{\gamma_W^\oplus \gamma_W^\ominus} = 51$ mJ/m², it follows from eq. [IV-5] that it should be assumed that $\gamma_W^\oplus = \gamma_W^\ominus = 25.5$ mJ/m², to be used as the reference value.

Thus the interfacial tensions and free energies given in this work are expressed in absolute S.I. units, but one must remain mindful of the fact that the $\gamma^\ominus$ and $\gamma^\oplus$ values, even though also expressed in S.I. units, are based on the assumption that for water, at 20°C, $\gamma^\oplus = \gamma^\ominus$.

It would seem, for the moment, that it is impossible to determine $\gamma_i^\oplus / \gamma_i^\ominus$ with the available techniques. The electron-acceptor parameter $\gamma_i^\oplus$ and the electron-donor parameter $\gamma_i^\ominus$ are only determined by equations [IV-3] and [IV-5], and, by an extension of these, by the Young-Dupré equation (eq. [IV-13]). In all cases, $\gamma_i^\oplus$ manifests itself only in conjunction with either $\gamma_i^\ominus$ or $\gamma_j^\ominus$, and $\gamma_i^\ominus$ manifests itself only in conjunction with either $\gamma_i^\oplus$ or $\gamma_j^\oplus$. The thermodynamically measured entities γ_i^{AB} and ΔG_{ij}^{AB} thus can only yield products of the general form $\gamma_i^\oplus \gamma_i^\ominus$. The separate values for the subfactors $\gamma^\oplus$ and $\gamma^\ominus$ of these products remain inaccessible, because their determination requires one more equation than is obtainable by currently available methods in any given case. It may, however, not be impossible for entirely different types of physical measurements to allow the independent determination of the separate $\gamma^\oplus$ or $\gamma^\ominus$ values of given compounds, in the future.

At this point it is desirable to set an arbitrary level of $\gamma^\oplus$ (or $\gamma^\ominus$) values, below which one may consider a compound to be a "monopole" of the opposite sign. One can state that in terms of the postulate of $\gamma^\oplus = \gamma^\ominus$ for water equals 25.5 mJ/m², it is reasonable to estimate a compound to be monopolar for, e.g., $\gamma^\ominus$, when its $\gamma^\oplus$ value is significantly less than 1 mJ/m². Alternatively, without postulating anything at all as to absolute $\gamma^\oplus$ and $\gamma^\ominus$ values, a strongly polar compound i may be taken to be preponderantly a $\gamma^\ominus$ monopole when its $\gamma_W^\oplus / \gamma_i^\ominus$ ratio is more than 25.5, while a strongly polar compound would be preponderantly a $\gamma^\oplus$ monopole when its $\gamma_W^\ominus / \gamma_i^\ominus$ ratio is greater than 25.5. In other words, a substance i is preponderantly a $\gamma^\ominus$ monopole when its $\delta_{iW}^\oplus \ll 0.2$, and mainly a $\gamma^\oplus$ monopole when its $\delta_{iW}^\ominus \ll 0.2$; see eqs. [IV-19A] and [IV-19B].

An equally important point is to ascertain how "monopolar" molecules or particles of a substance i must be (in absolute terms) to be capable of repelling each other, while immersed or dissolved in a polar, or more specifically, an aqueous medium.

---

* Gutmann (1978) has given an estimate for the "electron-acceptor number"/"electron-donor number" ratio of approximately 3. However there is no stringent quantitative connection between a number of this type, and γ and γ, because of a disparity of dimensions. The donor numbers in Gutmann's Table 2.1 have the dimensions of energy per mole, and were obtained from the enthalpy of the Lewis bases with $SbCl_5$ in dilute solution. The ΔH of benzene + $SbCl_5$ apparently is very small, vs. a free energy of 10 kJ/mole for the benzene-water hydrogen bond (Gutmann, 1978; Good and Elbing, 1971). Thus, $SbCl_5$ does not appear to be a broadly satisfactory Lewis acid for establishing a scale of donor numbers. The electron-acceptor numbers in Gutmann's Table 2.3 are dimensionless, being based on the relative chemical shift of ^{31}P in Et_3PO in the solvent under test, and are also on an arbitrary scale. No scale of acceptor numbers with units of energy per mole is available. The fact that an acceptor number of 8.9 is given for benzene, in Gutmann's Table 2.3, vs. 10.1 donor number, in his Table 2.1, indicates the irrelevance of the quantitative values of these numbers.

This clearly can only occur when the (mono)polar (AB) repulsion begins to exceed the LW attraction, in other words when the γ_{iw} begins to assume a negative value. To that effect one must look again at eq. [IV-9], and insert the known values for water ($\gamma_W^{LW} = 21.8$ and $\gamma_W^{\oplus} = \gamma_W^{\ominus} = 25.5$ mJ/m²), as well as the γ_i^{LW} for most biological and many other organic materials (it has been found that $\gamma_i^{LW} \approx 40$ mJ/m² for a considerable number of such substances). Thus:

$$\gamma_{iw} = (\sqrt{40} - \sqrt{21.8})^2 + 2(0 + 25.5 - 0 - \sqrt{\gamma_i^{\ominus} \cdot 25.5}) \quad\text{[IV-9A]}$$

where we assume that we are dealing with a $\gamma^{\ominus}$ monopolar substance i. This becomes:

$$\gamma_{iw} = 2.74 + 51 - 10.1 \sqrt{\gamma_i^{\ominus}} \quad\text{[IV-9B]}$$

For $\gamma_{iw} = 0$, $\gamma_i^{\ominus}$ then equals 28.3 mJ/m². In other words for a typical monopolar substance i to display a negative interfacial tension with water, its $\gamma_i^{\ominus}$ value must exceed 28.3 mJ/m² (or, in order for $\gamma_i^{\ominus}$ repulsion to occur, in water, $\delta_{iw}^{\ominus}$ must exceed 1.054, for γ_i^{LW} mJ/m²). Higher values for $\gamma_i^{\ominus}$ than 28.3 mJ/m² are not at all unusual (see Chapters XVII and XIX) so that negative interfacial tensions between various monopolar surfaces and water, and thus, "monopolar repulsions" between particles or molecules of material i in water, are commonly encountered. These "monopolar repulsions" favor the solubility of (macro) molecules, the dispersion and stability of particles, and the formation and maintenance of microemulsions (see Chapters XIX, XXI–XXIII).

It will be clear from the example illustrated in eq. [IV-9A] that eqs. [IV-8A] and [IV-8B] are best obeyed when i is a strongly monopolar substance.

Meanwhile it has become increasingly obvious that in the dry state virtually all dry *solid* surfaces of polar compounds are *monopolar electron-donors* (van Oss *et al.*, 1997), which manifest a sizeable $\gamma^{\ominus}$ and a $\gamma^{\oplus}$ which is either zero or very small. This also holds for complex mineral materials, such as clay particles and other solid metal oxides (Giese *et al.*, 1996; Giese and van Oss, 2002); see also Chapter XVII. When nevertheless a small but non-negligible $\gamma^{\oplus}$ value appears in addition to a large $\gamma^{\ominus}$, this is practically always due to hydration or residual wetness. With talc, which is quite heat-resistant, heating to 400°C reduced its $\gamma^{\oplus}$ value from 2.4 to 0.1 mJm² (van Oss *et al.*, 1997)

Polar compounds, such as sugars or their polymers, whose surfaces are monopolar electron-donors in the dried state, manifest dipolarity when in aqueous solution (Docoslis *et al.*, 2000; van Oss *et al.*, 2002) with however, still a significantly larger $\gamma^{\ominus}$ than their $\gamma^{\oplus}$ value. The reason for the $\gamma^{\ominus}$ monopolarity (in the dried state) of such polar solutes is that their excess of electron-donicity over electron-acceptivity causes, upon drying, a neutralization of all $\gamma^{\oplus}$ by the excess $\gamma^{\ominus}$, so that on a dried surface of such a solute one measures only the residual $\gamma^{\ominus}$ (Docoslis, 2000).

As far as polar *liquids* are concerned, there are *monopolar organic liquids*, such as benzene, toluene, xylene (all Lewis bases) and chloroform (Lewis acidic). And there are *dipolar organic liquids*, such as the lower alcohols, nitriles, amides and dimethylsulfoxide. The *monopolar organic liquids* among the above are miscible with the higher alkanes, whilst the dipolar ones (which are more strongly self-associated due to their dipolarity) are not significantly soluble in the higher alkanes (van Oss *et al.*, 1997).

DIFFERENT MODES OF INTERACTION BETWEEN TWO POLAR SUBSTANCES

The frequent occurrence of monopolarity among polar compounds makes it important to realize that the interaction between two randomly chosen polar compounds may either have a very pronounced polar component, or it may have no polar component at all! In principle, any statement concerning 1 and 2, in the list following, is also valid, in a symmetrical manner, for 2 and 1. However, when substance 1 is a liquid, forming a drop on a solid 2, a certain asymmetry may occur in a few cases, which gives rise to a total of ten different variants, illustrated in Fig. IV-1. The solid arrows indicate LW and the interrupted arrows, AB interactions. Horizontal arrows designate cohesive, and vertical arrows, adhesive interactions. Cohesive interactions in the *solid* are not indicated, as they do not influence the shape of the drop. Cohesive interactions in the liquid, on the other hand, play an important role in the shape of the drop (and thus in the value of the constant angle θ), in the competition between the 0adhesion of the liquid to the solid which tends to flatten the drop, and the cohesion of the liquid itself, which tends to make the drop revert to the spherical shape. Thus it is useful to enumerate the different possible classes of binary systems, composed of substances 1 and 2 (van Oss *et al.*, Langmuir, 1988); see also Figure IV-1:

   I.   1 and 2 are apolar.
   II. 1 is apolar and 2 is monopolar.
   IV. 1 is apolar and 2 is bipolar.
   IV. 1 and 2 are monopolar in the same sense.
   V.   1 and 2 are monopolar in the opposite sense.
   VI. 1 is monopolar and 2 is bipolar.
   VII. 1 and 2 are bipolar.

In class I, polar interactions are of course totally absent. Classes IV and VI can be subdivided into subclasses A and B (see below). In IV-A and IV, polar properties do not contribute to the interaction across the interface, because monopolarity does not affect the cohesion of a liquid (II), or because polar solids do not influence apolar liquids (II and IV-A) or because monopoles of the same sign do not influence each other (IV). In other words, in classes I, II, IV-A, IV-B and IV, eq. [IV-13] reduces to:

$$(1+\cos\theta)\gamma_L = 2\sqrt{\gamma_S^{LW}\gamma_L^{LW}}$$ [IV-28]

which, except for class IV-B, may be further reduced to:

$$(1+\cos\theta) = 2\sqrt{\gamma_S^{LW}/\gamma_L}$$ [IV-28A]

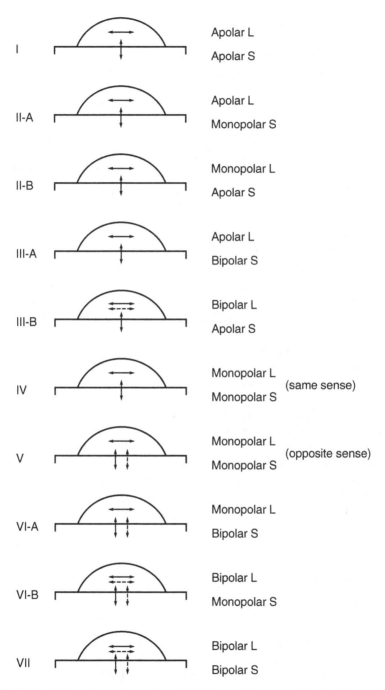

FIGURE IV-1   Schematic representation of the interactions governing the shape of liquid drops on solid surfaces. Roman numerals: classification according to the categories described in the text. LW interactions: solid arrows; AB interactions: arrows connected by interrupted lines. Horizontal arrows indicate cohesion; vertical arrows signify adhesion.

Examples are (see Fig. IV-1):

Class I: Alkanes (heptane to hexadecane), on Teflon.

Class II: Acetone and other ketones, or tetrahydrofuran, on Teflon, or on polyethylene.

Class IV-A: Hexadecane on water encased in an agarose gel (van Oss, Roberts *et al.*, 1987).

Class IV-B: Water or glycerol on Telfon.

Class IV: Dimethylsulfoxide* on surfaces of: dry dextran, polyethyleneoxide, cellulose acetate, cellulose nitrate, polymethylmethacrylate (PMMA), various dried proteins (van Oss *et al.*, J. Protein Chem., 1886).

Class V: Dimethylsulfoxide on dried RNA (van Oss *et al.*, J. Chromatog., 1987), or on PMMA.

Class VI-A: Dimethylsulfoxide on various hydrated proteins (van Oss *et al.*, J. Protein Chem., 1986).

Class VI-B: Water or glycerol on dry dextran, polyethyleneoxide (van Oss *et al.*, Advan. Colloid Interface Sci., 1987), cellulose acetate, cellulose nitrate (van Oss *et al.*, J. Chromatog., 1987), PMMA (van Oss, Good and Busscher, 1990), various dried proteins (van Oss *et al.*, J. Protein Chem., 1986).

Class VII: Water or glycerol on various hydrated proteins (van Oss *et al.*, J. Protein Chem., 1986), or on ethylene glycol or formamide encased in polyacrylamide gels (van Oss, Ju *et al.*, 1989).

The situation depicted in class IV-B (Fig. IV-1) shows that it is possible to obtain γ_S^{LW} (i.e., γ_S) of an apolar solid surface with a bipolar liquid (e.g., water). The polarity of the liquid influences only its cohesion and not its adhesion to the solid. For class IV-B, eq. [IV-28] must be used. An example is water on Teflon ($\theta = 118°$).

For classes V and VI-A (Fig. IV-1), eq. [IV-13] applies, with the slight modification that, here, $\gamma_L = \gamma_L^{LW}$, so that one can use a somewhat simplified version of eq. [IV-13]:

$$(1 + \cos\theta)\gamma_L = 2(\sqrt{\gamma_S^{LW}\gamma_L^{LW}} + \sqrt{\gamma_S^{\oplus}\gamma_L^{\oplus}}) \qquad \text{[IV-29A]}$$

in the case of a $\gamma^{\oplus}$ monopolar liquid, or:

$$(1 + \cos\theta)\gamma_L = 2(\sqrt{\gamma_S^{LW}\gamma_L^{LW}} + \sqrt{\gamma_S^{\ominus}\gamma_L^{\ominus}}) \qquad \text{[IV-29B]}$$

in the case of a $\gamma^{\ominus}$ monopolar liquid. For class VI-B eq. [IV-29A] or eq. [IV-29B] is also used, depending on whether the solid is $\gamma^{\ominus}$ or $\gamma^{\oplus}$ monopolar. For class IV-B the original Young equation (eq. [IV-10]) may be used to determine the interfacial

---

* The monopolarity of dimethylsulfoxide is an oversimplification. DMSO is predominantly monopolar with a strong $\gamma^{\ominus}$ parameter, but it also has a small $\gamma^{\oplus}$ parameter; see Chapter XVII.

tension between a bipolar liquid (e.g., water) and a monopolar solid (e.g., zein, sucrose, dextran, polymethyl-methacrylate, cellulose acetate, etc.; see van Oss *et al.*, Advan. Colloid Interface Sci., 1987). In using the original Young equation in this situation, the existence of negative interfacial tensions can be demonstrated without the need for any quantitative data with respect to γ^{LW}, $\gamma^{\oplus}$ or $\gamma^{\ominus}$ of either solid or liquid; just the contact angle, the surface tension of the liquid and the qualitative knowledge of monopolarity of the solid suffice. For class VII the full eq. [IV-13] must be employed. In the situations depicted in class IV-B (Fig. IV-1), using eq. [IV-28], it clearly is also possible to determine γ_L^{AB} whose γ_L is known, having obtained γ_{LW}^{LW}, on an apolar solid (e.g., Teflon) with a known γ_S^{LW}, by making use of eq. [IV-2] (see Fowkes, 1987).

THE SURFACE TENSION OF LIQUIDS AND SOLIDS

The concept of surface tension of liquids (at the interface with air, or vacuum, expressed in mN/m, or preferably in mJ/m²), like that of interfacial tension between two liquids, or between a liquid and a solid (expressed in the same units), seems easy to grasp, and as a rule raises few intuitive problems. The concept of surface tension of a solid, however, in view of the relative indeformability of that state of matter, has on occasion encountered difficulties in its visualization, and has even, among some workers, given rise to incredulity as to its very existence. In order not to diverge into a philosophical disquisition into the subject, it suffices to define the surface tension of solids *and* of liquids by restating eq. [II-33] as:

$$\gamma_i \equiv -\frac{1}{2}\Delta G_{ii}^{coh} \qquad\qquad [\text{II-33A}]$$

i.e., as minus half the energy of cohesion of a solid, or a liquid material i. The major remaining difference between liquids and solids then lies in the deformability of the one and the lack of deformability of the other. The only places where that difference plays a role are (a) in the methodologies of measurement of γ_i of liquids and solids (see Chatpers VIII and IX), and (b) in the asymmetry of applicability of some of the seven classes of binary systems of polar compounds (listed in the preceding section) to the interpretation of contact angle measurements (see Fig. IV-1). Thus, the knowledge of γ_i of polar materials remains important, as far as *liquids* are concerned (see eq. [IV-13]), but is, by itself, of little use for *solid*, except for its rather limited utilization in Young's general equation [IV-10]. On the other hand, for both liquids and solids i, knowledge of their surface tension *components*, γ^{LW} and γ^{AB}, and of the latter's *polar parameters*, $\gamma^{\oplus}$ and $\gamma^{\ominus}$, is indispensable for the determination of γ_{ij}, ΔG_{ij}, ΔG_{iki} and ΔG_{ikj} (where the subscript j indicates other liquids and/or solids with which substance i may interact, and the subscript k stands for a liquid, through which i and i, or i and j, interact).

INTERFACIAL TENSIONS BETWEEN POLAR LIQUIDS

Contrary to interfacial tensions between liquids and solids (which can only be determined indirectly from the individual γ components and polar parameters of the individual substances, see eq. [IV-9]), interfacial tensions between immiscible liquids can be measured directly, e.g., by the drop shape or drop weight methods (Adamson, 1990). We know that these methods are only valid for *non-polar* organic liquids, e.g., for *alkane-water* interfacial tensions (van Oss *et al.*, J. Dispersion Sci. Tech., 2002); see Chapter XIV. Girifalco and Good (1957) collected measured γ_{12} values of 98 organic liquid-water interfaces and compared the experimentally found γ_{12} with theoretical γ_{12} values obtained *via*:

$$\gamma_{12} = \gamma_1 + \gamma_2 - 2\sqrt{\gamma_1\gamma_2} \qquad\qquad [\text{IV-30}]$$

and found, in most cases, large discrepancies between $\gamma_{12}^{\text{exp.}}$ and $\gamma_{12}^{\text{theor.}}$ (eq. [IV-30]). Most of the organic liquids had γ_1 values between 20 and 29 mJ/m², so that their interfacial tension with water ($\gamma_2 = 72.8$ mJ/m²) should have been in the range of $10 < \gamma_{12} < 17$ mJ/m², if eq. [IV-30] were applicable. In reality, however, the measured γ_{12} values for these liquids ranged from 1.7 to 51 mJ/m². We now know that eq. [IV-30] is only valid for exclusively apolar (LW) interactions (see eq. [IV-35A]). Good and Girifalco (1960) introduced their proportionality factor Φ (see also Girifalco and Good, 1957) into eq. [IV-30]:

$$\gamma_{12} = \gamma_1 + \gamma_2 - 2\Phi\sqrt{\gamma_1\gamma_2} \qquad\qquad [\text{II-41A}]$$

to adjust for the discrepancies between Berthelot's geometric combining rule and the actual value for the interaction energy:

$$\Phi = \frac{-\Delta G_{12}}{\sqrt{\Delta G_{12}\Delta G_{22}}} \qquad\qquad [\text{IV-31}]$$

(cf. eq. [II-42]). In so doing, it emerged that when the anomalies are caused by the polar contributions to the energy of *cohesion* of one of the polar liquids (especially water), $\Phi < 1$; and when the anomalies are caused by polar interactions of *adhesion* between two polar liquids, $\Phi > 1$. It is only when the interactions are totally apolar (LW) (see the last section of Chapter II), *or* when the polar cohesion of the polar liquids, and the polar adhesion between the liquids cancel each other, that $\Phi = 1$. Subdividing the right side of eq. [IV-9] into three sections:

$$\gamma_{12} = \underbrace{\gamma_1^{\text{LW}} + \gamma_2^{\text{W}} - 2\sqrt{\gamma_1^{\text{LW}}\gamma_2^{\text{LW}}}}_{\text{A}} + \underbrace{2\sqrt{\gamma_1^{\oplus}\gamma_1^{\ominus}} + \sqrt{\gamma_2^{\oplus}\gamma_2^{\ominus}}}_{\text{B}}$$

$$\underbrace{-\sqrt{\gamma_1^{\oplus}\gamma_2^{\ominus}} + \sqrt{\gamma_1^{\ominus}\gamma_2^{\oplus}}}_{\text{C}} \qquad\qquad [\text{IV-9C}]$$

It is easily seen (cf. eq. [II-41A]) that when A is the only right-hand term, the interaction is completely apolar (LW), and $\Phi = 1$. When polar *cohesion* of one or both liquids is the major factor, and B is therefore preponderant, $\Phi < 1$. When polar *adhesion* between the liquids (C) is dominant, $\Phi > 1$. This is a fairly rare occurrence in liquid-liquid interactions, and will not readily give rise to very high Φ values, because in such cases, strong polar adhesion among liquids tends to lead to miscibility between the liquids (see also Chapter XII), in which case their interfacial tension is not directly measurable. However, in liquid-solid interactions, in the cases where a negative interfacial tension occurs (see Chapter XV), the occurrence of $\Phi > 1$ is quite common. Finally, when the polar cohesion of the liquids and the polar adhesion between the liquids are about of equal value, these two effects balance out (i.e., B $\approx$ C), and only A remains, causing Φ again to revert to unity; see, e.g., cyclohexanoland water, heptanoic acid and water, methylpropyl ketone and water, monochloroacetone and water (Girifalco and Good, 1957). (In subsequent papers by Good, the parameter Φ was calculated from the assumption of pairwise additivity of all components of intermolecular forces, which is not valid in the condensed state, for Keesom and Debye interactions; see Good, 1977).

Whilst the introduction of the phenomenological factor Φ did not directly elucidate the physical causes underlying the considerable deviations from the values expected via eq. [IV-30], it has been an important incentive for the development of the beginning of solutions for the problem (Fowkes, 1962, 1967).

EARLIER APPROACHES TO CORRELATE CONTACT ANGLES AND SURFACE TENSIONS IN POLAR SYSTEMS

Zisman (1964) noted a correlation between γ_S and the values of γ_L, found by an extrapolation of cos θ to cos $\theta = 1$, when θ was measured with a number of liquids L, giving rise to Zisman's designation of γ_S thus obtained, as γ_C, for "critical surface tension". Zisman's work was the forerunner of all later studies on the determination of γ_S; see Chapter XIV for limitations and corrections regarding Zisman's approach.

More recently an "equation of state" was developed (Neumann *et al.*, 1974) by means of which it was hoped that γ_S could be derived from contact angle measurements with only one single liquid, for all possible materials. This claim is based on the premises that the surface tension of liquids or solids may not be divided into components and that γ_{12} cannot be negative. In Chapter XIV the fallacies inherent in the "equation of state" are examined in more detail.

COMPARISON BETWEEN REPULSIVE LIFSHITZ– VAN DER WAALS AND REPULSIVE ACID-BASE INTERACTIONS

Both LW and AB interactions between two identical bodies, or between two different bodies *in vacuo*, are always attractive. However, when immersed in a liquid, conditions readily occur under which these interactions can give rise to a net repulsion.

With respect to van der Waals interactions, only those which occur between two *different* materials, immersed in a third (liquid) medium can be repulsive. This was already implied by Hamaker (1937) and reaffirmed by Derjaguin (1954). Fowkes (1967) gave instances of combinations of materials which might undergo a van der Waals repulsion. Visser (1972) outlined the precise conditions under which van der Waals interactions must be repulsive (see also Schulze and Cichos, 1972; Derjaguin *et al.*, 1972; Sonntag *et al.*, 1972; Wittman *et al.*, 1971; Churaev, 1974; Krusly-akov, 1974). Van Oss, Omenyi and Neumann (1979) gave the first experimental demonstration of actual van der Waals repulsions, giving rise to phase separation in polymer solutions in various apolar systems (see also Neumann, Omenyi and van Oss, 1979; Omenyi *et al.*, 1981; Neumann *et al.*, 1982; van Oss *et al.*, 1983, for descriptions of repulsions which, however, in most instances occurred in at least partly polar systems). The van der Waals interactions mentioned above were, at the time, held to be applicable to van der Waals–London or dispersion forces only. However, as shown in Chapter II, in condensed media, *on a macroscopic level*, all three apolar interfacial interactions [i.e., dispersion (London) and induction (Debye) and orientation (Keesom) interactions] obey the same laws, so that what was initially proven only for dispersion interactions remains valid for all three types of van der Waals forces, which are now grouped together as LW forces. Following Visser's treatment (1972), which was experimentally confirmed by van Oss, Omenyi and Neumann (1979) and van Oss, Absolom and Neumann (1979), the conditions for the occurrence of a repulsive Lifshitz–van der Waals interaction can be predicted to exist when the apolar component of the surface tension of the liquid medium (γ_3^{LW}) has a value which is intermediary between those of the apolar component of the surface tensions of the two different components γ_1^{LW} and γ_2^{LW}.

Exactly as LW interactions between two bodies are always attractive *in vacuo*, so do AB interactions between two bodies *in vacuo* necessarily give rise to an attraction. However, when immersed in a polar liquid medium, the net AB interaction between two polar bodies can be repulsive, *even when the two bodies consist of the same material* (van Oss *et al.*, Advan. Colloid Interface Sci., 1987; Chem. Rev., 1988). This is due to the fact that even a single, pure, homogenous polar compound can (and usually does) have two different and independent polar properties, i.e., it will have a hydrogen-donor parameter of a given value, and a hydrogen-acceptor parameter of quite another value. It can be shown that the polar interactions between moieties of the *same* material, immersed in a polar liquid such as water, are repulsive when the value of the $\gamma^\oplus$ and $\gamma^\ominus$ parameters of that liquid ($\gamma^\oplus$ and $\gamma^\ominus$ are taken to be equal for water; see above) is intermediate between the $\gamma^\oplus$ and $\gamma^\ominus$ values of the polar material.

The crucial difference between apolar (LW) and polar (AB) interactions between molecules of material (subscript 1) with the molecules of the liquid (subscript 3) in which that material is immersed lies in the respective combining rules. The difference between these combining rules is greater than might at first sight appear. For apolar materials, or for the apolar part of the interaction energy of polar materials:

$$\Delta G_{13}^{LW} = -2\sqrt{\gamma_1^{LW}\gamma_3^{LW}} \qquad [\text{II-39}]$$

whilst the polar interaction energy is described as:

$$\Delta G_{13}^{AB} = -2(\sqrt{\gamma_1^{\oplus}\gamma_3^{\ominus}} + \sqrt{\gamma_1^{\ominus}\gamma_3^{\oplus}}) \qquad \text{[IV-3A]}$$

It should be clear that the simple geometric mean combining rule, applicable to LW interactions (eq. [II-39]), is quite different from the combining rule that applies to AB interactions, which is the sum of *two* square root terms (eq. [IV-3]). The difference becomes more obvious when expressing the LW and AB components of the interfacial tension γ_{13}:

$$\gamma_{13}^{LW} = (\sqrt{\gamma_1^{LW}} - \sqrt{\gamma_3^{LW}})^2 \qquad \text{[II-35A]}$$

whilst:

$$\gamma_{13}^{AB} = 2(\sqrt{\gamma_1^{\oplus}\gamma_1^{\ominus}} + \sqrt{\gamma_3^{\oplus}\gamma_3^{\ominus}} - \sqrt{\gamma_1^{\oplus}\gamma_3^{\ominus}} - \sqrt{\gamma_1^{\ominus}\gamma_3^{\oplus}}) \qquad \text{[IV-7B]}$$

whereas γ_{13}^{LW} must always be positive, or zero, γ_{13}^{AB} can readily be negative; see eqs. [IV-8A, 8B]. The total interfacial tension can be negative, when $\gamma_{13}^{AB} < 0$ and $|\gamma_{13}^{AB}| > |\gamma_{13}^{LW}|$; see eq. [IV-9]. This can be of considerable importance in arriving at the interaction energy between two molecules of material (1), immersed in liquid (3):

$$\Delta G_{131} = -2\gamma_{13}$$

because, when $\gamma_{13} < 0$, $\Delta G_{131} > 0$, which then denotes a *repulsion* between molecules of material (1), when immersed in liquid (3); see Chapters XIX–XXIII.

The expressions for the apolar (LW) *and* the polar (AB) components of the free energies of interaction between:

Molecules of the same material (1), *in vacuo* (energy of cohesion), i.e., ΔG_{11}
Molecules of material (1), with molecules of another material (2), i.e., ΔG_{12}, or
      with molecules of the liquid medium (3), *in vacuo* (energy of adhesion),
      i.e., ΔG_{12} or ΔG_{13}
Molecules of the same material (1), immersed in liquid medium (3), i.e., ΔG_{131}
Molecules of two different materials (1) and (2), immersed in liquid medium (3),
      i.e., ΔG_{132}

are listed in Table IV-1, A and B.

From Table IV-1, A and B it can be seen that it is less appropriate to compare ΔG_{131}^{LW} (which must be negative, or zero) with ΔG_{131}^{AB} (which can be positive) than it is to compare ΔG_{132}^{LW} (which can be negative, zero, or positive), with ΔG_{131}^{AB} (which also can be negative, zero, or positive), because both ΔG_{132}^{LW} and ΔG_{131}^{AB} can be expressed as the product of the differences between positive entities. Thus $\Delta G_{132}^{LW} > 0$, when:

$$\gamma_1^{LW} > \gamma_3^{LW} > \gamma_2^{LW}$$

TABLE IV-1A
Apolar Interactions[a,b]: Expressions for ΔG_{11}^{LW}, ΔG_{12}^{LW}, ΔG_{131}^{LW} and ΔG_{132}^{LW}, and Conditions at which ΔG_{132}^{LW} is Positive, i.e., Repulsive

Free energy of cohesion *in vacuo*	$\Delta G_{11}^{LW} = -2\gamma_1^{LW}$ ΔG_{11}^{LW} always is negative, i.e., attractive
Free energy of adhesion between two different materials, 1 and 2, *in vacuo* (material, 2, may also be a liquid, 3)	$\Delta G_{12}^{LW} = -2\sqrt{\gamma_1^{LW}\gamma_2^{LW}}$ ΔG_{12}^{LW} always is negative, i.e., attractive
Free energy of interaction between two indentical materials, 1, immersed in a liquid, 3	$\Delta G_{131}^{LW} = -2(\sqrt{\gamma_1^{LW}} - \sqrt{\gamma_3^{LW}})^2$ ΔG_{131}^{LW} always is negative, i.e., attractive
Free energy of interaction between two different materials, 1 and 2, immersed in a liquid, 3	$\Delta G_{132}^{LW} = -2(\sqrt{\gamma_1^{LW}} - \sqrt{\gamma_3^{LW}})(\sqrt{\gamma_2^{LW}} - \sqrt{\gamma_3^{LW}})$[c] ΔG_{132}^{LW} is positive, i.e., repulsive when: $\gamma_1^{LW} < \gamma_3^{LW} < \gamma_2^{LW}$ or when: $\gamma_1^{LW} > \gamma_3^{LW} > \gamma_2^{LW}$

[a] It should be kept in mind that in all cases the total interfacial (IF) interaction is expressed as: $\Delta G^{IF} = \Delta G^{LW} + \Delta G^{AB}$.
[b] See van Oss, J. Dispersion Sci. Tech., 1990.
[c] Note the similarity with the equation for ΔG_{131}^{AB} in Table IV-1B.

TABLE IV-1B
Polar Interactions: Expressions for ΔG_{11}^{AB}, ΔG_{12}^{AB}, ΔG_{131}^{AB} and ΔG_{132}^{AB}, and Conditions at which ΔG_{131}^{AB} and ΔG_{132}^{AB} is Positive, i.e., Repulsive

$\Delta G_{11}^{AB} = -4\sqrt{\gamma_1^{\oplus}\gamma_1^{\ominus}}$
ΔG_{11}^{AB} always is negative, i.e., attractive

$\Delta G_{12}^{AB} = -2(\sqrt{\gamma_1^{\oplus}\gamma_2^{\ominus}} + \sqrt{\gamma_1^{\ominus}\gamma_2^{\oplus}})$
ΔG_{12}^{AB} always is negative, i.e., attractive

$\Delta G_{131}^{AB} = -4(\sqrt{\gamma_1^{\oplus}} - \sqrt{\gamma_3^{\oplus}})(\sqrt{\gamma_1^{\ominus}} - \sqrt{\gamma_3^{\ominus}})$[a]
ΔG_{131}^{AB} is positive, i.e., repulsive, when
$\gamma_1^{\oplus} > \gamma_3^{\oplus}$ and $\gamma_1^{\ominus} < \gamma_3^{\ominus}$ or when
$\gamma_1^{\oplus} < \gamma_3^{\oplus}$ and $\gamma_1^{\ominus} > \gamma_3^{\ominus}$. In water at 20°C (where it is assumed that $\gamma_3^{\oplus} = \gamma_3^{\ominus}$), ΔG_{131}^{AB} is positive when
$\gamma_1^{\oplus} > \gamma_3^{\oplus} = \gamma_3^{\ominus} < \gamma_1^{\ominus}$ and when $\gamma_1^{\oplus} < \gamma_3^{\oplus} = \gamma_3^{\ominus} > \gamma_1^{\ominus}$

$\Delta G_{132}^{AB} = 2[(\sqrt{\gamma_1^{\oplus}} - \sqrt{\gamma_2^{\oplus}})(\sqrt{\gamma_1^{\ominus}} - \sqrt{\gamma_2^{\ominus}})$

$\quad - (\sqrt{\gamma_1^{\oplus}} - \sqrt{\gamma_3^{\oplus}})(\sqrt{\gamma_1^{\ominus}} - \sqrt{\gamma_3^{\ominus}})$

$\quad - (\sqrt{\gamma_2^{\oplus}} - \sqrt{\gamma_3^{\oplus}})(\sqrt{\gamma_2^{\ominus}} - \sqrt{\gamma_3^{\ominus}})]$

ΔG_{132}^{AB} most usually is positive, i.e., repulsive,
when $(\sqrt{\gamma_1^{\ominus}} + \sqrt{\gamma_2^{\ominus}}) > 2\sqrt{\gamma_3^{\ominus}}$ and $\gamma_1^{\oplus}$ and $\gamma_2^{\oplus}$ are very small,
or (more rarely) when $(\sqrt{\gamma_1^{\oplus}} + \sqrt{\gamma_2^{\oplus}}) > 2\sqrt{\gamma_3^{\oplus}}$, and $\sqrt{\gamma_1^{\ominus}}$ and $\sqrt{\gamma_2^{\ominus}}$ are very small

[a] Note the similarity with the equation for ΔG_{132}^{LW} in Table IV-1A.

or when:

$$\gamma_1^{LW} < \gamma_3^{LW} < \gamma_2^{LW}$$

Similarly (in aqueous media, where by convention $\gamma_3^\oplus = \gamma_3^\ominus$, see above) $\Delta G_{131}^{AB} > 0$, when:

$$\gamma_1^\oplus > \gamma_w^\oplus = \gamma_w^\ominus > \gamma_1^\ominus$$

or when:

$$\gamma_1^\oplus < \gamma_w^\oplus = \gamma_w^\ominus < \gamma_1^\ominus$$

Thus it is seen, for strictly *apolar* interactions, that a van der Waals repulsion occurs between *two different* materials, when the (apolar component of the) surface tension of the liquid in which the two materials are immersed has a value which is intermediate between the values of the (apolar components of the) surface tensions of the two materials. For *polar* interactions, a repulsion occurs between particles or molecules of the *same* material, immersed in water, when the electron-acceptor and electron-donor parameters of water have a value which is intermediate between the values of the electron-acceptor and the electron-donor parameters of that material.

The crucial observation here is that to achieve repulsion, the applicable physical property of the liquid must have a value intermediate between the values of the same physical property of the material suspended or dissolved in that liquid. In the case of apolar interactions, each material has only one physical property that is applicable, i.e., its van der Waals, or Hamaker coefficient, which is proportional to the apolar component of its surface tension, γ^{LW} (Chapter II). Thus, a van der Waals repulsion is only possible between *two different* materials. Each polar compound, on the other hand, itself has two different polar parameters, an electron-acceptor and an electron-donor parameter, and it is perfectly feasible for the electron-acceptor and the electron-donor parameters of a polar liquid (e.g., water) to have a value intermediate between the two different polar parameters of a single material. When using water as a liquid, the reasoning used above is of course facilitated by the fact that, by a convention established for water, it is assumed that $\gamma_3^\oplus = \gamma_3^\ominus$. But for other polar liquids, for which $\gamma_3^\oplus \neq \gamma_3^\ominus$, the conditions for $\Delta G_{131}^{AB} > 0$ are only slightly more complicated than for water (eqs. [IV-8A,B]). Thus the conditions for $\Delta G_{132}^{AB} > 0$ and for $\Delta G_{131}^{AB} > 0$ are quite similar, provided each set of conditions is expressed in terms of the surface tension components or parameters which are appropriate for apolar (LW) or polar (AB) interactions, as the case may be.

THE MECHANISMS ALLOWING ΔG_{132}^{LW} AND ΔG_{131}^{AB} TO BE POSITIVE

For a description of the *mechanisms* behind the repulsive action, linked to positive values of ΔG_{132}^{LW} and ΔG_{131}^{LW}, one can express these entities in terms of free energies of the type of ΔG_{ij}^{LW} and ΔG_{ij}^{AB}. The apolar free energy of interaction between two different materials immersed in a third, liquid medium then is:

$$-\Delta G_{132}^{LW} = \Delta G_{12}^{LW} + \Delta G_{33}^{LW} - \Delta G_{13}^{LW} - \Delta G_{23}^{LW} \qquad [IV\text{-}32]$$

and the polar free energy of interaction between two particles or molecules of the same material immersed in a liquid medium is:

$$\Delta G_{131}^{AB} = \Delta G_{11}^{AB} + \Delta G_{33}^{AB} - 2\Delta G_{13}^{AB} \qquad [IV\text{-}33A]$$

However, the expression for ΔG_{131}^{LW} can be similarly written, as:

$$\Delta G_{131}^{LW} = \Delta G_{11}^{LW} + \Delta G_{33}^{LW} - 2\Delta G_{13}^{LW} \qquad [IV\text{-}33B]$$

and the latter (ΔG_{131}^{LW}) certainly never can become positive (see Table IV-1A).

Thus, verbal description of the mechanisms, paraphrasing eqs. [IV-32, IV-33A,B], do *not* allow any prediction as to whether ΔG_{131}^{LW} or ΔG_{131}^{AB} can assume a positive value, without further decomposing the various ΔG_{ij}^{LW} and ΔG_{ij}^{AB} components into their (quite distinct) apolar, respectively polar surface tension components and/or parameters. Thus, if one rewrites eqs. [IV-32] and [IV-33A], by decomposing the various ΔG_{ij}^{LW} and ΔG_{ij}^{AB} components as described above, one obtains:

$$\Delta G_{132}^{LW} = 2(\sqrt{\gamma_1^{LW}\gamma_3^{LW}} + \sqrt{\gamma_2^{LW}\gamma_3^{LW}} - \sqrt{\gamma_1^{LW}\gamma_2^{LW}}$$
$$- \sqrt{\gamma_3^{LW}\gamma_3^{LW}}) \qquad [IV\text{-}32A]$$
$$= -2(\sqrt{\gamma_1^{LW}} - \sqrt{\gamma_3^{LW}})\,(\sqrt{\gamma_2^{LW}} - \sqrt{\gamma_3^{LW}})$$

and:

$$\Delta G_{131}^{AB} = 4(\sqrt{\gamma_1^{\oplus}\gamma_3^{\ominus}} + \sqrt{\gamma_1^{\ominus}\gamma_3^{\oplus}} - \sqrt{\gamma_1^{\oplus}\gamma_1^{\ominus}} - \sqrt{\gamma_3^{\oplus}\gamma_3^{\ominus}})$$
$$= -4(\sqrt{\gamma_1^{\oplus}} - \sqrt{\gamma_3^{\oplus}})\,(\sqrt{\gamma_1^{\ominus}} - \sqrt{\gamma_3^{\ominus}}) \qquad [IV\text{-}34]$$

Now the strong analogy between ΔG_{132}^{LW} and ΔG_{131}^{AB} becomes obvious, when one likens γ_1^{LW} to $\gamma_1^{\oplus}$ and γ_2^{LW} to $\gamma_1^{\ominus}$. On the other hand eq. [IV-33B] can also be written as:

$$\Delta G_{131}^{LW} = -2(\sqrt{\gamma_1^{LW}} - \sqrt{\gamma_3^{LW}})^2 \qquad [IV\text{-}35]$$

the right-hand part of which clearly cannot become positive (see Table IV-1).

Thus, the mechanism of ΔG_{132}^{LW} (eq. [IV-32A]) may be described as follows: "The apolar free energy of interaction between two different materials, 1 and 2, immersed in a liquid, 3, is the sum of the apolar free energies of adhesion between 1 and liquid 3 and between 2 and liquid 3, minus the sum of the apolar free energy of adhesion between 1 and 2 and the apolar free energy of cohesion of liquid 3." When the apolar component of the surface tension of the liquid, γ_3^{LW}, has a value intermediate between the values of the apolar components of the two solutes or solids, γ_1^{LW} and γ_2^{LW}, that interaction will be repulsive; see Table IV-1. Eq. [IV-32A], and the above expression for the apolar interaction between two different materials immersed in a third liquid, can also be given in terms of the Hamaker constants of these materials (Omenyi *et al.*, 1981; Neumann *et al.*, 1982; van Oss *et al.*, 1983). For the link between the apolar component of the surface tension of pure compounds, and their Hamaker constants, see Chapter III.

Similarly to ΔG_{132}^{LW}, the mechanism of ΔG_{131}^{AB} (eq. [IV-34]) may be described as follows: "The polar free energy of interaction between molecules or particles of the same material, 1, immersed in a liquid, 3, is (2 ×) the sum of the polar free energies of adhesion between the electron acceptor of 1 and the electron donor of the liquid, and of the electron donor of 1 and the electron acceptor of the liquid, minus the sum of the polar energies of cohesion of material, 1, and of liquid, 3." When the electron acceptor and the electron donor parameters of the liquid have a value intermediate between the values of the electron acceptor parameter of material 1, $\gamma_1^{\oplus}$ and its electron donor parameter, $\gamma_1^{\ominus}$, that interaction will be repulsive; see Table IV-1B. It should be kept in mind that the values of $\gamma_1^{\oplus}$ and $\gamma_1^{\ominus}$, which, together, give rise to the polar energy of cohesion of material 1, are almost always different.

As ΔG_{131}^{AB} may be likened to ΔG_{132}^{LW}, it is not surprising that the expression for ΔG_{132}^{AB} is more complicated than either of these (see Table IV-1A and B), so that there are many more combinations of conditions at which ΔG_{132}^{AB} can be positive, or negative.

IMPLICATIONS OF $\Delta G_{132}^{AB} > 0$

When $|\Delta G_{131}^{AB}| > |\Delta G_{131}^{LW}|$ and the interfacial (IF) free energy of interaction, $\Delta G_{131}^{IF} = (\Delta G_{131}^{LW} + \Delta G_{131}^{AB}) > 0$, a repulsion must occur between the particles or molecules of material, 1, immersed or dissolved in liquid, 3. (Electrostatic interactions between particles, 1, are not considered here, but when they exist, they must contribue further to the mutual repulsion, as they then occur between particles of the same sign of charge.) The occurrence of a positive ΔG_{131}^{AB} is the sole, or a strongly contributing cause of the solubility in water of most biopolymers and other polar polymers (Chapter XXII); it is also the cause of the high osmotic pressure (at high concentrations) and molecular weight independence of that osmotic pressure of water-soluble oligomers and polymers, such as polyethylene oxide and dextran (Chapter XIX). A positive ΔG_{131}^{AB} also is the major factor in the stability of red blood cells and of polyethyleneoxide-stabilized particle suspensions (Chapter XXIII).

IMPLICATIONS OF $\Delta G_{132} > 0$

When the interfacial (IF) free energy of interaction, $\Delta G_{132}^{IF} = (\Delta G_{132}^{LW} + \Delta G_{132}^{AB}) > 0$, a net repulsion must occur between the particles or molecules of materials 1 and 2, immersed or dissolved in liquid, 3 (provided there is no over-riding electrostatic attraction between materials 1 and 2). The occurrence of a positive ΔG_{132} is a necessary condition for phase separation in apolar, polar and aqueous polymer solutions (Chapter XXI).

Finally, in the case of suspended particles, 1, in the presence of a dissolved polymer, 2, in an aqueous medium, 3, when $\Delta G_{132} > 0$, as well as $\Delta G_{131} > 0$ and $\Delta G_{232} > 0$, then, when $\Delta G_{132} > \Delta G_{131}$, there will be "depletion flocculation", and when $\Delta G_{132} < \Delta G_{131}$, there will be "depletion stabilization" of the suspension (Chapters XX and XXIII).

CONNECTION BETWEEN "HYDROPHOBIC" AND HYDROPHILIC INTERACTIONS

In water "hydrophobic" attractions are solely due to the hydrogenbonding free energy of cohesion of the water molecules (equal to -102 mJ/m^2) (cf. eq. [IV-17B] discussed in Chapter XVIII, and eq. [IV-17], above). That attractive AB energy of cohesion of water does not disappear when a net hydrophilic repulsion occurs. A hydrophilic repulsion can only prevail when it is quantitatively stronger than the always present "hydrophobic" attraction caused by the free energy of cohesion of water due to hydrogen-bonding; see Table IV-2.

The *conditio sine qua non* for a positive value of $\Delta G_{131}^{AB}(= -4(\sqrt{\gamma_1^\ominus} - \sqrt{\gamma_3^\oplus})(\sqrt{\gamma_1^\ominus} - \sqrt{\gamma_3^\ominus})$ in eq. [IV-34]) is for one of the polar surface tension parameters (usually $\gamma_1^\oplus$) to be smaller than that of water, i.e., $\gamma_1^\oplus < \gamma_W^\oplus$, while the other parameter ($\gamma_1^\ominus$) must be larger than that of water, i.e., $\gamma_1^\ominus > \gamma_W^\ominus$; see above. It is clear that with a constant value of one polar surface tension parameter (e.g., $\gamma_1^\ominus$, which is higher than $\gamma_3^\ominus$), the highest positive value of ΔG_{131}^{AB} can only be obtained when the other polar surface tension parameter (e.g., $\gamma_1^\oplus$) is equal to, or close to zero. For instance, maximum solubility in water is best attained by monopolar polymers, and the highest suspension stability can be reached by particles with a monopolar surface, see Chapters XXII and XXIII, respectively. As shown earlier, a monopolar surface tension component (in practice, usually $\gamma_1^\ominus$) should have a value of at least 28.3 mJ/m^2, to overcome not only the usual apolar attraction of about $-2 \times 2.74 \approx -5.5$ mJ/m^2 (for a γ_1^{LW} of ≈ 40 mJ/m^2), but especially the background of *polar* ("*hydrophobic*") *attraction*, due to the hydrogen-bonding energy of cohesion of water of $-4 \times 25.5 = -102$ mJ/m^2; see Table IV-2.

CODA

Contrary to apolar interfacial Lifshitz–van der Waals (LW) interactions, which have only one aggregate property, exemplified by the Hamaker constant, *polar* or (Lewis) acid-base (AB) interactions are governed by two different (and independently variable) properties, exemplified by the electron-acceptor ($\gamma^\oplus$) and the electron-donor ($\gamma^\ominus$) surface tension parameters. The total polar (AB) surface tension component, γ^{AB}, equals twice the geometric mean of $\gamma^\oplus$ and $\gamma^\ominus \cdot \gamma^\oplus$ and $\gamma^\ominus$ thus are not additive, but γ^{LW} and γ^{AB} are: $\gamma^{LW} + \gamma^{AB}$ make up the total surface tension, γ, of non-metallic condensed-phase materials. Also contrary to the apolar, LW interfacial tension, γ_{ij}^{LW}, the *polar*, AB interfacial tension, γ_{ij}^{AB}, comprises four interaction terms and can assume a positive as well as a negative value. Thus the total interfacial tension, γ_{ij}, of polar materials can be positive as well as negative, which corresponds to the possibility of

TABLE IV-2
Comparison Between the Interfacial Mechanisms of "Hydrophobic" Attraction and Hydrophilic Repulsion for Typical "Hydrophobic" and Hydrophilic Entities

$\Delta G_{1W1} =$	$-2(\sqrt{\gamma_1^{LW}} - \sqrt{\gamma_W^{LW}})^2$	$-4\sqrt{\gamma_1^\oplus \gamma_1^\ominus}$	$-4\sqrt{\gamma_W^\oplus \gamma_W^\ominus}$	$+4\sqrt{\gamma_1^\oplus \gamma_W^\ominus}$	$+4\sqrt{\gamma_1^\ominus \gamma_W^\oplus}$		Eq. [IV-17][a]
	I	II	III	IV	V		
	(LW attraction between 1 and water)	(Polar cohesion of 1)	(Polar cohesion of water)	(Polar adhesion between el. acceptor of 1 and el. donor of water)	(Polar adhesion between el. donor of 1 and el. acceptor of water)		
	——— LW ———	————————————————————— AB —————————————————————					
ΔG_{1W1} of "hydrophobic" attraction between apolar particles or molecules, such as octane, of $\gamma_1^{LW} \approx 22$ mJ/m2, immersed in water, at 20°C	0	0	-102 mJ/m²	0	0	$=$	-102 mJ/m²
ΔG_{1W1} of hydrophilic repulsion between typical hydrophilic particles or molecules, of $\gamma_1^{LW} \approx 40$ mJ/m2, $\gamma_1^\oplus = 0$ and $\gamma_1^\ominus \approx 50$ mJ/m2, immersed in water, at 20°C	-5.5 mJ/m²	0	-102 mJ/m²	0	$+127.8$ mJ/m²	$=$	$+20.3$ mJ/m²

[a] Subscript 3 of eq. [IV-17] is here designated by subscript, W, for water.

having either attractive or repulsive interfacial free energies of interaction between *identical* polar surfaces, in polar, and especially in aqueous media. (Between two *different* surfaces, immersed in a liquid, both LW and AB interaction energies can be attractive or repulsive.)

The Young equation linking contact angles of liquid drops on solid surfaces comprises three unknowns: γ_S^{LW}, $\gamma_S^{\oplus}$ and $\gamma_S^{\ominus}$. To solve for these three unknowns for any given polar solid surface, S, contact angles must be measured with (at least) three liquids, L, each with known γ_L^{LW}, $\gamma_L^{\oplus}$ and $\gamma_L^{\ominus}$ (two of which liquids must be polar); to obtain finite contact angles it is a prerequisite that $\gamma_L > \gamma_S$; see Chapter XII. Examples are given of apoalr and polar (or mixed) liquid-solid interactions (e.g., in the guise of contact angles), and the unusual intractions of monopolar liquids or solids (e.g., with $\gamma^{\oplus} = 0$ and a significant $\gamma^{\ominus}$, or *vice versa*) are illustrated.

V Electrical Double Layer Interactions

ELECTROKINETIC POTENTIAL AND IONIC DOUBLE LAYER

The electrokinetic, or ζ-potential, of macromolecules, particles or larger surfaces, dissolved or immersed in an appropriate liquid (preferably containing dissolved ions), is the potential measured at the slipping plane by electrokinetic methods, such as electrophoresis, electroosmosis, or streaming potential. The potential of the particle itself, called the ψ_o-potential, is situated at the precise particle-liquid interface; the ψ_o-potential is not directly measurable, but it is the potential from which the free energy of electrostatic interaction traditionally is derived, *via*:

$$\psi_o = \zeta(1 + z/a)\exp(\kappa z) \qquad \text{[V-1]}^*$$

where z is the distance from the particle's surface to the slipping plane, a distance that is generally of the order of $\approx$3 to 5 Å, a is the radius of the particle, and $1/\kappa$ is the thickness of the diffuse ionic double layer, or Debye length:

$$1/\kappa = [(\varepsilon kT)/(4\pi e^2 \Sigma v_i^2 n_i)]^{1/2} \qquad \text{[V-2]}$$

where ε is the dielectric constant of the liquid medium (for water, $\varepsilon \approx 80$), k is Boltzmann's constant ($k = 1.38 \times 10^{-23}$ J per °K), T the absolute temperature in °K, e the charge of the electron ($e = 4.8 \times 10^{-10}$ e.s.u.), v_i the valency of each ionic species and n_i the number of ions of each species per cm^3 of bulk liquid. Table V-1 lists some values for the Debye length $1/\kappa$ for a number of electrolyte concentrations.

Thus, the lower the ionic strength, the thicker (but also the less substantial) the ionic double layer and the higher the ionic strength, the thinner (but also the denser) the ionic double layer. At a constant ψ_o, ζ clearly is considerably smaller than ψ_o at high ionic strengths (and thus at small $1/\kappa$). An important problem connected with the extrapolation from ζ (which is measurable) to ψ_o (which is not directly

---

* A list of symbols can be found on pages 399–406.

TABLE V-1

Values for the Debye Length $1/\kappa$ for a Number of Aqueous 1-1, 1-2 and 2-2 Electrolyte Solutions

Solution	$1/\kappa$ (nm)
H_2O	1,000
10^{-5}M NaCl	100
10^{-3}M NaCl	10
10^{-1}M NaCl	1
10^{-5}M Na_2SO_4	56
10^{-3}M Na_2SO_4	5.6
10^{-1}M Na_2SO_4	0.56
10^{-5}M $MgSO_4$	48
10^{-3}M $MgSO_4$	4.8
10^{-1}M $MgSO_4$	0.48

Source: Gouy, 1910; Henry, 1931.

measurable), particularly with a view to the calculation of the electrostatic free energy of interaction, ΔG^{EL} (see below), is that in aqueous systems, inside the slipping plane, the dielectric constant ε may be much smaller than that of the bulk liquid, in cases of strong orientation of the water molecules closest to a polar surface, due to hydration (Vaidhyanathan, 1986). A pronounced orientation of the water molecules in the first and to a lesser extent in the second layer of hydration of serum albumin at pH 7.2 was demonstrated by van Oss and Good (1988a). From the slipping plane outward that orientation decays quite rapidly, which makes it often more reasonable to consider, for practical purposes, the surface of the particle as situated at the slipping plane, and thus to use the measured ζ-potential, rather than the ψ_o-potential extrapolated from it, in equations [V-3–V-6], used below Zukoski and Saville (1986) showed that when a particle's potential is measured by conductivity determinations, a higher potential is found than the particle's ζ-potential measured by electrokinetic methods. Zukoski and Saville's potential includes potentials originating within the second ionic (or Stern) layer, in other words, within the slipping plane. However, for electrostatic interactions between mobile macromolecules or particles, it may be a more realistic approach to situate the operative boundary of the macromolecule, particle or surface, at the slipping plane and not inside of it.

The ζ-potential of a surface is rarely directly related to its γ^{AB} (see Chapter IV) but it is strongly linked to its $\gamma^{\ominus}$ (see Chapter XIX). The measurement of contact angles on amphoteric surfaces may also be influenced by the ζ-potential of the surface, when these contact angles are measured with drops of buffered water of different pH values. This is because the water-surface interaction then can influence the degree of dissociation of amphoteric groups, as well as of electron-accepting or electron-donating groups (Holmes-Farley *et al.*, 1985). Also, electrokinetic migration in polar organic liquids appears to be at least qualitatively linked to electron-acceptor or electron-donor properties (Labib and Williams, 1984, 1986, 1987; Fowkes *et al.*, 1982; Fowkes, 1987; see also Chapter XIX).

For a detailed treatment of ζ- and ψ_o-potentials and the electrical double layer, see e.g., Overbeek (1952), Shaw (1969), Sparnaay (1972), Hunter (1981), and Kitahara and Watanabe (1984).

FREE ENERGY OF ELECTROSTATIC INTERACTIONS

For moderate values of ζ (i.e., not higher than $\approx$ 25 to 50 mV), the electrostatic interaction energy between two flat plane parallel plates, at a distance ℓ, is expressed as (Verwey and Overbeek, 1948; Overbeek, 1952; Hiemenz, 1986):

$$\Delta G^{EL} = 1/\kappa \cdot 64 \, nkT\gamma_o^2 \exp(-\kappa\ell) \qquad \text{[V-3]}$$

where n is the number of counterious in the bulk liquid per cm^3 and:

$$\gamma_o = [\exp(ve\psi_o/2kT) - 1)] / [\exp(ve\psi_o/2kT) + 1] \qquad \text{[V-4]}$$

where v is the valency of the counterions in the electrolyte (for the other symbols see eq. [V-2]). For the goodness of fit of eq. [V-3], see p. 52 of Kitahara and Watanabe (1984). Under similar conditions the interaction energy between two equal spheres of radius R can be expressed as (Verwey and Overbeek, 1948; Overbeek, 1952; Kitahara and Watanabe 1984):

$$\Delta G^{EL} = 0.5 \, R\psi_o^2 \, \varepsilon \ln [1 + \exp(-\kappa\ell)] \qquad \text{[V-5]}$$

and for a sphere of radius R and a flat plate (Visser, 1976):

$$\Delta G^{EL} = R\psi_o^2 \, \varepsilon \ln [1 + \exp(-\kappa\ell)] \qquad \text{[V-6]}$$

where ℓ is measured from the outer edges of the spheres. Overbeek (1952) indicated the delimitations of the use of eq. [V-5], as a function of κR. Generally, eq. [V-5] is valid for $\kappa R > 10$; for lower values of κR, Overbeek gives tables and graphs, allowing better approximations than eq. [V-5]. Eq. [V-6], for a sphere and a flat plate, is also valid for two (perpendicularly) corssed cylinders of radius R, which is applicable to the configuration used in Israelachvili's force balance; see Chapter XVI. All of the above approximations (eqs. [V-3, 5, 6]) wre derived for a constant surface *potential* model. For the computation of ΔG^{EL} according to a constant surface *charge* model, see Kitahara and Watanabe (1984, pp. 70ff.). To derive the aggregate, ψ_{o_a} value from two interacting values ψ_{o_1} and ψ_{o_2}, use the geometric mean between the two, i.e., $\psi_{o_a} = \sqrt{\psi_{o_1} \psi_{o_2}}$. This applies to eqs. VII-3, 5 and 6.

ΔG^{EL} is function of the square of the electrokinetic potential (eqs. [V-3, 5, 6]) and thus in practice becomes rapidly negligible below a certain threshold. In biological systems, where, at 0.15 M NaCl, $1/\kappa \approx 8$ Å, and using ζ as the operative potential, ΔG^{EL} is reduced to rather low values, when ζ is $\approx$ 14 mV or lower. Even for originally somewhat more highly charged surfaces, an increase in ionic strength (see eq. [V-2]) would strongly reduce ζ (eq. [V-1]), and thereby also depress ΔG^{EL} (eqs. [V-3, 5, 6]). This explains why high salt concentrations are so detrimental to electrostatic particle repulsion, and thus to the stability of those particle suspensions which are mainly stabilized through electrostatic repulsion. (R and a, for radius, are used interchangeably in this chapter.)

ELECTROKINETIC PHENOMENA

The five main electrokinetic phenomena and the connection between them are schematically indicated in Fig. V-1. *Electrophoresis* is the transport of charged particles, in an electric field, toward the oppositely charged electrode, in a polar liquid (Fig. V-1A). *Electroosmosis* is the transport of the liquid medium alongside a surface that is electrically charged but immovable, in the direction of the electrode with the same sign of charge as that of the immovable surface (Fig. V-1B). *Electroosmotic counter-pressure* is the pressure that builds up against one extremity of a chamber when electroosmotic flow is prevented by enclosure of the chamber (Fig. V-1C). *Streaming potential* is the electrical potential difference caused by forcing a polar liquid to stream under pressure alongside an electrically charged surface (Fig. V-1D). Finally, *migration potential* is the electrical potential difference caused by the migration (in practice virtually always through sedimentation) of charged particles in a polar liquid (Fig. V-1E).

From Fig. V-1, it is easily seen how these five electrokinetic phenomena are interrelated: Electrophoresis (Fig. V-1A) reverts to electroosmotic flow (Fig. V-1B) when the charged particles are made immovable. Electroosmotic flow (Fig. V-1B) clearly can only manifest itself as electroosmotic counter-pressure when liquid flow alongside a charged surface is forcibly prevented (Fig. V-1C). Conversely, when a pressure is applied on the liquid, making it stream alongside a charged surface, an electrical potential difference called streaming potential is created (Fig. V-1D). Finally, conversely to electrophoresis (where a charged particle migrates under the influence of an electric field, Fig. V-1A), when a charged particle migrates through a polar liquid with, e.g., sedimentation as a driving force, an electrical potential results, called a migration potential (Fig. V-1E). Thus a simple relationship can be perceived between any of these five electrokinetic phenomena and its closest neighbor, when classed in the order:

Electrophoresis,
Electroosmosis,
Electroosmotic counter-pressure,
Streaming potential,
Migration potential,
Electrophoresis, etc., (see Fig. V-1).

For a more formal treatment of the interrelationship of a number of these phenomena, see Mazur and Overbeek (1951). Not discussed here is, e.g., streaming current, which

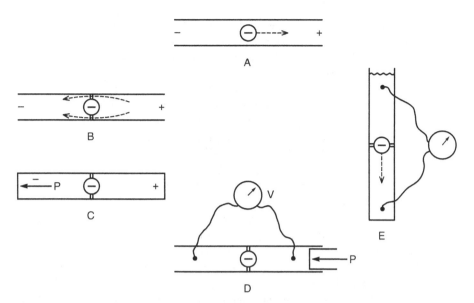

FIGURE V-1 Schematic representation of five electrokinetic phenomena: A) electrophoresis, B) electroosmosis, C) electroosmotic counter-pressure, D) streaming potential, and E) migration potential; for details see text. In a clockwise as well as in a counter-clockwise direction each one of these five electrokinetic phenomena depicted above stands in a simple relationship to its closest neighbors; – and + indicate the polarity of an applied electric field; P indicates pressure; – stands for a particle with a negatively charged surface; ‖ indicates that such a particle is immobilized by attachment inside the chamber; and V designates a voltmeter indicating an electrical potential difference resulting from the last two electrokinetic phenomena (D and E). Not drawn here is a model of streaming current, which is closely related to streaming potential and has analogies with electroosmosis. (From van Oss, 1979.)

is closely related to streaming potential, and has analogies with electroosmosis (Overbeek, 1951).

ζ-POTENTIAL, IONIC STRENGTH AND ELECTROKINETIC MOBILITY

As a rule the electrophoretic mobility of molecules and particles is solely governed by the potential at their surface of shear (ζ-potential) and does *not* depend on their size or shape. There are, however, two sets of conditions under which the size or shape of molecules and particles may influence their electrophoretic mobility. These are conditions under which the ratio between the dimensions of the molecules or particles and the thickness of their diffuse ionic double layer is such that the distortion of the electric field by the particles is neither at its minimum nor at its maximum. A few limitations, of a practical order, will be imposed in advance (Overbeek and Wiersema, 1967):

1. Only rigid molecules or particles are considered here. Very rarely, if ever, does one encounter molecules or particles that are sufficiently deformable under the influence of an electric field, to influence their electrophoretic mobility (Abramson *et al.*, 1942, 1964).

2. Only non-conducting molecules or particles are considered here. Even normally fairly conducting particles behave as insulated particles when subjected to electrophoresis. The electrical double layer surrounding the particles offers such a strong resistance to an electric current, that it transforms conducting particles into insulated ones.

3. The particles' movement inside the liquid remains laminar. The electrophoretic mobilities that most particles can attain (a few μm volt^{-1} sec^{-1} cm) cannot normally give rise to turbulence (van Oss, 1955).

4. Electroosmotic counter-flow of the bulk liquid along the walls of the vessel [which is governed by the ζ-potential of the walls (Overbeek, 1952)], although in practice frequently an important factor, will not be taken into account here, as it can always (at least theoretically) be dissociated from the real electrophoretic mobility of molecules or particles.

5. The influence of Brownian movement of the particles, being quite slight (Wiersema, 1964), will not be taken into account.

INFLUENCE OF THE DOUBLE LAYER

THICK DOUBLE LAYER

The simplest situation that may be considered is the one in which one electrophoreses a small particle, with a very thick but unsubstantial electrical double layer, which offers no significant impediment to the electric field (see Fig. V-2). Under these conditions the three forces (Hückel, 1924), acting on the particle (see Fig. V-3), ought, once a uniform velocity is attained, to cancel one another out:

$$F_1 + F_2 + F_3 = 0 \qquad\qquad [V\text{-}7]$$

These forces are:

The attractive force of the electric field for the charged particle:

$$F_1 = + ne\chi \qquad\qquad [V\text{-}8]$$

(in which ne is the total effective electric charge of the molecule or particle and χ the electric field strength).

The hydrodynamic friction force that the particle undergoes in its movement through the liquid:

$$F_2 = -6\pi\eta aU \qquad\qquad [V\text{-}9]$$

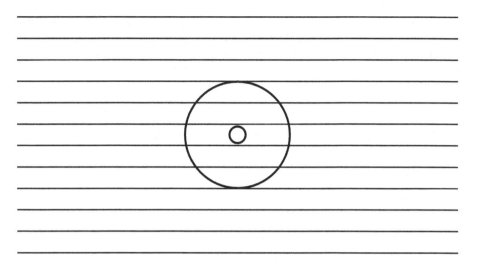

FIGURE V-2 Small particle surrounded by a thick electrical double layer (small κa; the lines of force of the electric field are virtually unhindered by the particle or its double layer). (From van Oss, 1975.)

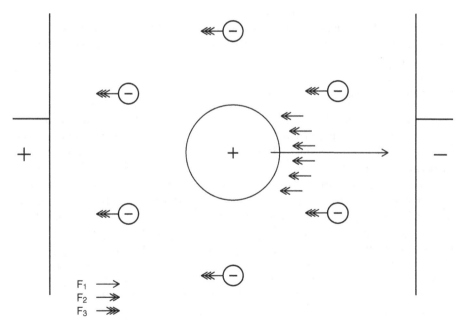

FIGURE V-3 The three forces acting on particle in an electric field. F_1 is the force the electric field exerts directly on the charge(s) of the particle. F_2 is the hydrodynamic friction force the particle undergoes, caused by its movement through the liquid. F_3 is the electrophoretic friction force, caused by the oppositely charged ions moving in the direction opposed to that of the particle. (From van Oss, 1975.)

(in which η is the viscosity of the medium, a the radius or other expression for the dimension of the particle and U the electrophoretic velocity the particle attained).

The electrophoretic friction force, caused by the movement in the direction opposed to that of the particle, of the oppositely charged ions:

$$F_3 = \chi(\varepsilon\zeta a - ne) \qquad [V\text{-}10]$$

(in which ε is the dielectric constant of the medium and ζ the potential of the particle at its surface of shear).

From eqs. [V-7, 10] it follows that:

$$u = \frac{\zeta\varepsilon}{6\pi\eta} \qquad [V\text{-}11]$$

where:

$$\frac{U}{\chi} = u \qquad [V\text{-}12]$$

(u being the electrophoretic mobility).

Thus, particles that are small compared to the thickness of their electric double layers have an electrophoretic mobility u that is solely proportional to their electrokinetic ζ-potential.

THIN DOUBLE LAYER

The other situation that must be considered is the one in which one electrophoreses a somewhat larger particle (relatively speaking), with a thin but quite dense electrical double layer. The electric field will be considerably distorted, on account of both the particle size and the compactness of the double layer (von Smoluchowski, 1921) (see Fig. V-4). In this case the oppositely charged ions that are directly in the path of the moving particle will oppose little or no electrophoretic fraction force to the particle. There remains an electrophoretic friction force by oppositely charged ions that are more tangential to the moving particle, but this effect is fairly small. Thus the electrophoretic mobility of particles in this situation is greater than that of the particles considered earlier. Under these conditions, the equation for the electrophoretic mobility (which results from the interaction between the flow of electricity and the flow of liquid in the double layer) (Overbeek, 1952) is derived with the help of Poisson's equation:

$$\frac{d}{dx}\left(\varepsilon\frac{d\psi}{dx}\right) = -4\pi\sigma \qquad [V\text{-}13]$$

(x is the distance of any given point from the surface of shear,   the electric potential of the double layer and σ the surface charge density), and integration by parts of

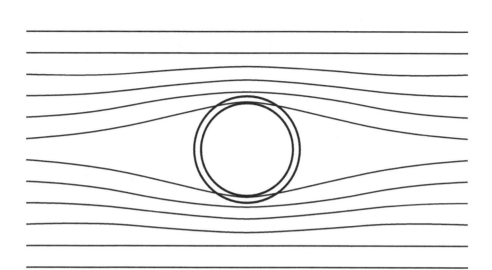

FIGURE V-4 Large particle surrounded by a thin electrical double layer (large κa); the lines of force of the electric field are strongly deflected by the particle. (From van Oss, 1975.)

the equation for the forces acting on all volume elements concerned. This yields (Overbeek and Wiersema, 1967; von Smoluchowski, 1918, 1921; Shaw, 1969), after equating ζ with , at the surface of shear:

$$u = \frac{U}{\chi} = \frac{\zeta \varepsilon}{4 \pi \eta} \qquad \text{[V-14]}$$

The two discordant equations, that of Hückel (1924):

$$u = \frac{\zeta \varepsilon}{6 \pi \eta} \qquad \text{[V-11]}$$

for the electrophoretic mobility of particles that are relatively small compared to the thickness of their electrical double layer and that of von Smoluchowski (1918, 1921):

$$u = \frac{\zeta \varepsilon}{4 \pi \eta} \qquad \text{[V-14A]}$$

for the electrophoretic mobility of particles that are large compared to the thickness of their double layers, were reconciled by Henry (1931) into one equation, by taking both the radius a of the particle and the thickness of its electrical double layer, expressed as 1/κ, into account (eq. [V-2]):

$$u = \frac{\zeta \varepsilon}{6 \pi \eta} f(\kappa a) \qquad \text{[V-15]}$$

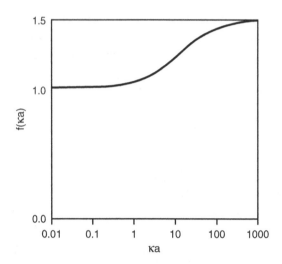

FIGURE V-5 Graph of f(κa) from Henry's equation [V-15] versus κa (the latter is on a logarithmic scale, for spherical particles. (Adapted from Overbeek, 1952; p. 209; see van Oss, 1975.)

when κa → ∞, or, in practice, when κa > 300, f(κa) becomes = 1.5, which makes eq. [V-15] revert to von Smoluchowski's equation [V-14]. When κa → 0, or, in practice, when κa < 1, f(κa) becomes = 1.0, which makes eq. [V-11] revert to von Smoluchowski's equation [V-11]. Figure V-5 shows the variation of f(κa) with κa for spherical particles (Overbeek, 1953; Dukhin and Derjaguin, 1974). For a long cylinder oriented *perpendicular* to the electric field, f(κa) becomes = 0.75 at low values of κa, while at high values of κa this function remains = 1.5. For a long cylinder, oriented *parallel* to the electric field, and for all cases of electroosmosis and streaming potential, f(κa) remains 1.5 for all values of κa (Overbeek, 1952).

RELAXATION

In all the situations treated above, it had been assumed, by Hückel (1924), by von Smoluchowski (1921), as well as by Henry (1924), that the initial symmetry of the electrical double layer surrounding the particle remained unimpaired during the electrophoretic movement. However, this is not always the case: under certain quite common conditions the diffuse electrical double layer may be more impeded in its electrophoretic progress through the liquid than the particle it surrounds. This is called relaxation. Relaxation forces (see Fig. V-6) can appreciably reduce the electro-phoretic mobility. Overbeek (1943) and Overbeek and Wiersema (1967) have indicated the conditions under which the relaxation effect may be neglected:

1. At ζ < 25 mV, for all values of κa;
2. At κa → 0 for all values of ζ, and
3. At κa → ∞ for all values of ζ.

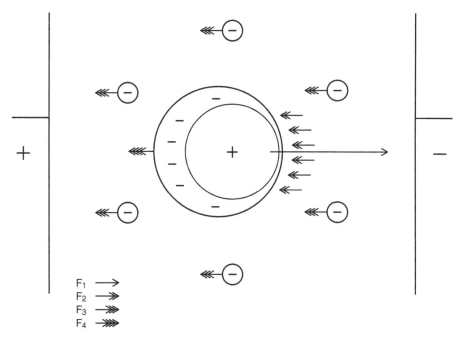

FIGURE V-6   The same diagram as Fig. V-4, with, in addition, the influence of the relaxation force F_4, which distorts the ionic double layer. (From van Oss, 1975.)

In cases 2 and 3, in practice it generally suffices that $\kappa a < 0.1$ or > 100 for the relxation effect to be quite small. In other words: relaxation is of little import when:

1. ζ potentials and thus electrophoretic mobilities are low, because the actual electrophoretic movement necessarily remains slow enough for double layer retardation to be minimal (at $u < 2$ μm volt^{-1}sec^{-1} cm);
2. The electrical double layer is very thick but also very unsubstantial, and
3. The electrical double layer is very thin and dense.

   In all other cases, i.e., when $\zeta > 25$ mV and $0.1 < \kappa a < 100$, the relaxation effect seriously influences the electrophoretic mobility (Overbeek, 1943; Overbeek and Wiersema, 1967; Wiersema, 1964; Booth, 1948, 1950; Derjaguin and Dukhin, 1974). A number of complicated equations taking this effect into account have been given; as they are not readily soluble without the help of a computer they are not repeated here. Figure V-7 illustrates the effect of relaxation on the electrophoretic mobility at various values of κa as a function of the ζ-potential, in a 1-1 electrolyte (Overbeek, 1943). For $0.3 < \kappa a < 20$ there appears to be a maximum in the attainable electrophoretic mobility, roughly between 3.8 and 6 μm volt^{-1}sec^{-1}cm, which agrees well with the "maximum" electrophoretic mobility of 5 μm volt^{-1}sec^{-1}cm found by von Hevesy (1913, 1917). The available aids for the calculation of ζ-potentials from electrophoretic mobilities under a variety of conditions of κa are discussed below (see also van Oss, 1975).

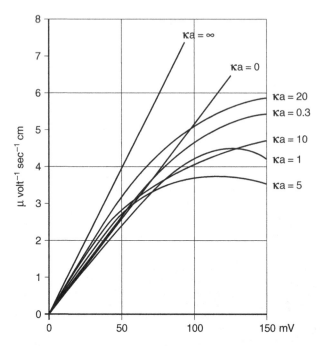

FIGURE V-7   Graph of the electrophoretic mobility u (in μ volt⁻¹sec⁻¹cm) versus ζ (in mV) for a number of values of κa. A maximum electrophoretic mobility between 3.8 and 6 μm volt⁻¹sec⁻¹cm for $0.3 < \kappa a < 20$ is quite apparent. (From Overbeek, 1943; see van Oss, 1975.)

VALIDITY RANGES OF THE EQUATIONS

1. $\kappa a < 0.1$

When the ionic strength of the medium is very low and when the molecules or particles are quite small, Hückel's equation [V-11] is valid.

2. $\kappa a > 300$

When the molecules or particles are large compared to the thickness of the electrical double layer, von Smoluchowski's equation [V-14] is valid. In Table V-2, some minimum dimensions of a are given for a number of values of $1/\kappa$, corresponding to a number of electrolyte solutions, see Table V-1 (Henry, 1931; Gouy, 1910).

3. $0.1 < \kappa a$ 300

From Table V-2 it also becomes obvious that with most biological or biochemical substrates (e.g., proteins, nucleic acids, subcellular particles), the dimensions (a) are such that at the usual concentrations of electrolytes $\kappa a < 300$.

In addition, with respect to the lower end of this range, with most buffers normally used for electrophoresis, $1/\kappa < 10$ μm, so that a must be <1 μm for the ratio between the double layer and the particles' dimension to be small enough to use the unmodified Hückel equation [V-11]. Proteins and nucleic acids, however, generally have dimensions > 1 μm, so that it must be concluded that for virtually all electrophoresis work with biological or biochemical substrates small than whole cells, the conditions are bound to be such that $0.1 < \kappa a < 300$. As in the *entire* range of $0.1 < \kappa a < 300$ relaxation effects make themselves felt (at least at ζ-potentials > 25 mV) (Overbeek, 1943), and as Hückel's equation [V-11] evolves into von Smoluchowski's [V-14] in the range of $1 < \kappa a < 300$ (at *all* values for ζ), in this range electrophoretic mobilities can only be translated into ζ-potentials after considerable computational effort. Fortunately, however, there exist tables that facilitate that effort (Wiersema *et al.*, 1966).

TABLE V-2
Ionic Strengths[a] and Minimum Dimensions of a (in nm)
for a Number of Values[b] of $1/\kappa$ (in nm) so that $\kappa a > 300$

Ionic strengths ($\Gamma/2$)	$1/\kappa$	a
0	1,000	300,000
10^{-5}	100	30,000
10^{-3}	10	3,000
0.1	1	300
3×10^{-5}	56	17,000
3×10^{-3}	5.6	1,700
0.3	0.56	170
4×10^{-5}	48	14,000
4×10^{-3}	4.8	1,400
0.4	0.48	140

[a] The ionic strength, $\Gamma/2 = \Sigma (v_i^2 m_i)/2$, where v_i is the valency of each ionic species and m_i the molar concentration of the ions of each species in the bulk liquid; cf. Table V-1.
[b] See eq. [V-2].

CONFRONTATION OF THE EQUATIONS
WITH EXPERIMENTAL DATA

1. SMALL κa

Few experimental data are available on the constancy or variability of the electrophoretic mobility of small molecules. Although in principle the verification of the independence of electrophoretic mobilities of a at low κa seems quite feasible, no report of such a verification has been published to date. With a view to the separation of bio-polyelectrolytes according to size, gel electrophoresis techniques have been developed, using either compounds that differ only in size but have approximately the same charge (nucleic acids) (Loening, 1967), or compounds of differing sizes and charges such as proteins that can be *given* identical charges by the admixture of sodium dodecyl sulfate (Stegemann, 1979). The same compounds might be used in unimpeded electrophoresis to accomplish this verification.

2. LARGE κa

Micro-electrophoresis of a variety of particles of different sizes, coated with given proteins, has demonstrated conclusively that, at least in the large κa range, the electrophoretic mobility is independent of particle size (Abramson et al., 1942).

3. TRANSITION 0.1 < κa < 300

The first published work on the verification of the transition of $\zeta\varepsilon/6\pi\eta$ to $\zeta\varepsilon/4\pi\eta$ with increasing κa was by Mooney (1931), who electrophoresed emulsified oil droplets of different sizes in water with a low electrolyte content. The electrophoretic mobility indeed gradually increased by 50% as the droplet size increased from 0.1 μm to 150 μm, to remain constant with further increases in droplet size. These and other verifications are discussed by Wiersema et al. (1966). The availability of monodisperse polystyrene latex particles of various diameters (Vanderhoff et al., 1956) has opened new possibilities for testing the degree of applicability of the electrophoresis equations.

Shaw and Ottewill (1965) were the first to use polystyrene latices of different diameters and at different ionic strengths for this purpose. Their mobilities conformed qualitatively to the theoretical curves, but they were in certain cases quantitatively higher than the ones calculated by Wiersema et al. (1966) and Overbeek and Wiersema (1967).

The difference in electrophoretic mobility caused by differences in the range of 0.1 < κa < 300 has been put to use for the determination of latex particle size distributions by McCann et al. (1973) and the phenomenon has been utilized for the electrophoretic separation of mixtures of polystyrene latex particles of 0.8 μm and 0.2 μm diameter, under conditions of 0 gravity, during the flight of Apollo 16 to the Moon. Under these conditions, where $1/\kappa = 10$ nm (the κa of the larger and smaller particles thus being respectively 40 and 10), the larger particles migrated approximately 42% faster than the smaller ones (Snyder et al., 1973).

CALCULATION OF ζ-POTENTIALS FROM MOBILITY DATA

It is important to realize that the potentials alluded to in all equations used above are expressed in electrostatic units (esu) of potential difference (Shaw, 1969). In order to convert esu potential differences to volts, ζ as well as χ must be divided by a factor 300. Thus, for example, in:

$$u = \frac{U}{\chi} = \frac{\zeta\varepsilon}{4\pi\eta} \qquad\qquad [V\text{-}14]$$

converted to:

$$\zeta = \frac{4\pi\eta U}{\chi\varepsilon} = \frac{4\pi\eta u}{\varepsilon} \qquad\qquad [V\text{-}16]$$

the value found for $4\pi\eta u/\varepsilon$ has to be multiplied by a factor 90,000, in order to express ζ in volts (where $u = U/\chi$ with u expressed in cm^2 volt^{-1}sec^{-1}).

Table V-3 gives the conversion to ζ-potentials (in mV), for u = 1 μm volt^{-1}sec^{-1}cm, at different temperatures, and thus at different values for ε and η (for water), for both the Hückel and the von Smoluchowski equations. For the range 0.1 < κa < 300 it is best to consult the graphs and tables, giving the variation of the electrophoretic mobility with κa, elaborated by Overbeek and Wiersema (1967), Wiersema *et al.* (1966), Overbeek (1966) and Wiersema (1964) (see also Loeb *et al.*, 1961). These authors have first converted the principal parameters to dimensionless variables, in order to make the results applicable to various temperatures, viscosities and other solvent properties. Thus, as a measure of the electrophoretic mobility, the entity E is introduced (Wiersema *et al.,* 1966; Wiersema, 1964):

$$E = \frac{6\pi\eta e}{\varepsilon kT} \frac{U}{\chi} \qquad [V-17]$$

where e is the elementary charge, k the Boltzmann constant and T the absolute temperature.

Instead of ζ, Y_o is given (Wiersema *et al.*, 1966; Wiersema, 1964):

$$Y_o = \frac{e\zeta}{kT} \qquad [V-18]$$

TABLE V-3
Conversion of u = 1 μm volt^{-1}sec^{-1}cm[a] to ζ-Potentials in Millivolts at Various Temperatures[b]

Temperature (°C)	η (Poise)[c]	ε (Dielectric constant of water)[c]	ζ in mV (after Hückel: $\zeta = 6\pi\eta u/\varepsilon$) (eq. [V-11])	ζ in mV (after von Smoluchowski: $\zeta = 4\pi\eta u/\varepsilon$) (eq. [V-14])
0	0.01787	87.90	34.49	22.99
5	0.01519	85.90	30.00	20.00
10	0.01307	83.95	26.42	17.61
15	0.01139	82.04	23.55	15.70
20	0.01002	80.18	21.20	14.13
25	0.008904	78.36	19.28	12.85
30	0.007975	76.58	17.67	11.78
35	0.007194	74.85	16.31	10.87

[a] The unit of μm volt^{-1}sec^{-1}cm has been used here because this unit is most commonly given in the literature to express electrophoretic mobilities. The values of ζ are given in mV, for the same reason of conformation to the most generally quoted unit for this entity. For the conversion of u to ζ, which entails a double conversion of stat volts to volts, a multiplication factor of 90,000 has been employed.
[b] See van Oss, 1975.
[c] From *Handbook of Chemistry and Physics*, Chemical Rubber Co., 51st Edition, Cleveland, 1970.

TABLE V-4
Values of E as a Function of Y_o and of κa, for 1-1 Electrolytes[a]

Y_o																		
	0	**0.01**	**0.02**	**0.05**	**0.1**	**0.2**	**0.5**	**1**	**2**	**5**	**10**	**20**	**50**	**100**	**200**	**500**	**1000**	**∞**
1	1.00	1.00	1.00	1.00	1.00	1.00	0.99	1.01	1.04	1.13	1.22	1.33	1.42	1.45	1.47	1.49	1.50	1.50
2	2.00	2.00	2.00	1.99	1.97	1.96	1.92	1.91	1.93	2.09	2.27	2.51	2.76	2.87	2.93	2.97	2.99	3.00
3	3.00	2.99	2.98	2.96	2.90	2.85	2.71	2.63	2.60	2.74	3.06	3.43	3.92	4.20	4.34	4.45	4.48	4.50
4	4.00	3.98	3.95	3.90	3.78	3.64	3.36	3.13	2.99	3.08	3.43	3.96	4.79	5.38	5.68	5.90	5.97	6.00
5	5.00	4.97	4.92	4.80	4.59	4.31	3.80	3.44	3.20	3.17	3.47	4.10	5.25	6.22	6.88	7.33	7.46	7.50
6	6.00	5.94	5.86	5.66	5.38	4.85	4.11	3.63	3.27	3.08	3.35	3.94	5.27	6.64	7.90	8.73	8.94	9.00

κa

[a] See Fig. V-8
Source: Wiersema, 1964.

Table V-4 gives the values of E versus Y_0 for a number of values for κa, for the electrophoresis of molecules or particles in a solution of 1–1 electrolytes in water (Wiersema *et al.*, 1966; Wiersema, 1964). Conversion factors for Y_0 to ζ in mV and for u (in μm volt^{-1}sec^{-1}cm) to E, at various temperatures, are given in Table V-5. For further refinement and for more complicated situations of ionic composition, etc., see Wiersema *et al.* (1966) and Wiersema (1964); see also Fig. V-8.

For $Y_0 < 1$, the top row of Table V-4 may be used to obtain the E/Y_0 ratio for the appropriate value of κa, as at these low values of Y_0 (and thus of ζ), relaxation plays no significant role.

TABLE V-5

Conversion of Y_0 to ζ in mV and of u (in μm volt^{-1}sec^{-1}cm) to E at Various Temperatures[a]

Temperature (°C)	Multiplying factor for u, to obtain E	Multiplying factor for Y_0, to obtain E
0	1.464	23.55
5	1.251	23.98
10	1.082	24.42
15	0.948	24.85
20	0.839	25.28
25	0.750	25.71
30	0.676	26.14
35	0.614	26.57

[a] The following values have been used: $k = 1.3806 \times 10^{-16}$ erg deg.$^{-1}$; $e = 4.8032 \times 10^{-10}$ esu; T = temperature in °C + 273.15; for ε and η, see Table V-3. In both conversions, from Y_0 to ζ, as well as from u to E, a factor 300 was used, for the transformation of esu into volts. From van Oss, 1975.

THE BROOKS EFFECT

Brooks and Seaman (1973) and Brooks (1973) calculated from observed electrophoretic mobilities of erythrocytes in solutions of neutral polymers (e.g., dextrans) that the presence of relatively high concentrations of such neutral polymers (2.5 to 10.0%, w/v) causes a marked increase in the erythrocytes' ζ-potentials. Arnold *et al.* (1988) noted that the same phenomenon appears to occur with the ζ-potential of liposomes (made up of mixtures of phosphatidylcholine and phosphatidic acid), in the presence of up to 10% (w/v) of polyethylene oxide (PEO). However, upon comparison with surface potentials of the liposomes, obtained by means of a positively charged spin probe and ESR spectroscopy, it became evident that in reality no change had occurred in the surface potential of the liposomes, in the presence of PEO (Arnold *et al.*, 1990). These authors then proposed that the PEO concentration in the near vicinity to the liposomes' surfaces is much smaller than the PEO concentration in the bulk liquid, so that the influence of the viscosity, η, on the ζ-potential, at a given

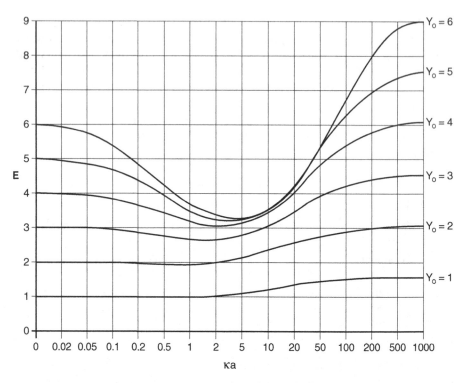

FIGURE V-8 E as a function of κa for different values of Y_0 (see Table V-4). (from Wiersema, 1964; see also van Oss, 1975.)

electrophoretic mobility u (see eq. [V-15]), is much smaller than was presumed by Brooks and Seaman (1973), who equated the value of the viscosity, η, at the shaer plane, with the value of η of the bulk polymer solution.

The depletion of the concentration of neutral polymers, which are, however, Lewis bases (or $\gamma^{\ominus}$ monopoles; see Chapter II and XVIII), in the vicinity of other $\gamma^{\ominus}$ monopolar surfaces such as erythrocyte or liposomal surfaces, can of course be easily predicted; see eq. [IV-16], by means of which the value of the repulsion energy between polymers and particles or cells, immersed in water, can be obtained. Arnold *et al.* (1988; 1990) discuss the important implications of this phenomenon of polar polymer depletion near polar surfaces, for cell and liposome aggregation and fusion processes (see also Chapter XXIII).

CODA

The theory of electrical double layer interactions is summarized, giving the connection between electrokinetic measurements and surface potential, as a function of ionic strength and of the size and shape of electrically charged macromolecules and particles. It must, however, be stressed that in principle, and other factors being equal, the electrokinetic mobility of particles and macromolecules is only indirectly linked to their size or shape.

For a description of electrokinetic methods, see Chapter XV; for the linkage between EL and AB interactions, see Chapter XIX.

VI Brownian Movement Forces — Osmotic Interactions of Polymers

BROWNIAN MOVEMENT FORCES

It should be realized that every single detached molecule or particle, immersed in a liquid medium, is endowed with a Brownian (BR) free energy of 1½ kT (Einstein, 1907), where k is Boltzmann's constant (= 1.3806×10^{-23} J/K) and T the absolute temperature in degrees K. This energy keeps it in solution or in suspension, provided the energy of attraction between similar molecules (or particles) immersed in that liquid is less than 1½kT per pair of molecules or particles. But while a very small molecule has a Brownian free energy of +1½kT, a large particle also has a Brownian free energy of only +1½kT. Thus, micron-sized particles, each pair of which would typically have a contactable surface area (S_c) of the order of 10^4 nm², will overcome the repulsive forces of Brownian motion and become destabilized even if their free energy of mutual attraction, in a given liquid at close range, is as small as -10^{-3} mJ/m². Whilst the energies of thermal motion, or diffusion, are relatively small, they are not always negligible; in apolar media they can be the major contribution to solubility or stability. As all single entities immersed in a liquid are endowed with +1½kT, this contribution clearly plays the greatest role in the case of the smallest molecules or particles, as these have the smallest surfaces of mutual contact.

Thus Brownian movement forces, which are always repulsive, together with the LW, AB and EL interactions (each of which can be either attractive or repulsive) discussed in the preceding chapters, comprise the physical forces which are the only primary non-covalent forces that normally* play a role in particulate or macromolecular interactions in liquid media; see also Chapter XX. In other words, the total free energy of interaction between particles and/or macromolecules can be described as $\Delta G^{TOT} = \Delta G^{LW} + \Delta G^{AB} + \Delta G^{EL} + \Delta G^{BR}$. Thus, if ΔG^{LW}, ΔG^{AB} and ΔG^{EL} of a given system have been measured separately, ΔG^{TOT} can be obtained by summing the values for these entities, and adding to these +1kT for ΔG^{BR} (for a system with two degrees of freedom). However, if the total free energy of interaction, ΔG^{TOT}, of a given system has been measured as a whole [e.g., by interfacial tension determination (see Chapters XIII and XXII, or through measurement of the system's

---

* In most cases gravitational and magnetic forces may be neglected.

equilibrium constant (see below: *Size and Shape of Polymer Molecules*), the ΔG^{BR} is already included in ΔG^{TOT}.

OSMOTIC INTERACTIONS

Osmotic (OS) interactions are secondary phenomena, caused by combinations of BR, LW, AB and EL forces in any number of proportions. However, BR forces are *always* operative in engendering osmotic interactions, and LW, AB or EL forces may or may not play a role, depending on a variety of circumstances. The influence of BR forces in osmotic interactions (relative to the other forces) is most pronounced at low solute concentrations and at low solute molecular weight (van Oss, Arnold *et al.*, 1990b).

The Brownian component of the osmotic pressure, Π, is, according to van't Hoff's law:

$$\Pi = RTC \qquad \text{[VI-1]}^*$$

where Π is the pressure differential in N/m^2 or $dynes/cm^2$, if the cgs system is adhered to, R the gas constant ($R = 8.3143$ J/degree/mole, or, in cgs, 8.3143×10^7 ergs/degree/mole). The absolute temperature in degrees K and C the concentration in moles. At 20°C ($=293°K$), C = 1 molar, $\Pi = 24.36 \times 10^5$ $N/M^2 = 2.436$ MPa = 24.36 $\times 10^6$ $dynes/cm^2 = 24.85$ Kg/cm^2.

For polar macromolecules in aqueous solution, and especially in the case of linear polymers, it is necessary to expand eq. [VI-1], to include the second and in some instances even the third virial coefficients B_2 and B_3 (Hermans, 1949; Wagner, 1945):

$$\Pi = 1,000 \ RT(c/M_n + B_2 \ c^2/d^2 + 1/3 \ B_3 \ c^3/d^3 + \ldots\ldots) \qquad \text{[VI-2]}$$

where c is the concentration (expressed as the weight fraction of solute per total volume) and M_n the (number-average) molecular weight of the polymer, while:

$$B_2 = v_o \left(\frac{n}{M_n} \right)^2 (\tfrac{1}{2} - \chi) \qquad \text{[VI-3]}$$

and

$$B_3 = v_o \left(\frac{n}{M_n} \right)^2 \qquad \text{[VI-4]}$$

where v_o is the molar volume of the solvent (for water at 20°C, $V_o = 18$ cm^3), n the number of monomeric units of the linear polymer, d the density of the polymer, and χ is a dimensionless parameter such that:

---

* A list of symbols can be found on pages 399–406.

$$-\chi = \Delta H^M / n_1 ckT \qquad \text{[VI-5]}$$

where ΔH^M is the enthalpy of mixing, n_1 the number of solvent molecules per unit volume, k is Boltzmann's constant (k = 1.38×10^{-23} J per degree K, or 1.38×10^{-16} ergs per degree K) and T the absolute temperature in degrees K. Thus, kT is the difference in energy of a solvent molecule immersed in pure polymer compared with that in pure solvent (Napper, 1983). χ is an important (dimensionless) parameter in the Flory-Huggins theory of polymer solubility (see Flory, 1953). $B_2 = 0$ when $\chi = \frac{1}{2}$, which occurs at the θ temperature (or θ-point) for the polymer-solvent systems (Flory and Krigbaum, 1950) which is the temperature at which the polymer just begins to precipitate. In "good" solvents, $\chi < \frac{1}{2}$, and $B_2 > 0$. The third virial coefficient B_3 only has a significant impact on the osmotic pressure when M_n/n is small, and then only at high concentrations (i.e., c > 0.2, and even then only reaches 5 to 6% of the total osmotic pressure, just in cases of water-soluble linear polymers with numbers of monomeric units, n > 100; see van Oss, Arnold *et al.*, 1990b, and Table XIX-7 in Chapter XIX.).

Thus when, in addition to Brownian interactions, interfacial interactions occur, χ has to be taken into account. Now the interfacial free energy per unit surface area between polymer molecules (1) immersed in solvent (2), times the unit surface area S_c (i.e., the minimum contactable surface area between two polymer molecules), may be equated with χ, when expressed in units of kT (Flory 1953; Hiemenz, 1984; Hermans, 1949):

$$S_c \Delta G_{121} = -\chi_{12} \qquad \text{[VI-6]}$$

or, in accordance with:

$$\Delta G_{121} = -2\gamma_{12} \qquad \text{[VI-7]}$$

$$\chi_{12} = 2S_c\gamma_{12} \qquad \text{[VI-8]}$$

The osmotic pressure of polymer (1), dissolved in water (2), can be expressed as:

$$\Pi = RT\left[\frac{c}{M} + \frac{c^2}{d^2}v_o\left(\frac{n}{M}\right)^2(\tfrac{1}{2}-\chi_{12}) + \frac{c^3}{3d^3}v_o\left(\frac{n}{M}\right)^2\right] \qquad \text{[VI-9]}$$

where χ_{12} is given in eq. [VI-8]. S_c can be estimated from the molecular structure of the polymer and γ_{12} is expressed as:

$$\gamma_{12} = (\sqrt{\gamma_1^{LW}} - \sqrt{\gamma_2^{LW}})^2 + 2(\sqrt{\gamma_1^{\oplus}\gamma_1^{\ominus}} + \sqrt{\gamma_2^{\oplus}\gamma_2^{\ominus}}$$

$$-\sqrt{\gamma_1^{\oplus}\gamma_2^{\ominus}} - \sqrt{\gamma_1^{\ominus}\gamma_2^{\oplus}}) \qquad \text{[IV-9]}$$

so that (cf. eq. VI-8):

$$\chi_{12} = 2S_c(\sqrt{\gamma_1^{LW}} - \sqrt{\gamma_2^{LW}})^2 + 2\,(\sqrt{\gamma_1^{\oplus}\gamma_1^{\ominus}} + \sqrt{\gamma_2^{\oplus}\gamma_2^{\ominus}}$$
$$- \sqrt{\gamma_1^{\oplus}\gamma_2^{\ominus}} - \sqrt{\gamma_1^{\ominus}\gamma_2^{\oplus}})$$

[VI-10]

for which all the parameters can be determined by contact angle (θ) measurements with a number of different liquids, using Young's equation [IV-13].

Verification of eq. [VI-9] with the results of experimental measurements on the linear polymer, poly(ethylene oxide) [PEO] of different molecular weights gives a close correlation between eq. [VI-9] and the experimental results from the lowest to the highest (c = 0.6) concentrations (van Oss, Arnold et al., 1990b; see also Chapter XIX). The results of Arnold et al. (1988) indicate that at higher PEO concentrations, the osmotic pressure becomes *independent of molecular weight*. Inspection of eq. [VI-9] shows that at the highest molecular weights the first term becomes very small, and the second and third term only depend on the inverse molecular weight (squared) of the repeating subunit of the polymer. Clearly, when χ_{12} is strongly negative, the second term of eq. [VI-9] becomes preponderant, which explains both the molecular weight independence of Π, and its exceedingly high values at high concentrations. Π is more than 300 atmospheres for 60% (w/v) aqueous PEO solutions (Arnold et al., 1988); see Chapter XIX, where experimentally observed osmotic pressures of aqueous solutions of the extremely hydrophilic monopolar electron-donating linear polymer, PEO, are compared with calculated values, as a function of concentration and molecular weight.

RADIUS OF GYRATION

To obtain the osmotic excess free energy, as a first approximation:

$$\Delta G^{OS} = R_g \Pi$$

[VI-11]

where (in an ideal solvent) the radius of gyration R_g of a dissolved macromolecule is (Hiemenz, 1986; Hiemenz and Rajagopalam, 1997):

$$R_g = r/\sqrt{6}$$

[VI-12]

where r is the mean-square end-to-end length of a random coil in solution. For asymmetrical polymers of known total length L (in solution), and of much greater length than width, it is more convenient to use (Hiemenz, 1986; Hiemenz and Rajagopalam, 1997):

$$R_g = L/\sqrt{12}$$

[VI-13]

and for spheres with radius R (Hiemenz, 1986; Hiemenz and Rajagopalam, 1997):

$$R_g = R\sqrt{0.4}$$

[VI-14]

The concept of radius of gyration is especially used in light scattering of polymer solutions; for a more complete treatment, see Hiemenz (1984).

POLYMER CONCENTRATION

The determination of polymer concentration (c) in the *bulk liquid* usually is a fairly simple procedure. It can be done, e.g., by refractometry, spectrophotometry, or by a variety of colorimetric determinations (for instance in the case of proteins).

It is the measurement of polymer concentration in the *bound*, or *adsorbed layer* on the surface of particles or cells, that is often a much more difficult matter. By radioactive tagging of the polymer molecules (see, e.g., Brooks, 1973; Chien *et al.*, 1977; Janzen and Brooks, 1988), one can measure the amount of polymer bound *per unit surface area*, but in the absence of knowledge of the thickness of the layer of bound or adsorbed polymer, its real concentration in terms of amount of adsorbed polymer per unit volume is still unknown. For determining the thickness of relatively thick adsorbed layers, viscometric or sedimentation methods may be utilized (Sato and Ruch, 1980), and under favorable conditions, ellipsometry may be used for the measurement of the thickness of relatively thin adsorbed layers (Stromberg, 1967; Sato and Ruch, 1980), although for the measurement of the thickness of adsorbed layers on particle suspensions, ellipsometry still is not very practical. (Although the measurement of the thickness of an adsorbed layer of polymer on a flat polymer surface, e.g., polystyrene, may to a certain extent reflect the thickness of polymer adsorption on polystyrene latex particles, under otherwise comparable conditions, see van Oss and Singer, 1966.) In many cases, however, educated guessing of the thickness of the adsorbed layer is the best one can do. But when, for instance, the amount of polymer bound per unit surface area (of particles) exceeds the possibility, for asymmetrical polymer molecules, that they all are adsorbed lying flat on the surface of the particles, one knows at least that the thickness of the adsorbed layer must lie between the lengths of the shortest and the longest dimensions of the polymer molecules. At low concentrations (in a "good" solvent) the thickness of the adsorbed layer is close to the radius of gyration (Rg) of the polymer molecules (Kawaguchi *et al.*, 1988).

In addition to ellipsometry, more recently other optical methods emerged for the measurement of the thickness and/or the increase in thickness as a function of time, of deposited (bio)polymer layers. The most utilized of these is surface plasmon resonance (Fägerstam and Karlsson, 1994), mainly for the determination of kinetic rate constants of biopolymer (e.g., protein) interactions. For the same purpose the present author and his collaborators employed fluorescence spectrometry (Docoslis *et al.*, 1999: see Chapter XXV).

SIZE AND SHAPE OF POLYMER MOLECULES

For the determination of the molecular weight and the size, and especially the shape, of strongly asymmetrical polymers in solution, the modern chromatographic methods (gel-filtration, pore exclusion chromatography, HPLC, polyacrylamide gel electrophoresis, etc.) are grossly inadequate. For such asymmetrical polymers, the only satisfactory manner for arriving at their molecular size and asymmetry factors is to determine their diffusion constant (D) and their sedimentation constant (s), or their (usually weight average) molecular weight (M_W). D can be determined in a variety of ways (see, e.g., Svensson and Thompson, 1961), using static means, or even by using the analytical ultracentrifuge (Beckman-Spinco, Model E, Palo Alto, CA). The latter instrument also serves to determine sedimentation constants (s) by measuring sedimentation rates, or directly, M_W, by sedimentation equilibrium determinations; see, e.g., Claesson and Claesson (1961) and Svedberg and Pedersen (1940, 1959). For M_W it is best to use the Svedberg equation:

$$M_W = \frac{RTs}{D(1-\bar{v}p)} \qquad [VI\text{-}15]$$

where s is the sedimentation coefficient, $\bar{v}$ the partial specific volume of the polymer and ρ the density of the solution. For dilute aqueous solutions, $\rho \rightarrow 1$ and for proteins, $\bar{v} \approx 0.75$. The asymmetry of polymer molecules is obtained from the friction factor ratio f/f_o, where the actual friction factor f is:

$$f = M_W(1-\bar{v}\,\rho)/s \qquad [VI\text{-}16]$$

or:

$$f = RT/D \qquad [VI\text{-}17]$$

while the friction factor f_o of a completely spherical molecule of the same M_W then would be:

$$f_o = 6\pi\eta\ N\ \sqrt[3]{3M_W\bar{v}\rho/4\pi N} \qquad [VI\text{-}18]$$

where η is the viscosity of the medium, and N is Avogadro's constant ($N = 6 \times 10^{23}$). The friction factor ratio f/f_o then is:

$$f/f_o = 10^{-8}\ \sqrt[3]{(1-\bar{v}\rho)/(D^2 s\bar{v})} \qquad [VI\text{-}19]$$

For the connection between the asymmetry factor ratios f/f_o and the actual asymmetry ratio y, for oblong and oblate ellipsoids (Svedberg and Pedersen, 1940; 1959), see Table VI-1.

One of the most-used biopolymers, involved in a variety of osmotic interactions, is the linear polymer of glucose, dextran. Table VI-2 gives the asymmetry ratios and the molecular dimensions, and radii of gyration of some of the most used molecular weight fractions, measured in the author's laboratory (Edberg et al., 1972).

Effects of the interdiffusion of linear polymers have recently been discussed by Klein (1990).

TABLE VI-1
Friction Factor (f/f$_o$) Ratios and Asymmetry Factors (y) for Oblong (cigar-like) and Oblate (flat disc-like) Ellipsoids

Oblong ellipsoids				Oblate ellipsoids			
$f/f_o = \dfrac{1-y^2}{y^{2/3}\log[(1+\sqrt{1-y^2})/y]}$				$f/f_o = \dfrac{y^2-1}{y^{2/3}\arctan\sqrt{(y^2-1)}}$			
1/y	f/f$_o$	1/y	f/f$_o$	y	f/f$_o$	y	f/f$_o$
1.0	1.000	12	1.645	1.0	1.000	12	1.534
1.2	1.003	14	1.739	1.2	1.003	14	1.604
1.4	1.010	16	1.829	1.4	1.010	16	1.667
1.6	1.020	20	1.996	1.6	1.019	20	1.782
1.8	1.031	25	2.183	1.8	1.030	25	1.908
2.0	1.044	30	2.356	2.0	1.042	30	2.020
3.0	1.112	35	2.518	3.0	1.105	35	2.119
4.0	1.182	40	2.668	4.0	1.165	40	2.212
5.0	1.255	50	2.946	5.0	1.224	50	2.375
6.0	1.314	60	3.201	6.0	1.277	60	2.518
7.0	1.375	70	3.438	7.0	1.326	70	2.648
8.0	1.433	80	3.658	8.0	1.374	80	2.765
9.0	1.490	90	3.867	9.0	1.416	90	2.873
10.0	1.543	100	4.067	10.0	1.458	100	2.974

From Svedberg and Pedersen, 1940.

TABLE VI-2
The Physical-Chemical Parameters of Dextran Fractions

Fraction	Mw	S_{20}^0 [a]	$D_{20}^{0-2\%}$ [b]	f/f$_o$	y[c]	Diameter[d] (Å)	Length[d] (Å)	Rg[e] (Å)
T 10	10,980	1.36	0.88	1.82	16.5	9.6	158	46
T 20	21,400	1.71	0.60	1.96	19	10.3	196	57
T 40	39,400	2.08	0.43	2.39	31	10.8	335	97
T 70	72,000	2.51	0.33	2.88	48	12.5	600	173
T 110	100,500	3.18	0.28	3.14	57	13.2	752	217
T 150	146,500	4.16	0.22	3.56	73	13.7	1,000	289
T 250	237,000	6.28	0.16	3.81	88	15.1	1,330	384
T 500	492,000	10.20	0.14	4.13	103	20.8	2,766	798

[a] S_{20}^0 in units of 10^{-13} sec.
[b] $D_{20}^{0-2\%}$ in units of 10^{-6} cm^2 sec^{-1}.
[c] Asymmetry ratio, for oblong ellipsoids; See Table VI-1; see also Svedberg and Pedersen, 1959, pp. 38–42.
[d] Calculated from Mw and asymmetry ratio, for oblong ellipsoids.
[e] Radius of gyration, calculated *via* Rg = L/$\sqrt{12}$; see eq. [VI-13]; see also Hiemenz, 1986, p. 266.
From Edberg, 1971.

CODA

A few fundamentals of Brownian motion and of osmotic pressure are stressed, and the influence is outlined of polar (AB) interactions on the second virial coefficient of the osmotic pressure equation pertaining to aqueous solutions of very hydrophilic polymers (see also Chapter XIX). A few other principles of colloidal hydrodynamics are outlined, e.g., the influence of the radius of gyration of dissolved polymer molecules on the free energy of osmotic interaction, determination of polymer concentrations, measurement of the size of polymer molecules by sedimentation methods, connection between the friction factor ratio and the asymmetry factor of polymers.

VII Rate of Decay with Distance

UNRETARDED LIFSHITZ–VAN DER WAALS FORCES

Table VII-1 shows the rate of decay with distance ℓ of LW interactions, in the cases of (semi-infinite) parallel flat plates, a sphere and a flat plate, and two spheres. Fairly thick flat plates may be considered to be semi-infinite, however, for the interaction of very thin flat plates, or layers, see eq. [II-15]; see also Nir, 1976. These equations are only valid for distances up to 50 to 100 Å; beyond that distance, retardation sets in (see below).

Computation of the Hamaker constant A may be circumvented by obtaining ΔG^{LW} from γ^{LW} and using:

$$-2\gamma^{LW} = \Delta G_{\ell_0}^{LW} = -A/12\pi\ell_0^2 \qquad\qquad [VII\text{-}1]^*$$

(see Chapters II and III). Comparison between values for ΔG_{ii}^{LW} and A_{ii} of a sizeable number of liquid and solid compounds yields an empirical but quite reliable value of $\ell_0 = 1.57 \pm 0.08$ (SD) Å (where SD stands for standard deviation of the mean), for use with eq. [VII-1]; see Chapter XI. One may thus state:

$$\Delta G_{\ell}^{LW} = \Delta G_{\ell_0}^{LW}(\ell_0/\ell)^2 \qquad\qquad [VII\text{-}2]$$

where $\ell_0 \approx 1.57 \times 10^{-8}$ cm. Eq. [VII-1] is valid for values of ℓ up to 100 Å. It thus is a fairly simple matter to determine ΔG^{LW} for values of ℓ, from ℓ_0, to $\ell \approx 100$ Å, of all substances i with a known, or a measurable γ_i^{LW}. Beyond that value, a decrease in A sets in, due to retardation. Table VII-1 shows the rate of decay with distance ℓ of LW interactions, in the case of parallel flat plates, 2 spheres, and a flat plate and one sphere.

---

* A list of symbols can be found on pages 399–406.

TABLE VII-1

Energies of Interaction (ΔG^{LW}) and Forces of Interaction of (F^{LW}) of Unretarded Lifshitz–van der Waals Interaction, for a Number of Configurations, as a Function of Distance, ℓ

Configuration	ΔG_{ℓ}^{LW}	F_{ℓ}^{LW}
(semi-infinite flat parallel slabs)	$\dfrac{-A}{12\pi\ell^2}$	$\dfrac{A}{12\pi\ell^3}$
(sphere of radius R and semi-infinite flat slab: also valid for two crossed cylinders at 90°)	$\dfrac{-AR}{6\ell}$	$\dfrac{AR}{6\ell^2}$
(two spheres of radius R)	$\dfrac{-AR}{12\ell}$	$\dfrac{AR}{12\ell^2}$

RETARDED LONDON–VAN DER WAALS FORCES

Retardation of London–van der Waals (or dispersion) forces (Casimir and Polder, 1948) is due to the fact that the electrodynamic interactions that give rise to dispersion forces are propagated at the speed of electromagnetic radiation, i.e., with the (finite) speed of light. Thus, when two atoms are about 1,000 Å apart, in the time it takes for the electric field of atom 1 to reach atom 2, and for the field caused by the induced dipole to return to atom 1, the trajectories of individual electrons will have changed direction, and the dipoles will attract each other considerably less strongly (Israelachvili, 1985).* At $\ell \approx 100$ Å, A can be significantly less than one half of its unretarded value (if one continues to use an equation of the type of eq. [VII-1]). There is, unfortunately, no single equation in closed form for calculating the influence of retardation as a function of ℓ, due to the fact that the effects of retardation vary with wavelength as well as with distance. However, Mahanty and Ninham (1976) give the full numerical methods for calculating retardation. Curiously, at large distances, the decrease of the interaction energy of dispersion forces, due to retardation, slows down, and eqs. [II-16A] and [VII-1] resume their validity

---

* Israelachvili (1985, p. 154; 1992, p. 198) seeks to explain the spreading of pentane ($\gamma_1 = 16.05$ mJ/m²) on water ($\gamma_W^{LW} = 21.8$ mJ/m2), and the non-spreading of dodecane ($\gamma_1 = 25.35$ mJ/m²) by retardation effects. However, there is no anomaly that needs explaining. The actual mechanism of spreading of pentane on water, and of non-spreading of dodecane on water is clear and straightforward: only liquids of a higher surface tension than the substrate material form a measurable contact angle. Dodecane on water yields cos $\theta = 0.85$, corresponding to a measurable contact angle $\theta \approx 32°$. In all these cases only LW interactions are relevant; see Fig. IV-1, category III-A. Actually, pentane, hexane, heptane, spread on water; octane is the limiting case (Hauxwell and Ottewill, 1970), and nonane, decane, . . . etc., form measurable contact angles. For simplicity's sake, water is here taken to be a solid (i.e., the water/vapor surface and the water/alkane surface are taken to form a single plane). If encased in a gel, water may even be formally treated as a solid; thus the γ_W^{LW} of water has been determined by contact angle measurements with hexadecane on water; a value for $\gamma_W^{LW} = 21.4 \pm 1.2$ mJ/m² was found (van Oss et al., 1987e). Richmond et al. (1973) studied the influence of Lifshitz–van der Waals forces (including retardation) on the adsorption of pentane, hexane, heptane, octane and dodecane on pure water, using Lifshitz's theory. Their study illustrates well why octane is the limiting case, with respect to the spreading or non-spreading of alkanes on water.

(Mahanty and Ninham, 1976). (At the height of the retardation regime, the rate of decay is proportional to ℓ^{-3}, instead of ℓ^{-2}). The reversal to the nonretarded regime at large distances has not yet been verified experimentally (probably hampered by the fact that at large distances dispersion forces are very weak indeed and thus difficult to measure with any degree of precision). Retardation effects at $\ell \approx 50$ to $1,000$ Å *in vacuo* have been confirmed experimentally by Derjaguin and Abrikossova (1951, 1954) and by Overbeek and Sparnaay (1952), and Israelachvili and Adams (1978), and in liquids (mica-coated cylinders in water) by Israelachvili (1985); see also Chapter XII.

POLAR (AB) INTERACTIONS

The rate of decay with distance ℓ, at $\ell > \lambda$ [λ is the correlation length, or decay length, of the molecules of the liquid medium (see Chan *et al.*, 1979)], can be expressed in terms of force (Parsegian *et al.*, 1985; Rand and Parsegian, 1984; Marcelja and Radic, 1983):

$$F_\ell = F_{\ell_0} \exp[(\ell_0 - \ell)/\lambda] \qquad\qquad \text{[VII-3]}$$

By integration one obtains (van Oss *et al.*, 1986b):

$$\Delta G^{AB}\ell = \Delta G^{AB}_{\ell_0} \exp[(\ell_0 - \ell)/\lambda] \qquad\qquad \text{[VII-4]}$$

for the parallel flat plate configuration. For pure water, it has been estimated that $\lambda \approx 0.2$ nm (Chan *et al.*, 1979; Parsegian *et al.*, 1985; Rand and Parsegian, 1984). In actual practice, however, the value of λ for water may be up to 1.0 nm, for positive (Israelachvili, 1985; Claesson, 1986) as well as for negative values of ΔG^{AB} (Israelachvili and Pashley, 1982, 1984; Israelachvili, 1985; see also marcelja, 1990) or even higher than 1.0 nm in some cases (Claesson *et al.*, 1986; Christensen and Claesson, 1988). The fact that the value of λ in liquid water is larger than the molecular dimension may be attributable to the fact that the water molecules tend to occur in (continuously associating and dissociating) clusters of 4 or more molecules, which form hydrogen bonds with each other (see, e.g., Eisenberg and Kauzmann, 1969; van Oss, Giese and Good, 2002a). Our best value for the value of λ for water in the repulsive mode* comes close to 0.6 nm (van Oss, Biophysics of the Cell Surface, 1990); see also Chapter XXIII.

---

* In the 1980s and early 1990s various results emerged from work with Israelachvili's force balance, indicating that whilst the decay length, λ, for water is of the order of 1 nm, in the *repulsive mode* (see, e.g., Claesson, 1986; Christenson, 1988), in many cases of "hydrophobic" *attraction*, λ may be much larger, e.g., of the order of 13 nm or more (Rabinovich and Derjaguin, 1988; Christenson, 1992; Kurihara and Kunitake, 1992). It turned out that these large values for λ, resulting from observations on hydrophobic attractions, can be ascribed to imperfect or incomplete attachment of hydrophobic compounds or materials to the extremely hydrophilic mica surface of the crossed cylinders of the force balance (see Chapter XIV and Figure XIV-1). Wood and Sharma (1995) showed that when for direct force balance measurements smooth hydrophobic monolayers are *robustly attached* to the mica surfaces, the bizarre long-range hydrophobic interactions mentioned above no longer manifest themselves. In other words, with properly prepared surfaces, repulsive as well as attractive AB forces are subject to only one decay length (λ), which is approximately 1.0 nm, for distances $1 \geq 1.0$ nm.

For a configuration of two spheres, with radii R, and a distance ℓ between the outer edges of the two spheres, one can arrive at the following approximation [by analogy with the calculations for electrical double layer interactions (Verwey and Overbeek, 1946; Hiemenz, 1986; see also Derjaguin, 1934; Lis *et al.* 1982)], for $\ell > \lambda$:

$$\Delta G_\ell^{AB} = \pi R \lambda \cdot \Delta G_{\ell_o}^{AB''} \exp[(\ell_o - \ell)/\lambda] \qquad [VII-5]$$

while for a sphere with radius R and a flat plate, as well as for two crossed (90°) cylinder with radius R, for $R > \ell$:

$$\Delta G_\ell^{AB} = 2\pi R \lambda \cdot \Delta G_{\ell_o}^{AB''} \cdot \exp[(\ell_o - \ell)/\lambda] \qquad [VII-6]$$

where $\Delta G_{\ell_o}^{AB''}$ stands for the interaction energy between two parallel flat surfaces, at the minimum equilibrium distance, ℓ_o. Table VII-2 shows the rate of decay with distance ℓ of AB interactions, in the cases of parallel flat plates, a sphere and a flat plate, and two spheres. Contrary to LW forces, with AB forces, even though $\Delta G_{\ell_o}^{AB''}$ can be measured directly, ℓ_o is not known precisely. It seems, however, reasonable to estimate ℓ_o for AB interactions to be of the same order of magnitude as the ℓ_o of LW interactions, i.e., approximately 1.6 Å, (see Chapter III).

In the presence of polymer molecules (2), dissolved in the bulk liquid (3), the dissolved polymer molecules interact on suspended particles (1) with a polar free energy ΔG_{132}^{AB}. Here the decay with distance ℓ, for a polymer concentration ϕ_2 (where ϕ_2 is the volume fraction occupied by polymer molecules) may be expressed as:

$$\Delta G_{132_\ell}^{AB} = \Delta G_{132_{\ell_o}}^{AB} \left[(1-\phi_2)\exp\frac{\ell_o-\ell}{\lambda} + \phi_2 \exp\frac{\ell_o-\ell}{Rg} \right] \qquad [VII-7]$$

see Chapter V for the values of the values of the radius of gyration, Rg). When Rg $\gg \lambda$ and $\ell > Rg$, eq. [VII-7] reduces to:

$$\Delta G_{132_\ell}^{AB} = \Delta G_{132_{\ell_o}}^{AB} \phi_2 \exp\frac{\ell_o-\ell}{Rg} \qquad [VII-8]$$

Thus, following eq. [VII-8], for particles suspended in a not too dilute polymer solution of high molecular weight, and thus with a rather high Rg value, the decay with distance of the polar interaction between particles and dissolved polymer molecules in solution can be exceedingly gradual. This extremely gradual decay can also influence the (polar) interaction between particles (1), suspended in a polymer (2) solution in a liquid (3). Such an interaction can be expressed (van Oss, Arnold and Coakley, 1990), as:

$$\Delta G_{131}^{AB} - \Delta G_{132}^{AB} = \Delta \Delta G^{AB} \qquad [VII-9]$$

where ΔG_{132}^{AB} decays as indicated in eq. [VII-4], and:

TABLE VII-2
Energies of Interaction (ΔG_ℓ^{AB}) and Forces of Interactions (F_ℓ^{AB}) of Polar [Electron-Acceptor–Electron-Donor, or Lewis Acid-Base (AB)] Interactions, for a Number of Configurations, as a Function of Distance, ℓ [a]

Configuration	ΔG_ℓ^{AB}	F_ℓ^{AB}
(flat parallel plates)	$\Delta G_{\ell_o}^{AB''} \exp[(\ell_o - \ell)/\lambda]$	$F_{\ell_o}^{AB''} \exp[(\ell_o - \ell)/\lambda]$ $= -\Delta G_{\ell_o}^{AB''}(1/\lambda)\exp[(\ell_o - \ell)/\lambda]$
(sphere of radius R and flat plate; also valid for two crossed cylinders at 90°)	$2\pi R\lambda\Delta G_{\ell_o}^{AB''} \exp[(\ell_o - \ell)/\lambda]$	$-2\pi R\Delta G_{\ell_o}^{AB''} \exp[(\ell_o - \ell)/\lambda]$
(two spheres of radius R)	$\pi R\lambda\Delta G_{\ell_o}^{AB''} \exp[(\ell_o - \ell)/\lambda]$	$-\pi R\Delta G_{\ell_o}^{AB''} \exp[(\ell_o - \ell)/\lambda]$

[a] $\Delta G_{\ell_o}^{AB''}$ is obtained from eqs. [IV-6], [IV-16], or [IV-17], and $F_{\ell_o}^{AB''} \approx -(1/\lambda)\Delta G_{\ell_o}^{AB''}$, or F_ℓ^{AB} is measured experimentally. The superscript " indicates that $\Delta G''$ or F'' were obtained at the plane parallel plate configuration, at the minimum equilibrium distance ℓ_o.

$$\Delta G_{131\ell}^{AB} = \Delta G_{131\ell_o}^{AB} \exp\frac{\ell_o - \ell}{\lambda} \qquad \text{[VII-4A]}$$

Combining eqs. [VII-4A and 9] into eq. [VII-7]:

$$\Delta\Delta G_\ell^{AB} = \Delta G_{131\ell_o}^{AB} \exp\frac{\ell_o - \ell}{\lambda}$$
$$- \Delta G_{132\ell}^{AB}\left[(1-\phi_2) \exp\frac{\ell_o - \ell}{\lambda} + \phi_2 \exp\frac{\ell_o - \ell}{Rg}\right] \qquad \text{[VII-10]}$$

At long range, i.e., for $Rg \gg \lambda$ and $\ell > Rg$, eq. [VII-8] becomes:

$$\Delta\Delta G_\ell^{AB} = -\Delta G_{132\ell_o}^{AB} \exp\frac{\ell_o - \ell}{Rg} \qquad \text{[VII-11]}$$

This implies, for positive values of $\Delta G_{132\ell_o}^{AB}$, that for particles suspended in a polymer solution, there will be a long-range attraction. At short range, i.e., when $\ell < Rg$:

$$\Delta\Delta G_{131\ell}^{AB} = (\Delta G_{131\ell_o}^{AB} - \Delta G_{132\ell_o}^{AB}) \exp\frac{\ell_o - \ell}{\lambda} \qquad \text{[VII-12]}$$

(As in a depletion layer of a thickness $\approx Rg$, occurring between the particles and the bulk polymer solution, no polymer molecules are found [See Chapters XX and XXIV], $\phi_2 = 0$ and the factor $(1 - \phi_2)$ in eq. [VII-10] becomes equal to one.) Thus, *at short range*, polar repulsion between particles suspended in a polymer solution occurs when $\Delta G_{131}^{AB} > \Delta G_{132}^{AB}$, and polar attraction prevails when $\Delta G_{131}^{AB} < \Delta G_{132}^{AB}$ (van Oss, Arnold and Coakley, 1990).

ELECTRICAL DOUBLE LAYER INTERACTIONS

The rate of decay with distance of the free energy of electrostatic interactions is expressed in eqs. [IV-3], [IV-5] and [IV-6], for flat plane parallel plates, two equal spheres, and a sphere and a flat plate; see Table VII-3. As electrokinetic potentials normally are determined in situations where there is a considerable distance ℓ between particles or surfaces, $\Delta G_{\ell_o}^{EL}$ is not known. In those cases where it is desirable to estimate $\Delta G_{\ell_o}^{TOT}$, an estimation must also be made for ℓ_o in the case of EL interactions. Here also, the same order of magnitude as the ℓ_o for LW interactions would appear the most reasonable estimation, i.e., $\ell_o \approx 1.6$ Å.

TABLE VII-3
Energies of Interaction (G_ℓ^{EL}) and Forces of Interactions (F_ℓ^{EL}) of Electrical Double Layer Interactions, for a Number of Configurations, as a Function of Distance, ℓ^a, for Relatively Weak Interactions, i.e., for $\zeta < 25$ mV

Configuration	$\Delta G_\ell^{EL\,b}$	F_ℓ^{EL}
(flat parallel plates)	$1/\kappa \cdot 64nkT\gamma_0^2 \exp(-\kappa\ell)^c$	$-64nkT\gamma_0^2 \exp(-\kappa\ell)^c$
(sphere of radius R and flat plate; also valid for two crossed cylinders at 90°)	$\varepsilon R\psi_0^2 \ln[1+\exp(-\kappa\ell)]$	$-\varepsilon R\psi_0^2 \ln[1+\exp(-\kappa\ell)]\kappa$
(two spheres of radius R)	$0.5\varepsilon R\psi_0^2 \ln[1+\exp(-\kappa\ell)]$	$-0.5\varepsilon R\psi_0^2 \ln[1+\exp(-\kappa\ell)]\kappa$

^{a} For explanation of the symbols used, see Chapter V, eqs. [V-2–V-6].
^{b} To derive the aggregate, ψ_{0_a} value from two interacting values, ψ_{0_1} and ψ_{0_2}, use the geometric mean between the two, i.e., $\psi_{0_a} = \sqrt{\psi_{0_1}\psi_{0_2}}$.
^{c} $\gamma_0 = [\exp(ve\psi_0/2kT) - 1]/[\exp(ve\psi_0/2kT) + 1]$ (eq. [V-4]).

BROWNIAN MOVEMENT INTERACTIONS

In *apolar media* (3), the interaction between particles with adsorbed or otherwise attached polymer molecules (1) obeys only LW and BR forces. The decay of BR forces with distance (ℓ) may then be expressed as:

$$\Delta G_{131\ell}^{BR} = \Delta G_{131\ell_o}^{BR} \exp\frac{\ell_o - \ell}{\lambda} \qquad [VII-13]$$

for short range interactions and/or in the case where the liquid consists of only a very dilute polymer solution.

When such particles are suspended in relatively concentrated polymer solutions, with a polymer volume fraction ϕ_2 (cf. eq. [VII-7]):

$$\Delta G_{132\ell}^{BR} = \Delta G_{132\ell_o}^{BR} \left[\phi_2 \exp\frac{\ell_o - \ell}{Rg} + (1-\phi_2) \exp\frac{\ell_o - \ell}{\lambda} \right] \qquad [VII-14]$$

For long-range interactions, eq. [VII-14] reduces to:

$$\Delta G^{BR}_{132\ell} = \Delta G^{BR}_{132\ell_o} \left[\phi_2 \, \exp \frac{\ell_o - \ell}{Rg} \right] \qquad \text{[VII-15]}$$

(cf. eq. [VII-8]). Thus at $Rg \gg \lambda$ and $\ell > Rg$, Brownian movement interactions, like AB interactions, also decay extremely gradually with distance.

COMBINED INTERACTION FORCES ACTING ON SOLIDS OR ON SUSPENDED PARTICLES IN POLYMER SOLUTIONS

In the case of particles (1) suspended in a polymer (2) solution with polymer concentration ϕ_2 in a liquid (3), it may be argued that *for long-range interactions*, i.e., when $Rg \gg \lambda$ and $\ell > Rg$, eq. [VII-11] applies, not only to $\Delta\Delta G^{AB}_\ell$, but also to $\Delta\Delta G^{BR}_\ell$, and even to $\Delta\Delta G^{LW}_\ell$, and $\Delta\Delta G^{EL}_\ell$. This is because eq. [VII-11] describes the decay of interactions transmitted over an appreciable distance by the subsequent interactions between large polymer molecules with a radius of gyration Rg. The nature of the fundamental triggering interaction force plays no further role in its steadily decaying propagation from one polymer molecule to the next polymer molecule, although of course the magnitude and the sign of each of the initial contributing fundamental forces must be taken into account. Thus, *for long-range interactions*, eq. [VII-11] may be generalized by the following approximation

$$-\Delta\Delta G^{TOT}_\ell = (\Delta G^{LW}_{132\ell_o} + \Delta G^{AB}_{132\ell_o} + \Delta G^{EL}_{132\ell_o}$$
$$+ \Delta G^{BR}_{132\ell_o}) \phi_2 \exp \frac{\ell_o - \ell}{Rg} \qquad \text{[VII-16]}$$

For short-range interactions, i.e., when $\ell < Rg$, the situation is quite different. Eq. [VII-12] is applicable, but must be expanded as follows:

$$\Delta\Delta G^{TOT} = (\Delta G^{LW}_{131\ell_o} - \Delta G^{LW}_{132\ell_o}) \left(\frac{\ell_o}{\ell} \right)^2$$
$$+ (\Delta G^{AB}_{131\ell_o} - \Delta G^{AB}_{132\ell_o}) \exp\left(\frac{\ell_o - \ell}{\lambda} \right)$$
$$+ (\Delta G^{EL}_{131\ell_o} - \Delta G^{EL}_{132\ell_o}) \exp(-\kappa\ell) \qquad \text{[VII-17]}$$
$$+ (\Delta G^{BR}_{131\ell_o} - \Delta G^{BR}_{132\ell_o}) \exp\left(\frac{\ell_o - \ell}{\lambda} \right)$$

Eq. [VII-17], as far as LW forces are concerned, is only valid at relatively short distances (ℓ), where the dispersion component of the LW forces is in the unretarded mode. When $\ell > 5$ to 10 nm, $(\ell_o/\ell)^2$ becomes $(\ell_o/\ell)^3$, for the dispersion component.

Finally, *within the homogeneous polymer solution*, the long-range inter-action energy between polymer molecules, $\Delta G^{TOT}_{232} = \Delta G^{LW}_{232} + \Delta G^{AB}_{232} + \Delta G^{EL}_{232} + \Delta G^{BR}_{232}$, represents the sum-total of the osmotic interactions, yielding the osmotic pressure, which is the same everywhere in the polymer solution.

THE DERJAGUIN APPROXIMATION

Whilst the contact angle method, used as a force balance, conveniently yields results in terms of *energy* per unit surface area (see Chapter VI, and eqs. [IV-13, IV-16–IV-18], the mechanical types of force balance more easily measure the *force*, as a function of distance, ℓ, either between a sphere and a flat plate (in earlier versions of the force balance) or between two crossed cylinders (see Chapter XIV). In order to express the measured *force* (F) between a sphere and a flat plate, which are a distance, ℓ, apart at their closest approach, in terms of *energy* per unit surface area between two flat parallel plates, which also are a distance ℓ apart, Derjaguin (1934) showed (in the case of additive inverse power potential) that:

$$F_\ell = -2\pi R \Delta G_\ell \qquad [VII\text{-}18]$$

This equation is also valid for two crossed cylinders. For two spheres the expression becomes:

$$F_\ell = -\pi R \Delta G_\ell \qquad [VII\text{-}19]$$

Equations [VII-18 and 19] also express the convention that, e.g., when the interaction is an attractive one, F is held to be positive, and G negative; see Tables VII-1 to 3. It can be shown that eqs. [VII-18 and 19] are valid not only for the special case of additive inverse power potentials (Derjaquin, 1934), but indeed for all types of force laws (Israelachvili, 1985).

ENERGY-BALANCE DIAGRAMS

It now becomes possible to construct interfacial energy vs. distances diagrams, taking LW, AB, EL, and where warranted, even BR forces into account. Classically, such energy vs. distance diagrams, or DLVO-plots (after Derjaguin, Landau, Verwey and Overbeek; see Chapter I), only considered LW and EL forces. However, as shown in Chapter IV, in polar, and especially in aqueous media, AB energies, whether repulsive or attractive, typically are up to 100 times greater than LW energies, and 10 to 100 times greater than EL energies, at close range. Thus, in aqueous media and therefore in all biological systems, the inclusion of AB forces is not a novel and supplementary refinement of the DLVO theory, but a drastic correction, for the first time allowing the calculation of aqueous interactions on a reasonably accurate scale. Only elastic interactions between soluble polymer molecules, attached or adsorbed onto opposing particles in aqueous suspension (see Chapter XVIII), will not be taken into account here, although the direct interfacial interactions between such dissolved polymer molecules are of course taken into account.* Nor will fluctuating structural forces (see also Chapter XVIII), due to the finite discreet size of molecules of the liquid medium be treated here; see Israelachvili (1985).

In the same manner as EL interactions, AB interactions can be attractive, or repulsive. Attractive EL interactions, which mainly occur in complex-coacervation (Chapter XXI) (or in flocculation: Chapter XXIII), in antigen-antibody interactions and, in general, in ion exchange processes, are rarely depicted in DLVO diagrams, because such diagrams are usually employed to delineate regimes of stability, and not of instability. Taking suspensions of human erythrocytes in saline water as an example, a classical DLVO-plot (featuring only LW and EL forces) is depicted in Fig. VII-1 (A and B) (van Oss, 1989a). Although there is a secondary maximum of repulsion of the order of $+5 \times 10^{-3}$ mJ/m^2, which would form a repulsive barrier of the magnitude of ≈ 1 to 2 kT (given the size of erythrocytes), this could easily be overcome by cells with slight surface irregularities, thus resulting in attachment at the primary minimum of attraction, which would lead to cell clumping. In practice, however, such a destabilization of erythrocytes does to occur, either *in vitro* or *in vivo*. Thus Figure VII-1 (C and D), which takes the polar (AB) repulsion between the polysaccharide strands of the erythrocytes' glycocalices into account (in addition to the LW attraction and the EL repulsion), gives a more realistic portrayal of the actual interaction between erythrocytes. Due to the AB repulsion, which at close range is a couple of orders of magnitude higher than the EL repulsion, there is no primary minimum of attraction, so that the cells should be hyper-stable. Experimental results showed that this is indeed the case: even centrifugal forces of $\approx 300,000 \times$ gravity applied for 10 minutes did not result in the attachment of erythrocytes to each other (van Oss, 1985).† It should be noted that in Figures VII-1 C and D the value for λ (applicable to the decay of the AB forces) was taken to be of the order of 0.2 nm (Leneveu *et al.*, 1977; Chan *et al.*, 1979). However, the fact that at 20°C at least 10% of the water molecules engage in mutual hydrogen bonding at any one time, the actual value of λ for water appears to be somewhat larger than 0.2 nm; it lies

---

* The elastic interaction effects of adsorbed polymer strands are visible on experimentally obtained forces vs. distance diagrams (see, e.g,. Israelachvili, 1985, p. 211, and Chapter XVIII). In these cases, the close-range repulsive interactions become very large at a distance ℓ corresponding approximately to the R_g value of the adsorbed polymer molecules, instead of at the distance $\ell = \ell_0$ (where the Born repulsion begins to prevail), which usually is observed in the interaction between smooth, "hard" surfaces. The fact that the minimum equilibrium distance between hard surfaces, coated with adsorbed polymer molecules, is of the order of the radius of gyration R_g of these polymer molecules would indicate that each *single layer* of polymer molecules can be either compressed, or interpenetrated by the other layer, to a thickness of $0.5 \times R_g$.

† Due to the still primitive technique and methods of interpretation of interfacial interaction energies of biological materials in aqueous media prevailing in 1983 (van Oss *et al.*, 1983), the quoted LW interaction energy between cells (van Oss, 1985) still was significantly underestimated, which led to the hypothesis (van Oss, 1985) that the electrostatic repulsion alone was responsible for the lack of cell adhesion at high g-forces. Figure VII-1 C and D depict the more accurate values for these energies, showing that the prevention of cell adhesion really is mainly due to the close-range AB repulsion; these are called Extended DLVO plots (or XDLVO plots); see Wu *et al.* (1999).

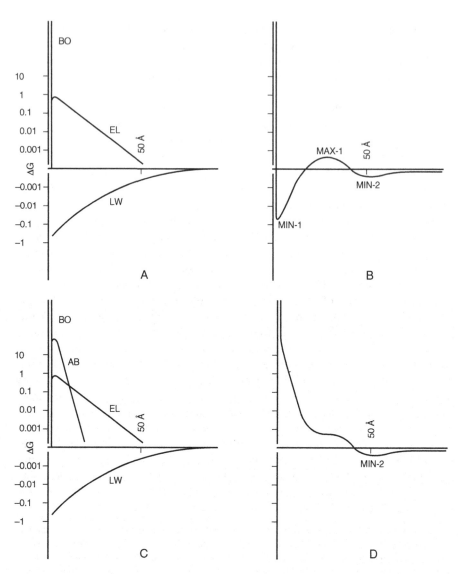

FIGURE VII-1   Energy balance diagrams (ΔG vs ℓ, with ΔG on a logarithmic scale, in mJ/m2) of erythrocyte suspensions. A: Electrostatic (EL) and Lifshitz-van der Waals (LW) interactions, combined in B: The classical DLVO plot with the primary minimum of attraction (MIN-1), the secondary maximum of repulsion (MAX-1) and the secondary minimum of attraction (MIN-2). BO indicates the Born repulsion, which prevents all atoms and molecules to approach each other to a distance smaller than ≈ 1.4Å. C: The AB interaction (in this case a repulsion) is depicted here, in addition to the EL repulsion, yielding in D: The combined curve that depicts the total interaction more realistically than the DLVO plot. The most important aspect of the complete curve shown in D, is the absence of primary minimum of attraction (still visible in B), owing to the AB-repulsion, which gives rise to superstability in vivo. From van Oss, Cell Biophys. 14:1 (1989), with permission. For interactions in aqueous media C and D depict Extended DLVO plots (or XDLVO plots); see Wu *et al.* (1999).

in between 0.2 and 1.0 nm, and the best approximation for the value of λ for water probably, is close to 1.0 nm (van Oss, 1990). See also Chapter XXIII, and van Oss, Giese and Costanzo (1990).

It should be pointed out that the Born repulsion is also portrayed in Figs. XXIII-1 and 2. This is a universally occurring, *very* short-range, strong repulsive force, which, *inter alia*, is responsible for the fact that atoms or molecules cannot penetrate other atoms or molecules, under normal conditions of energy, velocity, pressure and temperature. Born repulsion forces decay so exceedingly steeply with distance, that it is most reasonable to depict the Born repulsion in plots such as those shown, e.g., in Figs. XXIII-1A, 1B, 2A, 2B, as totally vertical straight lines, about 0.136 nm from the origin. This is somewhat less than the 0.157 nm we found for the minimum equilibrium distance (ℓ_o) found for LW interactions (see Chapter III), following from eq. [III-1]. However, if one takes the Lennard Jones potential, governing the Born repulsion, into account (van Oss and Good, 1984), then one finds, $\ell_o \approx 0.136$ nm (van Oss *et al.*, Chem. Rev., 1988). For the purpose of calculating Hamaker constants from γ^{LW} values it remains, however, more convenient to use eq. [III-1], with the value $\ell_o = 0.157$ nm (Chapter III).

When relatively short-range specific attractions occur in aqueous media between, e.g., hydrophilic biopolymer molecules and specific attractive sites imbedded in a hydrophilic substratum, the microscopic-scale specific attractive energies are best treated with the XDLVO approach. At the same time the long-range macroscopic-scale aspecific repulsive energies between the overal hydrophilic substratum and the equally hydrophilic biopolymer molecules can also, separately, be treated via the XDLVO approach. The final outcome depends on the interplay between these two XDLVO analyses (Chapter XXIV), which also play a crucial role in the determination of the aggregate kinetic rate constants of such complex interactions (Chapter XXV).

CODA

The different equations are given for the modes of decay of ΔG^{LW}, ΔG^{AB}, ΔG^{EL} and ΔG^{BR}, as a function of distance, ℓ. The importance is discussed of the Derjaguin approximation, which allows the calculation of the free energy between flat parallel plates, once the force between curved surfaces (e.g. spheres, a sphere and a flat plate, or two crossed cylinders) is known. For AB interactions, the role of the decay length of water (λ) is discussed: For repulsive as well as for attractive AB interactions, the value for λ is about 1.0 nm.

Part II

Interfacial Properties and Structure of Liquid Water

VIII   Lifshitz–van der Waals and Lewis Acid-Base Properties of Liquid Water—Physical and Physico-Chemical Effects

DOMINANCE OF THE LEWIS ACID-BASE PROPERTIES OF WATER

According to:

$$\Delta G^{coh} = -2\gamma_w = -2\gamma_w^{LW} - 2\gamma_w^{AB} \qquad \text{[VIII–1]}$$

the free energy of cohesion of water, at 20°C equals $-2\gamma_w^{LW} - 2\gamma_w^{AB} = -43.6 - 102 = -145.6$ mJ/m². Thus the polar (AB) part of the free energy of cohesion of water is: $102/145.6 = 70\%$ of its total free energy of cohesion. However, as a consequence of the differences in combining rules pertaining to the determination of the interfacial tensions between compound 1 and water (γ_{1w}) with respect to γ_{1w}^{LW} and γ_{1w}^{AB} (where $\gamma_{1w} = \gamma_{1w}^{LW} + \gamma_{1w}^{AB}$, cf. Chapter IV), the predominance of the role of $\Delta G_{ww}^{AB(coh)}$ over $\Delta G_{ww}^{LW(coh)}$ becomes even more dramatic when applied to the total free energy of interaction between two molecules or surfaces of a low-energy, non-polar compound 1 (where compound 1 is, for instance, octane), immersed in water (cf. eq. IV-17 and Table IV-2):

$$\Delta G_{1w1} = -\Delta G_{1w1}^{LW*} - \Delta G_{11}^{(coh)AB} - \Delta G_{ww}^{(coh)AB} + \Delta G_{1w}^{(adh)AB\dagger} \qquad \text{[VIII-2]}$$

---

* $\Delta G_{1w1}^{LW} = \underset{(coh)}{-2\gamma_1^{LW}} \underset{(coh)}{- 2\gamma_w^{LW}} + \underset{(adh)}{4\sqrt{\gamma_1^{LW}\gamma_w^{LW}}}$, the total of which terms becomes practically zero in the case of

1 = octane, because $\gamma_{octane}^{LW} = 22.6$ mJ/m² and $\gamma_w = 22.8$ mJ/m², so that $\Delta G_{1w1}^{LW} = 0.0004$ mJ/m².

† For octane, $\Delta G_{1w}^{AB} = +4(\sqrt{\gamma_1^{\oplus}\gamma_w^{\ominus}} + \sqrt{\gamma_1^{\ominus}\gamma_w^{\oplus}}) = 0$, because $\gamma_1^{\oplus} = \gamma_1^{\ominus} = 0$, as octane is non-polar.

The predominant Lewis acid-base properties of water make it essential to take the AB interaction energies into account in all approaches involving energy vs. distance analyses of hydrophobic as well as of hydrophilic materials, immersed in water, using the extended DLVO theory, incorporating van der Waals, electrical double layer *and Lewis acid-base interactions*; see Chapters VII and XXIII.

For octane (cf. table IV-2): $\Delta G_{1w1} = 0 - 0 - 102 + 0 = -102$ mJ/m², which is for 100% of AB origin. For a hydrophilic, non-polar polymer with $\gamma_1 = 40$ mJ/m², $\gamma_1^{\oplus} = 0$ and $\gamma_1^{\ominus} = 50$ mJ/m² (cf. eq. IV-17 and Table IV-2): ΔG_{1w1} (see eq. VIII-2) becomes: $-5.5 - 0 - 102 + 127.8 = +20.3$ mJ/m², which is still for 73% of AB origin. Thus the LW free energy of interaction between two identical hydrophobic materials, immersed in water (ΔG_{1w1}^{LW}) usually is from zero to less than 10% of the total ΔG_{1w1}. The AB free energy of interaction between two identical hydrophilic materials, immersed in water (ΔG_{1w1}^{AB}), usually ranges between 100%, and at least 70% of the total ΔG_{1w1}. Conversely, the components of the free energies of hydration (ΔG_{1w}^{LW} and ΔG_{1w}^{AB}) are, respectively, 100% ΔG_{1w}^{LW} for completely non-polar (hydrophobic) materials and approximately 45% ΔG_{1w}^{LW} vs. 55% ΔG_{1w}^{AB} for extremely hydrophilic materials; see Chapters IX and X, below.

EFFECT OF TEMPERATURE ON THE γ-COMPONENTS OF WATER

From the data given by Jasper (1972) on the total surface tension of water, γ_w decreases rather linearly with increasing temperature, at the rate of about 1.08 mJ/m² per 10°C. If it may be assumed that between, say, 0°C and 60°C, γ_w^{LW} continues to represent 30% of γ_w, then both γ_w^{LW} and γ_w^{AB} vs. t may be depicted as in Fig. VIII-1. From contact angle measurements with water on surfaces of polymethylmethacrylate (PMMA), freshly cast from a 5% solution of PMMA in toluene, done at 20°C and at 38°C, some tentative conclusions may then be drawn about the values of $\gamma_W^{\oplus}$ and $\gamma_W^{\ominus}$ at temperatures higher than 20°C.

At 20°C the PMMA surfaces in question had the following properties: $\gamma^{LW} = 43.6$ mJ/m², $\gamma^{\oplus} \approx 0$; $\gamma^{\ominus} = 20.4$ mJ/m² (from $\theta_{DIM} = 30°$, $\theta_{\alpha\text{-Br. naphth.}} = 14.5°$, $\theta_{H_2O} = 60.75°$ and $\theta_{GLY} = 59.25°$). Assuming that for solid PMMA, γ^{LW} and $\gamma^{\ominus}$ remain practically unchanged at 38°C (e.g., $\theta_{DIM} = 26.13°$ and $\theta_{\alpha\text{ Br. naphth.}} = 12.5°$) and given that at 38°C, θ_{H_2O} had decreased to 52.44°, and also given that at 38°C $\gamma_W^{AB} = 49$ mJ/m², then $\gamma_W^{\oplus}(38°) = 32.4$ mJ/m² and $\gamma_W^{\ominus}(38°) = 18.5$ mJ/m² (γ_W^{LW} at that temperature is 20.9 mJ/m²).

Thus, compared with the standard values of $\gamma_w^{\oplus} = \gamma_w^{\ominus} = 25.5$ mJ/m² for water at 20°C (corresponding to a γ_w/γ_w ratio of 1.0), when raising the temperature to 38°C that $\gamma_w^{\oplus}/\gamma_w^{\ominus}$ ratio increases by 75% to 1.75 (van Oss and Giese, 2004); see Table VIII-1. It should be noted however, that with the same increase in temperature from 20° to 38°C $|\Delta G_{ww}^{Cohesion(AB)}|$ only decreases by about 3%, from -102 to -99 mJm²; see Fig. VIII-1. This agrees well with a continued elevated free energy of cohesion of water, even when hot, whilst the strongly increased Lewis acidity of water at higher temperatures most likely correlates with a decrease in cluster size of the water molecules, because the increase in electron-acceptivity of warmer water concomitantly increases the availability of unbound aqueous H atoms, which occurs when clusters become smaller. Smaller cluster size also agrees well with the pronounced decrease in the viscosity of water with an increase in temperature, see below under *Cluster Formation in Liquid Water*.

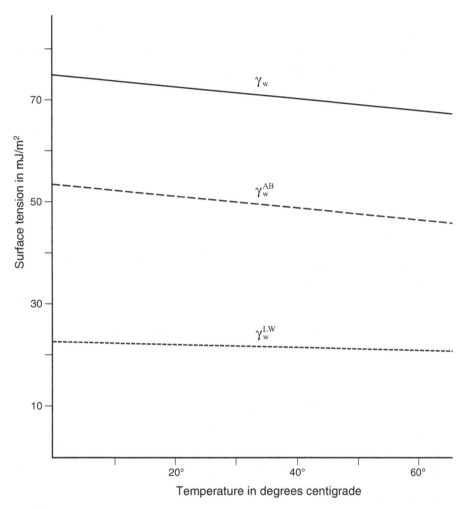

Figure VIII-1 Graph of the variation of the surface tension components of water with temperature. Data for γ_w derived from Jasper (1972). The plot of γ_w *vs.* t is virtually linear. To obtain γ_w^{LW} and γ_w^{AB}, it is assumed, for lack of more precise data, that γ_w^{LW} remains at 30% of γ_w for the temperature range shown. For the variation of $\gamma_w^{\oplus}/\gamma_w^{\ominus}$ with temperature, see text.

The increased electron-accepticity of warmer water also explains why warm water (even without soap) is more efficacious than cold water in washing dirty dishes. (The surfaces of accumulated dirt as well as of the dishes themselves are predominantly electron-donors which, due to the increase in aqueous electron-accepticity with temperature, become more hydrated with a raise in temperature, which favors a repulsion between the dishes' surfaces and dirt).

Upon cooling water below 20°C it becomes more Lewis basic with a $\gamma_w^{\oplus}/\gamma_w^{\ominus}$ ratio of 0.514 (by extrapolation of the $\gamma_w^{\oplus}/\gamma_w^{\ominus}$ ratios at 38° and 20°C); see also Table VIII-1.

TABLE VIII-1

Surface Tension Components for Water at Different Temperatures as Well as for (Hydrated) Ice, at 0°C; all γ-values in mj/m²

	Temperature (°C)	$R = \gamma_w^\oplus/\gamma_w^\ominus$	γ	γ^{LW}	γ^{AB}	$\gamma^\oplus$	$\gamma^\ominus$
Water	0°	0.5143[a]	75.8[b]	22.8[i]	53[i]	19[d]	37[d]
	20°	1.0[c]	72.8	21.8	51	25.5[c]	25.5[c]
	38°	1.75[e]	70.0	21.0	49	32.4[e]	18.5[e]
Ice (hydrated)	0°	0.25[f]	70.6[g]	28.5[h]	42.05	10.5[f]	42.1[f]

[a] Extrapolated from the values at 20° and 38°C (see text)
[b] Extrapolated from Jasper (1972)
[c] See Chapter IV
[d] From $R = \gamma_w^\oplus/\gamma_w^\ominus$ and γ^{AB}
[e] See text
[f] From the repulsion between an advancing ice front and hydrophilic clay mineral (SWy-1) particles and the attraction between the advancing ice front and hydrophic SWy-1-HDTMA particles (van Oss *et al.*, J. Adhesion, Sci. Tech., 1992).
[g] From the water contact angle on ice at 0°C, (Ketcham and Hobbs, 1969; van Oss *et al.*, Cell Biophysics, 1992).
[h] From the contact angle of *cis*-decalin, extrapolated to 0°C (van Oss *et al.*, J. Adhesion, Sci. Tech., 1992).
[i] See Fig. VIII-1.

THE SURFACE PROPERTIES OF ICE AT 0°C

The surface properties of (hydrated) ice at 0°C can be derived via: a) the contact angle with *cis*-decalin of 35° (van Oss *et al.*, J. Adhesion Sci. Technol., 1992), as far as γ_{ice}^{LW} is concerned; b) for $\gamma_{ice}^{AB} = 2\sqrt{\gamma_{ice}^\oplus \times \gamma_{ice}^\ominus}$ from the contact angle with water of 20°C (Ketcham and Hobbs, 1969) and: c) for $R = \gamma_{ice}^\oplus/\gamma_{ice}^\ominus$ from the fact that an advancing freezing front of hydrated ice (in water) repels (rejects) particles of the hydrophilic smectite (Swy-1), whilst it attracts (engulfs) Swy-1 particles that have been hydrophobized with hexadecyl trimethyl ammonium cations (van Oss *et al.*, J. Adhesion Sci. Technol, 1992); see Table VIII-1.

The surface tension properties of both ice and water at 0°C are data that are essential for determining, e.g., whether living blood cells are going to be engulfed by an advancing freezing front, or whether these cells are going to be rejected (i.e., pushed forward) by such an advancing front. Virtually all living mammalian cells are significantly hydrophilic, as are all water-soluble proteins (e.g., blood plasma proteins), polysaccharides, as well as salt ions. All of these are repelled by advancing freezing ice fronts, so that when freezing blood, blood cells, blood proteins, as well as the ambient salt ions, are pushed forward and thus become highly concentrated into a very small volume, which causes most of the salt ions to diffuse inside the cells. After thawing and concomitant re-dilution, the cells, which now contain high concentrations of salt ions, undergo severe osmotic damage. On the other hand, when upon freezing the cells become immediately engulfed by the advancing freezing front, they remain undamaged during freezing as well as after thawing. This will not happen however as long as the cells are suspended in pure water or saline, and it is only when the water is admixed with, e.g., high concentrations of glycerol (van Oss *et al.*, Cell Biophys., 1992) that the cells can become engulfed by the advancing freezing front, on account of the different $\gamma_{wg}^{\oplus}$ and $\gamma_{wg}^{\ominus}$ values of the water/glycerol mixture, which differ significantly from the $\gamma_w^{\oplus}$ and $\gamma_w^{\ominus}$ values of water; see also Chapter XIV, under *Advancing Solidification Fronts*. Frozen cells which are uniformly distributed inside a frozen water/glycerol mixture, do not undergo osmotic damage during freezing, nor during thawing. However, after thawing the admixed glycerol should be very gradually replaced with water, to avoid osmotic damage caused by the high glycerol concentrations upon sudden exposure to pure water.

In this context it would be useful to know the surface properties of glycerol at 0°C, but even its total surface tension appears to be unknown at that temperature This is probably a consequence of the fact that glycerol is exceedingly viscous at 0° (i.e., 12,100 cP) which is 6,777 times higher than the viscosity of water at 0°C. Still, whilst the surface properties of glycerol at 20°C are: $\gamma_g = 64$, $\gamma_g^{LW} = 34$, $\gamma_g^{AB} = 30$, $\gamma_g^{\oplus} = 3.92$ and $\gamma_g^{\ominus} = 57.4$ mJ/m^2 (see Table XII-4), and assuming that these values increase with a decrease in temperature in a manner analogous to those of water, one can at least estimate as a first approximation that for glycerol at 0°C: $\gamma_g = 66.3$, $\gamma_g^{LW} = 36$ and $\gamma_g^{AB} = 30.3$ mJ/m^2. To conform with the known inclusion of human erythrocytes by an advancing freezing front (van Oss *et al.*, Cell Biophys., 1992, see above), an optimal correlation would exist for glycerol at 0°C when $\gamma_g^{\oplus} = 3.0$ and $\gamma_g^{\ominus} = 76.7$ mJ/m2, with a $\gamma_g^{\oplus}/\gamma_g^{\ominus}$, R, of 0.039 (as compared to R = 0.068 for glycerol at 20°C). For water, R = 0.514 at 0°C, compared to R = 1.0 at 20°C; see Table VIII-1.

CLUSTER FORMATION IN LIQUID WATER

The strong hydrogen-bonding (AB) free energy of cohesion of water, apart from being the driving force of the hydrophobic effect, is also the cause of the continuous formation of clusters of water molecules, of an average size of about 4.5 water molecules per cluster, at 20°C (Rao, 1972; van Oss, Giese and Good, 2002; van Oss and Giese, 2004), which are however short-lived as they continuously form and break down again within about a picosecond (Luzar and Chandler, 1996).

Using the solubility equation, also given in Chapter XIII as eq. XIII-3:

$$-2\gamma_{iw}^{o}S_c = kT \cdot \ln s \qquad \text{[VIII-3]}$$

the contactable surface area (S_c) of a solute can be obtained when the aqueous solubility of that solute (s, expressed in mol fractions) and its zero time dynamic interfacial tension with water (γ_{iw}^{o}) or its free energy of interaction (ΔG_{iwi}), when immersed in water are known, using (see also eq. II-3):

$$\Delta G_{iwi} = -2\gamma_{iw}^{o} \qquad \text{[VIII-4]}$$

Now, the correct γ_{iw}^{o} values, as well as the contactable surface areas have been derived from their aqueous solubilities and their molecular dimensions (yielding S_c)

for cyclohexane, benzene, toluene and xylene (van Oss, Giese and Good, 2002; Table XIII-2), using eq. VIII-3, above. Meanwhile, again using eq. VIII-4, which expresses the free energy of interaction of two molecules, i, immersed in water, w, in terms of the zero time dynamic interfacial tension, γ_{iw}^o, one may state with equal validity that the free energy of interaction between two molecules of water, w, immersed in, e.g., organic liquid, i, is similarly related to γ_{iw} (van Oss, Giese and Good, 2002):

$$\Delta G_{wiw} = -2\gamma_{iw}^o \qquad \text{[VIII-4A]}$$

Thus, eq. VIII-4A, which is an equally valid expression for γ_{iw}^o as eq. VIII-4, shows that one can obtain the contactable surface area of water from its solubility in an organic solvent, i, using:

$$-2\gamma_{iw}^o \cdot S_{c(w)} = kT \cdot \ln s_{(w)} \qquad \text{[VIII-3A]}$$

One can therefore also obtain the contactable surface area, S_c, for *water*, from its solubility in an organic solvent. Thus, from the aqueous solubilities of water in a number of organic solvents, i.e., cyclohexane, benzene, toluene and xylene, at 20°C, S_c for water could be established at 0.358 nm^2 (± 0.013, SD) (van Oss, Giese and Good, 2002). This corresponds to a contactable surface area for water (at room temperature), which is much greater than the S_c for a single water molecule. It corresponds, instead, to an Sc value of a water *cluster* comprising approximately 4.5 water molecules per cluster, at 20°C; see also Rao (1972).

In contact with low-polarity compounds such water clusters would tend to form a fairly thin, probably curved sheet-like array of water molecules, similar to those found in water "clathrates" that surround other, small, usually low-polarity atoms or molecules (Davidson, 1973; see also Chapters IX and XVIII). Finally, the size of water clusters of about 4.5 water molecules per cluster at 20°C correlates much more closely with a radius of gyration (R_g) compatible with a decay length (λ) for water with a value close to 1.0 nm (see Chapters VI, VII, XVIII, XIX and XXIII).

The self-hydrogen-bonding water molecules which are associating, then dissociating, then re-associating, are responsible for the high $\Delta G_{ww}^{(AB)Cohesion} = -102$ mJ/m^2 of water at 20°C, decreasing mildly to about −87.8 mJ/m^2 at the boiling point of water. This indicates that even when reaching 100°C, $\approx 86\%$ of the free energy of cohesion of water still persists although, when heating water to 100°C, its viscosity decreases a drastic 3.5 fold (from 1.00 cP at 20°C to 0.284 cP at 100°C). This strongly points to a concomitant decrease in cluster size from about 4.5 water molecules per cluster at 20°C, to probably mainly single water molecules, at 100°C. The reduction in cluster size with an increase in temperature is further corroborated by the marked increase in the electron-acceptivity (i.e., in the Lewis acidity) of water, upon heating; see the Section on the *Effect of Temperature on the γ-Components of Water*, above.

The influence of the pH of water on its interfacial properties and on the surface properties of electrically charged and/or amphoteric or amphiphilic surfaces in contact with water of different pH's, is treated in Chapter XII; see also van Oss and Giese (2005).

CODA

This Chapter treats the influence of physical and physico-chemical phenomena on the interfacial properties of water. These include the dominant influence of the Lewis acid-base (AB) properties of water on its free energy of cohesion, which in turn is the driving force of hydrophobic attraction (the "hydrophobic effect"), as well as of hydrophobic repulsion ("hydration forces") exerted on or by molecules or particles immersed in water.

The influence of temperature (T) of water on its AB surface tension properties (where γ_w^{AB} is proportional to $\sqrt{\gamma_w^{\oplus} \cdot \gamma_w^{\ominus}}$) is slight, but the influence of T on the $\gamma_w^{\oplus}/\gamma_w^{\ominus}$ *ratio* is large. Data are given on the γw, γw and w values for water, as a function of T, as well as on the surface properties of ice at 0°C. Cluster formation of water molecules is discussed, as well as the influence of T on cluster size. At 20°C the average number (n) of single water molecules per cluster is found to be about 4.5, but n decreases with an increase in T. The connection between the cluster size of water at room T and the radius of gyration (R_g) of water is discussed, as is the connection between R_g and the decay length of water (λ).

IX Role of Water in Hydrophobic Attraction

THE HYDROGEN-BONDING (LEWIS ACID-BASE) FREE ENERGY OF COHESION OF WATER AND "THE HYDROPHOBIC EFFECT"

The driving force of hydrophobic attraction between (especially) low-polarity molecules or particles, immersed in water, is not a mystery: it is simply the hydrogen-bonding (Lewis AB) free energy of cohesion of the water molecules that surround these molecules or particles, see Table IV-2. A synonym for "hydrophobic attraction" is "The Hydrophobic Effect," although the "hydrophobic" part remains a misnomer (see Chapter XVIII). The expression: "hydrophobic effect" sounds more impressive and also more mysterious than "hydrophobic attraction," but it is the same thing and it is really solely caused by Lewis acid-base force-induced (which in the case of water, means hydrogen-bond-driven) cohesion between the water molecules. A similar polar cohesion of liquid molecules also occurs in other dipolar liquids, such as glycerol, formamide, ethylene glycol, etc., but the free energy of polar cohesion in these non-aqueous dipolar liquids is much weaker than the polar cohesion of water. In such non-aqueous dipolar organic liquids the interaction is not called "hydrophobic" but, equally unfelicitously, "solvophobic."

Obviously, the hydrophobic attraction phenomenon, caused as it is by the hydrogen-bonding free energy of cohesion between the water molecules of the aqueous medium is always present in water. The hydrophobic attraction therefore acts on all molecules and particles, whether they be apolar or polar. In the latter case however, the omni-present hydrophobic attraction can be attenuated, or even reversed, by a hydration-driven hydrophilic repulsion, see Chapter X, below.

The hydrophobic attraction between identical molecules, 1, immersed in water, w, is best described by the polar part of eq. IV-17, in the form displayed on top of Table IV-2, terms II to V. The upper one of the two examples displayed in that Table features an apolar compound of $\gamma^{LW} = 22$ mJ/m², for instance, octane. (As water at 20°C also has a γ^{LW} which is close to 22 mJ/m², the LW term (term I) is zero in this case). It is clear from the example shown in Table IV-2 that in the case of octane only term III is operative, which is the term representing the polar free energy of cohesion of water at 20°C, so that $\Delta G_{1w1}^{AB} = -102$ mJ/m². ΔG_{iwi}^{AB} represents the best quantitative definition for hydrophobic attraction as well as for hydrophilic repulsion (van Oss and Giese, 1995; see also Chapter XIX), where $\Delta G_{iwi}^{AB} < 0$ indicates hydrophobicity and $\Delta G_{iwi}^{AB} > 0$ hydrophilicity, both in quantitative terms.

Hydrophobic attractions are not limited to interactions between identical molecules, parts of molecules, particles or surfaces: they equally readily occur between two *different* entities, 1 and 2, immersed in water, w. Their free energies of interaction are then expressed as ΔG_{1w2} (see eq. IV-16, whose subscript, 3, is here replaced by subsript w, for water). Significant hydrophobic attractions between two different moieties, 1 and 2, immersed in water, can occur between two apolar molecules, or parts of molecules, but also between one low-polarity molecule or site and one hydrophilic molecule or site. Hydrophobic attractions are among the most important modes of interaction between polar, hydrophilic antigenic sites, or epitopes, situated distally on antigen molecules, cells or particles on the one hand, and hydrophobic antibody-active stites or paratopes on the other hand, which have the function to interact specifically with the aforementioned epitopes. Usually hydrophobic paratopes are situated in a concavity or cleft of the antibody molecules, which prevents them from causing undesirable aspecific interactions with other circulating proteins or cells. This type of specific interaction mechanism between antigen and antibody molecules also applies to most other specific biological ligand-receptor interactions; see Chapter XXIV.

HYDROPHOBIC HYDRATION

Whilst the free energy of *hydrophobic attraction* (ΔG_{1w1}) is due to Lewis AB intertactions [see above and Table IV-2 (top)], the free energy of *hydration* (ΔG_{1w}) (cf. eq. IV-6B):

$$\Delta G_{1w} = \gamma_{1w} - \gamma_1 - \gamma_w \qquad [\text{IX-1}]$$

of hydrophobic surfaces is caused solely by Lifshitz–van der Waals interactions. It can be seen from both eq. IX-1, and from (cf. eq. II-33A):

$$\Delta G_{1w1} = -2\gamma_{1w} \qquad [\text{IX-2}]$$

that the interfacial tension, γ_{1w}, plays a crucial role (for the determination and role of γ_{1w}, see Chapter XIII). Using the full non-polar/polar equation (eq. IV-18), the free energy of hydration of completely apolar commpounds is:

$$\Delta G_{1w} = -2[\sqrt{(\gamma_1^{LW} \cdot \gamma_w^{LW})} + \sqrt{(\gamma_1^{\oplus} \cdot \gamma_w^{\ominus})} + \sqrt{(\gamma_1^{\ominus} \cdot \gamma_w^{\oplus})}] \qquad [\text{IX-3}]$$

When $\gamma_1 = \gamma_1 = 0$, the resulting remnant of eq. IX-3 is:

$$\Delta G_{1w}^{LW} = -2\sqrt{(\gamma_1^{LW} \cdot \gamma_w^{LW})} \qquad [\text{IX-4}]$$

indicating that the free energy of hydration of non-polar compounds is indeed exclusively of LW origin, as far as the attraction of the first layer of hydrophobic hydration is concerned. Because of the virtually complete saturation of the hydrogen bonds in the thin network of the water of hydration around apolar entities, the distal surface of the first layer of such water of hydration is probably one of the most hydrophobic structural forms of which water is capable.

Given that for water, γ_w^{LW} = 21.8 mJ/m2, and that with most apolar condensed-phase materials γ^{LW} usually only varies between 17 and 35 mJ/m2, the free energy of hydration of such apolar materials would tend to range between –38.5 and –55.2 mJ/m2. These values, while not exceedingly elevated, still represent sizeable bond energies of "hydrophobic" hydration, thus contradicting the designation "hydrophobic" for such apolar or "water-fearing" materials; see also Hildebrand (1979).

For the water-air interface however, i, for air, is zero, so that the hydration energy, ΔG_{iw} (see eq. IX-4) is also zero, which makes air at the water-air interface the only known "real hydrophobic" surface in the strictest sense of the word. On the other hand, at room temperature air is not a condensed-phase material and it therefore continues to be true that there exists no condensed-phase material that is truly "hydrophobic", in the sense that it totally repels (fears) water.

"Hydrophobic" hydration being, as shown above, solely or mainly of LW origin, does not significantly attract one part of a cluster of water molecules more than the other. It therefore causes the first layer of the water of hydration on apolar surfaces to assume the appearance of a thin, relatively hydrophobic network of water clusters which closely resemble the water molecules of clathrate hydrates [van Oss and Giese (2004); see also Israelachvili (1992, pp. 129-130); Evans and Wennerström (1999); Davidson (1973)]; see also the following sub-section. In contrast with the flattening and thus to a certain extent, the densification of the water of hydrophobic hydration as indicated above, the putative decrease in density of hydrophobic hydration layers proposed by Besseling (1998) would not appear likely, especially in view of the strong similarity with clathrate-forming water layers; see below.

CLATHRATE FORMATION

Clathrate formation strongly resembles the formation of agglomerates of hydrophobic molecules immersed in water, surrounded by a thin layer of completely hydrogen-bonded water molecules. Clathrates (from the Greek: κλειω, to confine) is a designation for (usually) *single* atoms or molecules, encaged in a network of water molecules. When atoms or molecules that are apolar, or of low polarity, are immersed in water at a sufficiently low concentration to keep a large enough inter-atomic or inter-molecular (or, effectively, inter-clathrate) distance between one another to obviate hydrophobic agglomeration, such single atoms or molecules can become individually enclosed in cages consisting of thin layers of hydrogen-bonded water molecules (cf. Davidson, 1973). Clathrates of noble gases such as argon, krypton and xenon can also form with non-aqueous liquids, for example hydroquinone. Such clathrates can be used for the isolation, purification and handling of some of the rare gases (van Thoor, 1968).

ACTION AT A DISTANCE

The free energies of hydrophobic attraction in water, i.e., ΔG_{1w1}^{AB}, as well as ΔG_{1w2}^{AB}, decay exponentially as a function of distance, ℓ, following the equations given in Table VII-2, which is the same rate of decay vs. distance as also observed with hydrophilic repulsion. The basis for the exponential decay in both modes is the decay length of water, λ, of approximately 1.0 nm. Mechanistically speaking, the extension at a distance of hydrophobic attraction energies between hydrophobic molecules or particles is a consequence of the properties of the different successive layers of water of hydration which surround such molecules or particles when they are immersed in water. The distal surface of the first layer of water of hydration of hydrophobic molecules or particles is also rather hydrophobic (see the Section on *Hydrophobic Hydration*, above), but less hydrophobic than the surface proper of such molecules or particles, although it still gives rise to a significant degree of hydrophobic attraction, caused by the AB free energy of cohesion between the surrounding water molecules. Then, the distal surface of the second layer of water of hydration surrounding the first layer of hydration is also to a certain degree still somewhat hydrophobic and although less hydrophobic than the first layer of water of hydration, it is still subject to some hydrophobic attraction, and so on, to an exponentially decreasing extent, as the distance increases distally into the bulk liquid. With drops of octane (in water) of R = 1 nm, ΔG_{1w1} reaches a value of –0.5 kT at $\ell \approx 5.0$ nm, which –0.5 kT per drop of octane becomes counter-balanced by the Brownian motion energy, in one direction, of +0.5 kT. This occurs when such octane drops of R = 1 nm become mono-dispersed in water at a dilution greater than about 0.5% (v/v). Then, at $\ell \leq 10$ nm, the agglomeration of two such octane drops in water becomes improbable and only clathrate formation of single octane drops with R = 1 nm or less, should occur. However, once agglomeration of octane drops or molecules has taken place, the reversion of such agglomerates into a stable suspension of single spheres of R $\leq$ 1 nm, without the admixture of surfactants, is not favored.

MODULATION OF THE HYDROPHOBIZING CAPACITY OF WATER

As the hydrophobizing capacity of water (which is equal to ΔG_{iwi}^{AB}, cf. eq. IV-34) is based upon the self-hydrogen-bonding (AB) free energy of cohesion between the water molecules, the hydrophobizing capacity of water can be modulated by the admixture of solutes which either have a lower self-hydrogen-bonding free energy of cohesion than water (thus decreasing the hydrophobizing capacity of water), or which have a higher self-hydrogen-bonding free energy of cohesion than water (and thus increase the hydrophobizing capacity of water.

DECREASING $|\Delta G_{iwi}^{AB}|$

To decrease the hydrophobizing capacity of water one must decrease the (negative) value of ΔG_{iwi}^{AB} (cf. the polar part of eq. IV-17; see also Table IV-2). In water this value is quite high, due to the high value of the polar cohesion of water (term III of

ΔG_{iwi}^{AB}, displayed in Table IV-2), i.e., $\Delta G_{iwi} = -102$ mJ/m^2. To decrease $|\Delta G_{iwi}|$ one must lower γ_w^{AB} of water (which is equal to minus one half of term III (Table IV-2; see also eq. IV-5). To that effect one must add a solute, s, to the water, with a much lower γ_s^{AB} than that of water. Nonetheless, such a solute, to be soluble in water, still must be significantly polar. This can be achieved by using a strongly monopolar (or almost monopolar) compound such as one of the lower monohydric alcohols [e.g., methanol or ethanol, cf. Harrison *et al.* (2003); see also Washburn (1928)], or ethylene glycol or acetonitrile (see Chapter XXIV, under *Liquid Chromatography*). Such monopolar, or near-monopolar liquids usually have a high $\gamma_s^{\ominus}$, but a very low $\gamma_s^{\oplus}$, so that their γ_s^{AB} values are quite low (cf. eq. IV-5).

The admixture to water of such largely monopolar organic liquids results in a lowering of the γ^{AB} of their aqueous solution and thus of term III of its $|\Delta G_{iwi}|$ (cf. Table IV-2), which decreases the polar cohesion of the aqueous mixture, which is tantamount to lowering its hydrophobizing capacity. This approach is widely used in the elution step of reversed-phase liquid chromatography of, e.g., proteins; see Chapter XXIV, and Harrison *et al.* (2003). Using too high a concentration of the above-mentioned solvents should however be avoided as they diminish the γ_w value in the hydration term (term V of ΔG_{iwi} in Table IV-2), which then decreases a protein's solubility in the aqueous mixture.

INCREASING $|\Delta G_{iwi}^{AB}|$

To increase the hydrophobizing capacity of water one can dissolve solutes in it that have a higher polar free energy of cohesion than water, as exemplified by term III of ΔG_{iwi}; see Table IV-2. The two principal classes of polar cohesion of water enhancing solutes are simple sugars (not polysaccharides) and inorganic salts (van Oss and Giese, 2004).

Simple *sugars*, such as sucrose or glucose have a higher γ_s^{AB} than water, when in aqueous solution. Extrapolated to 100% sugar solutions, the γ_s^{AB} of sucrose is 100.2 mJ/m^2 and that of glucose 108.7 mJ/m^2, whilst the γ_w^{AB} of water is only about half that, at 51 mJ/m^2, all at 20°C. For instance at 30% sucrose in water (w/w), γ^{AB} of the solution is 60.3 mJ/m^2 and at 55%, γ^{AB} of the solution is 68.1 mJ/m^2. Fifty five percent aqueous sucrose solutions can destabilize kaolinite or silica suspensions which in distilled water form stable suspensions (Docoslis *et al.*, 2000). This hydrophobizing effect of sugars is only achievable with low molecular weight sugars, which have a high γ_s^{AB} owing to the OH- and O-groups of sugar molecules that can readily react with those of other sugar molecules via hydrogen-bonding, which results in a high free energy of cohesion. This effect does not occur with polymers of these sugars, such as ficoll or dextran. In such polysaccharides the strong self-hydrogen bonding occurring in monomeric or dimeric sugars has been inactivated by covalent polymerization. In contrast with, e.g., glucose, the linear polymer, dextran, has a γ_s^{AB} in aqueous solution equaling only 21.35 mJ/m^2, which is much lower than the γ_w^{AB} of water of 51 mJ/m^2. In contrast with their monomers, which are strongly dipolar in aqueous solution (thus causing their elevated γ_s^{AB} values), polysaccharides are close to being monopolar electron-donors (even in aqueous solution), which causes their rather low γ_s^{AB} (Docoslis *et al.*, 2000). Nonetheless, simple sugars as well as their polymers have one property in common, which is that they are both repelled by the air-side of the water-air interface, due largely to a sizeable net LW repulsion caused by the total absence of a γ^{LW} value for the air-side of the water-air interface (see. Chapters II and IV, above, for repulsive van der Waals interactions and see Chapter XI, below, which deals with the unusual physico-chemical properties of the water-air interface; see also Schematic Presentation XI-1)

Inorganic *salts*, such as Na_2SO_4, $(NH_4)_2SO_4$, K_2CO_3, can be used in the same manner as the sugars mentioned above, in molar or plurimolar aqueous solutions, e.g., for the insolubilization of proteins, caused by the enhanced hydrophobic attraction engendered by the free energy of cohesion (between the added inorganic cations and anions), which is stronger than that of water. For the surface tension-enhancing properties of various inorganic salts, when dissolved in water, see Washburn (1928).

One molar $(NH_4)_2SO_4$ solutions are used to immobilize very hydrophilic proteins, such as human serum immunoglobulin A (IgA), onto alkane (e.g., C_{18}) substituted hydrophobic beads, used in hydrophobic interaction liquid chromatography. This enhancement of the hydrophobizing capacity of water by the admixture of salt ions is necessary in the case of the exceedingly hydrophilic IgA molecules as, in contrast with most other blood serum proteins, they do not normally adhere hydrophobically to C_{18} substituted surfaces when IgA is dissolved in pure water. Elution of the IgA molecules from the C_{18} substituted beads is subsequently achieved with an inverted concentration gradient of $(NH_4)_2SO_4$, which decreases in concentration from 1 M to 0 M $(NH_4)_2SO_4$; see also Chapter XXIV. The precipitation of various proteins by the admixture of (usually plurimolar) amounts of $(NH_4)_2SO_4$ is called *salting-out**), see also Chapters XXI and XXII.

Finally, to enhance the hydrophobizing capacity of water, as far as partly polar solutes or particles are concerned, one can also decrease the temperature of the aqueous medium. A decrease in T only slightly increases the negative value of the part of ΔG_{iwi}^{AB} which represents the polar free energy of cohesion of water (term III of ΔG_{iwi}^{AB} shown in Table IV-2), while it more strongly decreases the (positive) value of term V, due to the pronounced decrease in γ_w, upon a decrease in T, e.g., from $20°$ to $0°C$; see Table VIII-1. Chapter X (below) treats the influence of a change in temperature from the point of view of hydrophilic repulsion. In the latter case the quantitatively more significant enhancement of hydrophilic repulsion energies among polar molecules is caused by an increase in T.

---

* The term *salting-out* is in contrast with a protein purification method, called *salting-in*, which is however an entirely unrelated method as far as the underlying mechanisms are concerned. Salting-in is used for the re-solubilization of *euglobulins* after these have been precipitated at a pH close to their isoelectric point, and in the absence of salt (usually after dialysis against distilled water to remove the salt ions). Euglobulins are "good" or "real" globulins, a term used to distinguish them from *pseudoglobulins*, which are reasonably soluble in distilled water. Examples of human blood serum euglobulins are: some of the immunoglobulins-G (IgG) and most of the Immunoglobulins-M (IgM) (van Oss and Buenting, 1967), and also subfractions of complement factors C1 and C3, see Kabat and Mayer (1961). Addition of physiological concentrations of NaCl to a euglobulin precipitate usually results in the re-dissolution of at least part of the protein; this is called salting-in. Salting-in often is significantly denaturing, in contrast with salting-out, which rarely causes much denaturation.

CODA

The hydrogen-bonding (Lewis acid-base, or AB) free energy of cohesion of water is discussed as the sole driving force of hydrophobic attraction between apolar entities when immersed in water. Hydrophobic hydration of apolar surfaces on the other hand is purely due to LW interactions. The first layer of hydration on, e.g., a non-polar sphere, immersed in water, is a thin, completely hydrogen-bonded network of water molecules which is relatively hydrophobic and is also observed in clathrates. This mode of hydrophobic hydration, which shows less and less organized layers of successively distal water of hydration, is the principal mechanism for the action at a distance of hydrophobic attraction of two entities immersed in water. The decay of hydrophobic attraction as a function of distance is exponential with as base the characteristic length (λ) of water, which is about 1.0 nm, corresponding to the radius of gyration of the average size of water clusters comprising 4 to 5 water molecules per cluster, at room temperature (see Chapter VIII).

Also discussed are the additives to water that can be used to alter its hydrophobizing capacity. To *decrease* the hydrophobizing capacity of water, monopolar or almost monopolar solvents one may, for instance, mix monohydric alcohols (e.g., methanol, ethanol) with water, whilst the addition of even small amounts of surfactants is even more effective. A decrease in temperature can also cause a decrease in the hydrophobizing effect of water, when acting on the polar moieties of partly polar molecules. To *increase* the hydrophobizing capacity of water, either simple *sugars* (e.g., sucrose or glucose) can be used, or one can employ inorganic *salts* [e.g., $(NH_4)_2SO_4$, Na_2SO_4, K_2CO_3] in molar or pluri-molar concentrations. The use of salts for the insolubilization of proteins through hydrophobic precipitation is called salting-out.

X Role of Water in Hydrophilic Repulsion

WATER AND HYDRATION FORCES

"Hydration Forces" (Pashley, 1982; Parsegian *et al.* 1985) is a term which alludes to the repulsion, in water, between two hydrophilic surfaces, such as two phospholipid layers, or two protein, or polysaccharide molecules or layers. Although the expression could also encompass various other non-covalent interaction forces[*]), we shall here concentrate our attention on those repulsive interactions that are neither due to van der Waals nor to electrical double layer forces (see also Pashley, 1982, and Gruen *et al.*, 1984). Hydration forces (sometimes also designated as "hydration pressure") which are the cause of hydrophilic repulsions in aqueous media are, like hydrophobic attractions (see Chapter IX), driven by Lewis acid-base forces (see also Chapters IV and XIX).

Referring to eq. IV-17 and turning to the examples given in Table IV-2, the hydrophilic example (bottom row) describes the free energy of polar (AB) adhesion between material, 1 (a monopolar, electron-donating hydrophilic molecule) and water, w, and is shown as term V: $+4\sqrt{(\gamma_1^{\ominus} \cdot \gamma_w^{\oplus})}$. In eq. IV-17 (repeated at the top of Table IV-2), term V has a positive value and therefore indicates a repulsive contribution. It is proportional to the free energy of hydration of entity 1 (cf. the third right hand term of eq. IV-18, which is the relevant term for an electron-donating monopolar entity 1), which at $+127.8$ mJ/m^2 amply suffices to overcome the always-present hydrophobic attraction (term III) of -102 mJ/m2, leaving a net hydrophilic (AB) repulsion of $+20.3$ mJ/m^2 (after also deducting a small van der Waals attraction). Being solely caused by the hydration of hydrophilic surfaces, the term "hydration forces" for its driving force is therefore entirely appropriate.

---

[*] See also Derjaguin *et al.* (1987) and Derjaguin (1989), who allude to what we would call *hydration forces*, as *disjoining pressure*. With time, however, the meaning of the term disjoining pressure as used by the above authors, has become increasingly synonymous with any non-covalent interaction, including, e.g., van der Waals-London attractions, which became designated as *negative disjoining pressures*, see, e.g.,Derjaguin (1989, p. 69).

109

NEGATIVE γ_{iw} VALUES AND HYDROPHILIC
REPULSION

It should be noted that, as $\Delta G_{iwi} = -2\gamma_{iw}$ (cf. eq. IV-17), when ΔG_{iwi} is positive (denoting a repulsion, occurring among very hydrophilic entities, i, immersed in water, even when i is electrostatically neutral) then the interfacial tension, γ_{iw}, between entities, i, and water, w, is *negative*; see Chapter XIX. This occurs in microemulsion formation and in other phase separation systems in aqueous media (Chapter XXI) as well as in the aqueous solubility of polymers (Chapter XXII) and in the stability of aqueous suspensions of electrically neutral, or close-to-neutral particles or cells (Chapter XXII).

In most cases a negative γ_{iw} is not thermodynamically stable and its decay to $\gamma_{iw} \rightarrow 0$ becomes the driving force of all the above-mentioned repulsion-driven separation phenomena. In some instances however, such as the interface formation with water adjacent to completely insoluble but very hydrophilic surfaces (e.g., a drop of water placed on clean glass, or a freshly separated layer of mica), a negative interfacial tension between water and such surfaces can be permanently maintained. Otherwise, after the dissipation of $\gamma_{iw} < 0$ to $\gamma_{iw} \rightarrow 0$, as in the cases of aqueous phase separations (Chapter XXI), the interfaces between two aqueous phases can maintain γ_{iw} values which are very close to zero and are usually expressed in units of $\mu N/m$ or $\mu J/m^2$, rather than as mJ/m^2 (Albertsson, 1986, pp. 33, 60; Zaslavsky, 1995, p. 172). Thus the visible phase separation boundary between two phases in aqueous systems, when contained in a glass cylinder, is precisely perpendicular to the glass surface and shows no sign of a curved meniscus at the glass edge.

HYDRATION ORIENTATION AND ACTION
AT A DISTANCE

In Chapter VII the rate of decay of polar (AB) forces was discussed, as a function of distance, ℓ. The fundamental reason for the possibility of propagating an at first sight short-range interaction, such as the hydrogen-bonding interaction, across a significant distance between the interacting surfaces, is a local change in the organization of the water molecules in the vicinity and under the influence of these surfaces (see also Chapter IX). The existence of a considerable degree of orientation of the water molecules of hydration of proteins and other macromolecules has long been surmised (reviewed by Ling, 1972). The manner of orientation of the water molecules adjoining polar surfaces was first illustrated schematically by Parsegian *et al.* (1985), without, however, indicating the gradual decay in orientation as a function of distance, ℓ. It is precisely this (exponential) decay in the orientation of water molecules, as a function of ℓ, which is described in eqs. [VII-3] to [VII-8]; see also Table VII-2. This direction of orientation of the water molecules near monopolar surfaces, and the decay of that orientation, are shown schematically in Fig. X-1. "Hydration forces" (Cowley *et al.*, 1978; Parsegian *et al.*, 1979, 1985, 1986, 1987; Marcelja and Radic, 1976; Pashley, 1982; Gruen *et al.*, 1984; Israelachvili, 1985) thus are the direct manifestation of the polar (AB) interaction forces, as propagated by the intermediary of the orientation

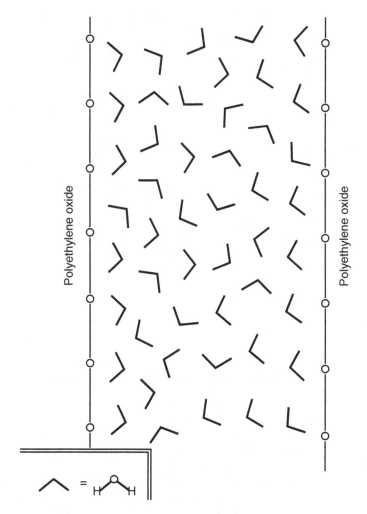

FIGURE X-1 Schematic presentation of two opposing monopolar polymer chains with strong electron donor parameters (e.g., poly(ethylene oxide): the circles represent the oxygen atoms in the polymer chains), immersed in water. The water molecules (see insert) are represented only as lines which denote the H-O bonds, at an angle of 104.5° (Pauling and Hayward, 1964).

In the first layer of hydration most of the water dispoles are strongly oriented, with the oxygen atoms pointing away from the polymer strands. The water molecules residing in layers which are more distant from the polar polymer strands are less oriented, and at even longer distances they resume a random orientation. However, when two such polymer chains are brought relatively closely together, e.g., within $\ell \approx 2$ to 3 nm, as illustrated here, there still is a measurable repulsion between them, brought about by the residual orientation of the water dipoles.

In this model the constantly occurring fluctuating complex formation between neighboring water molecules through hydrogen bonding is not depicted, but it is an additional factor that should be taken into account, as it does give rise to a characteristic decay length (λ) for water, which is somewhat larger than the radius of gyration of single water molecules, i.e., λ is somewhat larger than 0.2 nm (see text).

of neighboring water molecules, but mitigated by the decay of that orientation as a function of distance. Therefore, as has been most appropriately stated by Parsegian *et al.* (1987): "There is thus a neat separation between the respective behaviors of surface and medium." In:

$$\Delta G^{AB}\ell = \Delta G^{AB}_{\ell_o} \exp[(\ell_o - \ell)/\lambda] \qquad [VII-4]$$

$\Delta G^{AB}_{\ell_o}$ describes the (attractive or repulsive interaction energy of the polar surface, while the characteristic decay length, λ, of the liquid medium governs the rate of decay vs. distance.

HYDROPHILICITY AND HYDRATION ORIENTATION

It can easily be shown that although hydration orientation is qualitatively different from hydration, the two still are quantitatively linked. To begin with, a high $\gamma_i^\ominus$ value usually is accompanied by a γ_i^{AB} value which is zero or close to zero, in practically all cases listed in Table XIX-1, indicating monopolarity, which points to both a high degree of hydrophilicity and strong hydration orientation. Then a really strongly negative value for ΔG_{iw} (which indicates a high degree of hydrophilicity) can only be attained when $\gamma_{iw} < 0$ (cf. eq. [IV-17]), which again only occurs with monopolar or virtually monopolar compounds and thus also is linked to hydration orientation.

HYDRATION ORIENTATION OF PROTEINS

It is possible to measure the degree of hydration orientation of the water molecules of hydration of, e.g., a protein such as human serum albumin (HSA), by contact angle measurement (van Oss and Good, 1988a) on layers of concentrated hydrated HSA. the HSA is concentrated on an ultrafilter membrane, and the amount of residual water of hydration is determined by weighing the membrane before use and after accumulation of the (known) amount of protein, plus residual water. The contact angles obtained with apolar and polar liquids on layers of HSA with 1 and 2 layers of water of hydration per molecule are given in Table X-1. Using eq. [IV-13], the

---

TABLE X-1

Contact Angles Determined on Layers of Dry and Hydrated Human Serum Albumin

	Water	Glycerol	Hexadecane	α-Bromonaphthalene	Diiodomethane
Dry HSA[a]	63.5°	59.5°	—	23.2°	37°
HSA, 33% hydrated[b]	12°	48.5°	15°	—	—
HSA, 72% hydrated[b]	0	14.5°	13°	—	—

[a] From van Oss *et al.*, J. Protein Chem., 1986.
[b] From van Oss and Good, J. Protein Chem., 1988.

TABLE X-2

Surface-Tension Components and Parameters of Dry and Hydrated Human Serum Albumin and of Water and Interfacial Interaction Energy (ΔG_{121}) between HSA Molecules, in Water and the Degree of Orientation of the Water Molecules of Hydration (from data in Table X-1)

	$\gamma^{LW a}$	$\gamma^{\oplus a}$	$\gamma^{\ominus a}$	$\gamma^{AB a}$	$\Delta G_{1w1}{}^{a}$	Orientation[b]
Dry HSA	41	0.126	17.2	2.94	−23.0	NA
HSA, 33% hydrated	26.6	0.60	75.95	13.5	+62.2	73.5%
HSA, 72% hydrated	26.8	6.0	51.5	35.2	+21.6	31.0%
H$_2$O (bulk)	21.8	25.5	25.5	51.0	NA	0%

[a] In mJ/m^2.
[b] Expressed as % decrease in γ^{AB}, compared with the γ_w^{AB} of bulk water.
Note: NA, Not applicable.

contact angle data thus obtained yield the values of the surface tension parameters given in Table X-2, as well as the interfacial free energies (ΔG_{1w1}) between dry or hydrated HSA, and water. Water molecules in the bulk liquid are taken to be in a state of random orientation. We characterize the degree of orientation (in %) of water molecules in a given hydration layer as the decrease in γ_i^{AB} of the hydrated layer relative to γ_w^{AB} (= 51 mJ/m^2). Ultrafiltration of 25 ml 5% (w/v) HSA yielded a layer of hydrated protein containing 33% (w/w) water, and ultrafiltration of 25 ml 2% (w/v) HSA resulted in a layer of hydrated protein containing 72% (w/w) water (van Oss and Good, 1988a).

When the water of hydration is completely oriented, the γ^{AB} of such a hydrated surface becomes zero, as such a surface is then effectively monopolar (see above); i.e., it has a large $\gamma^{\ominus}$ but effectively no $\gamma^{\oplus}$. A 73.5% orientation for 33% hydrated HSA means that the water and glycerol in the drops with which the contact angles are measured virtually "see" only the oxygen atoms of the water molecules of hydration. Given the molecular dimensions of the HSA molecule (calculated from Sober, 1968) of a prolate spheroid of $\approx 40 \times 140$ Å, 33% hydration would correspond rather closely to just one molecular layer of hydration, assuming that the hydration is largely peripheral (see Ling, 1972). Similarly, 72% hydration corresponds to two layers of water molecules of hydration. At the isoelectric pH of 4.9 HSA is also still to some extent "hydrated," but a drastic change in its configuration has taken place, causing it to expose only its apolar moieties at the air (or initially, the nitrogen) interface, when highly concentrated by ultrafiltration (van Oss and Good, 1988a). A similar effect involving an increase in "hydrophobicity" of amphoteric materials accompanying (but only indirectly caused by) a decrease in ζ-potential was observed by Holmes-Farley *et al.* (1985); see also Chapter XIX.

It thus appears that there exists a fundamental difference between ordinary attraction of water molecules, a type of "hydration" which is exhibited by alkanes (see Chapter IX), and hydration accompanied by orientation of the water molecules as in the case of HSA at neutral pH (treated above) and other biopolymers. For instance, with HSA, its monopolarity is not strong enough to cause a repulsion when immersed in water in the dry state; as deduced from the strong negative value of

ΔG_{iwi} under these conditions (Table XIX-4). At a contactable surface area between two HSA molecules of 10 nm^2, for dry HSA, at $\Delta G_{iwi} = -5.75$ kT, this would mean virtually total insolubility in water, in that state. However, once hydrated (with strongly positive ΔG_{iwi} values), they become readily soluble in water due to the repulsive effect of the first layer(s) of oriented water of hydration. But this is possible only *either* because of an original relatively modest monopolarity of the protein in the dry state, which is strongly amplified by the orientation of the closest layers of water molecules, *or* because of a re-folding of the protein which then re-assumes its native hydrophilic conformation upon re-hydration, thus recovering from the reversible denaturation which had been caused by drying. From hydrodynamic measurements, it would follow that the first layer of hydration is the only layer sufficiently strongly bound to the HSA molecule to migrate with it (Ling, 1972); i.e., the slipping plane is situated at, or close to, the outer boundary of the first layer of hydration. As can be seen from the decrease in orientation in the second layer (Table X-2), the decay in orientation with distance appears to be quite steep; it correlates quite well with the exponential decrease proposed by Marcelja and Radic (1976) and Parsegian *et al.* (1985, 1986, 1987); cf. eq. [VII-4]. Some other proteins (e.g., IgG) also show the same solubility behavior; see van Oss *et al.* (1986b) and Chapter XXII. As with hydrophobic attraction, the decay length of water in hydrophilic repulsion, λ, is also about 1.0 nm, see Chapter XIX.

MODULATION OF THE HYDROPHILIC REPULSION THROUGH TEMPERATURE CHANGES

RAISING ΔG_{iwi}^{AB}

Increases in temperature (T) can cause a significant increase in hydrophilic repulsion, but it causes only a slight decrease in the hydrophobizing capacity of water, via a small decrease in the AB cohesion of water (term III in eq. IV-17; see Table IV-2). However, increases in T do cause a sizeable increase in $\gamma_w^\oplus$, which in turn causes a strong increase in the free energy of hydration of largely or solely monopolar electron-donating hydrophilic molecules, particles or cells, i.e., through an increase in the value of term V of eq. IV-17; see Tables IV-2 and VIII-1. Thus, heating water as a solvent increases the solubility of most polar solutes, as a consequence of the concomitant increase in $\gamma_w^\oplus$; see also Chapter XXII.

LOWERING ΔG_{iwi}^{AB}

Decreases in T have the opposite effect from increases in T. Lowering T causes a decrease in $\gamma_w^\oplus$ and therefore a decrease in the free energy of hydration of monopolar electron-donating hydrophilic entities (see term V of eq. IV-2, in Table IV-2).

Apart from modulating $|\Delta G_{iwi}^{AB}|$ through changes in T, one can also decrease the absolute value of ΔG_{iwi}^{AB} by increasing the hydrophobic attraction energy through the admixture of fairly high concentrations of salt ions, or sugars; see Chapter IX.

CODA

Like hydrophobic attractions in water (see Chapter IX) hydrophilic repulsions are also driven by polar (AB) free energies of interaction. Whilst for hydrophobic attractions this involves the (polar) AB free energy of cohesion of water, hydrophilic repulsions are due to the AB free energy of hydration (or "hydration force") which has to exceed the absolute value of the always-present AB free energy of aqueous cohesion, to give rise to a measurable net hydrophilic repulsion.

Thus, hydrophilic repulsion only occurs when the interfacial tension, γ_{iw}, between hydrophobic entity, i, and water, w, *has a negative value*. (It should be noted here that the γ_{iw} mentioned *passim* in this Chapter, is identical with the zero time dynamic interfacial tension with water, γ_{iw}^{o}, discussed in Chapter XIII, below). The negative value of γ_{iw} may dissipate, giving rise to: phase separation, microemulsion formation, solubility, or the formation of stable aqueous suspensions of hydrophilic particles or cells. It can also permanently maintain its negative value vis-à-vis solid hydrophilic surfaces, e.g., at the water-glass or the water-mica interface.

The action at a distance of hydrophilic repulsions in water is due to the orientation of water dipoles, which is strongest nearest the surfaces of immersed hydrophilic entities and decays exponentially as a function of distance away from these entities' surfaces. Finally, the influence of temperature (T) on hydrophilic repulsion is discussed: With increases in T hydrophilic repulsion can significantly increase. Decreases in T have the opposite effect, usually to a more modest extent.

XI | The Water–Air Interface

HYPER-HYDROPHOBICITY OF AIR AT THE WATER–AIR INTERFACE

Air, at the water-air interface, has the most unusual and extreme properties. It causes a zero-energy interface which can be attributed to the low density of gases, resulting in extremely weak interactions at the interface with another material (Docoslis *et al.*, 2000). The surface tension of air (A) is zero because the Lifshitz–van der Waals or apolar component of its surface tension (γ_A^{LW}) is zero, as well as its Lewis acid-base component (γ_A^{AB}), so that $\gamma_A^{TOT} = \gamma_A^{LW} + \gamma_A^{AB} = 0$, where $\gamma_A^{AB} = 2\sqrt{\gamma_A^{\oplus}\gamma_A^{\ominus}} = 0$, not because the interface is monopolar (as happens with many hydrophilic compounds) when $\gamma_A^{\oplus} = 0$ (van Oss, Chaudhury and Good, 1987a), but because *both* $\gamma_A^{\oplus}$ and $\gamma_A^{\ominus}$ are equal to zero.

Using the Young-Dupré equation, for instance for measuring contact angles of water somehow deposited upon a layer of air (cf. eq. IV-13):

$$\left(1 + \cos\theta_A\right)\gamma_W = 2\left(\sqrt{\gamma_W^{LW}\gamma_A^{LW}} + \sqrt{\gamma_W^{\oplus}\gamma_A^{\ominus}} + \sqrt{\gamma_W^{\ominus}\gamma_A^{\oplus}}\right) \qquad \text{[XI-1]}$$

where θ_A is the contact angle of drops of water in contact with air, at the W-A interface. Then, after inserting a zero for all symbols pertaining to air (A), what remains of eq. XI-1 is:

$$\left(1 + \cos\theta_A\right)\gamma_W = 0 \qquad \text{[XI-2]}$$

so that $\cos\theta_A = -1$, signifying that $\theta_A \rightarrow 180°$.

For practical reasons one probably can never attain a contact angle quite as high as 180° for water on air, because some minimal amount of solid apolar support material of non-zero surface tension still needs to be present. Nonetheless, it has been demonstrated that one can come remarkably close. Erbil *et al.* (2003) reached a contact angle of 160° with water on a very open polypropylene (PP) foam surface. On a slightly less open PP foam surface they obtained a contact angle with water of 149°. Earlier, Onda *et al.* (1996) obtained a contact angle with water as high as 174°, on a fractal surface of an alkylketene dimer (AKD); as shown below, it seems

117

unlikely that $\theta_W = 174°$ is going to be surpassed easily or by much. To illustrate this, it is useful to apply Cassie's equation to the porous surfaces utilized by Erbil *et al.*(2003) and Onda *et al.* (1996).

Using Cassie's equation (eq. XII-2; see also Cassie and Baxter, 1944 and Cassie, 1948) in the form:

$$\cos\theta_A = f_1\cos\theta_1 + f_2\cos\theta_2 \qquad [\text{XI-3}]$$

where θ_A is the experimentally measured contact angle of the supported water drop at the water-air interface and where f_1 and f_2 are the fractions of the total surface occupied (in the cases under study) by the apolar support material (f_1) and by the air at the W-A interface (f_2) such that:

$$f_2 = 1 - f_1 \qquad [\text{XI-4}]$$

Therefore eq. XI -3 can be expressed as:

$$\cos\theta_A = f_1\cos\theta_1 + \cos\theta_2 - f_1\cos\theta_2 \qquad [\text{XI-5}]$$

where, as shown above, $\cos\theta_2 = -1$. Thus:

$$\cos\theta_A + 1 = f_1\left(\cos\theta_1 + 1\right) \qquad [\text{XI-6}]$$

In the case of the 174° contact angle observed with water on a porous AKD surface by Onda *et al.* (1996), these authors found a contact angle for a flat slice of the AKD material to be 109°. Taking the latter value to represent θ_1 whilst $\theta_A = 174°$; then, using eq. XI-6, one finds $f_1 = 0.0081$, i.e., 0.8% of the surface under the water-air interface consisted of AKD and 99.2% of the interface contacted only air.

Similarly, applying eq. XI-6 to the results of Erbil *et al.* (2003) who used an open network of PP strands, where the PP polymer surface itself had a value for θ_1 of 110.47°, then for a medium high contact angle with water on a medium-porous PP of 149° a value for f_1 of 0.22 is obtained, i.e., 22% of the lower surface area of the drop contacted polypropylene and 78% of the drop interface contacted air. Using the most porous preparation of their PP foam surface, these authors achieved a contact angle of 160°. For the water drop on the latter porous surface, $f_1 = 0.093$, i.e., 9.4% of the interface with the water drop touched PP material and 90.7% of the interface only contacted air. See Table XI-1 for the values of the surface tension components used in the above calculations.

Busscher *et al.* (1992) also made hyper-hydrophobic surfaces, out of Teflon ($\theta_{water} = 109°$) by ion etching, as well as by ion etching followed by glow discharge. With ion etching alone, $\theta_{water} = 120°$ and with ion etching + glow discharge treatment, $\theta_{water} > 140°$. To calculate the resulting proportion of air interface obtainable with the latter treatment, we shall assume that in the latter case θ_{water} was 145°. Using again eq. XI-6, it follows that the $\theta_{water} = 120°$ surface presented 35% of its area as air, whilst with the $\theta_{water} = 145°$ surface 73% of its area was air.

TABLE XI-1
Values in mJ/m² of the Surface Tension Components Used in the Calculations; see Chapter XIX for PP and Chapter IV for the Surface-Thermodynamic Properties of Water at 20°C.

Component	γ^{LW}	$\gamma^{\oplus}$	$\gamma^{\ominus}$
Polypropylene (PP)	25.7	0.0	0.0
Water	21.8	25.5	25.5
Air	0.0	0.0	0.0

QUANTITATIVE EXPRESSION OF THE TOTAL HYDROPHOBIC ATTRACTION ENERGY AT THE WATER–AIR INTERFACE

The significantly higher water contact angle obtained by Onda *et al.* (1996) with a fractal foam support, compared to the contact angles obtained by Erbil *et al.* (2003) with a support of thin PP strands (both materials showing rather similar water contact angles of 109° and 110.5° respectively) would argue in favor of a polar foam support material that is macroscopically as well as microscopically porous (such as the fractal AKD support), rather than for a polar foam support material consisting of just macroscopic strands of an otherwise smooth, non-porous polymer.

The reason for the significantly greater hydrophobicity of air at the water-air interface than of polymer surfaces such as those of PP is not based on their two polar parameters: $\gamma^{\oplus}$ and $\gamma^{\ominus}$, *both* of which are zero in either case. The greater hydrophobicity of air lies in the fact that its Lifshitz–van der Waals component (γ_A^{LW}) is also zero, in contrast with the γ^{LW}-value of, e.g., PP, which equals 25.7 mJ/m²; whereas water has a γ_W^{LW}-value of 21.8 mJ/m2, at 20°C. The free energy of interaction between two particles or molecules of material, i, immersed in water, w, expressed as ΔG_{iwi} is (cf. eq. IV-17):

$$\Delta G_{iwi} = -2\left(\sqrt{\gamma_i^{LW}} - \sqrt{\gamma_W^{LW}}\right)^2$$

$$-4\left(\sqrt{\gamma_i^{\oplus}\gamma_i^{\ominus}} + \sqrt{\gamma_W^{\oplus}\gamma_W^{\ominus}} - \sqrt{\gamma_i^{\oplus}\gamma_W^{\ominus}} - \sqrt{\gamma_i^{\ominus}\gamma_W^{\oplus}}\right)$$

[XI-7]

For PP:

$$\Delta G_{iwi} = \Delta G_{iwi}^{LW} + \Delta G_{iwi}^{AB}$$

$$= -0.3 - 102$$

[XI-7A]

$$= -102.3 \text{ mJ/m}^2$$

ΔG_{iwi} is the most precise quantitative indicator of hydrophobicity (when negative) and hydrophilicity (when positive; see Chapter XIX and also van Oss and Giese, 1995).

For a PP surface, immersed in water, there is a slight LW attraction *and* a strong AB attraction (of 102 mJ/m², which is the attraction of the hydrophobic effect). For the water-air interface, $\Delta G_{iwi} = -43.6 - 102 = -145.6$ mJ/m², there is a significant net Lifshitz–van der Waals (LW) attraction plus the strong hydrophobic (AB) attraction, which combination, in aqueous media, is only attainable by the water-air interface (cf. the ΔG_{iwi} values given in Table XVIII-1). Thus, it is the additional 43.6 mJ/m² additional Lifshitz–van der Waals attraction which is responsible for the hyper-hydrophobicity of air, at the water-air interface and concomitant water contact angle of almost 180°, which is only approachable on a water-air interface with minimal solid support.

The presence of a rather high proportion of air surface between the rugosity-creating mini-protuberances which are the cause of the "roughness" of rough surfaces is the major reason for the increased contact angle values (see also Adamson, 1990, pp. 387–389). In the same manner the regularly repeated concavities occurring between the threads of tissues woven with relatively hydrophobic fibers, contribute significantly to the water-repellence of the woven tissues used for making raincoats, tents, etc. This property of water-repellant woven tissues inspired the research on contact angles formed on air-interspersed surfaces and tissues by Cassie and Baxter (1944) and by Cassie (1948); see also Chapter XII and eq. XI-3, above.

ABSENCE OF HYDRATION OF AIR AT THE WATER–AIR INTERFACE

Writing the Young-Dupré equation in terms of the free energy of hydration (eq. IV-11):

$$(1 + \cos \theta) \, \gamma_w = -2 \, \Delta G_{iw} \qquad \text{[XI-8]}$$

where (eq. IV-6B):

$$\Delta G_{iw} = \gamma_{iw} - \gamma_i - \gamma_w \qquad \text{[XI-9]}$$

ΔG_{iw} can be shown to be equal to zero, not only via eqs. XI-1 and XI-2, but also by analysis of γ_{iw} for the W-A interface (cf. eq. IV-9):

$$\gamma_{iw} = \left(\sqrt{\gamma_i^{LW}} - \sqrt{\gamma_w^{LW}} \right)^2 + 2 \left(\sqrt{\gamma_i^{\oplus}\gamma_i^{\ominus}} + \sqrt{\gamma_w^{\oplus}\gamma_w^{\ominus}} \right. $$
$$\left. - \sqrt{\gamma_i^{\oplus}\gamma_w^{\ominus}} - \sqrt{\gamma_i^{\ominus}\gamma_w^{\oplus}} \right) \qquad \text{[XI-10]}$$

which, for the water-air interface (for which all terms indicated by a subscript, i, are zero) becomes (at 20°C):

$$\gamma_{iw} = (\sqrt{21.8})^2 + 2(0 + 25.5 - 0 - 0) = 21.8 + 51 = 72.8 \text{ mJ/m}^2,$$

so that (using eqs. XI-9 and XI-10):

$$\Delta G_{iw} = 72.8 - 0 - 72.8 = 0 \qquad \text{[XI-9A]}$$

This means that the water-air interface is not hydrated at all. Indeed, of all hydrophobic surfaces it is probably only the water-gas or water-vacuum interface for which this is true. Even the completely non-polar polymer, poly(propylene) (PP), mentioned above still has a significant free energy of hydration (eq. XI-9):

$$\Delta G_{iw} = 51 - 25.7 - 72.8 = -47.3 \text{ mJ/m}^2 \qquad \text{[XI-9B]}$$

despite its designation as "hydrophobic" (which means: "water-fearing"). Thus the air side of the water-air interface actually is the only surface known which literally may be designated as "hydrophobic".

ATTRACTION OF PARTLY POLAR, OR AMPHIPHILIC SOLUTES TO THE WATER–AIR INTERFACE, WHICH DECREASES THE APPARENT SURFACE TENSION OF WATER

In aqueous media, water-soluble compounds that have an overt apolar (hydrophobic) moiety as well as a very polar (hydrophilic) one, combined within one "amphiphilic" molecule (i.e., one side of the molecule likes water and the other side likes oil), such as surfactant molecules, migrate preferentially to the water-air interface, with their hydrophobic moiety protruding into the air-side.

Aqueous solutions of amphiphilic molecules manifest a strong decrease in their apparent measured surface tension (Docoslis *et al.*, 2000). One can easily halve the apparent surface tension of water (γ_w of pure water at 20°C is 72.8 mJ/ m^2) by the admixture of only about 0.1% (w/w) of non-ionic surfactants such as various polyoxyethylene *n*-dodecanols (Lange, 1967). This is possible because of the migration of a major proportion of the dissolved non-ionic surfactant molecules toward the water-air interface, with the dodecyl chains of these solutes protruding into the air-side. This leaves only a small fraction of the surfactant molecules in the bulk solution, so that the apparent measured surface tension of the resulting strongly anisotropic solution has no longer any connection with the free energy of cohesion of the total liquid. One should therefore never measure contact angles with drops of such heterogeneous liquid mixtures for the purpose of determining the surface thermodynamic properties of a given solid, because homogeneity of the liquid of the drops is a *conditio sine qua non* for allowing its use for contact angle measurements while assuring that the cosine of the contact angle of the liquid drop is directly linked to the free energy of cohesion of the liquid drop as compared to its free energy of adhesion to the solid surface; see Figure XII-1 and, in general, Chapter XII, below. Figure XI-1 (left hand side) schematically illustrates the behavior of amphiphilic solutes dissolved in drops of water.

Amphiphilic compounds such as alcohols cause a considerable decrease in the surface tension of water, even when added at rather low concentrations. For instance the admixture of only 1% (w/w) *n*-propanol to water causes the apparent surface

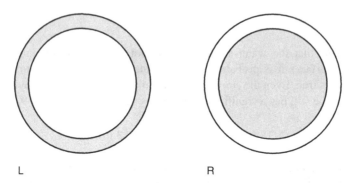

L   R

FIGURE **XI-I**   Schematic presentation of liquid drops of aqueous solutions, surrounded by air. Left drop (L): Drop of an aqueous solution of a partly hydrophilic-partly hydrophobic, or amphiphilic solute, with migration of most of the solute toward the aqueous solution- air interface, so that the solute is most concentrated closest to the air interface. Solutes of this type are, e.g. alcohols, surfactants, proteins. Such solutes *strongly decrease* the apparent surface tension of water. Right drop (R): Drop of an aqueous solution of a more homogeneously hydrophilic, or close-to-hydrophilic solute, with depletion of the solute molecules closest to the air interface. Solutes of this type are, e.g., sugars, salts, glycine, polysaccharides. [Sugars are close to being hydrophilic, but according to the definition given by van Oss and Giese (1995), their $\Delta G_{iwi} \leq 0$, which makes them midly hydrophobic, as also is witnessed by their finite aqueous solubility] ; see also Table XXII-1, as well as Docoslis *et al.* (2000) and van Oss *et al.* (2002). In this category of low molecular weight solutes are sugars and many inorganic salts which, when added to water, *increase* its surface tension. However, due to their depletion at the water-air interface, the *measured* increase in surface tension (e.g., by contact angle determination) is lower than the overall increase in surface tension of the bulk liquid, inside the drop. On the other hand, sugar *polymers* (polysaccharides) *decrease* the surface tension of the bulk aqueous solution, as compared to pure water. However, this decrease in surface tension of the bulk solution inside the drop is greater than the *measured* decrease in surface tension (e.g., by contact angle determination), due to the depletion layer at the water-air interface.

tension of the solution to decrease 32%, from 72.14 to 49.4 mJ/m^2, at 25°C (Weast, 1970/71, p. F-29). In the case of monohydric alcohols, the longer the alkyl chains, the lower the concentration needed to decrease the apparent surface tension of water but, concomitantly, the smaller their aqueous solubility.

Even less overtly amphiphilic molecules such as various globular proteins can, dependent on their secondary and tertiary structure, cause a significant decrease in the surface tension of water, by orienting their (normally internalized) hydrophobic moieties toward the water-air interface. This gives rise, for instance, to the hydrophobicity of air-dried globular plasma proteins, such as albumins or immunoglobulins, which is in contrast to the hydrophilicity of the same proteins in the hydrated (dissolved) state, tabulated in Chapter XVIII; see also Absolom, van Oss *et al.* (1981). However, some non-globular plasma proteins, such as fibrinogen, remain as hydrophilic after air-drying as they were when still dissolved in water, see Chapter XVII.

REPULSION OF HYDROPHILIC, OR NEAR-HYDROPHILIC SOLUTES BY THE AIR-SIDE OF THE WATER–AIR INTERFACE; THE INCREASE OF THE SURFACE TENSION OF AQUEOUS SOLUTIONS BY THE ADMIXTURE OF SUGARS AND SALTS

Completely hydrophilic compounds such as polysaccharides (e.g., dextran or ficoll), as well as near-hydrophilic compounds such as monomeric or dimeric sugars [e.g., glucose, sucrose[*])] are repelled by the water-air interface (Docoslis *et al.*, 2000)[†].] Thus, in exact opposition to amphiphilic compounds (see above), aqueous solutions of these hydrophilic, as well as very solule near-hydrophilic compounds, leave a solute-depleted layer at the water-air interface. This lends their aqueous solutions the appearance of having a lower measured surface tension than corresponds to the free energy of cohesion of the bulk solution, which begins a nanometer or so inside the water-air interface. Sugars such as the majority of glucose and sucrose, as well as the majority of water-soluble inorganic salts, when dissolved in water increase the surface tension of the bulk solution.

"Bulk" here signifies the main part of the aqueous solution, with the exception of the almost pure layer of water (of the order of one to a few nm thickness) adjacent to the water-air interface. The *increase* in the surface tension of the bulk solution by the admixture of small polar molecules such as sugars and salts is contributed by the elevated hydrogen-bonding (AB) free energy of cohesion of the sugar molecules in solution and by the electrostatic (EL) plus AB free energy of cohesion between the ions, in the case of dissolved salts. As a consequence of the solute-depleted zone at the water-air interface, the surface tension one *measures* with such sugar or salt solutions is significantly lower than the surface tension of the bulk solution, although both of them are higher than that of pure water (Docoslis *et al.*, 2000). Conversely, *polymers* of low-molecular weight sugars, such as dextran or ficoll *decrease* the surface tension of bulk water, but they also create a depletion layer near the water-air interface, so that their *measured* surface tension is somewhat higher than that of the bulk solution. The reason for the *decrease* in surface tension caused by, e.g., dextran molecules in aqueous solution, lies in the fact that the previously monomeric sugar molecules which could freely engage in the formation of hydrogen bonds with other single sugar

---

[*] Sugars such as glucose and sucrose are quite soluble in water, but their aqueous solubility is nonetheless finite, i.e., 47.7% for glucose and 66.7% for sucrose (Stephen and Stephen, 1963). Thus (see eqs. XIII-2A and XIII-3), their ΔG_{iwi} value is negative, which therefore makes these sugars (by definition) slightly hydrophobic (van Oss and Giese, 1995; see also Table XXII).

[†] It is noteworthy that there always is an AB *attraction* between these hydrophilic, or near hydrophilic compounds and the water-air interface which is, however, counter-balanced by a stronger LW *repulsion*. This is a consequence of the fact that LW forces acting between two different materials, 1 and 2, immersed in a third liquid, L, gives rise to negative Hamaker coefficients (A_{1L2}), i.e., to a repulsion, when $A_1 > A_L > A_2$, or when $A_1 < A_L < A_2$, where A_i is directly proportional to γ_i^{LW}. With poly- or oligo- or mono-saccharides this happens often, because usually the γ^{LW} of carbohydrates is of the order of 40 mJ/m², whilst γ^{LW} for water is 21.8 mJ/m² and γ^{LW} for air is zero. Thus $\Delta G_{iWA}^{LW} > 0$, because $\gamma_1 > \gamma_W > \gamma_A$, i.e., $40 > 21.8 > 0$, see Chapter II.

molecules, cease to contribute to the free energy of cohesion in aqueous solution when they become more rigidly bound together upon covalent polymerization. Then, once polymerized, they lose their capacity for self-hydrogen bonding with similarly polymerized sugar molecules, which renders them incapable of enhancing the free energy of cohesion of the solution (Docoslis *et al.*, 2000). Thus, both low molecular weight sugars (as well as neutral salts) *and* sugar polymers are repelled by and leave a depletion layer at the water-air interface, but *low-molecular weight sugars increase* the surface tension of their bulk aqueous solution, whilst the dissolution of *sugar polymers decreases* it. Figure XI-1 (right hand side) schematically illustrates the behavior of this class of solutes, when dissolved in drops of water.

THE ζ–POTENTIAL OF AIR BUBBLES IN WATER

One rarely finds reports of measurements of the ζ–potential of air (or other gas) bubbles suspended in pure water, in the absence of stabilizing agents such as surfactants or proteins. However, in 1995 Graciaa *et al.* measured the ζ–potential of air bubbles in deionized water, using a rotating tube device, and found a ζ–potential of –65 mV. In view of the rheological problems of measuring the velocity of air bubbles moving in an electric field while touching the upper part of the polymer-coated inner surface of a rotating glass cylinder, as well as many other experimental and theoretical quandaries, these –65 mV should at best be taken as just a rough order of magnitude.

Nonetheless, while it seems reasonable enough that air bubbles could well have a (negative) ζ-potential of about –65 mV in pure water, there is also little doubt that Craig *et al.* (1993) correctly stated that air (in their case actually N_2) bubbles in deionized water attract one another via strong (non-electrostatic, i.e., LW + AB) forces, which causes them to coalesce. It is easily shown that the total interfacial (LW + AB) ΔG_{iwi} of air bubbles with a radius, R = 1 mm, would equal -11.2×10^7 kT (see Table XI-2), whereas a ζ–potential of even –100 mV, at ionic strength, $\mu = 0.01$, would not quite suffice to help avoid inter-bubble coalescence *at contact*, but under these conditions contact would nevertheless be prevented by a net repulsive ΔG_{iwi} of $+1.4 \times 10^6$ kT occurring at a distance, ℓ, between bubbles, of 5 nm (see Table XI-2), which would prevent their coalescence.

However, in contradiction with the results of Craig *et al.* (1993), a higher salt content, at $\mu = 0.10$ (see Table XI-2), even at an initial ζ-potential of –100 mV (see also Table XI-2) would not be able to counteract the LW + AB inter-bubble attraction. In any case, a pre-existing gas bubble's ζ-potental in pure deionized water of about –65 mV would not suffice, if Garciaa *et al.* (1995) are right, to lend stability to the bubbles, as the addition of salt (as done by Craig *et al.*, 1993) to the deionized water would only *decrease* the absolute value of the ζ-potential of gas bubbles in water, which would make the inter-bubble coalescence even more pronounced. As an *increase* in the absolute value of the ζ-potential with added salt, as implied by the results of Craig *et al.* (1993), is against the laws of colloid behavior in aqueous media, there has to be something wrong with the experimental results of either Craig *et al.* (1993) or Graciaa *et al.* (1995) or both, unless it can be shown

TABLE XI-2

Extended DLVO Data on Air Bubbles of R = 1 mm, in Pure Water and in Water with Added NaCl, at $\mu = 0.01$ and at $\mu = 0.10$, at 20°C. All in kT Units.

$\dfrac{\ell}{\text{nm}}$	ΔG_{iwi}^{LW}	ΔG_{iwi}^{AB}	$\Delta G_{iwi}^{IF} = \Delta G_{iwi}^{LW} + \Delta G_{iwi}^{A}$ [a]	ΔG_{iwi}^{EL} $\mu = 0.01$ $1/\kappa = 3.16\ \mu m$ at $\Psi_o = -100$ mV	(ΔG_{iwi}^{EL}) [b] $\mu = 0.10$ $1/\kappa = 1.0\ \mu m$ (at $\Psi_o = -100$ mV)	$\Delta G_{iwi}^{TOT} = \Delta G_{iwi}^{LW} + \Delta G_{iwi}^{AB} + \Delta G_{iwi}^{EL}$ at $\mu = 0.01$; $1/\kappa = 3.16\ \mu m$ and $\psi_o = -100$ mV
$\ell = \ell_o = 0.157$	-5.3×10^7	-7.9×10^7	-1.3×10^8	$+1.1 \times 10^7$	$(+1.1 \times 10^7)$	-1.2×10^8
1.0	-8.3×10^6	-3.4×10^7	-4.2×10^7	$+6.0 \times 10^6$	$(+3.4 \times 10^6)$	-3.6×10^7
2.0	-4.2×10^6	-1.3×10^7	-1.7×10^7	$+4.7 \times 10^6$	$(+1.4 \times 10^6)$	-1.2×10^7
3.0	-2.8×10^6	-4.6×10^6	-7.4×10^6	$+3.6 \times 10^6$	$(+5.2 \times 10^5)$	-3.8×10^6
4.0	-2.1×10^6	-1.7×10^6	-3.8×10^6	$+2.7 \times 10^6$	(2.0×10^5)	-1.1×10^6
5.0[c]	-5.2×10^4	-6.2×10^5	-6.7×10^5	$+2.1 \times 10^6$	(7.2×10^4)	$+1.4 \times 10^6$
6.0	-3.6×10^5	-2.3×10^5	-2.7×10^5	$+1.5 \times 10^6$	(2.7×10^4)	$+1.2 \times 10^6$
7.0	-2.7×10^4	-8.2×10^4	-1.1×10^5	$+1.1 \times 10^6$	$(+1.0 \times 10^4)$	$+1.0 \times 10^6$
10.0	-1.3×10^4	-4.2×10^3	-1.7×10^4	$+4.6 \times 10^5$	$(+5.0 \times 10^2)$	$+4.4 \times 10^5$

[a] In pure, deionized water, $\Delta G_{iwi}^{EL} = 0$ and only $\Delta G_{iwi}^{LW} + \Delta G_{iwi}^{AB}$ determine the free energy of interaction, which is here attractive for both LW and AB forces.

[b] The figures in parentheses in this column indicate what the ΔG_{iwi}^{EL} values would be at an ionic strength of the bulk liquid, $\mu = 0.10$. At this ionic strength, at all distances, ℓ, the positive values of ΔG_{iwi}^{EL} remain so small that would remain negative, i.e., attractive, denoting instability, i.e. bubble coalescence.

[c] At about $\ell = 5.0\ \mu m$, LW retardation sets in and the LW free energy between two spheres no longer decays as a function of ℓ as $1/\ell$, but as $1/\ell^2$.

experimentally that the addition of small amounts of salt (e.g., of the order of 10^{-3} M) to a suspension of particles or bubbles in de-ionized water somewhat *increases* the particles' ζ-potential, even though further addition of salt would subsequently *decrease* the ζ-potential in the classical manner. [There is reason to believe that the higher salt concentrations also used by Craig *et al.* (1993) are quickly dissipated by the experimental condition of causing air bubbles to rise through the salt solution, which have to absorb significant amounts of salt in the process, thus continuously lowering the actual ionic strength of the aqueous medium surrounding any subsequent rising air (or rather N$_2$) bubbles].

CODA

It has become clear that the air-side of the water-air interface is probably the most hydrophobic, non-polar surface known to Man. This has been demonstrated, *inter alia*, by a water contact angle measured on a flat porous surface of which 99.2% was air (Onda *et al.*, 1996), which contact angle, at 174°, was only 6° removed from the theoretical, but unattainable 180°. Using the negative value of ΔG_{iwi}^{IF} of polypropylene (PP) as well as of the water-air interface (using the data given in Table XIX-1), it can be shown that whereas for PP, $\Delta G_{iwi}^{IF} = -102.3$ mJ/m^2, which is a very high (negative) value typical for completely hydrophobic condensed-phase materials, the value for ΔG_{iwi}^{IF} of the water-air interface is -145.6 mJ/m^2, which even exceeds some of the highest known (negative) ΔG_{iwi}^{IF} values [e.g., diiodomethane, at -112.1 mJ/m^2, cf. van Oss, Wu *et al.*, (2001)] by 30%. At the same time, whilst even the most hydrophobic condensed-phase materials still show a sizeable free energy of hydration, ΔG_{iw}, which is typically at least of the order of -40 mJ/m^2 (e.g., in the case of hexane or octane), whilst for the water-air interface, $\Delta G_{iw} = 0$, which shows that the air side of the water-air interface is the only real "hydrophobic" surface known, although air, of course, is not a condensed-phase material. Among condensed-phase matter there is no real "hydrophobic" material *sensu stricto*.

The hyper-hydrophobicity of a water-air interface makes it attract hydrophobic molecules, and/or hydrophobic moieties of amphiphilic solutes (e.g., alcohols, surfactants, proteins, etc.) toward that interface, with their most hydrophobic sites oriented into the air-side, which strongly decreases the apparent surface tension of water. Conversely, the total lack of any γ^{LW} property of the air-side at the water-air interface causes very hydrophilic solutes (such as mono-, oligo- and poly-saccharides) to be repelled away from the water-air interface, into the bulk liquid. Due to the formation of that depletion layer which is situated proximal to the water-air interface, both the *increase* in γ_w caused by simple sugars dissolved in the bulk liquid, as well as the *decrease* in γ_w caused by their polymers (i.e., polysaccharides) dissolved in the bulk liquid, are greater than the γ_w values one actually measures. Thus, contact angles on solid surfaces should never be measured with drops of solutions, or mixtures of liquids.

According to Craig *et al.* (1993) the admixture of salts (e.g., NaCl) changes the inter-bubble attraction in deionized water into a repulsion, and causes bubble stability, thus preventing bubble coalescence. However, in view of the ζ-potential found by Graciaa *et al.* (1995) for air bubbles in deionized water, the bubble stability, apparently caused by the admixture of salt (Craig *et al.* 1993), remains difficult to explain, unless it can be shown experimentally that when particles or bubbles (with a ζ-potential of, e.g., –65 mV) are immersed in de-ionized water and one adds small amounts of salt (e.g., $\approx 10^{-3}$ M), the absolute value of the ζ-potential then *increases* instead of decreasing.

Part III

---

*Experimental Measurement
Methods*

XII | Contact Angle and Surface Tension Determination and Preparation of Solid Surfaces

THE SESSILE DROP AS A FORCE BALANCE

Contact angle measurement, first described by Thomas Young in 1805, remains at present the most accurate method for determining the interaction energy between a liquid (L) and a solid (S), at the minimum equilibrium distance ℓ_o between S and L. The cosine of the contact angle θ (measured through the drop, at the tangent to the drop, starting at the triple point solid-liquid-air; (see Fig. XII-1) is a measure of the resultant of the energy of cohesion of the liquid, in the guise of γ_L and the energy of adhesion between liquid and solid, expressed as ΔG_{SL}; see eq. [IV-11] (see also Fig. IV-1 and de Gennes, 1985). Thus the left-hand side of eq. [IV-11] shows the part of the free energy of cohesion of the liquid, which is equal to the free energy of adhesion between the liquid and the solid, shown on the right-hand side.

It should be stressed that contact angle determinations, used as a force balance, can only serve to measure macroscopic-scale interaction energies, which is due to the macroscopic scale of the surface area of interaction between the liquid drop and the solid substratum, which is of the order of 0.3 to 1.0 cm.

SPREADING PRESSURE

To be assured that we may use the Young-Dupré equation in the general form given in eq. [IV-11], in all cases where $\gamma_L > \gamma_S$, we must investigate whether in these cases the general practice of neglecting the equilibrium spreading pressure π_e in the Young equation, expressed by Bangham and Razouk (1937):

$$(1 + \cos \theta)\, \gamma_{LV} = -\Delta G_{SL} - \pi_e \qquad \text{[XII-1]}^*$$

---

* A list of symbols can be found on pages 399–406.

(where V stands for the vapor given off by the liquid, L) should be continued. Fowkes *et al.* (1980) studied the possibility of spreading pressures arising with high-energy liquids deposited on low-energy solids, and found that this did not occur. On the other hand, when the vapor of a low-energy liquid could interact with a somewhat

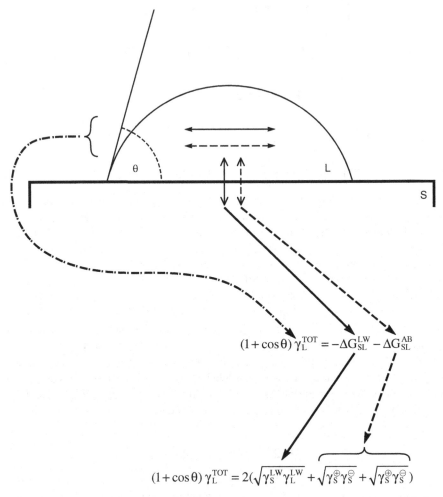

$$(1+\cos\theta)\,\gamma_L^{TOT} = -\Delta G_{SL}^{LW} - \Delta G_{SL}^{AB}$$

$$(1+\cos\theta)\,\gamma_L^{TOT} = 2(\sqrt{\gamma_S^{LW}\gamma_L^{LW}} + \sqrt{\gamma_S^{\oplus}\gamma_S^{\ominus}} + \sqrt{\gamma_S^{\oplus}\gamma_S^{\ominus}})$$

FIGURE XII-1 The contact angle (θ) as a force balance. Cos θ is a measure of the equilibrium between the energies of cohesion between the molecules of liquid L (horizontal arrows; left-hand side of the equation) and the forces of adhesion (vertical arrows; right-hand side of the equation) between liquid L and solid S. Apolar energies are indicated by uninterrupted horizontal or vertical arrows; polar (Lewis acid-base) energies are designated by interrupted horizontal or vertical arrows. Finite (non-zero) contact angles can only be obtained when the energy of cohesion of the liquid is greater than the energy of adhesion between liquid and solid. The manner in which these energies give rise to the Young equation for polar systems (see lower of the two equations; see also eq. [IV-13]) is also presented schematically.

higher-energy solid surface, the effect of the resulting positive spreading pressure caused an increase in the contact angle of water on that solid surface, which allowed the determination of π_e (Fowkes *et al.*, 1980). Busscher *et al.* (J. Colloid Interface Sci., 1986) on the other hand contend that even when $\gamma_L > \gamma_S$, spreading pressures can have a considerable influence on the contact angle. They base this on ellipsometric observations of the adsorption of water and of propanol on a variety of surfaces; they correlated the adsorption of these liquids with spreading pressures, which they found to be of the same order of magnitude for water and for propanol on both high- and low-energy surfaces. However, as the surface tensions of low energy liquids ($\gamma_{propanol} = 23.7$ mJ/m^2) and high-energy liquids ($\gamma_{water} = 72.8$ mJ/m^2) are known to be very different, especially on solids with a $\gamma_S \approx 35$ mJ/m$^2 \pm 10$ mJ/m^2 (cf. Fowkes *et al.*, 1980), the spreading behavior (vis-à-vis such surfaces) of propanol clearly differs fundamentally from the non-spreading behavior of water. Thus, the correlation between the ellipsometric results and the putative spreading pressures, advanced by Busscher *et al.* (J. Colloid Interface Sci., 1986), particularly when using propanol-water mixtures, must be regarded as tenuous.

From results obtained in wicking and thin-layer wicking (see below), it became evident that with all spreading liquids used (i.e., in all cases where $\gamma_L < \gamma_S$), pre-wetting occurred by the formation of a "precursor film" of the contact angle liquid on the solid surface, even before the bulk liquid actually reaches that part of the solid. This manifests itself in wicking by the production of a straight line with a strongly *positive slope*, when plotting $\eta h^2/t$ *vs.* γ_L (when η is the viscosity of the contact angle liquid and h the height of the liquid reached by the capillary rise, in a time t). However, with non-spreading liquids (i.e., $\gamma_L > \gamma_S$ and cos $\theta < 1$) neither spreading nor pre-wetting takes place, as evidenced by a strongly *negative slope* of plots of $\eta h^2/t$ *vs.* γ_L (van Oss, Giese, Li, *et al.*, 1992).

More recently we published an analysis of the degree to which putative "equilibrium spreading pressures" (π_e) need to be taken into account when measuring contact angles with non-spreading liquids i.e., liquids for which $\gamma_L > \gamma_S$ (van Oss *et al.*, 1998). It turned out that with all commonly used contact angle liquids, the maximum decrease in contact angle varied between 0.02° (for glycerol) to 1.55° (for water), when measured on a flat, smooth surface of poly(methylmethacrylate) (PMMA). These results are of the same order of magnitude as the standard deviation of most contact angle measurements, which lie typically between ±1° and ±2°. It appears therefore permissible, more than 65 years after the majority of Colloid and Surface scientists started following Bangham and Razouk's (1937) largely unfounded caveat, to stop worrying about the dreaded but largely imaginary "equilibrium spreading pressure" in all cases where $\gamma_L > \gamma_S$, and to drop the superfluous subscript "V" from γ_{LV} and γ_{SV}, and to resume designating all surface tensions measured in air, regarding liquids (L) as γ_L and those regarding solids (S), as γ_S.

We shall therefore also, with the majority of workers in this field, continue to neglect π_e in all cases where $\gamma_L > \gamma_S$, and where vapors of low-energy substances are absent. Thus eq. [IV-11] remains the basic Young-Dupré equation; see also Good (1979), van Oss *et al.* (1998), and the following section.

CONTACT ANGLE MEASUREMENT
IN AIR–HYSTERESIS

The Young equation is held to be valid for contact angles measured as the *advancing* angle, i.e., the angle the drop makes when it has just ceased advancing (e.g., for a few seconds; Chaudhury, 1984; Good, 1979). Whilst advancing contact angles (θ_a) have been held to be a measure of the apolar aspect of a surface, and retreating contact angles (θ_r) a measure of its polar aspects (Fowkes *et al.*, 1980), it would be hazardous to take that statement so literally as to attempt to derive the polar surface tension component of a surface by measuring θ_r.

The major reason for never using the retreating angle, θ_r, for any serious surface thermodynamic measurement purpose is that the retreating drop has (obviously) been retreating on that part of the surface to be measured, that had just been contaminated through wetting by the very liquid of which that retreating drop was composed. The retreating contact angle, θ_r, is therefore usually smaller than the advancing angle, θ_a. For that reason it is only *the advancing contact angle*, as defined earlier in this Section, for which the Young-Dupré equation is valid (cf. eq. IV-13; see also Figure XII-1). This is so because it is the very act of advancing that guarantees that the drop, with its three-phase (air-liquid-solid) line, touches a hitherto non-wetted part of the surface to be measured. For a method to determine the degree of contamination of the surface through wetting by the retreating drop, see the Section on Cassie's Equation, below.

The difference between θ_a and θ_r is called the contact angle hysteresis (ή ὑστέρησισ, deficiency, want, delay). Another important cause for hysteresis is surface roughness. Methods have been worked out for applying correction factors to the measured (advancing) contact angles, once the radius of roughness has been determined (Chaudhury, 1984). However, using smooth surfaces, i.e., surfaces with radii of roughness significantly smaller than 1 μm, is by far preferable (Chaudhury, 1984). Various other possible causes for hysteresis are discussed by Adamson (1982, 1990). For the main cause of the apparent hydrophobizing effect of roughness, see Chapter XI.

Usually hysteresis, when it occurs, is positive, i.e., $\theta_a > \theta_r$, indicative of complete or partial residual wetting of the solid surface by the liquid of the retreating drop. However, negative hysteresis (i.e., $\theta_a < \theta_r$), although rare, can take place. It usually occurs when a component of the liquid (or the liquid itself) can cause a change in the properties of the solid surface (other than just wetting). For instance, it is well known that plurivalent counterions (e.g., Ca^{++}) not only tend to neutralize the electric potential of (e.g. negatively) charged solid surfaces, but also (being electron acceptors) counteract the hydrophilicity of such a surface by depressing the surface's electron donicity parameter ($\gamma^{\ominus}$), thus not only rendering that surface electrically closer to neutral, but also making it more "hydrophobic" (Ohki, 1982; van Oss *et al.*, *Molecular Mechanisms of Membrane Fusion*, 1988; van Oss and Good, J. Protein Chem., 1988; Mirza *et al.*, 1998). Now, by using drops of water containing, e.g., 25 mM $CaCl_2$, for the measurement of contact angles on layers of negatively charged phospholipids, the Ca^{++} contained in the drop will make the phospholipid surface more "hydrophobic" and the low contact angles initially observed with such drops subsequently increase substantially. In such cases, $\theta_a < \theta_r$, that is to say, without any need to make the drop retreat by aspiration with a syringe, the drop "retreats" all by itself, while its contact angle increases. (It should be noted that the relatively low $CaCl_2$ concentrations used here do not significantly influence the surface tension properties of water.) In the case of amphoteric surfaces, negative contact angle hysteresis should also be possible, e.g., by using drops of water of different pH's (cf. Holmes-Farley et al., 1985; see also Mirza et al., 1998).

The much more common occurrence of positive contact angle hysteresis with polar liquids (e.g., water) on polar surfaces is an additional indication of the insignificance of the equilibrium spreading pressure in the cases where a finite (advancing) contact angle can be obtained, i.e., when $\gamma_L > \gamma_S$. In all cases where $\theta_a > \theta_r$, the liquid is clearly capable of partially wetting the solid surface, when in the retreating mode. However, the much higher value of θ_a than of θ_r shows that with the advancing drop no such wetting (or only a negligible degree of wetting) has occurred. Thus, in comparison with the effect of a retreating angle, no quantitatively significant amount of wetting is caused by evaporation of liquid from the drop, followed by re-condensation on the previously dry surface in the immediate vicinity of the drop (which is the putative cause of equilibrium spreading pressure).

The best way for measuring contact angles probably is by direct observation of the drop, and by determining the contact angles on either side of it, after having ascertained that, seen from above, the base of the drop is circular (i.e., not noticeably asymmetrical). The observation is done, e.g., by viewing the drop with a low-power telescope, with a goniometer attached to the rim of the ocular, which is provided with a cross-hair (Gaertner Scientific, Chicago, IL). It is useful to attach the viewing telescope/goniometer to a vertical X-Y track, manually adjustable in both directions with rotating knobs (Gaertner Scientific). The solid surface should be adjustable vertically on a mini-jack, and horizontally, away from and toward the viewer, on the tracks of an optical bench. A (diffuse) light source should be placed some distance behind the drop. It is of course also possible to photograph the drops, and measure the contact angles after wards. In any event, a fair number of drops of a given liquid should be measured on the left as well as on the right-hand side of each drop, and the average taken. When the reflection of the drop in the solid surface is visible, it becomes easier to determine the triple point (air-liquid-solid), which will improve the accuracy of the contact angle measurement. The best reproducibility is obtained with drops of a diameter of at least 5 mm (Good and Koo, 1979; Shu, 1991). Various other approaches to contact angle measurement, such as vertical plate (Wilhelmy-plate) methods, and tilting plate methods, spherical body methods, have been described by Neumann and Good (1979).

CONTACT ANGLES ON HETEROGENEOUS SURFACES—CASSIE'S EQUATION

On solid surfaces which are mosaics of different materials, e.g., an apolar material, 1, and a polar material, 2, Cassie and Baxter (1944) and Cassie (1948) showed that:

$$\cos \theta_A = f_1 \cos \theta_1 + f_2 \cos \theta_2 \qquad \text{[XII-2]}$$

where θ_A denotes the aggregate contact angle measured on the mosaic surface, and $f_1 + f_2 = 1$, while θ_1 and θ_2 are the contact angles one would find on a solid surface solely consisting of material, 1, and of material, 2, respectively. It can be stated that for material, 1 (cf. eq. [IV-13]):

$$f_1(1 + \cos\theta_1)\gamma_L = 2f_1(\sqrt{\gamma_1^{LW}\gamma_L^{LW}} + \sqrt{\gamma_1^{\oplus}\gamma_L^{\ominus}}$$
$$+ \sqrt{\gamma_1^{\ominus}\gamma_L^{\oplus}}) \qquad\qquad\text{[XII-3A]}$$

and for material, 2:

$$f_2(1 + \cos\theta_2)\gamma_L = 2f_2(\sqrt{\gamma_2^{LW}\gamma_L^{LW}} + \sqrt{\gamma_2^{\oplus}\gamma_L^{\ominus}}$$
$$+ \sqrt{\gamma_2^{\ominus}\gamma_L^{\oplus}}) \qquad\qquad\text{[XII-3B]}$$

As the interaction energies of the various classes of interaction (i.e., those of Lifshitz–van der Waals, electron-acceptor–electron-donor and electron-donor–electron-acceptor interactions) are *additive* (see Chapter IV), it may also be stated that, e.g., $f_1 = \sqrt{\gamma_1^{LW}\gamma_L^{LW}} + \sqrt{\gamma_2^{LW}\gamma_L^{LW}} = \sqrt{\gamma_A^{LW}\gamma_L^{LW}}$, where again, the subscript A, denotes the total, "aggregate" interaction energy and, similarly, $f_1\sqrt{\gamma_1^{\oplus}\gamma_L^{\ominus}} + f_2\sqrt{\gamma_2^{\oplus}\gamma_L^{\ominus}} = \sqrt{\gamma_A^{\oplus}\gamma_L^{\ominus}}$ and $f_1\sqrt{\gamma_1^{\ominus}\gamma_L^{\oplus}} + f_2\sqrt{\gamma_2^{\ominus}\gamma_L^{\oplus}} = \sqrt{\gamma_A^{\ominus}\gamma_L^{\oplus}}$. Eqs. [XII-3A] and [XII-3B] may be added together, to yield:

$$[1 + \cos\theta_A)\gamma_L = f_1(1 + \cos\theta_1)\gamma_L + f_2(1 + \cos\theta_2)\gamma_L \qquad\text{[XII-4]}$$

which, upon rearrangement, can be seen to be equivalent to Cassie's equation (eq. [XII-2]).

Israelachvili and McGee (1989) derived an equation which differs from eq. [XII-2], but they did so by designating a single (e.g., apolar) property to material, 1, and another single (e.g., polar) property to material, 2, etc., and using a single geometric mean combining rule for apolar as well as for polar materials. Two things are wrong with this approach. First, whilst in apolar systems one may solely deal with LW interactions, in polar systems one always has AB *as well as* LW interactions. Thus, neglecting the omnipresent LW forces in the interactions between a contact angle liquid and polar patches inevitably leads to erroneous results. In addition, as shown in Chapter XIV, with polar materials the assumption of the existence of only one polar property (the "γ^P" approach) leads to further erroneous results.

More recently Drelich and Miller (1993) proposed a modification of Cassie's equation, by incorporating a line-tension term. However, Cassie's equation, in the form of eq. [XII-2], correlates well with a variety of experimental situations (Cassie and Baxter, 1944; Baxter and Cassie, 1945; Wenzel, 1949; see also Adamson, 1990, p. 388) and it seems reasonable to continue to use it without modifications.

To determine the percentage of the surface area under the receded part of the drop which has been contaminated by the wetting caused by the retreating action, one can use Cassie's equation in the form:

$$\cos \theta_r = f_1 \cos \theta_a + (1 - f_1) \qquad \text{[XII-2A]}$$

where $(1 - f_1) = f_2 \cos \theta_2$, because in this case $\theta_2 = 0$ (cf. eq. XII-2), on account of the fact that the angle the contact angle drop makes on that portion of the solid surface which is completely wetted by the contact angle liquid is zero, so that $\cos \theta_2 = 1$ and $f_2 \cos \theta_2$ is equal to $(1 - f_1)$. The data published by Good and Koo (1979) show that for drops of water on Teflon FEP, $\theta_a = 116.8°$ and $\theta_r = 98.6°$, whilst for drops of water on PMMA, $\theta_a = 74.4°$ and $\theta_r = 54.3°$. Entering these data into eq. XII-2A yield that the % caused by θ_r with drops of water on Teflon, $f_2 = 20.8\%$, whilst the % wetting caused by θ_r with drops of water on PMMA, $f_2 = 43.0\%$. Both Teflon and PMMA are of course hydrophobic in the sense that for both polymers $\Delta G_{iwi} < 0$ (van Oss and Giese, 1995; see also Chapter XIX and Table XIX-1). More precisely however, for Teflon (subscript T), $\Delta G_{TWT} = -102.4$ mJ/m^2 and for PMMA (subscript P), $\Delta G_{PWP} = -37.8$ mJ/m^2 (using eq. IV-17 and based on the data listed in Chapter XVII), so that PMMA is significantly less hydrophobic than Teflon. It may be appropriate to look at the hydration energies (ΔG_{iw}; see eq. IV-18) of both polymers, indicating that $\Delta G_{TW} = -39.5$ mJ/m^2 and $\Delta G_{PW} = -94.5$ mJ/m^2, so that the ratio of f_2 (Teflon)$/f_2$ (PMMA), i.e. 0.48, is roughly comparable to the ratio of $\Delta G_{TW}/\Delta G_{PW}$, which is 0.42.

CONTACT ANGLE MEASUREMENT IN LIQUIDS

Young's equation is not only valid for contact angle measurements of drops of a liquid, L, on a solid, S, in air, but also for the measurement of contact angles of drops of a liquid, L, on a solid, S, immersed in a different liquid which is immiscible in liquid, L, e.g., an oil, O, using:

$$\gamma_{OL} \cos \theta_{OL} = \gamma_{SO} - \gamma_{SL} \qquad \text{[XII-5]}$$

(see, e.g., Chaudhury, 1984). It should, however, be noted that with the two-liquid approach, using the advancing angle with one liquid implies that one obtains a retreating angle with the other liquid. In principle the two-liquid approach, combined with contact angle measurements in air, can serve to determine γ_S^{LW} values of solids with polar liquids such as water. To that effect one also uses:

$$\gamma_L \cos \theta_L = \gamma_S - \gamma_{SL} \qquad \text{[XII-6]}$$

so that by subtracting eq. [XII-5] from [XII-6], one obtains:

$$\gamma_L \cos \theta_L - \gamma_{OL} \cos \theta_{OL} = \gamma_S - \gamma_{SO} \qquad \text{[XII-7]}$$

For a completely apolar hydrocarbon, O, in view of eq. [II-35A], eq. [XII-7] may be written as in Chaudhury (1984):

$$\gamma_L \cos\theta_L - \gamma_{OL} = -\gamma_O + 2\sqrt{\gamma_O\gamma_S^{LW}} \qquad \text{[XII-8]}$$

However, when the organic liquid, O, is polar eq. [XII-7] becomes (see eqs. [III-4 and III-7]:

$$\gamma_L \cos\theta_L - \gamma_{OL} \cos\theta_{OL} = -\gamma_O^{LW} + 2(\sqrt{\gamma_S^{LW}\gamma_O^{LW}}$$
$$-\sqrt{\gamma_O^{\oplus}\gamma_O^{\ominus}} + \sqrt{\gamma_S^{\oplus}\gamma_O^{\ominus}} + \sqrt{\gamma_S^{\ominus}\gamma_O^{\oplus}}) \qquad \text{[XII-7A]}$$

Thus, all three surface tension parameters (γ_O^{LW}, $\gamma_O^{\oplus}$, $\gamma_O^{\ominus}$) must be known for the polar organic liquid, still leaving the three unknowns (γ_S^{LW}, $\gamma_S^{\oplus}$, $\gamma_S^{\ominus}$) of the polar solid surface. These considerations become pertinent with respect to a seemingly extremely attractive application of the immersed-drop contact angle where liquid O is an aqueous solution of polyethylene oxide and the contact angle liquid, L, an aqueous solution of dextran (Podesta, et al., 1987; Schürch et al., 1981). Aqueous solutions of polyethylene oxide and of dextran, when mixed, separate into two immiscible phases (see Chapter XXI, and van Oss, Chaudhury and Good, Separ. Sci. Tech., 1987). The interfacial tension between these two phases is very low (Albertsson, 1986), which gives rise to large contact angles of drops of dextran solution on hydrated biological surfaces, immersed in polyethylene oxide solutions. Small differences in surface properties of hydrated biological surfaces give rise to large differences in contact angles of drops of dextran solution, immersed in a solution of polyethylene oxide (Schürch et al., 1981). However, the difficulties of interpretation of eq. [XII-4A] in such a system become considerable. As both aqueous dextran and polyethylene oxide solutions are mixtures even γ_L values are of doubtful applicability to the Young equation. The value of γ_{OL} is extremely close to zero, so that the large values of, and the large differences between values of θ_{OL} found on only slightly different surfaces are not as helpful for the interpretation of the surface properties of hydrated biological surfaces as might have been hoped. And, while γ_L^{LW} and γ_O^{LW} may, as a first approximation, be estimated from the properties of the solutes and the solutions' compositions, the $\gamma_O^{\oplus}$ and $\gamma_O^{\ominus}$ values of the aqueous surfactant solutions are hard to assess. Thus this application of the immersed contact angle method at present can only yield semi-quantitative results as to whether a given hydrated biological surface is more, or less, "hydrophilic" or "adhesive" than another such surface. Podesta et al. (1987) wisely chose the observed contact angle itself, as a measure of "adhesiveness," or "hydrophobicity" of male and female human parasite Schistosoma mansoni surfaces, at various stages of development. In their system, θ_{OL} values higher than 90° are held to correspond to higher "adhesiveness" and "hydrophobicity"; see also McIver and Schürch (1982).

CONTACT ANGLE MEASUREMENT BY WICKING

Hard particles, of a diameter of 1 μm or more, when spread into flat layers, will form a surface that is too rough for accurate contact angle measurement by the sessile (advancing) drop method (Chaudhury, 1984). However, by using a packed column of such particles, capillary rise velocity measurements of a liquid in that column can also yield the contact angle of that liquid with respect to the particles' surface. This "wicking" approach can also be used for contact angle measurement on the solid material of porous bodies, strands of fibers, etc. In these cases one uses the Washburn equation (Adamson, 1982; Ku *et al.*, 1985):

$$h^2 = \frac{tR\gamma_L \cos\theta}{2\eta} \qquad\qquad \text{[XII-9]}$$

where h is the height the column of liquid L has reached by capillary rise in time t, R the average radius of the pores of the porous bed, θ the contact angle and η the viscosity of liquid L.

By first measuring values of h, at a constant time t, for a number of low energy liquids L, which can be taken to spread on the particles' surface (e.g., hexane and octane), the value of R can be obtained for the type of column at hand, packed with a given kind of particle, because in these cases, $\cos\theta = 1$.* Once R is determined, $\cos\theta$ can be obtained with a number of well-characterized apolar and polar liquids, which then permits the solution of eq. [IV-13], used three times, for γ_S^{LW}, $\gamma_S^{\oplus}$ and $\gamma_S^{\ominus}$ of the particles in question.

To use the wicking approach for the determination of $\cos\theta$, it is essential to use exceedingly well-packed columns of rather monodisperse particles (Ku *et al.*, 1985). If, upon first contact with an ascending (and lubricating) liquid, tighter packing can occur locally, a gap will be created between particles, which causes a strongly asymmetrical rise of the liquid in the packed column. This makes it very difficult to measure the precise length of travel of the liquid column. However, in those cases where only poly-disperse suspensions of irregularly shaped particles are available, an extremely useful alternate approach is the coating of such particles into, e.g., glass surfaces, followed by the measurement of the capillary rise of various liquids, L, as outlined above, in the manner of thin layer chromatography† (see, e.g., van Oss, Giese, Li *et al.*, 1992).

---

* Some doubts could exist whether the Washburn equation [XII-9] remains valid even when $\gamma_L < \gamma_S$ (i.e., with spreading liquids). Specifically, it might be questioned whether for spreading liquids the condition $\cos\theta = 1$ continues to be applicable. However, it turns out that $\cos\theta = 1$ remains valid for all spreading liquids, because they were experimentally shown to pre-wet the surfaces of the particles over which they subsequently spread, in the form of a "precursor film" (van Brakel and Heertjes, 1975; de Gennes, 1990; van Oss, Giese, Li, *et al.*, 1992). If, however, the formation of a precursor film can be avoided (e.g., by "out-baking" the particles), the term $\gamma_L \cos\theta$ should, for the spreading regime, be replaced by $(S + \gamma_L)$, where S is a spreading coefficient $(S = \gamma_S - \gamma_L - \gamma_{SL})$, defined by Harkins and Feldman (1922) (see also Good, 1973; Good and Lin, 1976.)

† This approach was first suggested by Dr. Manoj K. Chaudhury (Chaudhury, private communication, 1986).

For the determination of the surface tension of and contact angles on high-energy carbon yarns, Chwastiak (1973) also uses wicking rate measurements, but developed equations to that purpose that appear more appropriate than Washburn's eq. [XII-9]. As with Washburn's equation, Chwastiak's approach can only accommodate values for $\theta < 90°$; at $\theta > 90°$ no capillary rise along the porous material can occur.

For the surface tension components and parameters, as well as the viscosities of the liquids used in wicking, see Table XVII-10. In wicking, as in direct contact angle measurement, the preferred order in which to use the contact angles obtained with these liquids in calculating the various surface tension components and parameters of polar solids is to calculate γ_S^{LW} first, and then $\gamma_S^{\oplus}$ and $\gamma_S^{\ominus}$ see the beginning of Chapter XVII. For wicking, the use of glycerol is not recommended, on account of its very high viscosity. Wicking is particularly useful in the characterization of nonswelling clay particles (van Oss, Giese, Li et al., 1992).

Instead of the Washburn approach, which uses the rate of capillary *rise*, one can also use the method of Bartell and Whitney (1934), expressing the capillary *pressure* (ΔP):

$$\Delta P = \frac{2\gamma_L \cos\theta}{R} \qquad\qquad \text{[XII-10]}$$

Apparatus for capillary pressure determination have been described by: Bartell and Whitney (1932), White (1982), Dunstan and White (1986) and Diggins et al. (1990).

White (1982) described a way to circumvent the need to determine R by means of a liquid which wets the particles (i.e., where $\cos\theta = 1$; see above) by proving that in eq. [XII-10] R may also be calculated as a function of the volume fraction (ϕ) occupied by the particles, the specific surface area (A) per unit weight of particles and the specific density of the particle material (ρ):

$$R = \frac{2(1-\phi)}{\phi\rho A} \qquad\qquad \text{[XII-11]}$$

In practice however, R, thus obtained, may not be quite equal to R obtained by wicking with a spreading liquid, especially in the case of porous particles (van Oss, Giese, Li, et al., 1992). The capillary pressure method cannot be used in the thin-plate configuration, described above, so that one must revert to the use of vertical tubes packed with powder, with all the drawbacks of that approach, described above. Thus the thin-layer rate of capillary rise method (using a spreading liquid to determine R) is by far the most convenient, and usually also the most accurate approach.

An experimental comparison between direct contact angle measurements and wicking (e.g., thin layer wicking) at first seemed to present a problem which arose from the fact that with large, flat, smooth surfaces one could only use the direct contact angle approach, whilst with rough surfaces formed by deposits of (non-swelling) particles or powders one could only use wicking, so that the method suitable for the first condition excluded using the second approach, and vice-versa. Thus a comparison between the two methods, using identical materials, appeared to be an unattainable aim. However, experiments done with synthesized, hydrophilic, monosized and cuboid Hematite particles, of about 0.66 µm (Costanzo et al., 1995), which could be deposited on a glass plate, permitted such plates, on account of the smoothness of the flat layers of cuboid Hematite thus formed, to be useable for direct contact angle measurements, as well as for thin layer wicking. Using such Hematite-coated plates, it could be shown that the contact angles as well as the surface tension components and parameters derived from them, when found by direct measurements were, within experimental error, comparable with those determined by thin layer wicking.

It should be noted that particles which swell when brought into contact with a liquid (e.g., water), e.g., clay particles such as smectites (e.g., montmorilonites), cannot be measured by wicking, as the swelling caused by a wicking liquid such as water, which occurs equally in all directions, invalidates the use of Washburn's equation (eq. XII-9) because the wicking method depends on a uni-dimensional liquid transport movement. However, swollen smectite particles, when deposited on a glass plate and dried, fortunately present an extremely smooth surface which is quite suitable for direct contact angle measurements.

SURFACE TENSION OF LIQUIDS

The total surface tension of liquids (γ_L) is most easily determined by the Wilhelmy plate method (Adamson, 1982, 1990). Briefly, a thin rectangular plate (e.g., a chromic acid cleaned glass microscope cover slip, or an acid cleaned platinum plate) is suspended vertically from the arm of a micro-balance, just above liquid L, contained in a small beaker. The beaker is slowly raised, until it just touches the hanging plate. Upon contact with the liquid, a small additional force, or additional weight (ΔW), is exerted on the plate, so that (Adamson, 1982, 1990):

$$\gamma_L \cos\theta = \frac{\Delta W}{p} \qquad \text{[XII-12]}$$

where p is the periphery of the plate.

Another rather reliable approach to measuring $_L$ is the hanging drop method (Adamson, 1982, 1990). After measuring the dimensions d_e and d_s (the maximum diameter d_e, and the diameter d_s measured a distance d_e above the bottom of the drop) from enlarged photographs of the drop, γ_L can be obtained as follows (Adamson, 1982, 1990):

$$\gamma_L = \frac{\Delta\rho g d_e^2}{H} \qquad \text{[XII-13]}$$

where $\Delta\rho$ is the difference in density between liquid and the gas phase (or air) and g the acceleration caused by gravity. H is a shape-dependent parameter; Adamson (1982) gives tables, correlating $S = d_s/d_e$ vs. $1/H$.

The γ_L values of 2,200 pure liquid compounds, usually as a function of temperature, have been published by Jasper (1972). These values are usually extremely reliable; only one error has been noted which is of some importance to the present work, i.e., for diiodomethane, at 20°C, $\gamma_L = 50.8$ mJ/m², and not 66.98 mJ/m² (Chaudhury, 1984).

APOLAR AND POLAR SURFACE TENSION COMPONENT OF LIQUIDS

γ_L^{LW} is best determined by contact angle measurement of liquid L on a purely apolar solid surface, of $\gamma_S = \gamma_S^{LW}$, using:

$$(1 + \cos \theta)\gamma_L = 2\sqrt{\gamma_S^{LW}\gamma_L^{LW}} \qquad\qquad [\text{IV-28}]$$

Once γ_L^{LW} is known, as well as γ_L, γ_L^{AB} can be derived from:

$$\gamma_L = \gamma_L^{LW} + \gamma_L^{AB} \qquad\qquad [\text{IV-2}]$$

Apolar solids which have been used for this purpose are polyethylene and polypropylene (Fowkes, 1987) and polytetrafluoroethylene (van Oss, Ju *et al.*, 1989).

An entirely different approach to measuring γ_L^{LW} and γ_L^{AB} (and even $\gamma_L^{\oplus}$ and $\gamma_L^{\ominus}$, see below) is to encase the liquid inside a gel, which allows one to treat the liquid as a *solid*, upon a surface of which drops of another liquid can be deposited, for contact angle measurement. The interior of a gel, especially when made at a low concentration of the gel-forming matrix, retains to a large extent the principal physico-chemical properties of the continuous liquid phase (Michaels and Dean, 1962). By using this approach, the γ_L^{LW} value of water could be determined, using agarose gels. When an aqueous agarose gel, consisting of X% agarose, is cut open,* the cut surface may be considered as a flat surface $(100 - X)\%$ "solid" water, at room temperature. By measuring contact angles with hexadecane $(\gamma^L = \gamma^{LW} = 27.5$ mJ/m² at 20°C) on these hydrous surfaces, formed with 2.0, 1.75, 1.50, 1.25 and 1.0% (w/v) agarose gellified with pure water we found, by extrapolation to 0% agarose, a value of 21.4 ± 1.2 (S.D.) mJ/m² for γ_L^{LW} of water (van Oss, Roberts *et al.*, 1987). The value

---

* Upon gel formation the molecules that constitute the gel matrix tend to orient themselves toward the interfaces with the air and with the walls of the vessel in which the gel was initially formed, so that the boundary surfaces of a gel may manifest certain properties (e.g., low surface tension) of the gel matrix more strongly than the overall proportion of gel matrix material to bulk liquid would warrant (van Oss *et al.*, 1977). However, when a gel is formed against a very hydrophilic surface, such as clean glass, or when a gel is cut open, the surface properties at the cut surface (or those formerly facing a clean glass surface) correspond much more closely to those of the liquid. In measurements of contact angles with hydrocarbons on such gel surfaces, there still may be a measurable effect caused by the interaction between the water molecules and the macromolecules of the gel matrix, as well as a residual van der Waals attraction exerted by the gel matrix macromolecules themselves. However, both of these effects diminish as the liquid content of the gel increases. It thus is possible to measure contact angles at the freshly cut surfaces of gels with a high water content, and by extrapolating to a water content of 100%, to obtain the apolar surface tension component of pure water.

for γ_L^{LW} found for water in this manner agrees well with the value of 21.8 ± 0.7 (S.D.) mJ/m² calculated by Fowkes (1963) from the interfacial tensions between water and various hydrocarbons given by Girifalco and Good (1957).

ESTIMATION OF THE POLAR SURFACE TENSION PARAMETERS $\gamma_L^\oplus$ AND $\gamma_L^\ominus$ OF LIQUIDS

GEL METHOD

The method of encasing polar liquids in gels of various concentrations, and measuring contact angles with various other liquids on cut surfaces of such gels (extrapolating to zero gel matrix concentration) (van Oss, Roberts *et al.*, 1987) (see above), can also be used for the determination of $\gamma_L^\oplus$ and $\gamma_L^\ominus$ by using well-characterized polar contact angle liquids, employing eq. [IV-13].

Agarose gels, of the type that earlier had been used for the measurement of contact angles of hexadecane on water (van Oss, Roberts *et al.*, 1987) can also be used for encasing glycerol. Agarose is suspended in distilled water, in concentrations (w/v) of 1.0, 1.25, 1.5, 1.75 and 2.0% and boiled) to dissolve the agarose. The precise final concentrations are then determined by weighing the solutions (containing a known weight of agarose). The solutions are cast in glass petri dishes and the gels allowed to set by cooling to room temperature. With the aid of a ruler and parallel scalpels, straight bars are cut out of the gels, and the contact angles measured (van Oss, Ju *et al.*, 1989). If the fluid under investigation is a liquid other than water, the bars are deposited into smaller glass petri dishes, which are filled with the fluid in question. That fluid is then replaced once a day, until all the water from the gel bars has been replaced, as monitored by refractometry of the equilibrium liquid. During the last few days before contact angle measurement, the petri dishes containing the bars with fluid must be kept in a vacuum desiccator over $CaSO_4$ (Drierite), in order to remove the last traces of water.

With glycerol encased in agarose gels, an unusual phenomenon was observed: drops of water *spread* on agarose gels encasing glycerol. And, vice versa, glycerol *spreads* on agarose gels encasing water (van Oss, Ju *et al.*, 1987). The $\gamma_L^\oplus$ and $\gamma_L^\ominus$ values for glycerol therefore cannot be ascertained by contact angle measurement on gels encasing glycerol. However, by using the $\gamma_L^\oplus$ and $\gamma_L^\ominus$ values obtained for glycerol by other means (see below), i.e., $\gamma_G^\oplus \approx 3.92$ and $\gamma_G^\ominus \approx 57.4$ mJ/m² (van Oss, Good and Busscher, 1990) and inserting these values into eq. [IV-13] (see also eq. [IV-1]) we obtain (a) for drops of water on glycerol ($\theta = 0$):

$$145.6 = 54.45 + 20 + 76.5 - \pi_e$$

and (b) for drops of glycerol on water ($\theta = 0$):

$$128.0 = 54.45 + 20 + 76.5 - \pi_e$$

showing that in both cases there must indeed be a positive spreading pressure π_e, which is highest for glycerol drops on water, as might be expected. In other words, the exceptionally high AB adhesion between glycerol and water, caused by the interaction between $\gamma_G^\oplus$ and $\gamma_W^\ominus$ ($2\sqrt{\gamma_G^\oplus \gamma_W^\ominus} = 20$ mJ/m^2) and between $\gamma_G^\ominus$ and $\gamma_W^\oplus$ ($2\sqrt{\gamma_G^\ominus \gamma_W^\oplus} = 76.5$ mJ/m^2) (eq. [IV-13]), gives rise to the spreading, not only of glycerol on water, but also of water on glycerol.

Dimethylsulfoxide (DMSO), ethylene glycol (EG) and formamide (FO) dissolve agarose gels. Therefore, a covalently crosslinked gel, polyacrylamide, must be used instead for these three liquids, in initial concentrations of 7, 8.5, and 10% (w/v). Lower concentrations of that material do not allow the formation of solid gels. Table XII-1 shows the polar surface tension data obtained by this method for DMSO (with water, glycerol and formamide), ethylene glycol (with water, glycerol and formamide), and formamide (with water and glycerol). As there is a strong miscibility between these contact angle liquids and the liquids encased in the gels, the results shown in Table XII-1 only reflect the order of magnitude of the $\gamma_L^\oplus$ and $\gamma_L^\ominus$ values found, so that attempts were made to obtain somewhat more precise values of these entities for glycerol and formamide, by using a different approach (van Oss, Good, and Busscher, 1990).

POLAR SOLIDS METHOD

From the spreading of water (W) on glycerol (G) encased in agarose gels (see above) it is easily seen that (compared with the reference values for water: $\gamma_W^\oplus = \gamma_W^\ominus = 25.5$ mJ/m^2) $\gamma_G^\oplus$ must be smaller than 4.83 mJ/m^2 and $\gamma_G^\ominus$ larger than 46.62 mJ/m^2 (utilizing eqs.

TABLE XII-1
Polar Surface Tension Parameters for DMSO, Ethylene Glycol (EG) and Formamide (FO), Estimated via the Gel-Method (van Oss, Ju *et al.*, 1989), in mJ/m^2. (For more accurate values for EG and FO, see Table XII-4.)

	Experimental values					
Liquid	$\gamma_L^{LW\,a}$	$\gamma_L^\oplus$	$\gamma_L^\ominus$	γ_L^{AB}	γ_L^{AB}	γ_L^b
DMSO	30.5^c–38.9^d	3.1^e	24.0^e		13.5^c–5.06^d	44
EG^f	29^a	4.18	41.78	26.43^g	19^a	48
FO^h	39^a	2.95	36.72	20.82^g	19^a	58

[a] Data from the literature; see, e.g., Chaudhury (1984).
[b] Jasper (1972).
[c] Fowkes (1987).
[d] From contact angle measurements of PTFE ($\gamma^{LW} = 19.17$) ($\theta_{DMSO} = 76°$).
[e] In view of other results (e.g., van Oss, Chaudhury and Good, Separ. Sci. Tech., 1987; Advan. Coll. Int. Sci; 1987) this value for γ_{DMSO}^+ is probably too high and the value for $\gamma_{DMSO}^\ominus$ probably too low.
[f] A better value for γ and γ of ethylene glycol, obtained in part from solubility data, is 1.92 and 47.0 mJ/m^2, respectively; see Table XVII-10, and Chapter XXII.
[g] Calculated from the $\gamma_L^\ominus$ and $\gamma_L^\oplus$ values experimentally obtained *via* the gel-method.
[h] A more accurate value for $\gamma^\oplus$ and $\gamma^\ominus$ of formamide was obtained by measurement of contact angles with formamide on monopolar surfaces: $\gamma^\oplus = 2.28$ and $\gamma^\ominus = 39.6$ mJ/m^2 (van Oss, Good and Busscher, 1990); see below.

[IV-4] and [IV-13]). Next to water, glycerol and formamide (F) are the highest energy hydrogen-bonded polar (non-toxic) liquids readily available for contact angle measurements on polar surfaces.

Glycerol has been used, together with water, to obtain the $\gamma_S^\oplus$ and $\gamma_S^\ominus$ values for a variety of polar surfaces (see e.g., van Oss *et al.*, J. Protein Chem., 1986; Separ. Sci. Tech., 1987; Advan. Coll. Int. Sci., 1987). However, a drawback of glycerol is its high viscosity (i.e., 1490 times that of water, at 20°C), which complicates standardization of the timing of contact angle measurement (an important consideration, particularly when there is any degree of solubility between solid and liquid). Another liquid with a surface tension that is almost as high as that of glycerol is formamide ($\gamma_F = 58$ mJ/m², $\gamma_F^{LW} = 39$ mJ/m² and $\gamma_F^{AB} = 19$ mJ/m²) (Chaudhury, 1984). Formamide has a viscosity which is much more comparable to that of water (i.e., ≈ 4 centipoises at 20°C) and, like glycerol, formamide also has much more pronounced (Lewis) base characteristics than water, see Tables XII-1 and XVII-10. Thus, while formamide still has a sizeable $\gamma_F^\oplus$ parameter, its $\gamma_F^\oplus/\gamma_F^\ominus$ ratio is sufficiently different from that of water to yield useful contact angle data, which in conjunction with contact angle data obtained with water, will yield reliable values for $\gamma_S^\oplus$ and $\gamma_S^\ominus$ for a variety of polar solids. Also, whilst $\gamma^\oplus$ and $\gamma^\ominus$ can be entirely defined with two polar liquids, the availability of a third polar liquid can yield a useful set of control values. In addition, the surface tension properties of a fourth polar liquid, ethylene glycol, have been estimated using its γ^{LW} and γ^{AB} values (Chaudhury, 1984), and its solubility data (Ch. XXII); see Table XII-4 (reproduced in part from Table XVII-10).

The $\gamma_W^\oplus/\gamma_G^\oplus$ and $\gamma_W^\oplus/\gamma_F^\oplus$ ratios were therefore determined by contact angle measurements on surfaces of a number of relatively strong Lewis bases, so as to be able to observe a significant degree of interaction between these base parameters, and the relatively minor acid parameters of glycerol and formamide. The monopolar (Lewis) basic solid surfaces used are: poly (methylmethacrylate) (PMMA), poly (ethylene oxide) (PEO), clay films (obtained by drying on a poly [vinyl chloride]) surface, corona-treated poly (propylene) (CPPL), dried agarose gel, dried zein (a water-insoluble corn protein), cellulose acetate film, dried film of human serum albumin (HSA); see Table XII-2. Some of the measurements were taken from earlier work, published (van Oss *et al.*, Advan. Coll. Int. Sci. 1987) or unpublished (see Table XII-3); other measurements were done more recently (PMMA, clay, PEO) and some of the data (on nylon 6,6) were furnished by Dr. R. E. Baier, SUNY-Buffalo. All of these surfaces were measured with diiodomethane and/or -bromonaphthalene for the determination of γ_S^{LW}. All surfaces were also measured with water and glycerol, to obtain the value of $\gamma_G^\oplus$ or $\gamma_F^\oplus$ according to a simplified version of eq. [IV-13]:

$$(1+\cos\theta)\gamma_L = 2(\sqrt{\gamma_S^{LW}\gamma_L^{LW}} + \sqrt{\gamma_S^\ominus\gamma_L^\oplus}) \qquad \text{[XII-14]}$$

which could be used in view of the monopolarity of the solid S, which implies the (virtual or total) absence of $\gamma_S^\oplus$. For the establishment of essential monopolarity of a solid surface, see van Oss *et al.* (Advan. Coll. Int. Sci,. 1987). In addition, contact angle data were used that had been obtained on nylon 6,6 (averaged over seven experiments), to calculate $\gamma_F^\oplus$. To that effect the value of 3.92mJ/m² for $\gamma_G^\oplus$, obtained

TABLE XII-2
Contact Angle Data Obtained with Five Liquids, on Various Polar Surfaces

	αBrN^a	DIM^b	H_2O^c	GLY^d	FO^e
PMMA 1	18°	30°	60°	60°	51°
PMMA 2	13°	40°	72°	65.5°	53°
Clay	20°	43°	63°	67.5°	57.25°
PEO	11°	18.25°	21.5°	43.5°	25°
CPPL	39°	56°	78.5°	78°	
CPPL	42°	54.5°	85.5°	78°	
CPPL	41°	53.5°	81°	78°	
Agarose	23°	37°	57.3°	57.3°	
Zein	24°	36°	65°	65°	
Cellulose acetate	32°	55°	54.5°	62°	
HSA	23.2°	34°	63.5°	62°	
Nylon 6,6 (Average of 7 observations)	36°	46°	64°	65°	55°

[a] $\gamma_L = \gamma_L^{LW} = 44.4$ mJ/m^2.
[b] $\gamma_L = \gamma_L^{LW} = 50.8$ mJ/m^2.
[c] $\gamma_L = 72.8$ mJ/m^2, $\gamma_L^{LW} = 21.8$ mJ/m^2, $\gamma_L^{\oplus} = \gamma_L^{\ominus} = 25.5$ mJ/m^2 (van Oss *et al.*, Advan. Coll. Int. Sci., 1987).
[d] $\gamma_L = 64$ mJ/m^2, $\gamma_L^{LW} = 34$ mJ/m^2, $\gamma_L^{AB} = 30$ mJ/m^2 (Chaudhury, 1984).
[e] $\gamma_L = 58$ mJ/m^2, $\gamma_L^{LW} = 39$ mJ/m^2, $\gamma_L^{AB} = 19$ mJ/m^2 (Chaudhury, 1984).
From van Oss, Good and Busscher, 1990.

from the average of the $\gamma_G^{\oplus}$ values shown in Table XII-3, was used to calculate the values of $\gamma_S^{\oplus}$ and $\gamma_S^{\ominus}$ for nylon 6,6, prior to determining the values for $\gamma_F^{\oplus}$ and $\gamma_F^{\ominus}$, using eqs. [IV-4] and [IV-13].

Table XII-2 shows the contact angles obtained with 5 liquids, on eleven monopolar surfaces used for obtaining $\gamma_G^{\oplus}$. The first four of these surfaces were also used to obtain $\gamma_F^{\oplus}$; for the other surfaces, which were measured at an earlier date, no reliable contact angle data with formamide on these materials are as yet available. $\gamma_G^{\ominus}$ and $\gamma_F^{\ominus}$ then follow from eq. [IV-4]; see Table XII-3. In addition, another important data point for $\gamma_F^{\oplus}$ and $\gamma_F^{\ominus}$ was obtained from the (averaged) contact angle measurements on nylon 6,6. The averages thus obtained were: $\gamma_G^{\oplus} = 3.92 \pm 0.7$ (SE) mJ/m^2 and $\gamma_F^{\oplus} = 2.28 \pm 0.6$ (SE) mJ/m^2. Using eq. [IV-4], the corresponding electron-donor parameters then are: $\gamma_G^{\ominus} = 57.4$ mJ/m^2 and $\gamma_F^{\ominus} = 39.6$ mJ/m^2 (van Oss, Good and Busscher, 1990).

For a list of the surface tension components and parameters of the liquids used in contact angle measurements on various solids, see Table XII-4, and for the order in which to use the contact angles obtained with these liquids in calculating the surface tension components and parameters of polar solids, see the beginning of Chapter XVII.

It should be noted that slight differences in contact angle measurement technique have a much stronger impact on the $\gamma^{\oplus}$ and $\gamma^{\ominus}$ values found than on the γ^{LW} values (see., e.g., the different values found for the same PMMA, in Table XII-3). Reactivity

TABLE XII-3

$\gamma_L^\oplus$ Values for Glycerol (G) and Formamide (F) Derived from Contact Angle Measurements on Monopolar Basic Solids[a]

	Solid			
	γ_S^{LW}	$\gamma_S^\ominus$	$\gamma_G^\oplus$	$\gamma_F^\oplus$
PMMA	43.2[b]	22.4[b]	4.18[g]	1.72[g]
PMMA	41.4[b]	12.2[b]	4.94[h]	3.24[h]
Clay	39.9[b]	21.5[b]	2.56	1.28
PEO	45.9[b]	58.5[b]	4.29	2.88
	33.0[c]	11.1[c]	2.39	
CPPL	32.7[c]	6.2[c]	4.56	
	33.0[c]	9.1[c]	2.91	
Agarose	41[d]	26.9[d]	5.32	
Zein	41.4[d]	18.7[d]	3.54	
Cellulose acetate	38[e]	32.3[e]	3.78	
HSA	41[e]	20.3[e]	4.62	
	γ_S^{LW}	$\gamma_S^\ominus$	$\gamma_S^\oplus$	
Nylon 6,6	36.4[f]	21.6[f]	0.02[fj]	2.8[ij]

AVE.: $\gamma_G^\oplus = 3.92 \pm 0.7$ (SE) and $\gamma_F^\oplus = 2.28 \pm 0.6$ (SE)

[a] All values for surface tension components and parameters are given in mJ/m².
[b] From recent measurements; see Table XII-2.
[c] Earlier unpublished measurements; see Table XII-2.
[d] van Oss *et al.* (Advan. Coll. Int. Sci., 1987).
[e] van Oss *et al.* (J. Chromatog., 1987).
[f] From unpublished data furnished by Dr. R. E. Baier; see Table XII-2.
[g] These data are from equilibrium contact angle measurements.
[h] These data are from still barely advancing contact angle measurements, on the same PMMA.
[i] Using eq. [IV-13] and $\gamma_G^\oplus = 3.92$.
[j] Not used in obtaining final averages.

of the contact angle liquids with the solid surface (Table XII-3: PEO, Agarose, HSA) tends to give rise to somewhat higher than average $\gamma_L^\oplus$ values. (Note that PMMA dissolves slowly in formamide.) Thus the values found for $\gamma_G^\oplus$ and $\gamma_F^\oplus$ are perhaps more likely to be somewhat on the high, rather than on the low side. The main choice is between strongly polar (e.g., PEO) and weakly polar (e.g., CPPL) surfaces, for this type of measurement. Weakly polar surfaces (generally weak Lewis bases) do not bind strongly with polar liquids that have only a small to moderate Lewis acid parameter, which leads to relatively inaccurate results. Strongly polar surfaces have other drawbacks. They rapidly attract all sorts of impurities, but they can of course be used for the purpose of determining polar γ_L parameters with impurities and all. However, such impure surface conditions may well lead to a decreased reproducibility of the measurements. Strongly polar solids also are more likely to be soluble in polar liquids, which is another cause for inaccuracies. Nevertheless, within

TABLE XII-4

Surface Tension Components and Parameters of Liquids Used in Direct Contact Angle Determination at 20°C, in mJ/m² (see also Table XVII-10)[a]

Liquid	γ	γ^{LW}	γ^{AB}	$\gamma^{\oplus}$	$\gamma^{\ominus}$
APOLAR					
cis-Decalin	32.2	32.2	0	0	0
α-Bromonaphthalene	44.4	44.4	≈ 0[b]	≈ 0[b]	≈ 0[b]
Diiodomethane	50.8	50.8	0	≈ 0.01[c]	0
POLAR					
Water	72.8	21.8	51.0	25.5	25.5
Glycerol	64.0	34.0[a]	30.0	3.92	57.4
Formamide	58.0	39.0[a]	19.0	2.28	39.6
Ethylene glycol	48.0	29.0[a]	19.0	3.0[d]	30.1[d]

[a] The viscosities of these liquids, at 20°C, are given in Table XVII-10.

[b] Not quite zero, but not precisely known

[c] see van Oss, Wu et al. (2001)

[d] These values for ethylene glycol are more accurate than the ones given in the 1994 edition; see van Oss, Ju et al. (1989).

the constraints of measurement conditions and techniques (of which the most crucial probably is the solubility of the solid in the liquid drops), the polar parameter values found for glycerol and formamide and ethylene glycol (see Table XVII-10) should be sufficiently reliable for use in obtaining a first approximation of the $\gamma_S^{\oplus}$ and $\gamma_S^{\ominus}$ values of polar solid surfaces.

PURITY OF CONTACT ANGLE LIQUIDS

Liquids used for contact angle measurements should be pure liquids and should not be mixtures of two or more or liquids, nor solutions of solutes in liquids. This is especially true for water, when used as a contact angle liquid. It is not necessary to use ultra-pure water; distilled water or reversed osmosis treated water usually are entirely satisfactory. As far as all solutes, dissolved in water are concerned, one must make a distinction between *two categories* of these (Docoslis *et al.*, 2000; van Oss, Docoslis and Giese, 2002):

Category A consists of usually very soluble solutes, most of which, when dissolved in water *increase* its surface tension. These are, e.g., sugars, some amino acids, such as glycine, most neutral salts (e.g., NaCl), as well as polysaccharides, such as dextran (which is a monopolar electron-donor and, in aqueous solution, a strong electron-donor and a fairly weak electron-acceptor). Category A solutes which are polymers of the dextran type, somewhat *decrease* the surface tension of water. All category A solutes, when dissolved in water undergo a local decrease in their concentration in the layer closest to the edge of the water-air interface by which they are repelled, but they still cause a net measurable decrease in the surface tension of water. Concentrations below 1% (w/v) of aqueous solutions of category A compounds do not increase the surface tension of water by more than 0.3 to 0.5%. Thus, e.g., isotonic physiological salt solutions of about 0.88% NaCl in water can safely be used for measuring contact angles on, e.g., layers of mammalian or microbial cells, as such isotonic solutions have an almost unchanged surface tension, at 20°C of about 72.5 mJ/m² (as compared to 72.8 mJ/m² for pure water at that temperature, cf. Jasper, 1972).

   Category B compounds on the other hand, *significantly decrease* the surface tension of water. This is because they strongly concentrate at the water-air interface, with their hydrophobic moieties orienting toward the air-side. Category B molecules comprise amphipathic or surface-active compounds, such as alcohols, peptides, proteins, surfactants [e.g., soaps, detergents, etc., including poly(ethylene oxide)]. Their proneness to concentrate at the water-air interface causes even very low concentrations of such amphipathic compounds to give rise to a *dramatic decrease* in the apparent surface tension of water. For instance many surfactants dissolved in water at 0.1% or less, may well lower the latter's apparent surface tension from 72.8 mJ/m² to about 40 mJ/m², just by orienting 90% to more than 99% of their apolar (i.e., low surface tension) moieties toward the air-side of the water-air interface, which causes a pronounced concentration anisotropy closest to that interface, exactly at the locus where the surface tension of liquids is measured and where it makes itself felt; see Chapters XI and XXII and Docoslis *et al.* (2000) and van Oss, Docoslis and Giese (2002). The use of category B solutes in aqueous solution, *at any concentration*, with the aim of measuring surface properties of solid surfaces is therefore strongly discouraged, as the results of such an endeavor are impossible to interpret quantitatively and usually even qualitatively.

SURFACE TENSION AND SURFACE FREE ENERGY

Finally, it should be noted that the surface *tension*, which is expressed as γ, is not the same thing as surface *energy* (or surface *free energy*), which one expresses as ΔG. These two differ numerically as well as by sign (cf. eq. II-33B):

$$\Delta G_{11}^{coh} = -2\gamma_1 \qquad \text{[XII-15]}$$

where ΔG_{11}^{coh} is the *free energy of cohesion* of condensed-phase material, 1, and γ_1 is its *surface tension*.

INFLUENCE OF TEMPERATURE ON THE PROPERTIES OF WATER AS A CONTACT ANGLE LIQUID

It should be noted that even pure water changes drastically in its surface-thermodynamic properties when cooled below, or heated above 20°C (see also Chapters VIII and XXII).

   By convention (see Chapter IV), at 20°C: $\gamma_W^\oplus = \gamma_W^\ominus = 25.5$ mJ/m², so that at that temperature the $\gamma_W^\oplus / \gamma_W^\ominus$ ratio equals unity. However at, e.g., 38°C, water becomes more Lewis acidic: $\gamma_W^\oplus = 32.4$ mJ/m² and $\gamma_W^\ominus = 18.5$ mJ/m², which increases the $\gamma_W^\oplus/\gamma_W^\ominus$ ratio from 1.0 to 1.75 at 38°C (see Chapter XXII). Conversely, upon cooling to 0°C, water becomes more Lewis basic, with $\gamma_W^\oplus = 19$ mJ/m² and $\gamma_W^\ominus = 37$ mJ/m², so that the $\gamma_W^\oplus/\gamma_W^\ominus$ ratio drops from 1.0 at 20°C to 0.514 at 0°C (van Oss and Giese, 2004; see also Chapter VIII). (All these $\gamma_W^\oplus$, $\gamma_W^\ominus$ and $\gamma_W^\oplus/\gamma_W^\ominus$ values are in comparison with values, standardized at 20°C).

Therefore, contact angles measured with pure water on a solid monopolar electron-donating surface such as poly(methyl methacrylate) (PMMA, cf. Chapter XXII), at a temperature (T) higher than 20°C, will be smaller than those measured at 20°C, thus giving the appearance that PMMA becomes less hydrophobic at higher T. This is not, however, what really happens: in reality it is just the water with which the contact angles are measured that becomes more Lewis acidic at higher T. The increase in the $\gamma_W^\oplus/\gamma_W^\ominus$ ratio of water with an increase in T is also the cause of the increase in aqueous solubility of electron-donating solutes with T, such as glucose and sucrose; cf. Chapter XXII.

INFLUENCE OF THE PH OF WATER AS A CONTACT ANGLE LIQUID ON THE SURFACE PROPERTIES OF ELECTRICALLY CHARGED SOLID SURFACES

The influence of the pH of the water used in contact angle measurements on homogeneously electrically charged or on amphoteric surfaces can be considerable. This is because the pH of the aqueous medium in contact with such surfaces can influence their surface (ζ-) potential, which in turn can change such a surface's hydrophilicity or hydrophobicity because, concomitantly with an increase in a surface's ζ-potential, its electron-donicity ($\gamma^\ominus$) tends to increase (and thus also to increase its hydrophilicity), whilst with a decrease in a surface's ζ-potential (and thus also in its electron-donicity) the surface will become less hydrophilic or may even become hydrophobic (Wu *et al.*, 1994; see also Chapter XIX, under "Linkage between EL and AB Forces").

Thus, the highest water contact angle (and therefore the greatest hydrophobicity) was found with drops of water of a pH corresponding to the point of minimum charge of an amphoteric surface (Holmes-Farley *et al.*, 1985). The same holds true for contact angle measurements with drops of water, as a function of their pH, on layers of dissolved proteins, where the contact angles of the water drops were at a maximum at the isoelectric point of the protein (van Oss and Good, 1988a; see also Chapter XXII).

Water contact angles on negatively charged vesicles consisting of phosphatidyl serine, or of phosphatidylic acid, reached a maximum value with drops of water acidified to pH 2, or below, where the ζ-potentials of such vesicles were at a minimum (Mirza *et al.*, 1998).

These phenomena are, however, strictly limited to the influence of pH on amphoteric or homogeneously electrically charged surfaces which have come in contact with water of different pHs and should not be interpreted as an indication of a possible change in the surface-thermodynamic properties of the water itself, when used as a contact angle liquid; see the following Section.

INFLUENCE OF THE PH OF WATER ON THE PROPERTIES OF WATER ITSELF, IN CONNECTION WITH ITS USE AS A CONTACT ANGLE LIQUID

In addition to the influence of the pH of drops of water on the contact angles they form on homogeneously electrically charged or on amphoteric solid surfaces (see the preceding Section, above), it is also of interest to ascertain to which (if any) extent the pH of water can influence, e.g., the $\gamma_W^\oplus/\gamma_W^\ominus$ ratio of the water itself.

Admixture of HCl slightly decreases γ_w, whilst addition of NaOH slightly increases it (Handbook of Chemistry and Physics, 1970/71), so that the difference in γ_w of water that has been acidified with a small amount of HCl, to attain a pH of about 2, compared with the γ_w of water made more basic with a small amount of NaOH, to attain a pH of about 13, caused a barely significant difference in contact angles, of 1.9°, measured on layers of dried dextran with drops of water at these different pHs. Differences in contact angles on dried dextran measured with close to neutral pure water, compared with either water of about pH 2, or with water of about pH 13, yielded differences in contact angles of 1.0° and 0.9° respectively, neither of which difference being statistically significant (van Oss and Giese, 2005). Now, the slight decrease in γ_w (and thus also in contact angle measured on dextran) caused by adding HCl is due to the fact that HCl has a lower surface tension than water. On the other hand, adding NaOH slightly increases γ_w (and thus also the contact angle measured on dextran), because NaOH has a higher surface tension than water (Handbook of Chemistry and Physics, 1970/71). The reason for the lower surface tension of HCl (compared to water) lies in its lower polar cohesive energy than that of water, so that as far as $\gamma_{HCl}^\oplus$ and $\gamma_{HCl}^\ominus$ are concerned, the *product*: $\gamma_{HCl}^\oplus \cdot \gamma_{HCl}^\ominus$ is smaller than the product: $\gamma_W^\oplus \cdot \gamma_W^\ominus$ (cf. Chapter IV). Similarly, the higher surface tension of NaOH (compared to water), is due to its higher polar cohesion energy than that of water, so that the *product*: $\gamma_{NaOH}^\oplus \cdot \gamma_{NaOH}^\ominus$ is larger than the product: $\gamma_W^\oplus \cdot \gamma_W^\ominus$. Now, crucial changes in the polar surface-thermodynamic properties of water would occur if the *ratio*: $\gamma_W^\oplus/\gamma_W^\ominus$ changed as a function of pH, as it does, for instance when the *temperature* of water is changed (see the section on "The Influence of Temperature," in this Chapter, above). However, although very small changes in the *product*: $\gamma_W^\oplus \cdot \gamma_W^\ominus$ occur (albeit only significantly so at the extremes of pH: at pH 2 and pH 13), which are due to differences in cohesive energy between HCl and NaOH, no significant changes in the $\gamma_W^\oplus/\gamma_W^\ominus$ *ratio* could be discerned, as a function of pH (i.e., through the addition of small amounts of HCl or NaOH) (van Oss and Giese, 2005). Thus, non-buffered water, either acidified with HCl, or made alkaline with NaOH, can be used as a contact angle liquid on electrostatically strictly neutral polar surfaces (see the preceding Section of this Chapter), at pH values between, e.g., pH 3 and pH 11, without concern that (non-buffered) acidified water might cause significant changes in Lewis acidity or that water made basic in an analogous manner could cause significant changes in its Lewis base properties. Indeed, the Lewis acid-base system (Lewis, 1923) has very little in common with the acidic or basic systems that are so designated in conjunction with their pH.

PREPARATION OF SOLID SURFACES

The easiest surfaces to prepare for contact angle measurement are smooth, flat surfaces of solid materials. These usually only need thorough cleaning, e.g., with chromic acid solution, followed by rinsing with distilled water, and extraction with acetone in a Soxhlet extractor for several hours, as applied to Teflon by Chaudhury (1984). Under certain conditions, other types of surfaces may need even less preparation. For instance, polystyrene surfaces on the inside of polystyrene Petri dishes which are hermetically sealed in sterile packs can be used without further preparation. Also, freshly split mica may be presumed to be clean, but to avoid almost immediate contamination of such high-energy surfaces by various air pollutants, mica surfaces may have to be kept, and measured, in a vacuum.

Various solids, usually present in powdered form, can be molten, by heating, and poured, e.g., onto microscope slides. An example is polyethylene oxide 6,000 (PEO) (which actually has a molecular weight of about 8,000), which generally is obtained in the form of a powder, or flakes. PEO can easily be molten on a low flame, and poured. PEO surfaces should be kept in a dessicator, as they are exceedingly hygroscopic.

PARTICLES AND CELLS

Contact angles on layers of bacteria and other cells were first measured in the author's laboratory in 1971 (Gillman and van Oss, 1971; van Oss and Gillman, 1972a; b). Two methods, both of which now widely utilized, were then used: 1) formation of layers by deposition of cells on a flat surface, or by growth of cells on a flat solid (nutrient) medium; and 2) formation of cell layers by suction-filtration of cell suspensions onto flat porous filters.

1. Many types of bacterial cells are readily grown as flat "lawn" on nutrient agar (van Oss and Gillman, 1972a; b). Cells capable of growing in tissue-culture are also readily grown in confluent layers on various flat surfaces (van Oss *et al.*, 1975). Fairly dense, if not entirely confluent layers of phagocytic cells (polymorphonuclear leucocytes, monocytes and macrophages) can be obtained thanks to their propensity for adhering to surfaces such as, e.g., siliconized glass (van Oss *et al.*, 1975). Cells suspended in aqueous media can, after concentration by centrifugation, be deposited on agar surfaces, or on moist membranes (van Oss *et al.*, 1975).

2. Cells which are not particularly "sticky," such as erythrocytes, and various bacteria grown as suspensions in broths, as well as phospholipid vesicles, are readily deposited on porous membranes by suction filtration (van Oss and Gillman, 1972; van Oss *et al.*, 1975). Membranes with pore diameters of about 0.40 to 0.45 μm, made of, e.g., cellulose-acetate (Millipore,

Bedford, MA) usually are quite suitable. For smaller particles or cells, pore sizes as small as 0.2 μm exist, and for larger particles, membranes with pore sizes up to several μm can be obtained. When a certain amount of drying of the cell layer, deposited on the membrane, is required, cellulose acetate membranes are wont to curl which is detrimental to effective contact angle measurement. In such cases porous silver membranes (existing in various poresizes) may be employed (Flotronics, Selas Corp., Dresher, NJ; continued as Hytrex membranes, Osmonics, Minnetonka, MN).

When cells are freshly deposited on a surface or a membrane (or have just been filtered on a membrane), a certain amount of drying is required, to remove the excess moisture. During studies on the standardization of the degree of required drying, it was found that in all cases, after a given minimum drying time, a period occurred (varying from 30 minutes to well over 1 hour), during which the contact angle has a constant "plateau" value, which is the value to be used (van Oss *et al.*, 1975).

DRIED SOLUTES

A large variety of micromolecular and macromolecular solutes can be transformed into smooth flat solid surfaces by deposition of a solution of such a solute in an appropriate solvent onto a glass surface, followed by evaporation of the solvent (in air or *in vacuo*). Table XII-5 lists a number of solutes that have been treated in this manner, together with the most appropriate solvents. In most cases it is advisable to keep the dried surfaces (on glass slides) in a vacuum dessicator. It usually is possible, upon drying, to obtain a very smooth, glassy surface with most polymers (e.g., water-soluble proteins and polysaccharides). However, with some organic polymers a fairly rough surface ensues upon drying from solution in a number of solvents. It is usually possible, however, to find a solvent from which a given polymer will dry as a very smooth surface. For instance, poly (methylmethacrylate) will yield rough surfaces when dried from solutions in methylethylketone or chlorobenzene, but it will give a very smooth glassy surface when dried from a solution in tetrahydrofuran, or in toluene. Only zein (a water-insoluble corn protein) has hitherto always yielded a relatively rough surface (e.g., upon drying from a 70% acetone–30% H_2O solution); see Table XII-5.

HYDRATED SOLUTES

To obtain flat layers of a hydrated biopolymer, membrane ultrafiltration can be used (in analogy with the deposition of layers of cells on microporous membranes; see above). To that effect, a concentrated solution of, e.g., a given protein is ultrafiltered on an anisotropic membrane, which is impermeable to the protein, under a few atmospheres' pressure, until no further ultrafiltrate ensues (van Oss *et al.*, 1975; van Oss, Moore *et al.*, 1985; van Oss *et al.*, J. Coll. Interf. Sci., 1986; J. Prot. Chem, 1986). Here also, after a certain initial drying time, a plateau value persists for the contact angle for 30 minutes to one hour. The contact angle at the plateau value is consistent for any given protein,

TABLE XII-5
Examples of Dried Solute Surfaces

Solute	Solvent	Reference
Human serum albumin	water	1
Human immunoglobulin G	water	1
Human immunoglobulin A	water	2
Egg white lysozyme	water	1
Corn zein[a]	acetone/water (70/30)	3
Gelatin	water	4
Dextran (polyglucose)	water	5
Ficoll (polysucrose)	water	4
Agarose (polygalactopyranoside)	water	3
Cellulose nitrate	methylethylketone	5
Cellulose acetate	methylethylketone	5
Sucrose	water	3
Glucose	water	4
Calf-thymus DNA	water	5
Yeast RNA	water	5
Guanidine hydrochloride	water	1
Urea	water	4
Ammonium sulfate	water	4
Poly (methyl methacrylate)	tetrahydrofuran	6
Poly (isobutylene)	tetrahydrofuran	6
Poly (styrene)	methylethylketone	6
Poly (propylene)	chlorobenzene	6
Poly (vinyl chloride)	chlorobenzene	6

[a] Fairly rough surface.
[1] van Oss et al. (J. Coll. Interf. Sci., 1986).
[2] van Oss, Moore et al. (1985).
[3] van Oss, Good and Busscher (1990).
[4] Hitherto unpublished results.
[5] van Oss et al. (Advan. Coll. Interf. Sci., 1987).
[6] van Oss et al. (Separ. Sci. Tech., 1989).

whereas plateau contact angles vary considerably among different proteins (van Oss et al., 1975; van Oss, Absolom et al., 1981). In all probability the contact angle at the plateau value, for a given hydrated protein, corresponds approximately to the contact angle of that protein, surrounded by, at most, one layer of water of hydration (van Oss and Good, J. Protein Chem., 1988). The most reasonable explanation for the constancy of the plateau contact angle of a given species of protein would be that after some initial drying, which removes the excess, loosely bound water, the last layer of hydration water is reached, which is much more strongly attached to the protein, and thus resists further removal by evaporation for a considerable time.

CODA

Methods are given for contact angle measurement, with a view to using the sessile (slightly advancing) drop as a force balance. *Inter alia*, discussions are given of: the role of a putative spreading pressure accompanying finite contact angles; positive (and negative) hysteresis; the correctness of Cassie's equation for contact angle interpretation on mosaic surfaces; the advantage and disadvantage of measuring contact angles immersed in other liquids; the use of wicking (permitting contact angle measurement in beds or columns of powders); measurement methods of the surface tension of liquids (including liquids encased in gels) and measurement of the interfacial tension between liquids; determination of the polar surface tension parameters of liquids. Surface tension properties are given of the most useful high-energy apolar and polar liquids best suited for contact angle measurement.

Finally, the influence is discussed of the temperature as well as of the pH of water on its properties as a contact angle liquid.

Methods are given for the preparation of flat solid surfaces, by melting, evaporation from solutions, or deposition on membranes (e.g., particles or cells). The need is stressed for first establishing a plateau value for the constant contact angle which prevails for a period of time *after* a certain time lapse of air-drying (which time lapse has to be established first). That plateau value then yields the operative contact angle for hydrated biopolymers or cells. The use of porous or semipermeable membranes in these techniques is also discussed.

XIII Interfacial Tension Determination — Influence of Macroscopic- and Microscopic-Scale Interactions

INTERFACIAL TENSIONS AND FREE ENERGIES OF INTERFACIAL INTERACTION

Virtually all the interfacial interaction energies occurring in liquids contain the interfacial tensions between condensed-phase molecules and those of the liquid, either as the sole term, or as one or several of the terms in one of the mathematical expressions for these molecules' free energies of interfacial interaction with or in the liquid, as exemplified by one or another version of the Dupré equation (see also the footnote thereto), such as eq. II-36, which pertains to the interfacial (IF) free energy of interaction (ΔG_{12}^{IF}) between two different condensed-phase materials, 1 and 2 in air or *in vacuo*, but which for compounds or materials, i, immersed in water, w, alludes to the interfacial free energy of hydration (ΔG_{iw}^{IF}). The simplest version of the Dupré equation (see eq. II-33A) shows the proportionality between γ_{12} and the interfacial free energy of cohesion of molecules, 1, immersed in a liquid, 2, i.e., ΔG_{121}^{IF}, which when liquid, 2, is water, is expressed as ΔG_{1w1}^{LW}. The latter is also the expression which quantitatively defines the degree of hydrophobicity (when $\Delta G_{1w1}^{LW} < 0$) or hydrophilicity (when $\Delta G_{1w1}^{LW} > 0$), of compound or material, 1; see also van Oss and Giese (1995) and Chapter XIX. Finally, eq. II-37 expresses the interfacial free energy of interaction between two different condensed-phase materials, 1 and 2, immersed in liquid, 3, where ΔG_{132}^{IF} is a function of all three pertinent interfacial tensions, γ_{12}, γ_{13} and γ_{23}. All the above equations for various forms of ΔG^{IF}, quoted from Chapter II, have LW as superscript in that Chapter, as it deals with LW interactions, but these equations are equally valid with superscript AB and, as used in the present Chapter, with superscript IF, where $\Delta G^{IF} = \Delta G^{LW} + \Delta G^{AB}$, cf. eq. IV-1.

Thus the interfacial tension, γ_{12}, plays a crucial role in all interfacial free energies of interaction in condensed-phase systems. It should again be pointed out that the expressions *interfacial tension* and *interfacial free energy are not interchangeable*. As should be clear from all the above-mentioned equations, γ_{12} and the various forms of ΔG of which γ_{12} is a component, always differ quantitatively and they often also differ in sign.

INTERFACIAL TENSION BETWEEN IMMISCIBLE LIQUIDS

Interfacial tensions (γ_{12}) between immiscible liquids can, in principle, be measured by the same methods with which liquid/vapor surface tensions are measured (Adamson, 1982, 1990; Neumann and Good, 1979). However, especially for the measurement of fairly small γ_{12} values, a few of these methods are better adapted for this particular purpose, e.g.: the pendant drop-shape method (see above), the simpler, but less accurate drop-weight method (Adamson, 1982, 1990), and a few other methods using deformed interfaces, such as the rotating drop method (Vonnegut, 1942; Princen *et al.*, 1967), and the method described by Lucassen (1979), using drops suspended in another liquid with a density gradient. Vonnegut's basic equation is:

$$\gamma_{12} = \tfrac{1}{4}\omega^2 \Delta \rho r_o^3 \qquad\qquad [\text{XIII-1}]$$

where ω is the speed of revolution around the long axis, $\Delta\rho$ the density difference and r_o the maximum radius of the drop at the axis of rotation.

For very samll values of γ_{12}, Vonnegut's (1942) rotating drop (or cylinder) method (which has been further elaborated upon by Princen *et al.*, 1967) is the most suitable. As a first approximation, the presence of a straight, flat meniscus at the interface between two liquids inside a vertical tube is an indication of a very low interfacial tension between two liquids.

For two-liquid systems, where at least one liquid is *non-polar*, these approaches may work, but for the determination of interfacial tensions between an organic liquid and a *polar* liquid, such as water, these approaches only yield correct results when the organic liquid is completely non-polar, i.e., for all practical purposes, an alkane; see the following Section.

DETERMINATION OF INTERFACIAL TENSIONS
BETWEEN CONDENSED-PHASE MATERIALS
AND WATER

One problem with two-liquid equilibrium drop-shape (or drop-weight) analysis of a drop of a polar organic liquid, immersed in water, is one of *scale*. With a drop of a polar organic liquid, such as octanol, immersed in water, in a matter of only one or two picoseconds a significant number of the OH groups of the alcohol orient themselves toward the drop's interface with the surrounding water. Thus, on a macroscopic scale, the interfacial tension, γ_{ow}^{mac}, between the drop of the organic phase (o), i.e., octanol in this case, and water (w), quickly becomes quite low, as it is heavily influenced by the dense layer of OH groups which had migrated to the water interface. Therefore, the macroscopic-scale interfacial tension with water (γ_{ow}^{mac}), as measured by drop-shape (or drop weight) analysis would only enable one to determine the macroscopic-scale free energy of interaction (ΔG_{owo}^{mac}) between *two adjoining macroscopic-scale drops* of octanol, brought closely together, when immersed in water, according to:

$$\Delta G_{owo}^{mac} = -2\gamma_{ow}^{mac} \qquad \text{[XIII-2]}$$

which is, however, an interaction energy which few scientific workers are likely to need to determine. What a greater number of such workers might instead aspire to ascertain is the microscopic-scale free energy of interaction (ΔG_{owo}^{mic}) between *two adjoining microscopic-scale molecules* of octanol, immersed in water, according to:

$$\Delta G_{owo} = -2\gamma_{ow}^{mic} \qquad \text{[XIII-2A]}$$

where the microcopic-scale interfacial tension, γ_{ow}^{mic}, is significantly larger than γ_{ow}^{mac} alluded to above (eq. XIII-2). Because γ_{ow}^{mic}, during the interaction between a drop of octanol and the water in which the drop had been immersed only exists during the first picosecond or so after this immersion (Luzar and Chandler, 1996) and thus is not directly measurable, we named that interfacial tension the "zero time dynamic" interfacial tension (van Oss *et al.*, 2002; van Oss and Giese, 2004; 2005), designated as γ_{ow}^{o}, which is equivalent with γ_{ow}^{mic} (eq. XIII-2A). Thus, although γ_{ow}^{o} is not practically measurable via drop-shape (or drop weight) analysis on account of the extreme shortness of time lapsed before it decays into its much smaller, macroscopic form, the microscopic-scale, zero time dynamic, γ_{ow}^{o}, can be determined from the *aqueous solubility of the polar organic liquid*, which is a microscopic (molecular) scale equilibrium reaction. Using:

$$-2\gamma_{ow}^{o} \cdot S_c = kT \cdot \ln s \qquad \text{[XIII-3]}$$

(van Oss and Good, 1996; van Oss *et al.*, 2002; van Oss and Giese, 2004; 2005) one can determine γ_{ow}^{o} from the organic compound's aqueous solubility (s, which must be expressed in mol fractions). Eq. XIII-3 is only valid when γ_{ow}^{o}, i.e., the zero time dynamic, or microscopic-scale interfacial tension with the liquid (here, water) is used. Also needed is the contactable surface area (S_c) between two molecules of the organic compound (o), immersed in the liquid (water). The symbol, k, stands for Boltzmann's constant (see the list of symbols, at the end of this book) and T is the absolute temperature in degrees Kelvin. For the determination of the contactable surface area, S_c, between two identical organic molecules, immersed in water, it is in many cases convenient to be guided by the interfacial tensions with water of *alkanes* (γ_{ow}), which *can* be determined via drop-shape (or drop weight) measurements (some of these experimentally obtained values have been listed by Girifalco and Good, 1957). In Table XIII-1 a number of the γ_{ow} values are given for pentane through octane, as well as for cyclohexane, together with the known aqueous solubilities of these compounds which, using eq. XIII-3, yielded the S_c values for these hydrocarbons. In addition, the interfacial tensions between water and alkanes (γ_{ow}) can also be calculated from their surface tensions, γ_o, because with *alkanes*, $\gamma_o = \gamma_o^{LW}$ (on account of their lack of polarity), using eq. IV-9. It can be seen in Table XIII-1 that the experimental γ_{ow} values and the γ_{ow}^{o} values calculated from

TABLE XIII-I

Surface Tensions, γ_o (= γ_o^{LW}) Values, Calculated Interfacial Tensions with Water (γ_{ow}^o) and Experimentally Derived Idem (γ_{ow}) by Means of Drop-Shape or Drop Weight Analyses (all in mJ/m²), of a Number of Alkanes, as Well as Their Contactable Surface Areas, When Immersed in Water, Derived from Their γ_{iw}^o-Values and Their Aqueous Solubilities (s).

Alkane	γ_o [a] (= γ_o^{LW}) (mJ/m²)	γ_{ow}^o [b] (mJ/m²)	γ_{ow} [c] (mJ/m²)	S_o [d] (nm²)	s [e] (Mol. fractions)
n-pentane	16.05	51.44	49.0	0.37	8.7×10^{-5}
n-hexane	18.40	51.14	51.1	0.41	3.1×10^{-5}
n-heptane	20.14	51.03	50.2	0.46	9.0×10^{-6}
n-octane	21.62	51.00	50.8	0.51	2.4×10^{-6}
cyclohexane	25.24	51.13	50.2	0.45	1.1×10^{-5}

[a] From Jasper (1972): for completely non-polar compounds, i, $\gamma_i = \gamma_i^{LW}$; see Chapter II
[b] Calculated from the s and γ_o^{LW} values, using eq. IV-9 or XIII-3
[c] Experimentally obtained, using drop-shape (or drop-weight) analyses, from Girifalco and Good (1957) (which is permissible with alkanes).
[d] Calculated, using γ_{ow}^o and eq. XIII-3.
[e] Aqueous solubilities from Stephen and Stephen (1963).

the alkanes' known apolar surface tensions, γ_o^{LW}, are quite comparable. They are virtually identical for hexane and octane and differ less than 5% for pentane. Thus the reasonable equality, *with alkanes*, between the experimentally measured γ_{ow} and the γ_{ow}^o calculated from the known γ_o^{LW} values, shows that one may with some confidence estimate the contactable surfaces (S_c) for a number of hydrocarbons, as shown in Table XIII-1. These S_c values are helpful in estimating the S_c values for the corresponding alkyl groups of polar organic compounds, thus permitting, *inter alia*, to obtain the contactable surface areas for, e.g., an alcohol such as octanol, using $S_c = 0.51$ nm² from octane (see Table XIII-1), which may be taken to be also valid for the octyl group, plus 0.10 nm² for the OH group, so that the total S_c value for octanol then amounts to 0.61 nm² (see van Oss and Good, 1996, and Table XIII-2).

For *polar* organic compounds, drop-shape (or drop weight) analyses should *never* be used for the determination of their molecular (i.e., microscopic-scale) γ_{ow}^{mic} (which term is interchangeable with γ_{ow}^o). However, this nevertheless leaves other approaches for obtaining γ_{ow}^o for polar organic liquids, *as well as for other polar solutes*.

THE AQUEOUS SOLUBILITY APPROACH
FOR DETERMINING γ_{ow}^o

There is for instance the *aqueous solubility* approach, using eq. XIII-3, which is an equilibrium method (van Oss and Good, 1996; van Oss *et al.*, 2002; 2005; van Oss and Giese, 2004). The solubility method was used for obtaining all the new, corrected

TABLE XIII-2

Updated List of the (new) Values of the Interfacial Tensions with Water (γ_{ow}^o) of Some Polar Organic Compounds, Compared with the Old (invalid) Data Derived from Drop-Shape (or drop weight) Data.

Polar organic compound	Old γ_{ow}[a] (mJ/m²)	New γ_{ow}^o[b] (mJ/m²)	Aqueous solubility (s)[c] (mMol fractions)	Contactable surface area (S_c) (nm²)	References
Chloroform	(31.6)	38.8	1.206	0.35	van Oss, Wu et al. (2001)
Diiodomethane	(48.5)	56.1	0.0806	0.34	van Oss, Wu et al. (2001)
Ethyl acetate	(6.8)	25.95	1.86	0.49	van Oss, Wu et al. (2001)
Ethyl ether	(10.7)	21.0	17.5	0.39	van Oss, Wu et al. (2001)
Benzene	(35.0)	41.6	0.404	0.38	van Oss, Giese and Good (2002)
Toluene	(36.1)	42.9	0.088	0.44	van Oss, Giese and Good (2002)
o-Xylene	(36.1)	44.0	0.019	0.50	van Oss, Giese and Good (2002)
Pentanol	4.4	22.5	5.32	0.47[f]	van Oss and Good (1996)
Hexanol	6.8	27.2	1.04	0.51[f]	Recent data (2004)
Heptanol	7.7	31.1	0.155	0.57[f]	van Oss and Good (1996)
Octanol	8.5	31.2	0.0810	0.61[f]	van Oss and Good (1996)
Cyclohexanol	3.9	18.5	6.48	0.55[f]	van Oss and Good (1996)
Sucrose[d]		5.65	35.1	1.20	Docoslis et al. (2000)
Glucose[d]		10.26	47.7	0.60	Docoslis et al. (2000)
Dextran[d] (T150)		−14.9	68.700[e]	14.0	Docoslis et al. (2000)

[a] Old values, from drop-shape or drop weight analyses (Girifalco and Good, 1957)

[b] New values, from aqueous solubilities and S_c-values, using eq. XIII-3.

[c] From Stephen and Stephen (1963), or from Handbook of Chemistry and Physics (1953)

[d] The data pertaining to these carbohydrates refer to the aqueous solutions, extrapolated to 100% concentration.

[e] From van Oss and Giese (2004)

[f] From the S_c values given for the corresponding alkanes, in Table XIII-1, plus 0 .10nm².

γ_{ow}^{o} values listed in Table XIII-2 (with the exception of dextran, which is also listed in Table XIII-2, for which γ_{ow}^{o} was derived using eq. IV-9. Here eq. XIII-3 was used to obtain an estimate of the exceedingly large theoretical solubility of this polymer, from its otherwise derived γ_{ow}^{o} and S_c values; cf. van Oss and Giese, 2004).

One of the more immediately useful results obtained with the solubility approach, is the new, corrected value for diiodomethane (DIM), see van Oss, Wu et al. (2001). DIM is one of the indispensable, almost completely apolar liquids, with a high enough surface tension ($\gamma_{DIM} = 50.8$ mJ/m², at 20°C), to be useful for measuring γ^{LW} on virtually all organic and bio-materials. Unfortunately DIM is ever so slightly polar (due to its iodine atoms), as can be deduced from its γ_{ow} value measured by drop-shape analysis (Girifalco and Good, 1957) of 48.5 mJ/m², which is distinctly somewhat on the low side, for its composition. For a monopolar, electron-acceptor liquid such as DIM this meant (also given its γ^{LW} of 50.8 mJ/m²), that its $\gamma_{DIM}^{\oplus}$ value should be about 0.7 mJ/m², which would be a non-negligible amount. However, using the corrected value for γ_{DIM-W}^{o} of 56.1 mJ/m², obtained from the aqueous solubility of DIM, and using eq. IV-9, one finds that in reality, $\gamma_{DIM}^{\oplus} = 0.009$ mJ/m², i.e., approximately 0.01 mJ/m² which, for most contact angle measurement purposes, is indeed negligible (van Oss, Wu et al., 2001). For instance for measurements of the value of γ^{LW} of PMMA (a monopolar electron-donating solid) with contact angles of DIM, and considering DIM to be non-polar, this would yield a contact angle, θ, which differs by only 1.3% from the "true value," which difference is of the same order as the normal margin of error of contact angle measurements and which would have yielded a value for γ_{PMMA}^{LW} of 42.0 mJ/m², instead of 41.4 mJ/m², a difference of 0.6 mJ/m2 (i.e., 1.4%), also within the normal error range (van Oss, Wu et al., 2001). For all practical purposes it is therefore permissible in most cases to use DIM as the contact angle liquid of choice for the determination of the LW component of solid surfaces, while considering it to be essentially apolar. If a greater accuracy than about ±1.4% is absolutely needed (which is normally reachable by averaging six or more contact angles), one can always take the γ_{DIM} to be equal to a γ^{LW} of 50.8 mJ/m² and in addition to have a $\gamma^{\oplus}$ of 0.01 mJ/m².

Table XIII-2 also shows that the use of the aqueous solubility approach for obtaining the interfacial tension with water of polar organic compounds is not limited to determining γ_{ow}^{o} of organic solvents that are only sparsely miscible with water, but it is equally applicable to very soluble organic molecules such as sugars (Docoslis et al., 2000; see also Chapter XXII). It should be mentioned here that dimeric or monomeric sugars have monopolar, electron-donating surfaces in the dry state, but that once they are dissolved in water they are dipolar, with a very elevated electron-donating, as well as a slightly lower but still quite respectable electron-accepting parameter, see Table XIII-2. For the polysaccharide, dextran (a polymer of maltose, which is itself a dimer of glucose) the same holds true, although its electron-donicity is somewhat smaller and its electron-acceptivity considerably smaller than those of its monomers, see also Table XIII-2.

USE OF THE CMC OF SURFACTANT MOLECULES FOR DETERMINING THEIR γ_{ow}^{o}

The aqueous solubility method also extends to surfactants, whose equivalent of aqueous solubility is their critical micelle concentration (CMC), as can be seen from the graph given by Adamson (1990, p. 509), so that for surfactants one may use (cf. eq. XIII-3):

$$-2\gamma_{ow}^{o} \cdot S_c = kT \cdot \ln CMC \qquad [\text{XIII-3A}]$$

see also Chapter XXII. It should however be realized that the nature of surfactant interactions with other condensed-phase materials or molecules is such that the use of the total interfacial tension between the complete surfactant molecule and water is *not* the most germane datum needed for the determination of the free energy of its interaction with another molecule or particle, when immersed in water. In all such cases one would, instead, have to know the interfacial tension with water of either the polar of the apolar moiety of the surfactant molecule, depending on whether the surfactant is needed for the hydrophobization of a hydrophilic, electrically charged particle, such as a clay particle (see Giese and van Oss, 2002; van Oss and Giese, 2003), or is required in order to obtain the solubilization or suspendability in water of a hydrophobic molecule or particle.

The γ_{ow} value of the apolar group is usually the easiest one to determine, from the known surface properties of, e.g., the alkane corresponding to the alkyl group of the surfactant, which is its apolar moiety. A later subsection of this Chapter treats the manner by which one obtains the $\gamma_{ow}^{o(apolar)}$ value from the γ_o^{LW} of an apolar organic compound. Then, once the $\gamma_{ow}^{o(apolar)}$ value of the apolar moiety is known (and expressed in kT units), it can be subtracted from the total [$\gamma_o^{o(total)}$] value of the complete surfactant (obtained from its CMC value, using eq. XIII-3A), to obtain the $\gamma_{ow}^{o(polar)}$ value of the surfactant's polar moiety:

$$\gamma_{ow}^{o(polar)} = \gamma_{ow}^{o(total)} - \gamma_{ow}^{o(apolar)} \qquad [\text{XIII-4}]$$

of which all terms should be expressed in kT units. This is the simplest procedure for obtaining $\gamma_{ow}^{o(polar)}$, in cases of non-ionic as well as ionic surfactants, because the direct derivation of $\gamma_{ow}^{o(polar)}$ is exceedingly laborious and fraught with assumptions. What severely complicates the direct derivation of the interfacial interaction energy between two polar moieties of an ionic (or of a non-ionic) surfactant, immersed in water, is the fact that due to, for instance, the electrical double layer repulsion (plus in addition the concomitant Lewis acid-base repulsion) between two electrically charged polar chains, these chains will separate from one another at its farthest at those parts of both chains that are the most distally situated with respect to their points of closest approach, which are next to their points of attachment to the apolar moieties of the surfactant molecules, which are bound to one another hydrophobically, at the minimum equilibrium distance (van Oss and Good, 1984). The same phenomenon exists between two non-ionic chains, which repel one another equally strongly by Lewis acid-base repulsion alone, when immersed in water. The analyses of these interactions between the polar chains of surfactant molecules entail the use of the extended DLVO (XDLVO) approach (see Chapters VII and XXIII), as well as a certain amount of guesswork as to the degree of oblique separation taking place between the two polar chains in each case. This is because for instance the ionic, polar moieties of two ionic surfactants (e.g. sulfate groups) branch sharply away from each other, to a distance which can only be roughly estimated, but which is needed to do an XDLVO analysis of the decay vs. distance of the polar chains (van Oss and Costanzo, 1992). With non-ionic surfactants a similar divergence occurs of the non-ionic polar chains [usually poly(ethylene oxide) or PEO chains], which divergence tends to be more gradual than is the case with ionic chains. This is because with ionic chains, one usually has a strong, practically single-point repulsion (e.g., between two SO_4 groups), whilst with non-ionic chains the regularly repeating electron-donating sites (e.g., oxygen atoms) on each one of the apposing chains repel each other over their entire length, making, e.g., PEO chains diverge at an angle, away from their point of closest attachment at their covalent links to the polar chains. At a certain distance from these points of attachment the polar chains will tend to cease diverging and ultimately to continue distally in a parallel manner (van Oss and Good, 1991a). However, as is visible in Figure XXII-1, a certain arbitrariness remains regarding the determination of the distance between the two parallel parts of the polar chains, as well as of the site on the polar chains where they cease to diverge and start running parallel to each other. These factors are difficult to estimate, as they require XDLVO analyses which have to be based in part on a number of unknown influences, such as, *inter alia*, the bending rigidity of the PEO chains, when dissolved in water.

CONTACT ANGLE-BASED METHODS
FOR DETERMINING γ_{ow}^o

Contact angle methods for determining interfacial tensions between condensed-phase materials or compounds (c) and water (w) can be based on *direct contact angle measurements*, combined with the Young-Dupré equation (eq. IV-13; see also Figure XII-1), as well as on *wicking*, also in conjunction with the Young-Dupré equation. In this manner the values for γ_C^{LW}, $\gamma_C^{\oplus}$ and $\gamma_C^{\ominus}$ can be obtained for any condensed-phase material or compound, C. From these three γ values the interfacial tension, γ_{CL}^o, with any liquid, L (including all cases where the liquid is water, w) can be obtained (cf. eq. IV-9), using the Young-Dupré equation three times, to solve for the three unknowns:

$$\gamma_{CL}^o = (\sqrt{\gamma_C^{LW}} - \sqrt{\gamma_L^{LW}})^2 + 2(\sqrt{\gamma_C^{\oplus}\gamma_C^{\ominus}} + \sqrt{\gamma_L^{\oplus}\gamma_L^{\ominus}} - \sqrt{\gamma_C^{\oplus}\gamma_L^{\ominus}} - \sqrt{\gamma_C^{\ominus}\gamma_L^{\oplus}}) \quad \text{[XIII-5]}$$

Finally, in such cases where γ_C is known (for instance where $\gamma_C = \gamma_C^{LW}$, when C is a *non-polar*, or a *monopolar solid*), one can use the *original Young equation* (see eq. IV-10):

$$\gamma_L \cos\theta = \gamma_C - \gamma_{CL}^o \quad \text{[XIII-6]}$$

thus permitting one, either by direct contact angle measurement with liquid, L, on condensed-phase solid, C, or via wicking, to determine γ_{CL}^o with only one measurement (which may however be repeated several times to attain greater accuracy, by averaging several observations).

XIV Different Approaches for Interpreting Contact Angles and Determining the Surface Tension and Surface Tension Components of Solids

THE CONCEPT OF SURFACE TENSION OF A SOLID

Even though the concept of surface tension of a solid is a much more difficult one than the concept of surface tension of a liquid, the definition of both of these is very elementary, i.e.:

$$\gamma_i \equiv -\frac{1}{2} \Delta G_{ii}^{coh} \qquad \text{[II-33]}^*$$

which simply defines the surface tension of any condensed material (i) as minus one half its free energy of (non-covalent) cohesion; see Chapter IV.

Apart from the obvious difference in deformability, an important difference between liquids and solids lies in the relative usefulness of γ_L compared to γ_S. The total surface tension of a liquid, γ_L, as a measure of its energy of cohesion is a crucial term in, and a *conditio sine qua non* for the solution of all variants of Young's equation (eqs. [IV-11, 12, 13, 28, 29 and XII-2–6]).

The total surface tension of a solid, γ_S, on the other hand, is usually irrelevant to most applications. This is not because γ_S does not exist, or is almost impossible to measure (Fowkes, 1988), but because, once γ_S has been determined, there is very little one can do with it. γ_S plays no role in obtaining values for γ_{SL} or ΔG_{SLS}^{TOT} (see Chapter IV) and it plays only a subsidiary role in obtaining ΔG_{SL}^{TOT} *via* the Dupré

---

* A list of symbols can be found on pages 399–406.

equation (eq. [IV-6B]), since the γ_{SL} term usually cannot be obtained from γ_S. One exception is the use of γ_S for obtaining γ_{SL}^o, by using Young's original equation (eq. XIII-6), discussed in Chapter XIII.

On the other hand, it should be understood that contrary to popular belief (see, e.g., Busscher *et al.*, 1986; Hamilton, 1974; Nyilas *et al.*, 1977; Andrade *et al.*, 1979; Dalal, 1987; Ström *et al.*, 1988; who use geometric means, or harmonic means, or multiplication factors, as combining rules for the polar surface tension component), the total surface tension of a solid generally has little correlation with its polarity or "hydrophilicity" (see, e.g., van Oss *et al.*, Chem. Revs., 1988). The problem with the use of geometric means, harmonic means, or multiplication factors for the polar surface tension component is that none of these treatments accounts for the common occurrence of a *zero polar surface tension component* among otherwise quite polar solids (or liquids), which are, in such cases, monopolar. Thus, according to:

$$\gamma_i^{AB} = 2\sqrt{\gamma_i^+ \gamma_i^-} \qquad\qquad [IV-5]$$

if $\gamma_i^\oplus = 0$ then $\gamma_i^{AB} = 0$ however large the value of $\gamma_i^\oplus$ may be. In all such cases which occur quite commonly, the contribution of γ_S^{AB} in:

$$\gamma_S = \gamma_S^{LW} + \gamma_S^{AB} \qquad\qquad [IV-2]$$

to γ_S then is non-existent, so that one frequently encounters polar solid materials, whose $\gamma_S = \gamma_S^{LW}$.

CRITICAL SURFACE TENSIONS OF SOLIDS

Zisman (1964) indicated a correlation between γ_S and the values of γ_L found by the extrapolation of $\cos \theta$ to $\cos \theta = 1$, when θ is measured with a number of liquids. However, Zisman was careful never to identify the value for the "critical surface tension", γ_c, thus found for the solid, with γ_S, as he was well aware of the potential difficulties encompassed in that approach. In the light of eqs. [II-39, IV-13, and IV-28] the validity of Zisman's approach holds true only for completely apolar liquids, in which case the determination of γ_S with only one apolar liquid is sufficient, yielding, however, only the value for γ_S^{LW}. As Good (1977) pointed out, the $\cos \theta$ values should be plotted vs. $\sqrt{\gamma_L^{LW}}$, and not vs. γ_L^{LW} as was done in the original Zisman approach (1964). Zisman plots retain a certain utility in the qualitative detection of polar properties of solid surfaces, by showing the unmistakable deviation from linearity among those points that have been obtained with polar liquids. Nonetheless, Zisman's pioneering work has been of importance in the general development of characterization methods of solid surfaces in many laboratories.

"EQUATION OF STATE" APPROACHES

Neumann *et al.* (1974), using Good's interaction parameter Φ (see eqs. [II-41 and IV-31] and Girifalco and Good, 1957) in the form:

$$\Phi = \frac{\gamma_S + \gamma_L - \gamma_{SL}}{2\sqrt{\gamma_S \gamma_L}} \qquad \text{[XIV-1]}$$

found that for a small number of fluorinated polymers, the straight line of plots of Φ vs. γ_{SL} was largely independent of the values of γ_S and γ_L, so that (for these cases) Φ could be empirically expressed as:

$$\Phi = 0.0075\gamma_{SL} + 1.00 \qquad \text{[XIV-2]}$$

Combining eqs. [XIV-1 and 2], Neumann now generalizes for all cases:

$$\gamma_{SL} = \frac{(\sqrt{\gamma_S} - \sqrt{\gamma_L})^2}{1 - 0.015 \ \sqrt{\gamma_S \gamma_L}} \qquad \text{[XIV-3]}$$

"which is an explicit formulation of (Neumann's) equation of state"; see Neumann (Chapter IX in van Oss et al., 1975). In conjunction with Young's equation:

$$\gamma_L \cos \theta = \gamma_S - \gamma_{SL} \qquad \text{[IV-10]}$$

eq. [XIV-3] yields:

$$\cos \theta = \frac{(0.015\gamma_S - 2)\sqrt{\gamma_S \gamma_L} + \gamma_L}{\gamma_L (0.015 \sqrt{\gamma_S \gamma_L} - 1)} \qquad \text{[XIV-4]}$$

Alternatively (Neumann et al., 1974; Neumann, Ch. IX in van Oss et al., 1975), using γ_S as an adjustable parameter (obtained, e.g., from modified Zisman plots), in conjunction with the corresponding values of γ_{SL}, values for Φ (eq. [XIV-1]) can be obtained, yielding plots of γ_L vs. Φ. From this, with the help of a computer program, a "best fit" set of values was generated, yielding, γ_S and γ_{SL} from any given value for Φ and γ_L, with only one liquid L, with the main rider that $\gamma_L > \gamma_S$ (Neumann et al., 1974). A strange requirement for the validity of this approach is the need to stipulate that *surface tensions may not be decomposed into different components* (i.e., denying the right to subdivide into apolar and polar components, as most other authors have been proposing, following Fowkes, 1964). A direct consequence of the negation of the existence of surface tension components is the necessity, perceived by Spelt et al. (1986), for the contact angles θ, measured on a polar surface with a set of liquids of different polarity, but with similar surface tensions γ_L, to be similar. Spelt (1985) and Spelt et al. (1986) tried to prove this experimentally, and believed that they succeeded. Using various pairs of liquids of identical γ_L, where one liquid was fairly apolar and the other polar,* they indeed found the same contact angle θ on a given

---

* e.g., 1-methyl naphthalene and methylsalicylate: ($\gamma_L \approx 39.0$); pentadecane and heptaldehyde: ($\gamma_L \approx 25.5$); benzaldehyde and dibenzylamide: ($\gamma_L \approx 42.0$).

(apolar) surface. However, as already briefly indicated by Fowkes (1987), these "polar" liquids used by Spelt *et al.* (1986) were in fact non–self-hydrogen bonding or, in our terms, monopolar, giving rise to interactions of classes I, II or IV (see Chapter IV and Fig. IV-1), which of course mimic completely apolar interactions. The papers by Spelt *et al.* (1985, 1986 and 1987) were discussed in detail by van Oss *et al.* (Langmuir, 1988).* Busscher *et al.* (1986, Table IV), showed discrepancies between γ eq. state (using the approach of Neumann *et al.*, 1974) obtained with water, and γ eq. state (using, again, the approach of Neumann *et al.*, 1974), ranging from +47% to –43%. A comparison can be made between the zero time dynamic interfacial tension between a number of organic liquids and water (γ_{iw}^{o}) and the same interfacial tension derived from these liquids' γ_i values and γ_w (of water), using the "equation of state" to compute γ_{iw}^{ES} [see also van Oss, Wu *et al.* (2001); van Oss, Giese and Good (2002)], showing discrepancies ranging between –86% to +102%, see Table XIV-1.

The computer program devised by Neumann *et al.* (1974) explicitly excludes negative values of γ_{12}, which flaw it has in common with the "γ^P" approach; see the following two sections, below. Thus the results obtained with the "equation of state," while fairly reasonable in some instances, are quite aberrant in others, especially in extremely polar systems; see Chapter XIX. However, the major reason for the impossibility of the task, set by the "equation of state" approach (i.e., to obtain γ_S [and γ_{SL}] values from contact angle [θ] measurements with only one liquid [L] on a solid surface [S]), lies in the fact that the approach attempts to solve for *three* independently variable *unknowns* (γ_S^{LW}, $\gamma_S^{\oplus}$ and $\gamma_S^{\ominus}$) with only *one equation*. Apart from the obvious futility of such an attempt, it should be realized that even *if* γ_S could be determined by such an approach, there is relatively little one would be able to do with that entity; see the beginning of this chapter. Table XIV-1 (see also van Oss *et al.*, Langmuir, 1988) illustrates strikingly the contrast between the measured interfacial tensions between water and various organic solvents and the interfacial tensions with water for these cases (calculated by means of the "equation of state").

Van de Ven *et al.* (1983) earlier demonstrated the lack of foundation for Neumann *et al.*'s "equation of state" (1974), on fundamental theoretical grounds. Another variant of an "equation of state" of this type was published by Gerson (1982); it

---

* For the measurement of surface tensions of living cells by contact angle determination with only one liquid, i.e., physiological saline water, the possibility of using an "equation of state" of the type described by Neumann *et al.* (1974) appeared at the time to be an attractive solution to safeguarding the physiological integrity of cells, while measuring their surface characteristics, and was therefore adopted by the present author and his collaborators during the early 1970s (see, e.g., van Oss *et al.*, 1975). Although Fowkes as early as 1963 had pointed the way to a more appropriate solution to the problem of polar surface tensions, it was only in the mid-1980's that the form which the precise solution to the problem was going to assume became clear, with the work of Chaudhury (1984) as a starting point. The author of this book and a number of his colleagues then embarked on a collaborative program with Dr. Chaudhury, which rapidly allowed the solution of the problem to emerge (van Oss *et al.*, Advan. Coll. Interf. Sci., 1987; Chem. Rev., 1988; van Oss and Good, J. Prot. Chem., 1988; J. Disper. Sci. Tech., 1988; van Oss, Roberts and Good, 1989; van Oss, Ju *et al.*, 1989; van Oss, Good and Busscher, 1989; van Oss, Cell Biophys., 1989; Biophys. of the Cell Surface, 1990; van Oss, Giese and Costanzo, 1990).

TABLE XIV-1
Zero Time Dynamic Interfacial Tensions (γ^o_{1w}) Between Water and a Number of Low-Aqneons-Solubility Organic Liquids, Obtained from Their Solubilities, Compared with the Interfacial Tensions for the Same Systems (γ^{ES}_{1w}) Derived from the Surface Tensions (γ_1) of These Organic Liquids and of Water, Using an "Equation of State" (Neumann et al., 1974; 1980). All γ-Values are Expressed in mJ/m^2. From van Oss, Wu et al. (2001); van Oss, Giese and Good (2002).

Organic liquid	γ_1 [a]	γ^o_{1w} [b]	γ^{ES}_{1w} ("eq. of state")	Difference between γ^o_{1w} and γ^{ES}_{1w}
Chloroform	27.3	**38.8**	30.3	−14%
Benzene	28.9	**41.6**	28.7	−31%
Diiodomethane	50.8 [c]	**56.1**	7.8	−86%
Ethyl acetate	24.0	**25.9**	33.5	+29%
Ethyl ether	17.1	**21.0**	40.5	+93%
n-Octane	21.6	**51.0** [d]	35.6	−30%

[a] From Jasper (1972)
[b] From aqueous solubilities; see Table XIII-2
[c] See Table XII-4
[d] See Table XIII-1

suffers of the same fatal flaws as the "equation of state" mentioned above, for the same reason; see also Johnson and Dettre (1989) for further experimental proof of the incorrectness of the "equation of state," and Morrison (1989) for an analysis of the errors in its thermodynamic assumptions.

THE CONCEPT OF A SINGLE POLAR SURFACE TENSION COMPONENT, "γ^P", IN CONJUNCTION WITH A GEOMETRIC MEAN COMBINING RULE

There are a number of authors (e.g., Owens and Wendt, 1969; Kaelble, 1970; Hamilton, 1974; Andrade et al., 1979; Janczuk et al., 1990) who, probably for intuitive reasons, implicitly postulated the existence of only *one* general property of polar molecules (designated by superscript, P), and who preferred a geometric mean combining rule for the polar interfacial tension, analogous to that used for γ^{LW} (eq. IV-2) so that:

$$\gamma^P_{12} = (\sqrt{\gamma^P_1} - \sqrt{\gamma^P_2})^2 \qquad \text{[XIV-5]}$$

As shown in Chapter II the geometric mean combining rule for γ^{LW} is accurate to within 2% (Chaudhury, 1984; van Oss et al., J. Colloid Interf. Sci., 1986). However, it can be demonstrated that the geometric mean combining rule (eq. [XIV-5]) is not only erroneous from a theoretical viewpoint (see Chapter IV) but that the application of the "γ^P" approach to various practical situations can be spectacularly unpredictive.

The crucial point to note is that, as the value of γ^P_{12} is supposed to be derived from the square of the difference between two entities (following the geometric mean rule; see eq. [XIV-5]), *it always must have a positive value*, so that ΔG_{121}, which is equal to $-2\gamma_{12}$, then must *always be negative* (cf. eqs. [II-35 and IV-2]). Thus if instead of γ^{AB} (eq. [IV-7]), γ^P, as defined in eq. [XIV-5], is used, the interaction between two molecules or particles (1) immersed in a liquid (2) *cannot be otherwise but attractive*, regardless of whether these molecules of particles (or the liquid) are polar or apolar. Especially in the case of water-soluble macromolecules, such an attraction must either give rise to insolubility, or to a very low solubility (in the limiting case where $\Delta G_{121} \rightarrow 0$), which is obviously not the case.

It should also be pointed out that if the geometric mean, "γ^P"-approach, is followed, Young's equation (cf. eq. [IV-1]) for polar materials becomes (instead of eq. [IV-13]):

$$(1+\cos\theta)\gamma_L = 2(\sqrt{\gamma^{LW}_S \gamma^{LW}_L} + \sqrt{\gamma^P_S \gamma^P_L}) \qquad [XIV-6]$$

COMPARISON BETWEEN THE γ^{AB}_{12} AND THE "γ^P_{12}" APPROACHES

Using the γ^{AB} approach (Chapter IV) and parallel to it, the "γ^P" approach to determine ΔG_{121} in polar media, it is useful to compare the applicability of both of these to the solubility of polymers in water, which is the most polar of liquid media. The solubilities in water (w) of most common water-soluble polymers, including, e.g., poly(ethylene oxide) (PEO), and dextran (a linear polymer of glucose) (DEX), may be expressed as

$$\Delta G_{iwi} S_c = kT \ln s \qquad [XIV-7]$$

where S_c is the contactable surface area between two adjoining solute molecules immersed in solvent, s the solubility (in mol fractions) and $kT = 4.05 \times 10^{-21}$ J, at 20°C (cf. Chapter XXII). Then, taking PEO and DEX specimens of various molecular weights, their solubilities can be calculated according to the γ^{AB}_{iw} approach (eq. [IV-7]), as well as according to the "γ^P_{12}" approach (eq. [XIV-5]), using the contact angles found experimentally for PEO and DEX, and the two different versions of the Young equation (eq. [IV-13] for the γ^{AB}_{iw} approach and eq. [XIV-6] for the "γ^P_{12}" approach); see Table XIV-2 (see also van Oss and Good, 1992).

It becomes clear from Table XIV-2 that of the two possibilities here considered, only the Lewis acid-base approach to the prediction of solubilities of polymers in water yields solubility values which are compatible with the solubilities found experimentally, for typical very water-soluble polymers such as poly(ethylene oxide) and dextran. The "γ^P" approach predicts solubility values which are up to several

TABLE XIV-2

Solubilities of Water-Soluble Polymers: Values Calculated _via_ the γ^p Approach and _via_ the Lewis Acid-Base Approach (using eq. [XIV-7] in both cases) Compared with Experimental Solubilities

Polymer	Degree of polymerization n	M_n	γ^p-approach				Lewis acid-base approach				s obtained experimentally (%)
			γ_{SW} (mJ/m²)	ΔG_{1w1} (kT)	S_c (nm²)	s (mol fractions)	γ_{SW} (mJ/m²)	ΔG_{1w1} (kT)	S_c (nm²)	s^h (%)	
PEO	23	1,000	5.77[a]	−10.8	3.8[b]	2.0×10^{-5}	−26.23[c]	+2.75	0.212[d]	>> 50	≥ 65
	136	6,000	5.77[a]	−64.5	22.6[b]	9.7×10^{-29}	−26.23[c]	+2.75	0.212[d]	>> 50	≥ 65
DEX	12	2,160	6.0[e]	−14.2	4.8[f]	6.8×10^{-7}	−11.3[g]	+2.23	0.4[f]	>> 50	> 50
	55.6	10,000	6.0[e]	−66.5	22.2[f]	1.3×10^{-29}	−11.3[g]	+2.23	0.4[f]	>> 50	> 50

[a] From $\gamma_{SW}^P = 32$ mJ/m², obtained from $\gamma_S^{LW} = 43$ mJ/m² and $\theta_{H_2O} = 18°$ (measurements done on PEO 6,000) (van Oss et al., Adv. Colloid Interface Sci., 1987; van Oss, Arnold et al., 1990a), using eqs. XIV-5 and 6.

[b] 0.166 nm² per monomeric subunit (van Oss, Arnold et al., 1990a).

[c] From $\gamma_S^\oplus = 0$, $\gamma_S^\ominus = 64$ mJ/m² and $\gamma_S^{LW} = 43$ mJ/m² (measurements done on PEO 6,000) (van Oss, Arnold et al., 1990a), using eqs. IV-9 and 13.

[d] See van Oss, Arnold et al., 1990a.

[e] From $\gamma_{SW}^P = 31$ mJ/m², obtained from $\gamma^{LW} = 42.9$ mJ/m² and $\theta_{H_2O} = 21.17°$ (measurements done on DEX 10,000).

[f] From S_c (monomeric unit) = 0.4 nm².

[g] From $\gamma_S^{LW} = 42.9$ mJ/m², $\gamma_S^\oplus = 1.0$ mJ/m² and $\gamma_S^\ominus = 47.4$ mJ/m², obtained by contact angle measurements on DEX 10,000: $\theta_{Diiodomethane} = 31.25°$; $\theta_{\alpha Bromonaphthalene} = 19.0°$; $\theta_{H_2O} = 21.17°$; $\theta_{Glycerol} = 30.75°$; $\theta_{Formamide} = 11.9°$

[h] In the case of extremely soluble solutes (with $\Delta G_{1w1} > 0$), i.e., with a solubility of 1 Mol fraction or greater, the precise aqueous solubility becomes too large to be meaningfully expressed in terms of Mol/L, or %; see Chapter XXII. From van Oss and Good, 1992.

TABLE XIV-3

Solubility of Polar Polymers-Comparison of the Acid-Base and the "γ^P"
Approaches

Lewis acid-base approach	**"γ^P" approach**
$\gamma = \gamma^{LW} + \gamma^{AB}$	$\gamma = \gamma^{LW} + \gamma^{P}$
Interfacial tension:	*Interfacial tension:*
$\gamma_{12}^{LW} = (\sqrt{\gamma_1^{LW}} - \sqrt{\gamma_2^{LW}})^2$	$\gamma_{12}^{LW} = (\sqrt{\gamma_1^{LW}} - \sqrt{\gamma_2^{LW}})^2$
$\gamma_{12}^{AB\,a} = 2(\sqrt{\gamma_1^{\oplus}} - \sqrt{\gamma_2^{\oplus}}) \times (\sqrt{\gamma_1^{\ominus}} - \sqrt{\gamma_2^{\ominus}})$	$\gamma_{12}^{P} = (\sqrt{\gamma_1^{P}} - \sqrt{\gamma_2^{P}})^2$
$\gamma_{12} = (\gamma_{12}^{LW} + \gamma_{12}^{AB})^{a}$	$\gamma_{12} = (\gamma_{12}^{LW} + \gamma_{12}^{P})^{b}$
[a] can be negative	[b] cannot be negative
The Young-Dupré equation:	*The Young-Dupré equation:*
$(1+\cos\theta)\gamma_L = 2(\sqrt{\gamma_S^{LW}\gamma_L^{LW}}$	$(1+\cos\theta)\gamma_L = 2(\sqrt{\gamma_S^{LW}\gamma_L^{LW}}$
$+\sqrt{\gamma_S^{\oplus}\gamma_L^{\ominus}} + \sqrt{\gamma_S^{\ominus}\gamma_L^{\oplus}})$	$+\sqrt{\gamma_S^{P}\gamma_L^{P}})$
$\Delta G_{121} = -2\gamma_{12}$	$\Delta G_{121} = -2\gamma_{12}$
ΔG_{121} can be positive, denoting the possibility of a repulsion; i.e., stability or solubility of 1, immersed in liquid, 2.	ΔG_{121} cannot be positive. This implies instability or insolubility of 1, immersed in liquid, 2; i.e., water-soluble polymers would have to be insoluble in water.

dozen decimal orders of magnitude (for PEO as well as for DEX) too small (van Oss and Good, 1992; van Oss and Giese, 2004).

The fundamental reason for the staggering inapplicability of the "γ^P" approach to the behavior of polar polymers in aqueous media lies in the constraint which makes it impossible for γ^P to assume a negative value, which in its turn excludes any possibility for ΔG_{121} to have a positive value,[*] i.e., the "γ^P" approach excludes any and all repulsive interactions. For neutral polymers to be soluble in water, a positive value for ΔG_{121} (or in the limiting case a value close to zero) is a *conditio sine qua non*. The differences between the Lewis acid-base and the "γ^P" approaches are summarized in Table XIV-3.

It may be useful to note that for electron-donor monopolar compounds (such as PEO), comparison between eqs. [IV-13] and [XIV-6] will show that (for $\gamma_i^{\oplus} = 0$) $\gamma_i^{\ominus} = 2\gamma_i^P$. For the purpose of denoting hydrophilicity (see Chapter XV), the difference between $\gamma_i^{\ominus}$ and "γ^P" may not appear terribly important as long as it is realized that under the conditions of monopolarity, γ^{AB} remains zero. The fatal error arises in using eq. [XIV-5] instead of eq. [IV-7] for determining stability or solubility (see also Table XIV-3 and van Oss and Good, 1992), and for determining free energies of interaction in polar systems in general. It is therefore advisable to avoid using the "γ^P" approach in all cases.

---

[*] For the same reason, the use of a harmonic mean (Andrade *et al.*, 1979) leads to equally erroneous results.

It should be noted that a modification of Cassie's equation (eq. [XII-2]), proposed by Israelachvili and McGee (1989) also involves a treatment comprising the "γ^P" approach; this is discussed in Chapter XII.

SEDIMENTATION OF PARTICLE SUSPENSIONS

By this approach one attempts to utilize the condition where ΔG_{131}^{TOT} reaches a maximum (positive value), in order to determine the γ_1 of solid particles (1), suspended in a liquid (3). Under certain conditions, ΔG_{131}^{TOT} is at a maximum, when ΔG_{131}^{LW} is at a maximum, i.e., when $\Delta G_{131}^{LW} \to 0$ (ΔG_{131}^{LW} cannot be positive; it should be remembered, however, that ΔG_{131}^{AB} *can be positive*). Such conditions can be observed to exist at the point of maximum sedimentation volume of suspensions of apolar particles in a series of apolar liquids of varying γ_L values. Such a series of liquids can be made, e.g., by mixing two apolar liquids in different proportions. The tube in which (for the same number of apolar particles) the largest sedimentation volume can be observed then contains the apolar liquid (3), for which $\gamma_3^{LW} = \gamma_1^{LW}$. Determinations of this type have been done with relatively (but not completely) apolar nylon 6,6 particles, suspended in mixtures of relatively (but by no means completely) apolar liquids, i.e., n-propanol + thiodiethanol (Neumann *et al.*, 1984). A γ_1^{LW} value of 39.6 mJ/m^2 was found for these particles, by that approach, which is fairly close to the best value of 36.4 mJ/m^2, based on many contact angle measurements done with various liquids by R.E. Baier and calculated, using eq. [IV-13]; see Tables XII-3 and XVII-5 to 7 and 8A. However, the difference of 3.2 mJ/m^2 between the two approaches is probably significant, and indicative of the influence of the polarity of the particles and of the liquid mixtures used. The two main reasons why this approach should not be used in polar systems are:

1. The distance ℓ between particles at maximum stability is usually of the order of $\ell \approx 5$ to 10 nm, where the rates of decay of ΔG_{131}^{LW} and ΔG_{131}^{AB} obey very different laws (see Chapter VII) so that at ΔG_{131}^{TOT} = max., at that equilibrium value ℓ, one has multiple unknowns and only one equation.
2. Mixtures of polar solvents have unreliable surface tensions, for utilization in Young's equation, because one of the two polar components invariably orients toward the liquid-air interface, so that the measurable liquid surface tension is no longer proportional to the energy of cohesion of the liquid (Good and van Oss, 1983; Docoslis *et al.*, 2000; see also Chapter XI).

Another varient of this approach was an attempt to learn the γ_1 value of living (or killed) cells (1), suspended in, or on, a liquid (3). Here, ΔG_{131}^{TOT} was held to have the lowest (negative) value when the highest cell concentration could be loaded onto a liquid density gradient, before "droplet sedimentation" was observed (Omenyi *et al.*, 1982, 1985, 1986; Omenyi, Snyder *et al.*, 1981; Neumann *et al.*, 1984) In all these cases the conditions of onset of "droplet sedimentation" were studied with glutaraldehyde-fixed erythrocytes layered on D_2O-H_2O gradients of various aqueous mixtures of dimethyl sulfoxide (DMSO). Maximum stability was usually observed at 12–15% (v/v) DMSO (depending on the species of origin of the erythrocytes used). While the results of these studies remain of interest because of various aspects of the observed influence of, e.g., added trivalent counterions (Omenyi et al., 1985) or cell ζ-potentials (Omenyi et al., 1982, 1986) on the stability of these cell suspensions, it probably is erroneous (as well as meaningless) to conclude that γ_1 of human erythrocytes is equal to 61 mJ/m^2, because that is the measured surface tension of a 14% v/v DMSO solution, in which these cells showed maximum stability. On the other hand, DMSO in aqueous solutions probably only orients slightly at the liquid-air interface, so that the γ_L^{LW} values may still be roughly estimated. At 14% (v/v) DMSO $\gamma_3^{LW} \approx 25$ to 26 mJ/m^2 (van Oss et al., 1986a), and measurements on hydrated human (native) erythrocyte stromata (van Oss and Cunningham, 1987, unpublished observations; van Oss, Cell Biophys, 1989) yield $\gamma_1^{LW} \approx 26.5$ mJ/m^2. Thus, the cell stability approach may possibly yield rough data for the γ_1^{LW} of cells. It would, however, be dangerous to try to extend this method to estimate polar surface tension parameters, because even though we know that, e.g., in the above case, ΔG_{131}^{AB} is at a maximum, we do not know precisely how high the value of that maximum is. We thus lack both of the two equations (i.e., eq. [IV-17]) needed for determining $\gamma_1^{\oplus}$ and $\gamma_1^{\ominus}$, even if ΔG_{131}^{AB} maxima could be established with two different, completely characterized polar liquids. Fuerstenau et al. (1991) and Diao and Fuerstenau (1991) attempted to ascertain the γ_S value of the solid surface of particles by flotation experiments in liquids of different surface tensions. This approach is analogous to the sedimentation stability experiments described in this section (above) and has the same drawbacks, for the same reason.

In summary, the sedimentation volume method is the only of this class of approaches that may be recommended, and that only for the determination of γ_1^{LW} of particles, by using completely apolar liquids. Nevertheless, it should be remembered that even in mixtures of totally apolar liquids, some of the molecules of the liquid with the lowest surface tension will preferentially migrate toward the air interface; see Chapter XIII.

ADVANCING SOLIDIFICATION FRONTS

This approach endeavors to utilize conditions where $\Delta G_{132}^{TOT} \rightarrow 0$, in order to ascertain the γ_1 of solid particles (1), immersed in a liquid (3), which is gradually being solidified into a solid (2) by cooling. Clearly, when such an advancing solidification front (2) pushes particles (1) in front of it, $\Delta G_{132}^{TOT} > 0$, and on the other hand when particles (1) are engulfed by the front as it advances, $\Delta G_{132}^{TOT} < 0$.

In apolar systems, where $\Delta G_{132}^{TOT} = \Delta G_{132}^{LW}$, we can use the Dupré equation:

$$\Delta G_{132}^{LW} = \gamma_{12}^{LW} - \gamma_{13}^{LW} - \gamma_{23}^{LW} \qquad \text{[II-37]}$$

in which ΔG_{132}^{LW} becomes equal to zero when either $\gamma_1^{LW} = \gamma_3^{LW}$ or $\gamma_2^{LW} = \gamma_3^{LW}$. In other words, in apolar systems, when particles (1) immersed in a liquid (3) are neither pushed nor engulfed by the advancing freezing front of the solidifying melt (2), $\gamma_1^{LW} = \gamma_3^{LW}$, which, in principle, is another method for determining the γ_1^{LW} of particles.* This

---

* This presupposes that $\gamma_1^{LW} \neq \gamma_2^{LW}$, which is practically always the case.

method was first tested in 1976 (see Omenyi and Neumann, 1976; and Neumann *et al.*, 1979), and described in detail by Omenyi (1978). Some of the materials studied by Neumann *et al.* (1979) can be reexamined, and recalculated *via* eqs. [IV-13 and IV-16] to circumvent the results calculated with the flawed "equation of state" approach. The rejection of nylon 6,6 (which was thought to be an anomaly, in the earlier work) is now easily accounted for, as a positive value for ΔG_{132}^{TOT} of 0.08 mJ/m^2 is obtained, using the surface tension data from Chapter XVII. However, this value was determined from data from a different nylon 6,6 sample than the one used by Omenyi in 1978. In the case of polystyrene, using the data from Chapter XVII, one obtains $\Delta G_{132}^{TOT} = -0.13$ mJ/m^2, which agrees with engulfment. Results on other particles studied by Neumann *et al.* (1979) are less easy to recalculate and compare, as insufficient data are available concerning them, at this time. It would appear, however, that this methodology might serve for the determination of upper and lower limits of the γ^{LW} values of apolar, or monopolar (electron-donor) particles using the equally monopolar naphthalene and other comparable systems. A varient of this approach is the phagocytosis by human polymorphonuclear leukocytes of non-opsonized bacteria, when suspended in liquid media of different surface tension (Neumann *et al.*, 1982). A minimum in bacterial uptake was observed to occur in the liquid which then was presumed to have the same surface tension as the bacteria. As these interactions occur in very polar media, this approach has the same drawbacks as those already outlined above.

The method has been expanded by Omenyi (1978), by using the concept of "critical velocity" of the advancing solidification front; see also Omenyi *et al.* (1981). This implies somewhat greater velocities than the relatively slowly advancing front approach described, e.g., by Neumann *et al.* (1979), and it appears to involve interactions between the advancing solidification front and the particles, at a distance, ℓ, where ℓ is of the order of 10 nm (Neumann and Francis, 1983). In that case, interpretation of polar interactions becomes exceedingly difficult, as ΔG^{LW} and ΔG^{AB} decay with distance ℓ according to totally different regimens (see the preceding section, and Chapter VII). Thus, the "critical velocity" advancing solidification front approach also is, at best, only applicable to the determination of the γ^{LW} value of particles, under carefully controlled conditions, using apolar liquids. However, determination of γ_1^{LW} values of particles is experimentally much simpler, and probably more accurate, by using the maximum sedimentation volume approach, described in the preceding section, or especially the wicking method, treated in Chapter XII.

Nonetheless, observations on either inclusion or exclusion of different well-characterized particles or macromolecules by (slowly) advancing icefronts can furnish indications on some of the surface properties of ice; see, *e.g.*, van Oss, Giese and Norris (1992); van Oss, Giese, Wentzek *et al.* (1992). From the partial inclusion and partial exclusion of hydrophilic montmorillonite (SWy-1) particles by a slowly advancing ice front it could be concluded that, for ice at 0°C, the value of $\gamma_i^{\oplus}/\gamma_i^{\ominus}$ could be pinpointed as being situated very close to 0.50 (as compared to the value for water at 20°: $\gamma_W^{\oplus}/\gamma_W^{\ominus} = 1.00$). That datum proved essential in determining the polar surface tension parameters of ice at 0°C (van Oss, Giese, Wentzek, *et al.*, 1992; see also Chapter XXIV and Table VIII-1 for different results).

ADHESION METHODS

Another approach to using $\Delta G_{132}^{TOT} \rightarrow 0$ is to determine the point of minimum adherence of particles (or adsorption of polymers) to:

1. a well-defined surface material in at least three different well-characterized polar liquids, *or*:
2. three different well-defined surface materials, in a well-characterized polar liquid.

A combination of approaches 1 and 2 was proposed by Neumann, Absolom *et al.* (1979), using cell adhesion onto various polymer surfaces, from suspensions in various DMSO-water mixtures, and by van Oss *et al.* (1981), using protein adsorption onto various polymer surfaces, from solutions in various DMSO-water mixtures. Although γ_1 determinations done by this approach are very laborious, one advantage is that, theoretically at least, a high accuracy might be reached thanks to the over-abundancy of data points. In a critique of the approach, in the light of the more recent studies of polar interactions (see Chapters IV and XXIV), the following considerations must, however, be taken into account:

1. The γ_2-values of the solid (polymer) surfaces were, unfortunately, obtained via the "equation of state," so that practically all of these would have to be determined anew. Also of course, knowing the γ_2 of these solid polymers is not helpful, because actually γ_2^{LW}, $\gamma_2^{\oplus}$ and $\gamma_2^{\ominus}$ must be determined, and not γ_2.
2. As already discussed in this chapter, in the section on Sedimentation of Particle Suspensions, the use of mixtures of polar liquids is fraught with pitfalls; see also Chapter XII.
3. A minimum in adhesion or adsorption may well not correspond precisely to $\Delta G_{132}^{TOT} = 0$, but to $\Delta G_{132}^{TOT} = x$, where x has a finite value.
4. The presence of progressively higher concentrations of DMSO is likely to have an increasingly dehydrating influence. Thus at higher DMSO concentrations cells as well as proteins no longer have the same γ_1^{LW}, $\gamma_1^{\oplus}$ and $\gamma_1^{\ominus}$ values as at low or zero DMSO.

Point #3 can be important, as this approach was geared to the proposition that when $\Delta G_{132} = 0$, $\gamma_1 = \gamma_3$, which allows one to obtain the value for γ_1, i.e., with the liquid with value γ_3, in which ΔG_{132}^{TOT} becomes zero (or is at a minimum) for all values of γ_2. If on the other hand, ΔG_{132}^{TOT}, at the minimum of adhesion or adsorption, is equal to a finite non-zero value, $\gamma_1 \neq \gamma_3$ (it should be kept in mind that γ_{13} can assume a negative value [see Chapter IV] and that this actually occurs with, and persists at, the interface between the liquid medium and, e.g., human blood cells; see van Oss, Cell Biophys., 1989). For the adhesive properties of surfactants, see van Oss and Costanzo (1992).

A *conditio sine qua non* for the utilization of the adhesion (or adsorption) approach is to be able to ascertain the exact value of ΔG_{132}^{TOT}. There are two approaches to this:

1. $\Delta G_{132}^{TOT} = 0$. This occurs in reversed-phase liquid chromatography of, e.g., proteins, at the point where the protein begins to appear in the eluate (van Oss, Isr. J. Chem., 1990; see also Chapter XXIV).

2. $\Delta G_{132}^{TOT} = x$, where x may have a positive or a negative value, it is of course easiest to determine a negative value of x, corresponding to a finite degree of adsorption. If one knows the total concentration of adsorption sites [S], the total concentration of free $[P_f]$ and of absorbed $[P_a]$ protein, $x = \Delta G = -RT.\ln K_{ass}$, where R is the gas constant (R = 8.314 J per mole per degree Kelvin), T the absolute temperature in degrees Kelvin, the association equilibrium constant, K_{ass}, can be determined as follows:

$$K_{ass} = \frac{[P_a]}{[P_f] \cdot ([S] - [P_a])} \qquad [XIV\text{-}8]$$

(cf. eq. [XXI-4]).

Especially the second approach may well prove feasible for the determination of the surface tension components and parameters of inert particles, using at least three different, well-characterized liquids, of which two must be polar, using eq. [IV-16] three times. For living cells or for proteins the method augurs less well, as in those cases the use of mixtures of polar solvents with water is virtually imperative, with all their drawbacks. It should, moreover, also be realized that in most cases, ΔG^{EL} is not negligible and thus must also be incorporated in ΔG_{132}^{TOT}.

Finally, see also Chapter XII, where the necessity is discussed for taking both macroscopic-scale repulsions and microscopic-scale attractions into account. Usually only the former interactions can be predicted via contact angle measurements. The latter interactions are more specific; their study needs a different approach.

OTHER APPROACHES

γ^{LW} AND HAMAKER CONSTANTS

Using eqs. [II-16 and II-33], one can derive γ^{LW} values for any compound, once its Hamaker constant is known; see Chapter III. Usually it is much more convenient to establish a compound's γ^{LW} value than to derive its Hamaker constants (see Chapter III), but in exceptional cases, the reverse process may be used. For instance, Hamaker constants of pure mica can be measured (Pashley, 1981), by force balance methods (see, e.g., Chapter XVI), and/or calculated (Israelachvili, 1985), yielding a value for $\gamma_{mica}^{LW} \approx 54$ mJ/m². This is somewhat higher than the value obtained by means of contact angle measurement with diiodomethane (DIM), which is one of the apolar liquids with the highest γ_L value. Even on extremely dry and freshly split mica, early contact angles with DIM of at least 20° are found, yielding $\gamma_{mica}^{LW} \approx 48$ mJ/m². Very soon thereafter, however, θ_{DIM} on mica increases to about 40°, which lowers γ_{mica}^{LW} to about 40 mJ/m²; this is probably due to the rapid rehydration of the mica surface, after initial drying. However, one must not lose sight of the fact that the Hamaker constant for mica found with the force balance was found with mica *immersed in water* (Pashley, 1981; Israelachvili, 1985), although the hydration layer of mica is perhaps not readily noticeable, under water, in the force balance.

PHASE SEPARATION

By means of phase separation observations between two polymer solutions, in the same solvent, or between a polymer solution and a particle suspension, in the same solvent, polar parameters of the surface tension of polymers, particles, or liquids, may be estimated (van Oss *et al.*, Separ. Sci. Technol., 1989); see also Chapter XXI.

ELECTROPHORETIC MOBILITY

The electrophoretic mobility of polar particles in (monopolar) *organic solvents* is a measure of the particles' polar parameter of the opposite sign of that of the solvent (Labib and Williams, 1983, 1986; Fowkes, 1987); see also Chapters XV and XIX. However, the correlation between the polar parameters found for various compounds by this approach is as yet more qualitative than quantitative.

CODA

A number of older interpretations of contact angle measurements are discussed. Some of these, like the Zisman and the particle sedimentation approaches, have been useful in their time, but are now somewhat limited in their possibilities. Other approaches, while seemingly logical in their day, are now known to be erroneous, e.g., the "equation of state" approach and the "γ^P" approach (which yields especially disastrous results when applied to the aqeous solubility of polymers). Also discussed are: methods using advancing solidification fronts (tricky, but sometimes useful); adhesion methods (dangerous when using mixtures of liquids); phase separation approaches, stability methods and electrophoretic mobility measurements in organic liquids.

XV Electrokinetic Methods

Each of the main five electrokinetic phenomena, i.e., electrophoresis, electroosmosis, electroosmotic counter pressure, streaming potential, and migration potential (see Chapter V, and Fig. V-1) can be used to determine the electrokinetic, or ζ-potential which, in its turn, serves to calculate ΔG^{EL}. However, only electrophoresis, electroosmosis and streaming potentials, in that order, are of practical importance in ζ-potential determination. For a detailed overview of electrokinetic methods, see Righetti *et al.* (1979).

MICROELECTROPHORESIS OF PARTICLES AND CELLS

Electrophoresis of particles, using direct microscopic observation of their electrophoretic mobility, is one of the oldest approaches to quantitative electrokinetic analysis. Until the development of moving boundary electrophoresis, and later of zone electrophoresis, microelectrophoresis was the only method available for the determination of the ζ-potential of dissolved polymers (e.g., proteins), which was measured after adsorption of the polymers onto various particles (Abramson *et al.*, 1942, 1964). Especially for the calculation of ζ-potentials of adsorbed proteins, one should not lose sight of the large difference in size between single protein molecules and particles with an adsorbed protein layer, which often gives rise to a significant difference in the κa ratio (see Chapter V), which can strongly influence the correlation between ζ-potential and electrophoretic mobility.

Microelectrophoresis is normally done inside a capillary or chamber which is closed at both ends. In such systems the electroosmotic backflow along the walls of the capillary or chamber, in conjunction with the forward electrophoretic movement of the particles, results in a parabolic particle migration profile, so that the true particle electrophoretic mobility is to be found only at the precise level of the "stationary layer" (Seaman and Brooks, 1979), which necessitates very precise focusing of the microscope. This causes severe problems with microelectrophoresis of larger particles, which fairly quickly sediment away from the stationary layer and thus tend to get out of focus, or disappear altogether from view. This effect may to some extent be countered by coating the capillary wall with a material of low ζ-potential (Seaman and Brooks, 1979), or by covering the capillary wall with a thin gel of low ζ-potential, inside of which the electroosmotic backflow is trapped, thus leaving the lumen free of any electroosmotic flow gradient, permitting one to focus at all levels inside the lumen left by the gel (van Oss, Fike *et al.*, 1974; van Oss and Fike, 1979).

Automated microelectrophoresis has now become possible, thanks to the development of laser Doppler velocimetry (Seaman and Brooks, 1979) and related methods.

MOVING BOUNDARY ELECTROPHORESIS

Moving boundary electrophoresis, described by Tiselius (1937), uses an open vertical "U"-tube (with sliding parts): an arrangement which essentially obviates all electroosmotic disturbances. The optical method best suited for the visualization of the change in concentration *versus* distance (van Oss, 1972) of the moving boundaries in electrophoresis, is the schlieren optical approach, as improved by Philpot (1938) and Svensson (1939); see also Svensson and Thompson (1961). The tremendous advances in our knowledge of the physicochemical properties of numerous proteins were all started by the moving boundary electrophoretic studies from the later 1930s to the mid-1950s. Then, however, the advent of the much more affordable, simpler and more convenient zone electrophoresis methods (see the following section) quickly made the large, costly and complicated moving boundary instruments obsolete. However, the schlieren optics of these machines alone still make them very useful for a variety of purposes. One of these is the accurate measurement of diffusion coefficients, although these can also be measured with the schlieren optics of the analytical ultracentrifuge (Edberg *et al.*, 1972). Another useful application is the observation by schlieren optics of the (otherwise invisible) concentration gradients and discreet layers in microemulsions, with or without electric field; see Chapter XXI. Unfortunately however, devices with working schlieren optics become increasingly difficult to find. Finally, for the accurate measurement of extremely small electrophoretic mobilities, e.g., of dextrans, moving boundary electrophoresis still is about the only applicable method (van Oss, Fike *et al.*, 1974). Moving boundary electrophoresis has also been applied to the determination of the ζ-potential of polystyrene latices (van Oss and Singer, 1966; Ottewill and Shaw, 1967).

ZONE ELECTROPHORESIS

Electrophoresis of dissolved biopolymers, deposited in a narrow band or "*zone,*" on strips of filter paper or cellulose acetate membranes, moistened with, and dipping with both extremities in, the electrophoresis buffer is called zone electrophoresis. Apart from cellulosic media, gels, thin layers of silica powder, blocks of starch or synthetic polymeric particles have been used in zone electrophoresis (Smith, 1979; Sargent and George, 1975). Paper electrophoresis, which was preponderant in the late 1950s and in the early 1960s, is now largely superseded by cellulose acetate and gel electrophoresis; see Fig. XV-1.

IMMUNOELECTROPHORESIS is an important variant of zone electrophoresis where the electrophoretically separated fractions are not visualized by staining, but by the very specific precipitation reaction engendered by immunodiffusion (Grabar and Williams, 1953; Grabar and Burtin, 1964; see also, Clausen, 1979; van Oss and Bartholomew, 1980). This method of visualization permits the differentiation between different, but electrophoretically overlapping protein fractions, which are indistinguishable from each other by conventional protein staining. Thus the five distinct mammalian blood serum proteins (albumin, α_1, α_2, β and γ globulins) detected with Tiselius's moving boundary method, which also are visible with ordinary zone electrophoresis, could be expanded to about 50 separate protein species by immunoelectrophoresis. Immunoelectrophoresis continues to be used in the diagnosis of human monoclonal gammopathies. Bi-dimensional methods have further increased the detectable number of different blood serum proteins by at least another decimal order of magnitude.

HIGH VOLTAGE ELECTROPHORESIS, at 40–20 V/cm, allows the electrophoretic separation of small molecules (e.g., amino acids, sugars, indoles, purines, pyrimidines, phenolic acids, keto acids, imidazoles, steroids) and can be carried out with a resolution impossible to achieve with the lower, more conventional voltage gradients (Smith, 1979). However, as high voltage electrophoresis requires potentials as high as 10,000 V and uses up to 2.5 kwh for a separation, special measures for

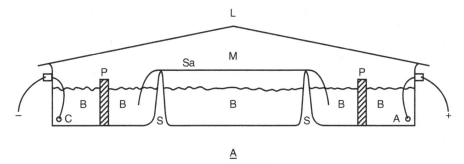

FIGURE XV-1   Cross section of a cellulose acetate strip zone electrophoresis device. M is a cellulose acetate membrane, soaked in buffer, of which both ends are immersed in the buffer B. Sa indicates the place where the sample is deposited. Each buffer compartment is divided into two sections by means of a porous (or perforated) barrier P, to prevent the membrane from being directly exposed to electrolysis products emitted by the electrodes C (cathode) and A (anode). The entire chamber should be closed by a lid L, when in use; it is desirable for that lid to have slightly sloping surfaces, so that droplets of condensed water will slide away to the sides of the chamber, without dripping on the membranes. Care should be taken that the levels of buffer B be equalized between the cathodal and the anodal side, to prevent untoward liquid migration inside the membranes through hydrostatic forces. Devices of this type are also used for paper electrophoresis, agar and agarose electrophoresis, and immunoelectrophoresis. For the latter applications the gel is confirmed to a tray that reposes with both ends on the supports S. Filter paper wicks steeped in buffer are then used to establish electrical contact between both sides of the gel and the buffer B. (From van Oss, 1979.)

safety as well as for cooling are necessary. The carrier that is most suited for this method is filter paper (preferably Whatman 3 MM). Various kinds of apparatus have been described. One type (capable of up to 10,000 V and 200 V/cm) comprised water-cooled metal plates between which paper insulated with polyethylene sheets is clamped (Gross, 1955, 1963; van Oss and Cixous, 1963). Another method, rarely capable of reaching more than 5,000 V, or 100 V/cm, is a cooled, water-immiscible liquid (such as toluene) that surrounds the wet filter paper (Smith, 1979; Katz et al., 1959). Both types of apparatus must be provided with adequate safeguards against the possibility of any kind of manual contact while the current is on. With the latter immersion-type apparatus, a graduate student was electrocuted in 1965, and it is recommended that safety measures developed since that occurrence (Spencer et al., 1966) be rigidly followed. High voltage electrophoresis of small molecules can be used in one dimension, on paper, with prior or subsequent partition chromatography in the dimension perpendicular to it. Usually high voltage electrophoresis is the first step, followed by chromatography. However, this author has obtained better results with amino acid separations by utilizing the two steps in the reversed order (van Oss and Cixous, 1963).

ZONE ELECTROPHORESIS OF PARTICLES AND CELLS

It should be realized that for the determination of any of the myriads of biological properties living cells possess, one usually needs to have *10^6 cells per isolated fraction* at one's disposal.* Thus to be able to determine which one of, say, ten different cell fractions contains cells capable of synthesizing compound X, *at least 10^7* cells of the original mixture have to be fractionated. The microelectrophoretic methods described above are quantitatively several decimal orders of magnitude removed from the capability of directly determining which biological properties accompany which electrophoretic mobility fraction in a given mixture of cells. For the electrophoretic separation of $> 10^7$ cells, a total separation distance between the slowest and the fastest group of cells of at least a few cm is generally required. Given the usual ζ-potentials of cells (generally between -10 and -20 mV), in electric fields that create no excessive Joule effects, times of the order of 10^3 to 10^4 seconds then are required, and during such time lapses most mammalian cells sediment about 1 cm in water. This means that, in the time required to separate mammalian cells in numbers sufficient to give separated fractions large enough for analysis, the cells will have sedimented to the bottom of most electrophoresis chambers, unless a way is found to prevent that occurrence. The following sections treat a number of approaches that have been devised to deal with that problem.

Descending Density Gradient Electrophoresis

Up to 4×10^7 cultured Chinese hamster bone marrow cells have successfully been subjected to downward electrophoresis into a Ficoll sucrose gradient by Boltz *et*

---

* When using a modern flow cytometry instrument, only 1 or 2×10^5 cells would suffice for analysis.

al. (1973). Similar separations were done by Griffith *et al.* (1975), with mixtures of erythrocytes from different species (rabbit and mouse, and human and rabbit); mouse spleen cells; mouse thymus cells and erythrocytes. The present author and his collaborators fractionated 10^8 cultured human lymphocytes by descending electrophoresis into a gradient of low molecular weight dextran (M_W = 10,000), which we found the least apt to clump the cells. A considerable enrichment in T cells was obtained in the fastest fractions (van Oss, Bigazzi and Gillman, 1974).

ASCENDING ELECTROPHORESIS

Cell aggregation by gradient-forming polymers such as Ficoll and dextran (Brooks, 1973; van Oss *et al.*, 1978) as well as cell deformation and other changes by sugars (van Oss *et al.*, 1978; van Oss, 1971) will remain a problem with all density gradient methods employing these substances. In addition, virtually uncharged polymers nevertheless can cause changes in the stability of cells; see Chapter XXIII. However, D_2O, which is 10% denser than H_2O, has no untoward physico-chemical or biological influence on cells. Methods to obviate stabilizing gradients in preparative cell electrophoresis by vertical upward electrophoresis into a D_2O gradient liquid column have been described by the present author and his collaborators. By ascending electrophoresis of 10^7 to 10^8 cells ("levitation"), practically pure T-cells were found in the fastest fraction of human peripheral lymphocytes (van Oss, Bigazzi and Gillman, 1974; Gillman *et al.*, 1974; van Oss and Bronson, 1979). This may result from the fact that human T-cells not only have higher electrophoretic mobilities but are also smaller than B-cells, and thus the least prone to sedimentation. Various cell types could be separated by ascending electrophoresis into shallow D_2O gradients, e.g., mixtures of 10^8 erythrocytes; granulocytes; lymphocytes; however, the fastest lymphocyte fraction always was closer to 100% T-cells, when no such shallow D_2O gradient was used (van Oss and Bronson, 1979). However, when a purely electrophoretic cell separation is required (without interference of cell size effects, advantageous or otherwise), ascending cell electrophoresis into a shallow D_2O gradient, with a cell-layer of $\approx 10^7 - 10^8$ cells deposited on top of a D_2O cushion as a starting point, appears to be one of the mildest and simplest methods available to date. The buffers used must nevertheless contain fairly large amounts of, e.g., glucose, to maintain isotonicity and to avoid high ionic strengths (and thus strong Joule heating). Boltz *et al.* (1976) also developed an ascending preparative cell electrophoresis method, into a (Ficoll) density gradient, albeit with relatively few ($10^5 - 10^7$) cells.

PACKED COLUMN CELL ELECTROPHORESIS

In columns packed with glass beads that are all approximately of the same size (0.1 mm diameter) it is possible to separate mixtures of 6×10^9 erythrocytes electrophoretically. The interstices between beads are small enough to prevent the migration into the column of erythrocytes by simple sedimentation at ambient gravity, but large enough to permit virtually unimpeded migration under the influence of an electric field. Complete separation was obtained between human and chicken erythrocytes, and between papain-treated and untreated human erythrocytes (Fike and van Oss, 1973). However, upon further experimentation with other types of mammalian cells, dimensional monodispersity of the cells proved to be essential for the success of the method. Cells with a fairly widespread range of sizes can at best be classified according to size with this method, but not according to charge. This method, therefore, appears to be useful mainly for the electrophoretic separation of erythrocytes, a separation for which there exists no urgent need at present.

For microgravity zone electrophoresis of particles and cells, see below.

ENHANCED ELECTROPHORETIC SEPARATION
OF SUBCLASSES OF CELLS THROUGH
RECEPTOR-TAGGING

The cells of one given class, e.g., human peripheral blood leukocytes, all tend to fall within a rather narrow range of electrophoretic mobilities (Ruhenstroth-Bauer, 1965) and, even though, e.g., a T-lymphocyte fraction may be electrophoretically enriched (see above), a complete isolation of any cell subclass by simple electrophoresis is probably impossible. However, by *tagging* the selected cells specifically with a polymer which has a markedly different ζ-potential, a more complete separation can be attained. This is best done by means of specific antibody molecules of the IgG-class, directed to an appropriate cell marker. IgG has a lower (slightly negative) ζ-potential than, e.g., lymphocytes, and a cell whose surface is specifically coated with a layer of IgG molecules therefore should have a markedly decreased electrophoretic mobility. In practice, however, hardly any significant decrease in electrophoretic mobility was noticeable after coating lymphocytes with a single layer of specific IgG molecules. But after coating such lymphocytes, which had already been coated with one layer of IgG, with a second layer of IgG-class antibodies, directed to the anti-lymphocyte antibodies, a 40% decrease in electrophoretic mobility was observed (Cohly *et al.*, 1985a; Cohly, 1986). The explanation of this phenomenon lies in the fact that the first layer of IgG molecules, attached to a ligand close to the cell membrane barely, if at all, protrudes beyond the tips of the cell's negatively charged sialoglycoprotein strands of its glycocalyx. The second IgG-class antibody, however, can protrude about 120 Å beyond the limit of the glycocalyx, and thus substantially influence the cell's -potential; see Fig. XV-2; see also Hansen and Hanning (1982). This difference between the influence of the first layer of attached IgG and a second layer, attached to the top of the first one, is confirmed by microelectrophoretic studies on murine lymphocytes and epithelial cells (Cohly *et al.*, 1985b), and on human erythrocytes treated with blood group antibodies (Cohly, 1986). An even more pronounced difference in electrophoretic mobilities could be achieved when the second antibody was cationized, according to the method of Gauthier *et al.* (1982); see Cohly *et al.* (1985a) and Cohly (1986).

MICROGRAVITY ELECTROPHORESIS

Microgravity, rather than zero gravity electrophoresis, is the more accurate way of describing this approach, as during space flights total zero gravity is never achieved,

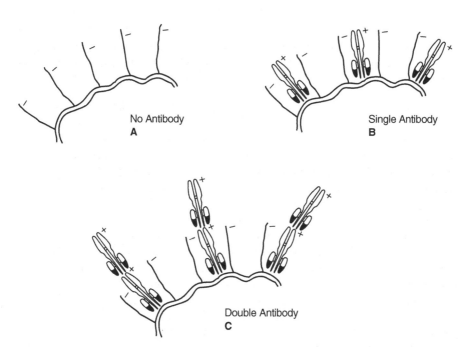

FIGURE XV-2 Schematic presentation of the influence of double antibody treatment on the effective charge of cells. A is a cell with its negatively charged glycoprotein strands sticking out. B is a cell which has IgG-class antibody molecules attached to the cell surface where the height of the antibody is similar to the height of the glycoprotein strands. C is a cell where the second antibody is attached to the "top" of the first antibody and thus extends beyond the glycocalyx. Only in the last case is an effective decrease in electrophoretic mobility observed. (From Cohly, 1986)

and very low (10^{-6} G) gravities only for relatively short periods of time. Nevertheless, during the greater part of most space flights gravity conditions would tend to fluctuate mainly between 10^{-3} and 10^{-5} G. During space electrophoresis experiments such microgravity conditions virtually abolish all sedimentation effects, as well as the fluid disturbances caused by local density differences created by the Joule effect (thermal convection).

The first successful electrophoresis separation at "0 G" was done during the flight of Apollo 16. Polystyrene latex particles of two different sizes and thus with different κa ratios (see Chapter V) (but with the same ζ-potentials; see Chapter V, Figure V-8) were subjected to electrophoresis together, as well as separately, in stationary tubes. The results showed the crucial importance of controlling the effects of electroosmosis (especially marked at 0 G, because under those conditions no stabilizing gradients can be formed to counter-balance these effects). They also showed that due to the absence of particle sedimentation and thermal convection, even in high potential fields, preparative electrophoretic separation of particles at 0 G is indeed feasible (Snyder et al., 1973; Micale et al., 1976). In the light of the experience obtained with the Apollo 16 experiment, experiments for the Apollo-Soyuz flight were designed, with living cells of various types. Successful electrophoretic separations were achieved during the Apollo-Soyuz flight with red cells of different species, as well as with human kidney cells, separated fractions of which were returned to earth. In addition, it was demonstrated during that flight that red cell separation by isotachophoresis at 0 G also is feasible (Allen et al., 1977); see the section on isotachophoresis, below. One of the most important results obtained was with the electrophoretic kidney cell separation experiment, which appeared to indicate that different subpopulations of cells (as separated by electrophoresis) produce different compounds, i.e., cells from one fraction produced urokinase, and those from other fractions produced human granulocyte conditioning factor, or erythropoietin (Allen et al., 1977). These results were made possible only by the use of low ζ-potential coatings of the inside of the electrophoresis tube (Vanderhoff et al., 1977), developed to obviate the type of electroosmotic backflow encountered on Apollo 16.

On the other hand, the use of microgravity conditions for continuous-flow electrophoretic systems, for which some enthusiasm appeared to exist in the early 1980's, seems to have little to commend it. The only (fairly slight) advantage of microgravity conditions for continuous-flow methods appears to be an enhanced throughput (possible on account of the decreased effect of thermal convection). That increase in throughput probably is of the order of a factor of 100; this is, however, far from sufficient to counterbalance the tremendous cost of space-processing. Especially for the electrophoresis of proteins, the use of microgravity conditions is totally superfluous, as there is an abundance of stabilizing materials which can easily counteract any detrimental effects caused by thermal convection. For cell electrophoresis, however, only stabilizing gradients can be of some use and among these, D_2O gradients give the most favorable results, with virtually no untoward osmotic or other side-effects; see the Section on Ascending Eletrophoresis, above.

MOLECULAR SIEVE ZONE ELECTROPHORESIS

Tiselius and Flodin (1953) observed that by means of electrophoresis in gels with very small pores or in very viscous polymer solutions, molecules may be separated according to their size or shape. They called the method "electrokinetic ultrafiltration." Electrophoresis through multiple layers of very dense gels has been used by the present author and his collaborators for the separation of stable lithium isotopes (van Oss et al., 1959). Most of the advances with this type of method occurred in the 1960s, after the development of starch gel electrophoresis (Smithies, 1955, 1959, 1962), with the widespread adoption of polyacrylamide gel electrophoresis (Raymond and Weintraub, 1959) and especially of "disc" electrophoresis (Ornstein, 1964; Davis, 1964).

STARCH GEL ELECTROPHORESIS

In 1955 Smithies demonstrated that electrophoresis in starch gels could yield a remarkable multitude of different protein fractions (Smithies, 1955). Smithies subsequently demonstrated that the separating effect was mainly due to differences in molecular size (Smithies, 1959). He observed that the denser the starch gel, the slower the migration of proteins in it (Smithies, 1962). Starch gel electrophoresis quickly became an important and powerful technique in the characterization and separation of proteins. However, the method is rather tedious, and was superseded by the much more convenient technique of polyacrylamide gel electrophoresis; see below.

Polyacrylamide Gel Electrophoresis

The use of concentrated polyacrylamide gels in electrophoresis was first described by Raymond and Weintraub (1959); they initiated the use of these gels as horizontal slabs. Ornstein (1964) and Davis (1964) subsequently developed "disc electrophoresis," with vertical cylinders of polyacrylamide gel. With that method exquisite resolution can be attained, in which the starting material can be resolved into manifold concentrated fractions, with the appearance of "discs," often no thicker than 10 µm each. Thus, disc electrophoresis has become a powerful analytical and preparative tool for obtaining maximum resolution with mixtures of biopolymers. The very popularity of the method, however, necessitates a *caveat*: molecular sieve electrophoresis is essentially based on two entirely different mechanisms that act simultaneously and in the same direction, one of them separating molecules according to their size and shape, and the other according to their electrophoretic mobility (which itself under common electrophoretic conditions may vary according to their size and shape; see Chapter V). Thus even in the simplest cases, the situation is analogous to one algebraic equation with two unknown variables. In other words, it is impossible to learn either the size and/or shape or the ζ-potential of a compound from its mobility in a dense gel. Of any given biopolymer isolated or characterized by such a method it is not only impossible to know (without further tests) if it has migrated to a certain point on account of its size, its shape or its charge (or any of these properties combined), but it is also impossible to know (without further tests) whether it is accompanied by other compounds that may have migrated to the same point on account of another combination of properties. What is worse, it is not only theoretically possible for two different compounds to migrate to the same point in a gel; a single homogeneous compound may, under favorable conditions, produce more than one band (Franglen and Gosselin, 1958; Cann, 1979). Polyacrylamide gel electrophoresis of larger molecules and viruses can be carried out by using more diluted gels (Edgell *et al.*, 1969), or by using mixed gels of polyacrylamide and agarose (Uriel, 1966).

Sodium Dodecyl Sulfate Polyacrylamide Electrophoresis

The advantage of having a constant charge/mass ratio, which is a natural attribute of nucleic acids, see below, can be artificially endowed upon other biopolymers (e.g., proteins), by treatment with a strongly negatively charged surfactant such as sodium dodecyl sulfate (SDS). This technique was first published by Shapiro *et al.* (1967). The reliability of the method has been tested on a large number of polypeptide chains (Weber and Osborn, 1969). The physicochemical conditions under which the molecular weights of SDS-polypeptides can be subsequently checked by analytical ultracentrifugation were elaborated upon by Barnett and Spragg (1971).

In recent years SDS polyacrylamide gel electrophoresis has been coupled to separations (by size) by other additional criteria (such as isoelectric pH of proteins or peptides) in two-dimensional gel electrophoresis allowing separation of individual components (or sometimes of different modifications of the same component) in complex protein mixtures (Gygi *et al.*, 2000).

GRADED POROSITY GEL ELECTROPHORESIS

Electrophoresis in gels of graded porosity allows macromolecules to migrate into a gel of which the pores continuously become smaller as a function of distance, until a pore size is reached that approaches their molecular size, which causes them to become trapped and prevents them from migrating any farther (Margolis and Kendrick, 1968). Thus this method of gel electrophoresis also separates solely according to the size of macromolecules and is independent of their charge, as long as they have enough surface charge to enable them to be transported electrophoretically (Gianazza and Righetti, 1979).

GEL ELECTROPHORESIS OF NUCLEIC ACIDS

The difficulties pointed out above, of having both size and charge contribute to the electrophoretic transport in dense gels, do not apply to molecules with a constant charge/mass ratio, such as RNA and DNA. Different RNA molecules were first successfully separated, according to molecular weight only, by polyacrylamide electrophoresis by Richards *et al.* (1965). Gel electrophoresis of (poly)nucleic acids has now become one of the most important tools in molecular biology. Agarose gels are generally used for the separation according to size of linear DNA molecules of M_W from 200,000 (using a 2% gel), up to M_W 40,000,000 (using a 0.3% gel); see Maniatis *et al.* (1982). For lower M_W DNA (i.e., fragments smaller than 1 kilobase (kb), or $M_W \approx 660,000$, down to 0.1 or 0.01 kb), polyacrylamide gels must be used, in gel concentrations ranging from 3.5% (0.1–1 kb) up to 20% (0.01 to 0.01 kb); both types of gels can also be used for strand-separation (Maniatis *et al.*, 1982). Al electrophoretically isolated DAN (or RNA) fractions can be, separately, electrophoretically eluted from the gel portion in which they are located.

For DNA sequencing one can also use gel electrophoresis. Briefly, DNA is degraded, either chemically or enzymatically. Using enzymatic degradation, a given enzyme is employed which breaks up DNA so that only strands with a terminal adenine (A) are produced. These are subjected to gel electrophoresis, which permits classification according to size of these terminal A strands. The same is done with thymidine (T), guanosine (G), and cytosine (C) strands. Comparison of the various sites of terminal-A, terminal-T, terminal-G, and terminal-C strands thus obtained then readily yields the complete DNA sequence (Ausubel *et al.*, 1987).

The use of gel electrophoresis involving radioactively labeled DNA for general DNA sequencing has been superseded by automated methods using specific fluorescent labeling of bases. However, it is still the technique of choice for analyses of DNA footprinting techniques that identify specific sequences of DNA which are protected from enzymatic and chemical cleavage by interaction with site-specific DNA-binding proteins (Schmitz *et al.*, 1978; Siebenlist and Gilbert, 1980).

More recently *pulsed field gel electrophoresis*, or electroporation was developed, for the separation of large DNA molecules (5 to 5,000 kilobases) (Schwartz *et al.*, 1983; Schwartz and Cantor, 1984; Carle and Olson, 1984). In this method two electrical fields in approximately perpendicular directions are applied alternatively. The time during which each field is applied is designated the pulse time. The pulse time determines the molecular size separation range. The pulse times are roughly related to the 4/3 power of the DNA size: they vary from 0.1 sec. for 5 kb to 1,000 sec. for 5,000 kb DN (Cantor *et al.*, 1986). Pulse field electrophoresis is used in analyses of DNA replication intermediates.

ZONE ELECTROPHORESIS OF ADSORBED PROTEINS

The high concentration of protein in the starting zone may be a potential cause of some inaccuracy in the measurement of electrophoretic mobilities of biopolyelectrolytes, such as proteins, when zone electrophoresis is used. For instance, diffusion, from a zone of protein adsorbed onto a glass surface, gives rise to a 100–fold increase in diffusion coefficient, due to a pronounced concentration effect caused by the protein-accumulation in the starting zone (Michaeli *et al.*, 1980). At the same time, no protein desorbed from the glass surface: the concentration-enhanced protein diffusion occurred entirely in the adsorbed state. However, in zone electrophoresis of proteins which are adsorbed onto glass, increases in electrophoretic mobility (u) due to high starting zone protein concentrations appear to be minor, at least at distances from the starting zone of at least 10 times the zone thickness. It is true that there are increases in u with highly concentrated protein in the starting zone, but these can, for the greater part, be accounted for by the higher values of u expected in the case of migration of cylindrical molecules or particles in a direction perpendicular to the electric field (Absolom, Michaeli and van Oss, 1981). On the other hand, when more randomly arranged (at lower protein concentrations in the zone), actually a somewhat lower electrophoretic mobility is observed for protein adsorbed onto glass, as well as for the same protein (bovine serum albumin) subjected to electrophoresis on cellulose acetate membranes (Absolom, Michaeli and van Oss, 1981). Thus, while zone-concentration effects do not appear to cause a significant increase in electrophoretic mobility, there is a noticeable decrease (from 20 to 40%) in the electrophoretic mobility of proteins adsorbed onto glass, as well as on cellulose acetate membranes, compared to moving boundary electrophoresis (Absolom *et al.*, 1981). Thus, in general, zone electrophoresis involving solid carriers (e.g., powders, glass surfaces, membranes) cannot be held to be a reliable approach to the determination of electrophoretic mobilities of solutes. If zone electrophoresis is to be used for the purpose of determining the electrophoretic mobility of a solute, only very dilute gels or relatively shallow density gradients should be considered as stabilizing media. For the mechanism of adsorption of albumin onto silica or glass, see Chapter XXIV.

CONTINUOUS FLOW ELECTROPHORESIS

Continuous flow electrophoresis is a procedure that started as continuous zone or curtain electrophoresis (see below) and that through many technical improvements and refinements outgrew the carrier and curtain stage, and has since been further developed as a free flowing liquid technique. In all cases the electric field is perpendicular to the direction of liquid flow. The method has three theoretical advantages of considerable importance:

1. It presents a simple built-in facility for continuous and automatic fraction collecting.
2. It allows the electrophoresis of fairly large quantities of solute which are directly proportional to the duration of electrophoresis; thus as long as a steady state is maintained, the output continues.
3. In the free-flowing liquid version it affords a possible solution to the intractable problem of sedimentation, which makes it especially attractive to cell electrophoresis.

Although all continuous flow electrophoresis methods are closely interrelated, it is convenient to subdivide them into the following sections: Curtain Electrophoresis, Free Flow Electrophoresis, Endless Belt Electrophoresis, Cylindrical Rotating Continuous Flow Electrophoresis, Stable Flow Electrophoresis, Electrophoretic Field Flow Fractionation, and Horizontal Rotating Cylinder Electrophoresis (see Fig. XV-3). Some of these methods may possibly fit as well under one of the other headings for reasons of taxonomical propinquity, but they all seem to have developed far enough into the direction of a well-defined separate method to merit treatment of their own.

CURTAIN ELECTROPHORESIS

Continuous zone electrophoresis was developed independently by Grassman and Hannig (1949) and Svensson and Brattsten (1949). The method quickly provided to be most practical when a vertical sheet of thick filter paper or cardboard was used as a carrier, hence the designation Curtain Electrophoresis (Grassman and Hannig, 1950; Durrum, 1951; Hannig, 1969; Vanderhoff *et al.*, 1979). The free-hanging sheet of filter paper carries a continuous downflow of buffer, while an electric field is applied in the plane of the filter paper, perpendicular to the buffer flow. The solute mixture that is to be separated is continuously applied to a point on the top part of the filter paper. The bottom of the filter paper is cut out in a pattern of multiple triangles, culminating in points at the bottom, under each of which a test tube is placed for the collection of the individual fractions (see Fig. XV-3A). Apparatus of this type can, once stable flow and field conditions are reached, be run for many days continuously, and fractionate, e.g., 200–1,000 mg of mixed proteins per 24 hours. The process should be done in a well-ventilated cold room, at +4°C. No curtain devices have been commercially available since 1971, although in some cases, parts may still be obtainable from the original manufacturers. The major drawback of curtain

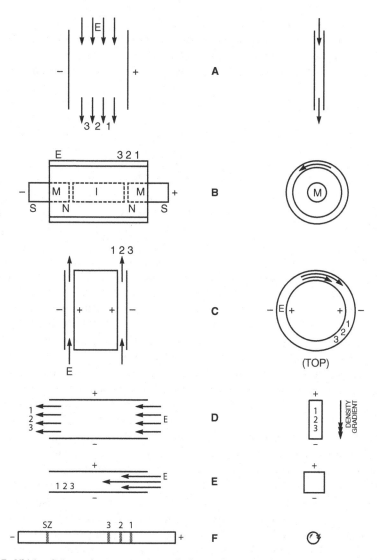

FIGURE XV-3 Schematic presentation of the various continuous flow electrophoresis methods; front views on the left and side views on the right (except for C, where the right-hand view is a top view). Polarities of the electric field are indicated by + and −. The direction of liquid flow is indicated by an arrow with a single arrowhead. The direction of rotation of a solid cylinder is indicated by an arrow with a double arrowhead; see C and F. The vertical (downward) direction of the density gradient in D is indicated by an arrow with a triple arrowhead. The approximate place of continuous entry of the sample mixture is indicated by E, and the places of three continuously emerging fractions are marked 1, 2 and 3, in the order of decreasing electrophoretic mobility. A: Curtain Electrophoresis and Free Flow Electrophoresis; B: Endless Belt Electrophoresis (M = magnets, N = north, S = south, I = soft iron cylinder); C: Cylindrical Rotating Continuous Flow Electrophoresis; D: Stable Flow Electrophoresis; E: Electrophoretic Field Flow Fractionation; F: Horizontal Rotating Cylinder Electrophoresis (SZ = starting zone).

electrophoresis was the aspecific adsorption of proteins onto the curtain material. However, once a steady state was reached and a certain load of protein had been adsorbed, no further protein appeared to accumulate and the separation became quite efficient when continuous runs of several days or more could be sustained.

FREE FLOW ELECTROPHORESIS

Continuous electrophoresis across freely flowing liquid between two closely spaced parallel flat plates is treated here (see Fig. XV-3A), endless belt electrophoresis is discussed in the next section, and cylindrical continuous flow electrophoresis, stable flow (Staflo) electrophoresis, electrophoretic field flow fractionation, and horizontal rotating cylinder electrophoresis are grouped in the sections following (see Fig. XV-3).

The development of free flow electrophoresis, or "nonstabilized continuous deflection electrophoresis," has been and remains linked with the pioneering work of Hannig (1961, 1964, 1969). The method has also been used successfully for the isolation of biopolymers that, due to adsorption and other problems, are not easily purified by means of carrier electrophoresis, such as acid mucopolysaccharides (Mashburn and Hoffman, 1966) and blood clotting factor VIII (Bidwell et al., 1966). Polystyrene latex particles of different sizes (and thus of different a) have also been separated by free flow electrophoresis (McCann et al., 1973; Hannig, et al., 1975).

The method still labors under many difficulties of electrokinetic as well as of hydrodynamic nature. Electroosmosis causes the distortion of the cross section of fraction streams to a crescent shape (Strickler and Sacks, 1973) and hydrodynamic considerations hamper the scaling-up of the process, particularly due to the thermal convection created by Joule-heating. The latter problem is still so severe that operation at 0 gravity has been considered as one of the better solutions (see above). Hannig et al. (1975) treated these and other theoretical and experimental aspects of free flow electrophoresis extensively. To obviate particle or cell sedimentation during continuous electrophoresis, Hannig (1969) placed the separation chamber vertically with the liquid flow in the downward direction; see also Hannig et al. (1975).

With free flow electrophoresis separations of cell mixtures consisting of several times 10^7 cells have become possible. Fractionations have been reported of T and B lymphocytes (Zeiller et al., 1974, 1975; von Boehmer et al., 1974; Hannig and Heidrich, 1977), rabbit kidney cells, rat liver lysosomes, as well as complete separation between inside-out and outside-in vesicles from human erythrocyte ghosts (Hannig and Heidrich, 1977).

ENDLESS BELT ELECTROPHORESIS

Kolin attained the same advantage by combining continuous flow with fluid rotation in a most ingenious manner: a continuously moving fluid belt is maintained by rotating it in a thin cylindrical or flattened cylindroid torus around a horizontal axis, by means of a double axial magnet that creates a radial magnetic field, and an electric field perpendicular to it (parallel to the axis). The electric field at the same time is the driving force for the elecrophoresis (Kolin, 1966, 1967; Kolin and Luner, 1969, 1971), see Fig. XV-3B. The method, of course, labors under the same difficulties as regular free flow electrophoresis between flat plates, with the added drawbacks of somewhat greater complexity of the apparatus, and of the lack of freedom to regulate the liquid flow velocity independently of the electric field used for the electric transport.

CYLINDRICAL ROTATING CONTINUOUS FLOW ELECTROPHORESIS

This type of continuous flow electrophoresis is a newer development, mainly geared at preparative separations; it is based on a concept by Philpot (1980). Like Kolin's endless belt electrophoresis (see above) it uses a rotating cylindrical torus of liquid, but its axis is vertical, and in addition to rotating, the entire cylindrical liquid vein also is forced to flow upward. The rotary liquid movement is maintained by rotating (at 150 rpm) the outer of the two cylinders that confine the liquid, which much enhances the cooling efficiency and thus allows high electric fields; see Fig. XV-3C. Also contrary to Kolin's method, the electric field is radial, which, of course, allows a very high field strength over a thin layer of liquid. Thus very large quantities of material may be continuously separated with this method, of the order of up to 100 grams of pretein per hour (Mattock *et al.*, 1980).

STABLE FLOW ELECTROPHORESIS

Stable flow (Staflo) electrophoresis was first proposed by Mel (1959); see also Mel (1960, 1964). This is horizontal fluid flow electrophoresis, with a vertical electric field, stabilized by a (sucrose) density gradient included in the fluid flow; see Fig. XV-3D. This method has been applied to a variety of separations (most of them done at the originating laboratory), e.g., of rat bone marrow cells (Mel *et al.*, 1965), spinach chloroplasts (Nobel and Mel, 1966), lipoproteins (Tippetts *et al.*, 1967), diploid yeast cells and spores (Resnick *et al.*, 1967), and sperm cells (Pistenma *et al.*, 1971). With stable flow (Staflo) electrophoresis separations of mixtures up to 10^8 blood cells are feasible. Mel obtained complete separation between rabbit and chicken erythrocytes and partial but significant separation of rat bone marrow cells (Mel, 1970). Pistenma *et al.* (1971) obtained a significant enrichment of viable and fertile cells by means of stable flow electrophoresis of fowl sperm cells. As one of the most intractable problems of free flow electrophoresis is stability (mainly against disturbances caused by thermal convection and flow), stable flow electrophoresis in which stability is maintained by means of a flowing *density gradient* seems to be a promising continuous flow approach for cell separation on a preparative scale. The one drawback, i.e., the use of high concentrations of sugars for the stabilizing gradient (which may adversely affect living cells) can possibly be obviated by the use of D_2O instead of sugar gradients (see above).

ELECTROPHORETIC FIELD FLOW FRACTIONATION

Based on general theories of field flow fractionation (FFF) developed by Giddings (1966, 1968), electrophoretic FFF (EFFF) was proposed by Caldwell *et al.* (1972) and applied to simple protein separations. EFFF is (exactly like Staflo; see above) horizontal free flow electrophoresis, with a vertical electrical field (but without a density gradient), spacer membranes, and with collection of fractions by their simple piling up in a given spot on the lower spacer membrane, once a steady state has been reached; see Fig. XV-3E.

HORIZONTAL ROTATING CYLINDER
ELECTROPHORESIS

Hjertén (1967) described a free zone electrophoresis method in a slowly rotating (at $\approx$ 40 rpm) horizontal tube, with the electric field in the direction of the axis; see Fig. XV-3F. In the same manner as Kolin's method (see above), rotating the fluid primarily serves to obviate sedimentation, and thus allows the electrophoresis of suspended particles and cells as well as of dissolved solutes. When a quartz tube is used, the method can be used for analytical purposes with the help of a UV scanning device. As in a number of the other methods described above, zone deformation by electroosmosis occurs, but it can to a certain extent be alleviated with the help of low ζ-potential coatings, e.g., with methyl cellulose. This procedure is essentially a micro-method, because thermal convection due to Joulian heating makes homogeneous cooling possible only with tubes of fairly small internal diameter, hence the advantage of working with toroid-shaped or flat liquid veins that can be cooled from both sides, as described above. Nevertheless the scale on which this type of electrophoresis can be applied surpasses that of microscope-electrophoresis (see above) by an order of magnitude, so that determinations could be done with the rotating tube method that were not possible earlier. It could, for example, be demonstrated that the highest insulin content within various fractions of β-cells of mouse pancreas islets of Langerhans are to be found in the fraction richest in spherical granules (Hjertén *et al.*, 1964). Virus isolation was also studied by this method (Fridborg *et al.*, 1965; Tiselius *et al.*, 1965).

CAPILLARY ELECTROPHORESIS

In 1985 Hjertén (1985a) introduced the technique of electrophoresis in tubes with an internal diameter of 0.05 to 0.3 mm, now known as capillary electrophoresis (CE), which is especially suited for the separation of non-particulate materials, e.g., for protein separations. Due to the large surface area-to volume ratio of capillary tubes, Joule heating is easily dissipated, thus favoring high-speed separations at high voltages. The only two remaining disadvantages of CE are a strong electro-osmotic backflow, and the undesirable adsorption of proteins onto the capillary's inner wall. However, coating of the inner capillary wall with, e.g., a monomolecular layer of non-crosslinked poly(acrylamide) could obviate both problems (Hjertén, 1985b). CE has become especially important as an analytical tool, in which the exit end of the capillary tube is made of quartz, to allow for continuous analysis of the exiting proteins or peptides via UV absorptivity monitoring, as also utilized in various analytical chromatography devices.

ISOELECTRIC FOCUSING

Both isoelectric focusing and isotachophoresis can be considered as electromigration in buffers of which the composition is non-constant with respect to location within the applied electric field (see Fig. XV-4; Righetti, 1975; Radola and Graeselin, 1977). While the electrophoretic transport of charged molecules in an electric field

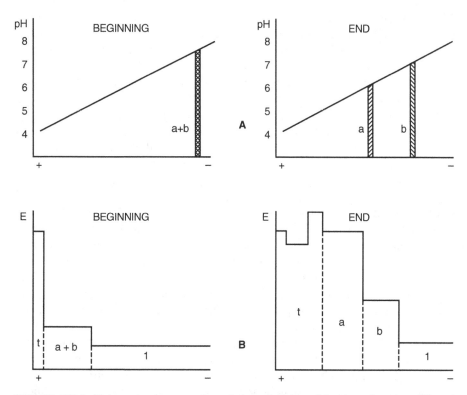

FIGURE XV-4 Schematic representation of the principles of isoelectrofocusing (A) and isotachophoresis (B). The beginning and end situations are depicted for both techniques. In isoelectric focusing (A) the pH is plotted *vs.* the path of migration. The pH gradient formed by the carrier ampholytes is ideally constant, although this is not always the case in actual practice (Fawcett, 1976). A mixture of two components a and b with different isoelectric points may be inserted at any place in the migration path, but is here shown to start near the cathodal and (a + b) (A, Beginning). When the separation is completed (in principle the separation continues to become sharper with time), when a and b have reached the locations in the migration path where the pH's correspond to their respective isoelectric points (6 for a and 7 for b) (A, End). In isotachophoresis (B) the electric field strength E (in V/cm) is plotted *vs.* the path of migration; see Everaerts *et al.* (1977). The sample mixture (a + b) is placed between the terminating electrolyte (t) in the cathode compartment, and the leading electrolyte (1) in the anode compartment. The effective mobility of the leading electrolyte (1) is the highest, and that of the terminating electrolyte (t) the lowest; the effective mobilities of materials a and b should be intermediate between these two. Once a steady state has been reached, the separation between a and b is complete and all four zones (t, a, b, 1) continue to move with equal velocity.

in a homogeneous buffer of a given pH is continuous and has to be stopped if one wishes to avoid losing the molecules through migration into the electrode compartment of the opposite charge, electromigration of a charged amphoteric molecule through a continuous pH-gradient automatically ceases when that amphoteric molecule has reached the place in the gradient where the pH is the same as its isoelectric point. Thus, isoelectric focusing is the sorting-out of amphoteric molecules according to their different isoelectric points, by electromigration through a pH-gradient (see Fig. XV-4A). Generally speaking, prolonged electromigration of a mixture of amphoteric molecules in a pH-gradient tends to enhance the resolution of each of its constituents, hence the designation of isoelectric *focusing*. The historical background has been given by Righetti and Drysdale (1974), who outlined the contributions by Kolin, Svensson (more recently called Rilbe) and Vesterberg, all of whom also described the origins of this method from their own particular viewpoints (Kolin, 1976, 1977; Rilbe, 1976, 1977). The development that finally made the general application of isoelectric focusing a practical reality was the synthesis of many different "carrier ampholytes," each having several acidic and basic groups with closely spaced pK values per molecule, with many different pI values (Vesterberg, 1968, 1969, 1976). Since the early 1970s, when these carrier ampholytes became generally available, a veritable publication explosion in the fields of analytical as well as preparative isoelectric focusing started (see e.g., Righetti and Drysdale, 1976; Radola and Graeselin, 1977; see also the review by Righetti, 1979). This is due to the fact that for, e.g., protein separation, isoelectric focusing allows an improvement in resolution of about an order a magnitude, compared to electrophoresis. This is because, with time, bands separated by isoelectric focusing reach and maintain an optimal sharpness (due to the focusing effect) while with electrophoresis separated bands continuously tend to broaden because of diffusion. For analytical purposes, and in particular for the separation of proteins by isoelectric focusing, one must be mindful of the fact that the carrier ampholytes most used are isomers and homologues of aliphatic polyamino-poly-carboxylic acids (Vesterberg, 1976), and are thus difficult to distinguish from proteins by the most commonly used colorimetric or spectrophotometric methods. However, as these carrier ampholytes have an average molecular weight of about 800, while that of most proteins is above 10,000, proteins generally can be easily separated from the ampholyte molecules by means of gel filtration (e.g., with Sephadex[R] G-50) (Vesterberg, 1969, 1976).

Various isoelectric focusing methods in completely free solution have been described by Quast (1979), but more generally stabilization by means of density gradients, gels or granulated gels is practiced (Righetti and Drysdale, 1976; Catsimpoolas, 1976). Isoelectric focusing can also be combined with continuous flow of the ampholine system (Just and Wermer, 1979).

Continuous flow isoelectric focusing has become an extremely powerful approach for the highly efficient fractionation of proteins, at a higher resolution than is attainable with ordinary (zone, or continuous flow) electrophoretic methods, on a micro-scale as well as at a large preparative level (Bier *et al.*, 1979; Bier and Egen, 1979); Bier's method has also been described by Maugh (1983).

It is not always necessary to attain equilibrium for isoelectric focusing to be useful. Catsimpoolas (1976) studied isoelectric focusing phenomena prior to reaching equilibrium, which he called "transient state isoelectric focusing," by means of optical scanning (Catsimpoolas, 1973). More recently, an approach was described by which the pH gradient could be permanently immobilized in the carrier gel by means of copolymerization of the pH-forming buffering species ("immobilines') with the polyacrylamide gel matrix (Bellquist, 1982; Righetti, 1984; Righetti *et al.*, 1987).

CELL ISOELECTRIC FOCUSING

It has been known for over thirty years that at least some blood cells (e.g., lymphocytes, platelets) have true isoelectric points (Bangham *et al.*, 1958). It would thus seem logical to endeavor to separate them by isoelectric focusing. One of the first attempts at preparative cell isoelectric focusing was reported by Leise and LeSane (1974), who fractionated up to 5×10^7 lymphocytes (human as well as rabbit). Cell viability (as measured by trypan blue exclusion) of fractions focusing at non-alkaline pHs was fair, but pronounced cell growth after subculturing the various fractions was found to be significant only with the fraction focusing at pH 4.2. In the second half of the 1970s great improvements in preparative cell isoelectric focusing were obtained in a number of laboratories. Manske *et al.* (1977) separated various kinds of native and modified cells (5×10^6 at a time), with little loss in viability among the electrofocused fractions. Boltz *et al.* (1977) studied the electrofocusing of a wide variety of cells (up to 5×10^6 per sample) and retained excellent viability and (with sperm cells) reproductive capacity of the fractionated cells. Finally, preparative cell isoelectric focusing by a continuous flow method has been achieved with erythrocytes (Just *et al.*, 1975) and with rat liver organelles (Just and Wermer, 1979).

ISOTACHOPHORESIS

The history and background of isotachophoresis has been described by Haglund (1970). Like isoelectric focusing, isotachophoresis is practiced in a buffer system of which the composition is non-constant with respect to location, but while isoelectric focusing is best done in a buffer system consisting of an essentially continuous pH gradient, isotachophoresis needs a buffer system of which the components have markedly discontinuous properties. The term isotachophoresis, meaning "transport at the same velocity," has become generally accepted for this method. However, Swedish workers from the Uppsala group feel that the term "Column Displacement Electrophoresis" is a more accurate description (Johansson *et al.*, 1988). In isotachophoresis the sample mixture is placed in the sample compartment, "terminating electrolyte" in this cathode compartment, and the "leading electrolyte" in the anode compartment (see Fig. XV-4B; Everaerts *et al.*, 1977). The "effective mobility" of the leading electrolyte is the highest, and that of the terminating electrolyte the lowest; the effective mobilities of the components in the sample mixture should be intermediate between those extremes. Because of the possibility of achieving a total separation between different ionic species, once a steady state is reached, isotachophoresis affords an extremely high resolution. The method is therefore increasingly employed for analytical purposes (Everaerts and Verheggen, 1975), e.g., by using capillary tubes (Arlinger, 1976; Delmotte, 1977). At the same time isotachophoresis is becoming useful for preparative separations (Lancaster and Sprouse, 1976); which also can include the capillary tube approach.

BIDIMENSIONAL METHODS

The combination of high-resolution isoelectric focusing in one dimension in a gel, with size-discriminating gel electrophoresis in a direction perpendicular to the isoelectric focusing direction, has made it possible to discern up to 1,500 blood serum proteins (Fawcett and Chrambach, 1986). O'Farrell (1975) used isoelectric focusing in the one, and sodium dodecylsulfate (SDS) gel electrophoresis in the other direction. Anderson and Anderson (1982) used isoelectric focusing in the one, and electrophoresis into a gradient gel of decreasing pore size as the other direction; this method is alluded to as "Iso-Dalt." Gianazza et al. (1986) and Dunn and Patel (1986) adapted gel isoelectric focusing with immobilized pH gradients to these techniques; other refinements to these two-dimensional techniques were published by Marshall and Williams (1986).

ELECTROOSMOSIS

It has been long established that there essentially is total equivalence between the electrokinetic phenomena (see Fig. V-1) and that ζ-potentials can theoretically be obtained through electroosmosis or streaming potentials as well as through electrophoretic determinations (Mazur and Overbeek, 1951; Overbeek, 1952); see Chapter V. Among others, Abramson et al. (1936) demonstrated that the ζ-potential of gelatin could be measured via the electroosmotic profile in a chamber coated with gelatin as well as through the electrophoretic mobility of gelatin-coated particles in the same chamber. Fike and van Oss (1976) have devised a method for the measurement of the ζ-potentials of living mammalian cells by determining the electroosmotic velocity of buffer through glass capillaries, of which the inner walls were entirely covered with monolayers of cultured cells. That method permits the determination of cell surface potentials of cell layers without the need for first dispersing the cells by chemical, enzymatical or mechanical means, required for the measurement of cell electrophoresis. It is known that all of these cell dispersion methods tend to alter the cells' ζ-potential more or less drastically (Häyry et al., 1965; Fike and van Oss, 1976). Doren et al. (1989) used electroosmosis to determine the ζ-potential of flat plates, in a modified microelectrophoresis device.

Synge and Tiselius (1950) and Mould and Synge (1954) first proposed to make use of electroosmotic flow in gel membranes or other porous media, to transport solventsthrough such porous media and thus to obtain separation according to the site of solutes transported with that solvent, the smaller solute molecules being able to go through smaller pores than the larger ones (see above). Pretorius et al. (1974) used electroosmosis as a means of transporting solvents, in thin-layer and in high-speed chromatography. The advantages of electroosmosis over other means of transporting solvents are two-fold: (1) With a reasonably high ionic strength (obtained by addition of electrolytes), and thus with a compressed electric double layer (see Chapter V), virtually ideal plug-flow can be achieved throughout the column, with a concomitant reduction in theoretical plate height (Pretorius *et al.*, 1974); (2) In open-ended tubes electroosmosis generates no noticeable hydrostatic pressure, and the rate of solvent transport is relatively independent of the size of the pores or channels. The author and his colleagues have used electroosmosis and electrophoresis combined and in the same direction through stacks of extremely dense membranes, to achieve enrichment of ^{7}Li (van Oss *et al.*, 1959).

Barry and Hope (1969) determined membrane potential differences across plant cell membranes *via* electroosmosis and Fensom *et al.* (1967) studied electroosmotic flwo across membranes of Nitella.

Thus, electroosmosis can be used as a means of solvent transport, as a method for determining cell surface potentials, and also as a method for determining cell membrane potentials (in a direction perpendicular to the cell surface). Electroosmotic solvent transport *cannot* be used to bring together two solutes; for that purpose only electrophoresis can be used, at a pH intermediate between the isoelectric points of the two solutes.

STREAMING POTENTIAL AND SEDIMENTATION POTENTIAL

ζ-Potentials of minerals and other granular materials are most conveniently determined by streaming potential measurements (Fuerstenau, 1956; Somasundaran and Kulkarni, 1973); such measurements are of considerable importance in the mining industry because of a strong correlation between the flotation properties of minerals and their ζ-potentials (Somasundaran, 1968, 1972). Other devices for the study of streaming potentials have been developed by Rutgers and de Smet (1947) and by Boumans (1957) who by means of turbulent flow in metal capillaries used streaming potentials to build up extremely high voltages (70,000 V) in a fluid-flow type of van de Graaff generator, developing, however, only 10^{-7} amps. Tuman (1963) measured streaming potentials through sandstone (up to several V) at very high pressure (up to 2,000 p.s.i.). Improvement on existing pressure ζ-potential measuring methods have been discussed by Korpi and de Bruyn (1972) and the interference of streaming potentials in the results obtained in automatic potentiometric systems was reported by Van den Winkel *et al.* (1974). The present author has described the role of streaming potentials elicited by the pressure difference used in ultrafiltration through charged membranes, on the ion retention by these membranes (van Oss, 1963; van Oss and Beyrard, 1963).

Streaming poential measurements have also been used to obtain information on the surface potentials of cells (van Wagenen *et al.*, 1976) and of various biopolymers (DePalma, 1976; DePalma *et al.*, 1977). As the pressures (and the concomitant shear forces) used in streaming potential measurements are considerable, the use of the method, in particular for the determination of cell surface potentials, is open to criticism, and without much doubt the use of the much gentler method of electroosmosis (see the preceding section) is to be preferred with living cells.

Streaming potentials *across* membranes (e.g., rabbit gall bladders) can be elicited by establishing differences in osmotic pressure on different sides of the membranes (Pidot and Diamond, 1964) as well as by mechanically applied pressures.

Sedimentation potentiasl elicited by the settling of charged particles under the influence of a gravitational field (Dorn effect) have been little studied by direct measurements. The influence of sedimentation potentials on the sedimentation coefficients of polyelectrolytes has been studied by Mijnlieff (1958). Flotation potentials caused by gas bubbles rising through a liquid are somewhat easier to measure; determinations were done at the beginning of this century by McTaggart (1914, 1922) and were continued by Alty (1924, 1926). More recently Usui and Sasaki (1978) reported on the "sedimentation" potentials resulting from the rise of small argon gas bubbles through aqueous solutions of cationic, anionic and non-ionic detergents.

ELECTROPHORESIS IN NON-AQUEOUS MEDIA

Lyklema (1968) gave one of the earlier reviews of this subject. Fowkes *et al.* (1982) described the electrophoresis of electron-accepting ($\oplus$) and of electron-donating ($\ominus$) particles, suspended in organic liquids which are electron-donating (for $\oplus$ particles) or electron-accepting (in the case of $\ominus$ particles). The rules are as follows:

$\oplus$ particles suspended in $\ominus$ liquids are negatively charged;
$\ominus$ particles suspended in $\oplus$ liquids are positively charged.

In addition, Fowkes *et al.* (1982) observed that $\oplus$ particles, suspended in $\oplus$ liquids in which $\ominus$ polymers have been dissolved, adsorb such $\ominus$ polymers on their surfaces (provided the $\oplus$ liquid has a dielectric constant $\varepsilon > 20$), and then become effectively $\ominus$ particles, with a positive charge. In 1982, Fowkes *et al.* used the classic (Rank) microelectrophoresis apparatus, necessitating observation at the stationary layer (see Seaman and Brooks, 1979). In 1987 Fowkes reported the use of a Pen Kem 3000 apparatus with liquids of low conductivity, and in 1988 Fowkes *et al.*, added the Coulter Delasa and the Matec instruments to devices with which electrohoresis in non-aqueous media can be effected.

Labib and Williams (1984) endeavored to arrive at a correlation between electron donicity values in (Kcal/mole) and ζ-potential values of various inorganic particles, by means of electrophoresis in a number of organic, $\oplus$, as well as $\ominus$ liquids. The same authors (1986) studied the connection between the aqueous pH scale and the electron donicity scale. And, returning to non-aqueous solvents Labib and Williams (1987) studied the effect of moisture on the ζ-potential or inorganic particles, in n on-aqueous liquids; it was observed that moisture can reverse the sign of charge of the solid surface and shift its electron donicity, making it more basic.

Labib and Williams (1984, 1986, 1987) could only indicate a qualitative correlation between ζ-potential and "donicity," with respect to the Gutmann scale (Gutmann, 1976, 1978). It may, however, be proposed that it is feasible to obtain a quantitative expression for the "donicity" (or "accepticity") of polar materials (1) from their ζ-potential in non-aqueous solvents, in terms of their $\gamma_1^\ominus$ (respectively, $\gamma_1^\oplus$) value, by using eqs. [V-1, 3, 5] to obtain ΔG_{131}^{EL}. One may then equate ΔG_{131}^{EL} with $\Delta G_{131}^{AB} = -2\gamma_{13}^{AB}$(eq. [II-33]), and using eq. [IV-7] one can then obtain, e.g., $\gamma_1^\ominus$, provided $\gamma_3^\oplus$ of the organic liquid (3) is known, and provided that material (1) and solvent (3) are monopolar (in the opposite sense, with respect to each other). The value of $\gamma_1^\ominus$, thus obtained, remains of course linked to the assumed reference value of $\gamma_W^\oplus = \gamma_W^\ominus$, for water at 20°C, but, as shown in Chapter IV, γ_{ij}^{AB} and the various ΔG^{AB} values derived from such a $\gamma_1^\ominus$ value are *not* linked to the assumed reference values for water.

It is of course necessary to know the dielectric constant (ε) of the liquid, and to determine the Debye length $1/\kappa$ (eq. [V-2]) for the particle in the liquid, and to know the κa ratio (Labib and Williams, 1984, 1986, 1987; Fowkes *et al.*, 1982), as well as the concentration of "counter ions" per unit volume of liquid (see Fowkes *et al.*, 1982). In connection with the low values for ε of most non-aqueous solvents, the ζ-potentials found are usually quite high (of the order of 100 mV); see Fowkes *et al.*, 1988. These authors also point out that the concentration of "counter ions" in the solvent is a function of the surface area of the particles per unit volume of the liquid.

For the linkage between ζ-potential and Lewis acid-base properties, see also Chapter XIX, and Schematic Chart XIX-1.

CODA AND LIMITATIONS

Analytical and preparative electrokinetic methods are described. Especially the latter approaches are treated in considerable detail.

It should be noted, however, that there are severe limits to the scaling-up possibilities of all preparative applications of, e.g., electrophoresis, due to the sharply increasing Joule heating with any increase in volume or thickness of electrophoresis chambers. The moment the dimension (perpendicular to the electric field) of such chambers surpasses about 1 cm in thickness, the excess heat developed by the applied electric field becomes increasingly difficult and/or expensive to dissipate by cooling. The same is true for High Voltage Electrophoresis; see under Zone Electrophoresis, above.

XVI Direct Measurement Methods, Treating the Force Balance in Particular

FORCE BALANCE

The contact angle method still remains the most accurate force balance methodology for measuring the interaction between two surfaces at the minimum equilibrium distance, $\ell = \ell_o$ (i.e., at molecular contact). But for determining the rate of decay of the interaction with distance, ℓ, one has to apply the various theoretical rate of decay rules which apply to each different interaction mode. However, one can measure the force of attraction between two surfaces, in air, or *in vacuo*, or immersed in a liquid, as a function of distance, ℓ, by means of a mechanical force balance.

This type of force balance was pioneered by Derjaguin and Abricossova (1951, 1953) in Russia, and by Sparnaay (1952), Overbeek and Sparnaay (1954), de Jongh (1958), and Black *et al.* (1960) in the Netherlands. These early devices were *inter alia* designed to verify the power mode of the decay with distance of retarded van der Waals–London forces (see Chapter VII). Derjaguin and Abricossova (1951, 1953) were the first to measure the interaction between a flat and a spherical quartz body as a function of distance. Sparnaay (1952), Overbeek and Sparnaay (1954), and at first also de Jongh (1958) tried to measure the force between two parallel flat quartz plates, but this conformation presented many drawbacks (difficulty of maintaining strict parallelism of the plates, difficulty of removing dust particles, difficulty of electrostatic effects). Most of these drawbacks could be overcome by adopting Derjaguin's sphere and flat plate configuration (de Jongh, 1958); see also Derjaguin *et al.* (1954), and van Silfhout (1966). At first the problem of build-up of electrostatic forces was combatted by the insertion of radioactive salts into the device (Derjaguin *et al.*, 1954; Overbeek and Sparnaay, 1954), but an even better measure appeared to be the periodic introduction of water vapor into the device followed by re-evacuation (de Jongh, 1958). Thus Derjaguin *et al.* (1954) and especially de Jongh (1958) and Black *et al.* (1960) were finally able to measure, even in the retarded mode, a Hamaker constant of quartz, that was at least of the same order of magnitude as the theoretically derived value. In 1954, Derjaguin *et al.* also described the first force balance capable of measuring interactions between two crossed cylinders immersed in a liquid, using two crossed quartz fibers. This crossed-fiber method was further improved (Derjaguin *et al.*, 1978, 1987; Rabinovich and Derjaguin, 1988).

Meanwhile, a number of British workers further improved the methodology by discarding glass or quartz (and other silica) surfaces, which have surface asperities of the order of 10 nm (van Blokland and Overbeek, 1979), replacing them with transparent hemicylindrical bodies coated with freshly cleaved mica, which is virtually molecularly smooth, thus allowing measurements to within about one Ångström (= 0.1 nm); see Tabor and Winterton (1969),* Israelachvili and Tabor (1972) and White *et al.* (1976); this device was still only used in air or *in vacuo*. This force balance was further developed by Israelachvili and Adams (1978), for use in liquids; it became the forerunner of the force balance now in use in laboratories all over the world. Anutech at Canberra (Australia) markets a version of this device, as described by Parker *et al.* (1989); see also Claesson (1986), Russell (1987), Christenson (1988) and Israelachvili and McGuiggan (1988). Figure XVI-1 shows a schematic presentation of Israelachvili's device, as illustrated by Israelachvili (1987); see also Israelachvili and McGuiggan (1988). Cain *et al.* (1978) developed a similar device, using, however, silicone rubber rather than mica-coated surfaces. For the use of a quartz plate and lens, in liquid, see Rouweler (1972), and van Blokland (1977), who also discuss some of the pitfalls of this general method, for determining unretarded and retarded Hamaker constants; see also White *et al.* (1976).

OTHER APPROACHES

Israelachvili (1985) briefly describes a number of other approaches for measuring surface forces, including the contact angle method (more extensively treated in Chapters IV and XVI, above), particle detachment experiments (Visser, 1968, 1970, 1976a, b), peeling experiments (Good, 1967b; van Oss *et al.*, 1977), measurement of the thickness of soap and other liquid films, determination of dynamic interparticle separations in liquids by means of light, X-ray (Cowley *et al.*, 1978; Parsegian *et al.*, 1979, 1986, 1987), or neutron scattering, and flocculation studies of particle suspensions. Direct osmotic pressure measurement methods on latex particle suspensions as a function of concentration was first done by Ottewill and colleagues (Barclay *et al.*, 1972; Ottewill, 1976; Cairns *et al.*, 1976). Here various pressures could be applied hydraulically, and the final latex concentration measured subsequently. Homola and Robertsen (1976) used a similar approach but they varied the latex concentration and subsequently measured the pressure (Napper, 1983). Using both approaches the repulsive force between latex particles could be measured, as a function of interparticle distance. The methods developed by Cowley *et al.* (1978) and Parsegian *et al.* (1979) also are essentially osmometric methods; see also Evans and Needham (1988).

---

* Tabor and Winterton (1969) thus were the first to give experimental proof that at $\ell \leq 10$ nm no retardation is apparent in van der Waals–London attractions, while at $\ell \geq 20$ nm the retardation regime completely governs the van der Waals–London decay.

In addition, mention should be made of the cell-pipetting approach used by Evans *et al.* (1988), which measures deformation of red cells or vesicles as a function of forces applied from the outside; see also Evans and Needham (1988). This approach is to some extent comparable to the method of measurement of the disjoining pressure between two bubbles, described by Derjaguin *et al.* (1954).

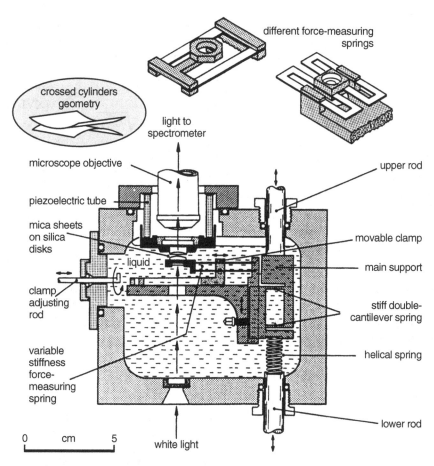

FIGURE XVI-1 Cross-section of Israelachvili's surface force apparatus for measuring the forces between two curved molecularly smooth surfaces in liquids, showing some of the alternative (interchangeable) force-measuring springs suitable for different types of experiments. The separation between the surfaces is measured (to better than 1 Å) by use of an optical technique using multiple beam interference fringes. The distance between the two surfaces is controlled by use of a three-stage mechanism: (i) coarse control (micrometer-driven upper rod), (ii) medium control (differential spring driven by lower rod), and (iii) fine control (to better than 1 Å via the voltage-driven piezoelectric crystal tube supporting the upper mica surface). The stiffness of the force-measuring spring can be varied by a factor of 1,000 by shifting the position of the dove-tailed clamp with the adjusting rod. (From Proc. Natl. Acad. Sci. USA, *84*, 4722–4724 (1987), Courtesy Professor Jacob N. Israelachvili.)

INTERFACIAL ATTRACTION EFFECTS
(HYDROPHOBIC INTERACTIONS) IN WATER

By coating the mica surfaces of the crossed cylinders of the force balance (see Fig. XVI-1) with a surfactant, consisting of alkane chains and a quaternary ammonium base (of which the positive charge combines with the negatively charged mica surface), one effectively coats the mica with a fairly homogeneous alkane monolayer facing outward.

In a number of force balance measurements of hydrophobic interactions, the attraction between layers of hexadecyl groups, in water, has been measured (Israelachvili, 1985; see also Herder, 1990). Combining eqs. [IV-34 and IV-35]:

$$\Delta G_{131}^{IF} = -2(\sqrt{\gamma_1^{LW}} - \sqrt{\gamma_3^{LW}})^2 + 4(\sqrt{\gamma_1^{\oplus}\gamma_3^{\ominus}}$$
$$+ \sqrt{\gamma_1^{\ominus}\gamma_3^{\oplus}} - \sqrt{\gamma_1^{\oplus}\gamma_1^{\ominus}} - \sqrt{\gamma_3^{\oplus}\gamma_3^{\ominus}})$$

[XVI-1]*

where IF stands for the total interfacial interaction, 1 for the solid and 3 for the (aqueous) liquid. Taking the solid to be hexadecane, for the hexadecyl-coated mica surfaces, where $\gamma^{LW} = \gamma = 27.5$ mJ/m^2, the appropriate values have been entered into eq. [XVI-1]:

$$\Delta G_{131} = -0.66 - 4 (0 + 25.5 - 0 - 0)$$
$$= -102.66 \text{ mJ/m}^2$$

[XVI-1A]

However, in experiments described by Pashley et al. (1985), a contact angle with water of only 95° was reported on the hexadecyl surfaces used. Hence these surfaces were not completely apolar. When employing the version of Young's equation which incorporates the polar interaction terms:

$$0.5 (1 + \cos\theta)\gamma_L = \sqrt{\gamma_S^{LW}\gamma_L^{LW}} + \sqrt{\gamma_S^{\oplus}\gamma_L^{\ominus}} + \sqrt{\gamma_S^{\ominus}\gamma_L^{\oplus}}$$

[IV-13C]

we find, after insertion of the contact angle with water of 95°:

$$33.23 = 24.48 + 8.75$$

[XVI-1B]

Thus, in this particular case, the polar part of the right-hand term of Young's eq. [IV-13C] then equals 8.75 mJ/m^2. Upon insertion of that value into eq. [XVI-1], this becomes:

$$\Delta G_{131} = -0.66 - 4 (0 + 25.5 - 8.75)$$
$$= -67.66 \text{ mJ/m}^2$$

[XVI-1C]

---

* A list of symbols can be found on pages 399–406.

Thus, the only slightly smaller experimental value (at closest approach) of approximately -60 mJ/m^2 (Pashley *et al.*, 1985) conforms rather well to the value found in eq. [XVI-1C], considering that with the residual polarity of the low-energy surfaces, a residual non-zero surface potential probably also still was present. It is clear from eq. [XVI-1A] that, ideally, $\approx 99\%$ of the total interfacial attractive energy is imputable to the hydrogen-bonding part of the cohesive energy of water. However, in somewhat less than ideal cases, where the hydrophobic surface still has some polar properties, the polar interaction energy between the solid and water (the two far right-hand terms of eq. [IV-13C]) must be subtracted from the ideally obtainable hydrophobic attraction energy (eqs. [XVI-1, XVI-1A, XVI-1C]); see van Oss and Good (1988b).

To cite a number of other examples: Claesson *et al.* (1986) measured an attractive energy ΔG_{131} of about -58 mJ/m^2 between dioctadecyl layers, in water, at contact; Christenson and Claesson (1988) found $\Delta G_{131} \approx -50$ to -80 mJ/m^2 for fluorocarbon surfaces in water, at contact, and Rabinovich and Derjaguin (1988) found $\Delta G_{131} \approx -30$ to -37 mJ/m^2 for the attraction, in water, between cylinders of fused quartz, previously "hydrophobized" by treatment with dimethyl dichlorosilane vapor. All these data were obtained by force-balances of the Israelachvili (Fig. XVI-1) or of the Derjaguin type (Rabinovich and Derjaguin, 1988), which are rather accurate for force vs. distance measurements at distances greater than the equilibrium distance. However *at* the equilibrium distance, i.e., at contact, attraction energies derived from contact angle measurements may be held to be more accurate, due to the unavoidable distortion of the cylinders' surfaces upon contact (see, e.g., Blomberg *et al.*, 1990). Thus, there is an overall quite satisfactory correlation between the interaction forces observed between between two low-energy surfaces immersed in water, derived from contact angle data, and those obtained from force-balance observations.

A strong attractive force between two different surfaces (i.e., one hydrophobized, positively charged, and one untreated negatively charged mica surface) was observed by Claesson *et al.* (1986); in addition to an electrostatic attraction, an interfacial attractive energy, $\Delta G_{132}^{\mathrm{IF}} \approx -35$ mJ/m^2 must have been involved here, which probably accounts for the rather high total attractive energy; $\Delta G_{132}^{\mathrm{TOT}} \approx -100$ mJ/m^2.

INTERFACIAL REPULSION EFFECTS
(HYDRATION REPULSION) IN WATER

The most-studied coating material which is capable of inducing non-electrostatic repulsion between particulate and other solid surfaces is, beyond doubt, polyethylene oxide (PEO) (or polyethylene glycol or polyoxyethylene). It is the major non-electrostatic stabilizing agent utilized in particle stabilization (see, e.g., Napper, 1983), and it also has become the principal coating material used in the measurement of repulsive forces in force-balance experiments (Israelachvili *et al.*, 1980; Claesson, 1986; Claesson and Gölander, 1987). The free energy of repulsion, at closest approach, between two layers of PEO (at $\gamma^{LW} = 43$ $\gamma^{\oplus} = 0$ and $\gamma^{\ominus} = 64$ mJ/m^2) (van Oss, *et al.*, 1987a); see also Table XVII-6), $\Delta G_{131} = +52.5$ mJ/m^2, using eq. [XVI-1]. Israelachvili *et al.* (1980) found a repulsive force of ≈ 80 mN/m between crossed mica cylinders, in the presence of 0.01% PEO; this corresponds to $\Delta G_{131} \approx +13$ mJ/m^2; see also Klein and Luckham (1984). Claesson and Gölander (1987) found, with mica surfaces electrostatically coated with PEO, forces F > 100 mN/m, i.e., $\Delta G_{131} >$ $+16$ mJ/m^2; see also Claesson (1986).* The value of $\Delta G_{131} = +52.5$ mJ/m^2 agrees well with an osmotic pressure $\Pi \approx 31$ MPa found for 60% PEO solutions by Gawrisch (1986); Arnold et al. (1988); see van Oss, Arnold et al. (1990), and Chapter XIX.

Observations on the repulsion between uncoated mica surfaces, under various conditions, have been reported since 1981 (Pashley, 1981, 1984). The repulsions were stronger than would be expected from just electrostatic interactions. In 5×10^{-4} M NaCl, repulsions of the order of $\Delta G_{131} \approx +16$ mJ/m^2 were observed (Pashley, 1981), but much smaller repulsions were measured in the presence of alkaline earth metal chlorides (Pashley and Israelachvili, 1984a). The latter observation correlates well with the known attenuating action of divalent cations on negatively charged electron-donating surfaces (Ohki et al., 1988). Pashley and Israelachvili (1984b) showed that, at close approach, in dilute potassium salt solution, repulsive energies of $\Delta G_{131} \approx +150$ mJ/m^2 (and sometimes even higher) could be observed. Oscillatory forces also were apparent in these observations.

INTERACTIONS BETWEEN ADSORBED POLYMER LAYERS IN APOLAR MEDIA

One of the early reports on the behavior of non-polar polymers adsorbed onto the mica half cylinders in Israelachvili's force balance was by Klein (1980). He used coatings of polystyrene, immersed in cyclohexane, and observed a long-range attraction (here mainly LW), and a short range repulsion (at distances $\ell \leq$ Rg). The short-range repulsion is entirely understandable, as the (LW) attraction is only of the order of -0.4 kT, so that the opposing polymer strands, as soon as they are brought within one another's sphere of action (i.e., to within 1 Rg), will repel each other by the energy of their Brownian motion (the same energy which allows the solubility of this polystyrene preparation in cyclohexane in the first place). Similar observations were reported by Luckham and Klein (1985) for layers of adsorbed PEO, in toluene. Both of these phenomena are relatively rare instances of steric stabilization *sensu stricto*; see Chapter XX.

INTERACTION BETWEEN ADSORBED BIOPOLYMERS AND BETWEEN ADSORBED PHOSPHOLIPIDS IN AQUEOUS MEDIA

It is a logical step to extend the use of the Israelachvili force balance to measurement of the interaction between adsorbed layers of polymers, phospholipids, etc. The

---

* These authors give only the F vs. ℓ plots; the steep, almost asymptotic increase in repulsive force at $\ell < 10$ nm makes it difficult to estimate the value of the repulsive force at the minimum equilibrium distance $\ell = \ell_o$.

fundamental problem that arises from this type of application lies in the growing uncertainty about the intermolecular distance, ℓ, with increasing polymer molecular weight, as well as in the uncertainty as to the perfection, homogeneity and thickness of the adsorbed layer, all of which also influence the value of ℓ.

Luckham *et al.* (1986) described the mutual repulsion between (myelin-like) layers of mixed cerebroside sulphate and cholesterol, and the attraction between one such lipid layer and an opposing layer of myelin basic protein. Lee and Belfort (1989) report the interaction between adsorbed layers of the basic protein ribonuclease A. They found that the (asymmetrical) RNase molecules were adsorbed onto the mica surface with their long axis perpendicular to the mica surface, and they describe a long range (non-DLVO) repulsion and a short-range attraction between RNase layers. Perez and Proust (1987) measured the interaction between opposing layers of the glycoprotein, bovine submaxillary mucin, adsorbed onto the two mica surfaces. At low surface coverage of mucin an attraction occurred, ascribed to polymer cross-binding. At higher surface coverage of the mica surfaces by mucin, a repulsion was observed. Curiously, the attraction due to cross-binding is a longer-range interaction than the repulsion, which the authors impute to "steric" effects. This paper illustrates the difficulties inherent in measuring the interactions, as a function of distance, ℓ, between coatings of undetermined thickness and orientation, of polymer molecules with one dimension which is greater than ℓ. However, the greater size of the cross-binding macromolecule is indubitably the reason for the longer range of this interaction and at higher mucin concentration, when all mucin receptors are occupied, an acid-base type repulsion (and not a "steric" one) is to be expected; see also Mirza *et al.* (1998).

Marra and Israelachvili (1985) and Marra (1985, 1986) studied the interaction, as a function of ℓ, between opposing phospholipid bilayers, coated onto mica. Usually a repulsion was observed; however, in the presence of Ca^{++} ions, an attraction prevailed. Helm *et al.* (1989) showed an interesting effect occurring upon close approach between two opposing phospholipid bilayers (within $\ell \approx 1$ nm), which has similarities to the interaction between two such bilayers in cell or vesicle fusion.

MEASUREMENT OF THE DECAY-LENGTH OF WATER

In view of the uncertainty about the precise value of ℓ, as well as about the strict homogeneity of the thickness of coated layers, attempts to measure the characteristic decay-length, λ, of water with Israelachvili's force balance, using various materials to coat the mica surfaces, would seem unadvisable. For some of the unusually large values of λ that have been reported, see, e.g., Christenson (1988), Rabinovich and Derjaguin (1988), and Derjaguin and Churaev (1989). For the influence of dissolved polymers on λ, see Chapter VI. A discussion on the largely artefactual nature of the unusually high values reported for λ_{H_2O} in the attractive modes can be found in Chapter VII.

ELECTROSTATIC EFFECTS

Electrostatic effects, i.e., the influence of ionic strength, ionic species, pH, etc., have been thoroughly studied with Israelachvili's force balance. On the whole, the DLVO theory (see Chapters VII and XXIII) only seems fit to describe the effects of electrical double layer forces observed with the force balance (Pashley and Israelachvili, 1984a; Israelachvili, 1985), as long as strong hydrophobic attractions (Christenson, 1988; Claesson *et al.*, 1987) or strong hydrophilic repulsions (see, e.g., Claesson, 1986) do not predominate. Also, the interaction of Ca^{++} ions tends to have an influence beyond that predicted by DLVO considerations (Marra, 1986; see also Prévost and Gallez, 1984; Ohki *et al.*, 1988; Mirza *et al.*, 1998).

OSCILLATORY EFFECTS

When the liquid medium consists of relatively large and rigid molecules, such as octamethylcyclotetrasiloxane (Horn and Israelachvili, 1981; Israelachvili, 1985), or benzene or cyclohexane, at short range force balance measurements divulge oscillating interactions, tending to fluctuate in a regular fashion between attraction and repulsion; see also Christenson (1988). Israelachvili (1985) ascribes these oscillatory interactions to solvation forces. Pashley and Israelachvili (1984b) also observed short-range ($\ell \leq 5$ nm) oscillatory effects in the repulsion between two mica surfaces, in aqueous media containing 10^{-3} M to 1 M KCl.

ATOMIC FORCE MICROSCOPY

Whilst Force Balance analysis, as well as contact angle determinations, only measure macroscopic-scale interactions (see Chapters XVI, XIII and XXIV), with the more recently developed methodology of Atomic Force Microscopy (AFM) (Binnig *et al.*, 1986), one can study macroscopic-scale as well as microscopic-scale interactions, or even a mixture of the two modalities.

Using AFM, Lee *et al.* (1994) measured the forces between complementary strands of DNA, which is a purely microscopic-scale interaction. On the other hand, clearly macroscopic-scale hydrophobic as well as hydrophilic interactions between a small sphere and a flat plate have been measured by Ducker *et al.* (1992), by means of AFM. A hybrid mixture of macroscopic and microscopic-scale interactions between poly(ethylene oxide) layers adsorbed onto glass surfaces was studied with AFM, by Braithwaite *et al.* (1994). Furthermore, AFM also permits the visualization of various other hybrid phenomena occurring during the adhesion or adsorption of different biological entities, as well as of the crystal growth of, e.g., viruses. These observations include the AFM visualization of surfactant micelles adhering to hydrophobic surfaces (Manne *et al.*, 1994).

CODA

An historical overview is given of the development of the force-balance. Recent results with Israelachvili-type balances are briefly discussed (e.g., "hydrophobic" attraction, hydration repulsion, influence of adsorbed polymers, electrostatic interactions, oscillatory effects), followed by a brief discussion on applications of atomic force microscopy.

Part IV

Associated Phenomena
and Applications

XVII Surface Tension Components and Parameters of Liquids and Solids

COMPLETELY APOLAR LIQUIDS

As can be seen from Table XVII-1, if one needs to use an apolar liquid to measure the γ^{LW} component of a solid surface (using eq. IV-28A), as soon as one deals with solids with γ_S^{LW}-values greater than about 28 mJ/m^2, the choice, among the alkanes, is rather poor because these are close to the freezing point, at 20°C. In practice, however, there are a few higher-energy liquids which are apolar for all practical purposes (see also Table XVII-10), e.g.:

Diiodomethane, for which $\gamma_L^{LW} \approx \gamma_L = 50.8$ mJ/m^2 and $\gamma^\oplus \approx 0.01$ mJ/m^2.

TABLE XVII-1

Interfacial Tensions with Water and Surface Tension Properties of Alkanes at 20°C in mJ/m^2

Liquid	γ_{iw}^{o} [a]	γ [b]	γ^{LW}	γ^{AB}	$\gamma^\oplus$	$\gamma^\ominus$
Pentane	51.4	16.05	16.05	0	0	0
Hexane	51.1	18.40	18.40	0	0	0
Heptane	51.0	20.14	20.14	0	0	0
Octane	51.0	21.62	21.62	0	0	0
Nonane	51.0	22.85	22.85	0	0	0
Decane	51.0	23.83	23.83	0	0	0
Undecane	51.1	24.66	24.66	0	0	0
Dodecane	51.1	25.35	25.35	0	0	0
Tridecane	51.2	25.99	25.99	0	0	0
Tetradecane	51.2	26.56	26.56	0	0	0
Pentadecane	51.3	27.07	27.07	0	0	0
Hexadecane	51.3	27.47	27.47	0	0	0
Nonadecane	51.5	28.59	28.59	0	0	0
Eicosane	51.5	28.87	28.87	0	0	0
Cyclohexane	51.1	25.24	25.24	0	0	0

[a] Calculated using eq. IV-9; see also Chapter XIII and Table XIII-1.
[b] From Jasper, 1972.

With diiodomethane it is possible to obtain γ_L^{LW} values for most solid surfaces. It is, incidentally, advisable to calculate the γ_S^{LW} values first when using eq. [IV-28A], even for polar solids because contact angle measurements with the two apolar liquids mentioned above (which usually do not react with the solid, so that the contact angle does not tend to decay) are somewhat more accurate than those obtained with the more polar (and thus more reactive) liquids. The apolar and the polar liquids most used in direct contact angle determinations, or in wicking, are grouped in Table XVII-10.

In TableXVII-1 the γ_{iw}^o values (i.e., the zero time dynamic interfacial tensions with water; see van Oss, Giese and Good, 2002) have been derived (using eq. IV-9) from the surface tensions of a number of alkanes, which for alkanes are equal to their apolar, Lifshitz–van der Waals surface tensions (given that with alkanes the $\gamma^\oplus$, the $\gamma^\ominus$ and hence the γ^{AB} values are all zero, so that here, $\gamma = \gamma^{LW}$; see eqs. IV-2 and IV-5). The γ_{iw}^o values thus derived are extremely close to the experimentally obtained values published by Girifalco and Good (1957). These experimental values had been obtained earlier in the 20th century by drop-weight or drop-shape measurements on the largely water-immiscible drops of these alkanes, when immersed in water. With this approach however, it has more recently been shown by van Oss and Good (1996) that it only produces correct results when applied to completely non-polar liquids, whereas when applied to even slightly polar organic liquids, this experimental approach produces γ_{iw} values that are considerably lower than the correct ones; see Chapter XVII.

MONOPOLAR LIQUIDS, IMMISCIBLE WITH WATER

For those water-immiscible partly polar organic liquids of which it may reasonably be assumed that they are monopolar, the knowledge of their interfacial tension with water allows the estimation of the value of the monopolar $\gamma^\ominus$ (or $\gamma^\oplus$, as the case may be), using eq. [IV-9].

However, with partly polar liquids, monopolar or otherwise, the values of their interfacial tensions with water, which had been earlier obtained experimentally by drop-shape or drop-weight measurements, as given in Table XVII-2 of the 1994 edition of this book, have since been shown to be much too low, due to the almost instantaneous spatial anisotropy arising among the polar and the non-polar moieties of these polar liquids, at the organic liquid-water interface (van Oss and Good, 1996). Some work was done more recently to determine the correct interfacial and surface tension properties of a number of predominantly monopolar, water-immiscible organic liquids, where the zero time dynamic interfacial tension with water (γ_{iw}^o) (see Chapter XVII) could be obtained from these liquids' aqueous solubilities (van Oss, Wu et al., 2001; van Oss, Giese and Good, 2002). The $\gamma^\oplus$ or $\gamma^\ominus$ parameters of these monopolar organic liquids could then be derived from their γ^{LW} and γ_{iw} values, using eq. IV-9; see Table XVII-2.

DIPOLAR LIQUIDS, IMMISCIBLE WITH WATER

Knowledge of the interfacial tension with water alone does not suffice to determine the $\gamma^\oplus$ as well as the $\gamma^\ominus$ values of dipolar liquids, but the interfacial tension with

TABLE XVII-2

Interfacial and Surface Tension Properties of Some Partly Polar (Predominantly Monopolar) Liquids at 20°C in mJ/m^2

Liquid	γ_{iw}^{0} [a]	γ	γ^{LW}	γ^{AB} [b]	$\gamma^{\oplus}$	$\gamma^{\ominus}$
Ethyl acetate [c]	25.95	23.9	23.9	≈ 0	≈ 0	6.2
Ethyl ether [c]	21.0	17.0	17.0	≈ 0	≈ 0	9.0
N-Octanol	31.2	27.5	27.5	≈ 0	≈ 0	3.97
Benzene	41.6	28.85	28.85	≈ 0	≈ 0	0.96
Toluene	42.9	28.5	28.5	≈ 0	≈ 0	0.72
o-Xylene	44.0	30.1	30.1	≈ 0	≈ 0	0.58
Chloroform [c]	38.8	27.15	27.15	≈ 0	1.5	≈ 0
Diiodomethane [c]	56.1	50.8	50.8	≈ 0	≈ 0.01	≈ 0

[a] Zero time dynamic interfacial tension (van Oss, Giese and Good, 2002); see also Table XIII-2. The γ_{iw}^{0} values of these organic liquids have been determined from their aqueous solubilities; see eq. XIII-3 and Chapter XIII in general. Then, assuming monopolarity for all these compounds, their $\gamma^{\oplus}$, or their $\gamma^{\ominus}$ values (as the case may be) could be determined from the values of their γ_{iw}^{0} and their γ^{LW}, using eq. IV-9.

[b] As all these partly polar liquids are predominantly monopolar, *i.e.*, they are either mainly electron-acceptors (with a finite $\gamma^{\oplus}$ value) or electron-donors (with a finite $\gamma^{\ominus}$ value), so that their γ^{AB} is zero or very close to zero (van Oss, Chaudhury and Good, 1987).

[c] See also van Oss, Wu, *et al.* (2001).

water, *and* the γ^{AB} of such liquids (using eqs. [IV-5] and [IV-9] will allow the determination of both $\gamma^{\oplus}$ and $\gamma^{\ominus}$; see Table XVII-3. γ^{AB} is determined by measuring the contact angle of such liquids on a low-energy apolar surface, such as Teflon or Parafilm, using eqs. [IV-2] and [IV-28A]; see also Chapter XVI. This has been done, to date, with only a few liquids, but can be done in principle, with all dipolar water-immiscible liquids. The final results may not always be obtainable with a high degree of accuracy, but one can get at least the orders of magnitude with this approach.

NAPHTHALENE

The surface tension components and parameters of naphthalene are given in Table XVII-3.

A curious case is that of naphthalene: it was earlier supposed that the surface tension of solid naphthalene was lower than that of liquid naphthalene (Neumann *et al.*, Colloid Polymer Sci., 1979). That supposition, however, was based on the known surface tension of liquid naphthalene (Table XVII-3; see also Jasper, 1972), which is correct, and the surface tension of solid naphthalene, obtained only from the contact angle with water, and the utilization of the "equation of state," now known to be flawed (see Chapter XIV). In reality the surface tension of solid naphthalene is higher than that of liquid naphthalene, thus following a general rule that denser materials tend to have a higher surface tension than less dense materials.

MONOPOLAR AND DIPOLAR WATER-MISCIBLE LIQUIDS

The γ-components and polar γ-parameters of a number of water-miscible liquids are given in Table XVII-4. In many cases these data are not easily obtainable. With ethanol and methanol, the γ^{AB} values could be measured, by contact angle measurement on, e.g., Teflon, but the $\gamma^{\oplus}/\gamma^{\ominus}$ ratios had to be estimated, as contact angle measurements

TABLE XVII-3
Surface Tension Components and Parameters of Solid and Liquid Naphthalene, in mJ/m²

	γ	γ^{LW}	γ^{AB} [b]	$\gamma^{\oplus}$	$\gamma^{\ominus}$
Naphthalene (liquid) [a]	32.8	32.8	0	≈ 0	≈ 1.0
Naphthalene (solid) [b]	42.7	42.7 [b]	0	≈ 0	1.36 [c]

[a] From Neumann, Omenyi and van Oss, 1979, measured at 80°C; at 20°C, by extrapolation, $\gamma \approx 39.4$ mJ/m².
[b] At 20°C; from θ (DIM) = 31.7° and θ (αBrN) = 19.5°.
[c] From θ (H$_2$O) = 90°, θ (Gly) = 76.5° and θ (FO) = 61°, and presuming that this compound is a monopolar electron donor; however, there may be a small $\gamma^{\oplus}$ value, which might explain its pronounced solubility in ether.

TABLE XVII-4
Surface Tension Compounds and Parameters of Water-Miscible Dipolar Liquids, at 20°C in mJ/m²

Liquid	γ	γ^{LW}	γ^{AB}	$\gamma^{\oplus}$	$\gamma^{\ominus}$
Methanol	25.5 [a]	18.2 [b]	4.3	≈ 0.06 [b]	≈ 77 [b]
Ethanol	21.4 [a]	18.8 [b]	2.6	≈ 0.019 [b]	≈ 68 [b]
Ethylene glycol	48	29 [d]	19 [d]	3.0 [c]	30.1 [c]
Glycerol	64	34 [d]	30	3.92 [f]	57.4 [f]
Formamide	58	39 [d]	19	2.28 [f]	39.6 [f]
Dimethylsulfoxide	44	36	8	0.5 [g]	32 [g]
Water	72.8 [a]	21.8 [h]	51	25.5 [i]	25.5 [i]

[a] Jasper, 1972.
[b] From van Oss et al., Chem. Rev., 1988; the $\gamma^{\oplus}$ and $\gamma^{\ominus}$ values are estimated from the high solubility of these alcohols in water and from the γ^{AB} values.
[c] From van Oss, Ju et al., 1989.
[d] From Chaudhury, 1984.
[e] From solubility data (Chapter XXII).
[f] From van Oss, Good and Busscher, 1990; see also van Oss et al., J. Colloid Interf. Sci., 1989.
[g] Estimated average from van Oss et al., J. Colloid Interf. Sci., 1989.
[h] From Fowkes, 1963.
[i] From van Oss et al., Advan. Colloid Interface Sci., 1987; J. Colloid Interface Sci., 1989; see Chapter IV.

on polar (usually high-energy) surfaces are excluded, due to the spreading of these low-energy liquids on higher-energy surfaces. Inclusion of polar liquids in a gel has worked relatively well for, e.g., formamide and ethylene glycol (van Oss, Ju *et al.*, 1989), but ethanol and methanol evaporate so rapidly that the gel method is not very promising for these last two liquids. It also is not easy to obtain very reliable polar data for dimethylsulfoxide (DMSO), due to its extraordinary hygroscopicity. The best values for glycerol and formamide were obtained *via* contact angle measurement, on a variety of known monopolar polymer surfaces (van Oss, Good and Busscher, 1990). For ethylene glycol (EG) the values for $\gamma_{EG}^{\ominus}$ and $\gamma_{EG}^{\oplus}$ are derived from its known γ_{EG} and γ_{EG}^{AB} (Chaudhury, 1984), and measurements by Professor H.J. Busscher; see van Oss, Ju *et al.*, (1989, note added in proof) replacing earlier values from the 1994 edition of this book, which are now known to be erroneous. They were based on an incorrect value of the solubility of ethylene glycol (Stephen and Stephen, 1963). In reality there is no finite aqueous solubility for ethylene glycol at room temperature; it is completely miscible with water (see, e.g., the Merck Index, 11th ed., 1989).

SYNTHETIC POLYMERS

With the usual array of contact angle liquids, i.e., diiodomethane, α-bromonaphthalene, water, glycerol and often, formamide (see also Table XVII-10), contact angles have been measured on a variety of polymer surfaces; see Table XVII-5. It can be seen that most polar organic polymers have a γ^{LW} of ≈ 40 mJ/m^2 $\pm$ 10%. Also, most of these polar polymers are mainly, if not solely, electron-donors. The strongest electron-donor of these is, without doubt, polyethylene oxide.

BIOPOLYMERS

The surface tension data of various biopolymers are given in Tables XVII-6 to 9. It can be seen that all these polar biopolymers (in the dry state) also have γ^{LW} values close to 40 mJ/m^2, $\pm$ 10%, with the exception of fibronectin, which possibly still contained residual water (Table XVII-6). All (dry) biopolymers studied appear to be preponderantly electron donors.

PLASMA PROTEINS

As seen in Table XVII-6, the plasma proteins studied appear to fall into two categories: those that have a relatively low $\gamma^{\ominus}$-value, even in the dry state (e.g., fibrinogen, fibronectin, low density lipoprotein), and those that have a relatively low $\gamma^{\ominus}$-value in the dry state (which would preclude their solubility in water in that state), but in the hydrated state have a much higher $\gamma^{\ominus}$ (which then would allow their ready dissolution in water): e.g., HSA and IgG. The most likely explanation is that the latter proteins become reversibly denatured upon drying, but revert to their native configuration when rehydrated (van Oss *et al.*, J. Prot. Chem., 1986; van Oss and Good, J. Prot. Chem., 1988; van Oss, J. Prot. Chem., 1989). It can indeed be readily observed, when immersing dried HSA or IgG in water, that solubilization of these proteins

TABLE XVII-5
Surface Tension Components and Parameters of Various Synthetic Polymers, at 20°C in mJ/m²

Polymer	γ	γ^{LW}	γ^{AB}	$\gamma^{\oplus}$	$\gamma^{\ominus}$
Teflon FEP	17.9	17.9[a]	0	0	0
Polyisobutylene	25.0	25.0[b]	0	0	0
Polypropylene	25.7	25.7[b]	0	0	0
Corona-treated polypropylene	33.0	33.0[c]	0	0[c]	11.1[c]
Polyethylene	33.0	33.0[d]	0	0	0
Nylon 6,6	37.7	36.4[c]	1.3	0.02[c]	21.6[c]
Polymethyl-methacrylate (PMMA)	40.0	40.0[e]	0	0[e]	14.6[e]
PMMA	40.6	40.6[b]	0	0[b]	12[b]
PMMA	41.4	41.4[c]	0	0[c]	12.2[c]
Polystyrene	42	42[f]	0	0	1.1[f]
"Cell culture" polystyrene[g]	46.8	43.7	3.1	0.12	20.0
Polyvinylalcohol	42	42[e]	0	0[e]	17–57[e,h]
Polyvinylpyrrolidone	43.4	43.4[i]	0	0[i]	29.7[i]
Polyvinylchloride	43.8	43.0[b]	0.75	0.04[b]	3.5[b]
Polyethyleneoxide (PEG—6,000)	43.0	43.0[e]	0	0[e]	64[e]
Polyethyleneoxide (PEG—6,000)	45.9	45.9[e]	0	0[c]	58.5[c]
Polyoxytetramethylene glycol[j] (MW ≈ 2.000)	44.0	41.4[k]	2.6	0.06[l]	27.6[l]
Co-poly(ethylene glycol, propylene glycol)[j] (MW ≈ 2.000)	47.5	42.0[m]	5.5	0.13[n]	58.8[n]
Co-poly(ethylene glycol, propylene glycol)[j] (MW ≈ 1.000)	47.9	40.9[o]	7.0	0.22[p]	55.6[p]

[a] From Chaudhury, 1984.
[b] From van Oss et al., J. Separ. Sci. Tech., 1989.
[c] From van Oss, Good and Busscher, J. Disp. Sci. Tech., 1990.
[d] From θ (DIM) = 52°: Fowkes, J. Adh. Sci. Tech., 1987.
[e] From van Oss et al., Advan. Colloid Interface Sci., 1987.
[f] From van Oss et al., J. Colloid Interface Sci., 1986.
[g] "Cell culture quality" polystyrene dishes, Nunc, Roskilde, Denmark; from θ (H_2O) = 58.9°, θ (DIM) = 28.7°, θ (αBrN) = 16.5°, θ (Gly) = 60.8°, θ (FO) = 36.8°.
[h] Depending on the sample used.
[i] Unpublished results; from θ (DIM) = 30°, θ (αBrN) = 17°; $\sqrt{\gamma^{\ominus}}$ averaged from: θ (H_2O) = 56.6°, θ (Gly) = 51.8; θ (FO) = 49.5°.
[j] Courtesy of Prof. E. Dellacherie, Laboratoire de Chimie-Physique Macromoléculaire, CNRS, Nancy France.
[k] From θ (αBrN) = 21.5°.
[l] From θ (H_2O) = 52°, θ (Gly) = 49.75°, θ (FO) = 44.5°.
[m] From θ (αBrN) = 19°.
[n] From θ (H_2O) = 19°, θ (Gly) = 48°, θ (FO) = 21.5°.
[o] From θ (αBrN) = 23°.
[p] From θ (H_2O) = 23°, θ (Gly) = 41°, θ (FO) = 22°, θ (EG) = 26°.

appears to be a two-step process, the first one being one of swelling and hydration, followed by actual dissolution; see also Chapter XXII. It should be noted that the surface of a non-globular plasma protein, $i.e.$, fibrinogen, remains hydrophilic, even when dried (Table XVII-6).

The orientation of the water of hydration of HSA, as a function of the thickness of the hydration layer, has been described earlier (van Oss and Good, J. Prot. Chem., 1988a) Briefly, the first layer of water appears to be ≈98% oriented (with the oxygen atoms outward), and the second layer ≈30% oriented in the same manner. The very similar (and rather low) γ^{LW} values of the surface of HSA hydrated with one layer and with two layers of water most likely is an indication of the fact that in both cases the contact angle liquid mainly "sees" a more distal surface of bound water of which the average Lifshitz–van der Waals properties are relatively independent of the orientation of the water molecules (Table XVII-6).

TABLE XVII-6
Surface Tension Components and Parameters of Plasma Proteins at 20°C, in mJ/m²

Plasma protein	γ	γ^{LW}	γ^{AB}	$\gamma^{\oplus}$	$\gamma^{\ominus}$
Human serum albumin (HSA), dry, pH 4.8	45.0	44.0[a]	1.0	0.03[a]	7.6[a]
HSA, dry, pH 7	41.4	41.0[b]	0.4	0.002[b]	20[b]
HSA, hydrated, 1 layer of hydration water, pH 7	27.6	26.6[b]	1.03	0.003[b]	87.5[b]
HSA, hydrated, 2 layers of hydration water, pH 7	62.5	26.8[b]	35.7	6.3[b]	50.6[b]
Human immunoglobulin-G (IgG) dry pH 7	45.2	42[a]	3.2	0.3[a]	8.7[a]
IgG, hydrated, pH 7	51.3	34[a]	17.3	1.5[a]	49.6[a]
Bovine fibrinogen, dry	40.3	40.3[c]	0	0	53.2[c,d]
Bovine fibrin, dry	44.0	40.2	3.8	0.3	12.0[e]
Human fibrinogen, dry	40.6	40.6[f]	0	0	54.9[f,g]
Human fibrinogen, hydrated	41.3	37.4[h]	3.9	0.1[h,i]	38.4[h,i]
Human fibronectin, dry	59.3	29.5[j]	29.8	4.3[j,k]	51.6[j,k]
Human low density lipoprotein, dry	41.1	34.45[l]	5.66	0.26[l,m]	30.8[l,m]
Human immunoglobulin-A (IgA), hydrated, pH 7	26.8	26.8	0[n]	0[n]	93.0[n]

[a] From van Oss, J. Protein Chem., 1989.
[b] From van Oss and Good, J. Protein Chem., 1988 cf. Table XIX-4.
[c] Unpublished results; from θ (DIM) = 39.25°, θ (α BrN) = 24°.
[d] From θ (H_2O) – 36°, θ (Gly) = 52°.
[e] From van Oss (Biofouling, 1991; J. Prot. Chem., 1990).
[f] Unpublished results; from θ (DIM) = 37.5°, θ (αBrN) = 25°.
[g] From θ (H_2O) = 31.8°, θ (Gly) = 53.25°, θ (Fo) = 37.6°.
[h] Unpublished results; from θ (DIM) = 43.7°, θ (αBrN) = 34°.
[i] From θ (H_2O) = 46.5°, θ (Gly) = 51.75°, θ (Fo) = 48°.
[j] Unpublished results; from θ (DIM) = 58°, θ (αBrN) = 51.25°.
[k] From θ (H_2O) = 11.2°, θ (Gly) = 19.5°, θ (Fo) = 12.5°.
[l] Unpublished results; from θ (DIM) = 48.5°, θ (αBrN) = 37.25°.
[m] From θ (H_2O) = 52.8°, θ (Gly) = 54.6°, θ (Fo) = 49°.
[n] From van Oss, Moore $et~al.$, 1985; given that θ (hexadecane) = 12.9° and θ (H_2O) = 0° and assuming $\gamma^{\ominus}$ monopolarity.

A curious phenomenon occurs at the isoelectric point of proteins, best illustrated with human serum albumin (HSA), at its isoelectric pH of ≈ 4.8. The $\gamma^\ominus$ value of dry HSA at pH 4.8 is much lower than at pH 7.0. At pH 4.8, dry HSA has much the same surface tension values as IgG at pH 7.0, which is close to the latter's *average* isoelectric point, but it must be kept in mind that (polyclonal) IgG is heterogeneous and has a whole array of isoelectric points, varying from ≈ 5.5 to ≈ 7.5. HSA, however, is much more homogeneous and has only one isoelectric point. In the hydrated form, isoelectric HSA remains in solution, but the molecules appear to turn their hydrophobic moieties inward, and orient their most hydrophobic aspects toward the solution/air interface (see also van Oss and Good, J. Prot. Chem., 1988; Table XVII-6)). The increased hydrophobicity close to the isoelectric pH (as manifested by an increase in contact angle values with water), of amphoteric solid surfaces, has been discussed by Holmes-Farley (1985), and was also observed by Fokkink and Ralston (1989). Thus, whilst ζ-potentials on the one hand, and $\gamma^\oplus$ and $\gamma^\ominus$ values on the other, are, in principle, independent variables, in practice, a change in ζ-potential nevertheless tends to be accompanied by a change in (usually) $\gamma^\ominus$. For the linkage between AB and EL forces, see Chapter XIX, and the end of Chapter XV.

OTHER PROTEINS

Lysozyme, as well as tobacco mosaic virus (Table XVII-7) also appears to fall in the category of globular proteins such as HSA and IgG, which in the dry state have become sufficiently (reversibly) denatured as to become insoluble in water (note their relatively low $\gamma^\ominus$ values, when dry). Myelin basic protein, on the other hand, is more akin to fibrinogen; it appears directly soluble in water, without the necessity to become hydrated prior to its dissolution. Fowkes (1987) also mentions the necessity for polyethylene oxide (PEO) (see Table XVII-5) to be hydrated in order to be soluble, but this may not be quite the same mechanism. At room temperature PEO clearly should be quite soluble in water, even without hydration. It is true that at temperatures near the θ-point PEO becomes insoluble, while its hydration should increase with an increase in temperature, but it is also likely that the surface properties of PEO change, when it approaches its θ point. The dramatic decrease in the solubility of PEO in water, with an increase in temperature of the order of only $30°C$, argues against the entropic nature of the "sterically" stabilizing properties of PEO; see also Chapters XVIII, XX and XXIII. Heating aqueous protein solutions to $\approx 60°C$ or beyond in many cases also results in (often irreversible) insolubilization, which is held to be due to structural changes ("denaturation"). A curious exception is Bence Jones protein, a light-chain dimer of human immunoglobulins which in about 16% of monoclonal dysproteinemias is produced in excessive quantities and then finds its way in the urine. When heated to 60°C, Bence Jones protein precipitates (as do many other plasma proteins), but upon continued heating, it redissolves at 95° to 100°C (Rose, 1979). In any event, there are many other examples of polymers which become insoluble at temperatures around 60°C. That insolubility is probably due to a reorientation ("denaturation") of the polymer chains, causing some of their more "hydrophobic" moieties to become exposed to the interface with water. It seems

TABLE XVII-7

Surface Tension Components and Parameters of Various Other Proteins, at 20°C, in mJ/m²

Protein	γ	γ^{LW}	γ^{AB}	γ	γ
Lysozyme, dry (egg white)	48.8	41.2[a]	2.6	0.07[a]	23.4[a]
Lysozyme, hydrated (egg white)	≈72.2	31.5[a]	≈40.7	≈4.5[a,b]	≈56.2[a,b]
Tobacco mosaic virus, dry	51.8	44.7[a]	7.1	0.66[a]	18.9[a]
Tobacco mosaic virus, hydrated	39.45	27.15[a]	12.3	0.66[a]	57.1[a]
Myelin basic protein, dry (rabbit brain)	37.0	37.0[c]	0	0	35.1[c,d]
Myelin basic protein, hydrated (rabbit brain)	56.4	26.0[e]	32.4	5.2[e,f]	50.5[e,f]
Zein (corn protein), dry	42.8	41.1[g]	1.7	0.04	18.4[g]
Gelatin, dry	37.6	37.6[h]	0	0	18.5[h,i]
Collagen, dry	48.9	42.0[j]	6.9	0.57[j]	21.1[j]

[a] From contact angles in van Oss *et al.*, J. Protein Chem., 1986.
[b] Extremely hydrophilic; these values are approximative.
[c] Unpublished results; from θ (DIM) = 48°, θ (αBrN) = 30.5°.
[d] From θ (H_2O) = 54°, θ (Gly) = 65.5°.
[e] Unpublished results; from θ (hexadecane) = 19.5°.
[f] From θ (H_2O) = 17°, θ (Gly) = 24.7°.
[g] From contact angles in van Oss, Good and Busscher, J. Dispersion Sci. Tech., 1990.
[h] Unpublished results; from θ (DIM) = 44°, θ (αBrN) = 32.5°.
[i] From θ (H_2O) = 67°, θ (Gly) = 67°.
[j] From θ (DIM) = 31.5°; θ (αBrN) = 25°, θ (H_2O) = 55°, θ (Gly) = 49.5°, θ (FO) = 37.5°.
Sample (Vitrogen, Lot: Hide AO 119) obtained through courtesy of Collagen Corporation, Palo Alto, California.

likely that this is also the principal mechanism responsible for the insolubilization of PEO at the cloud point, rather than a decrease in hydration; see above.

The corn protein, zein, is insoluble in water (hot or cold), but soluble in DMSO, formamide and ethylene glycol (van Oss *et al.*, J. Protein Chem., 1986; van Oss and Good, J. Macromolec. Chem., 1989). Zein appears to become feebly hydrated in water, but not sufficiently to become soluble; it is, however, partly soluble in 70/30 (v/v) acetone-water mixtures. When hydrated, zein becomes insoluble in formamide and in ethylene glycol (van Oss and Good, J. Macromolec. Sci. Chem., 1989).

Gelatin is insoluble in water at room temperature, but soluble at 100°C. Upon cooling down to room temperature, aqueous gelatin solutions gellify. At low concentrations, i.e., below approximately 3 (w/v), most types of gelatin solutions (formed at 100°C) remain liquid upon cooling.

CARBOHYDRATES

As with proteins, the γ^{LW} values of polymeric as well as of monomeric glucides appear to be around 40 mJ/m². Not surprisingly, the soluble polysaccharides tend to have $\gamma^{\ominus}$ values higher than 28 mJ/m², while the less soluble ones usually have a

$\gamma^\ominus < 28$ mJ/m^2; see Table XVII-8. In a number of cases, the $\gamma^\oplus$ values, while still small, are not negligible. Depending upon the circumstances, with some of the polysaccharides, one finds either a small $\gamma^\oplus$ value, or under slightly different conditions, $\gamma^\oplus \approx 0$; see, e.g., dextran (cf. van Oss *et al.*, Advan. Coll. Int. Sci., 1987). In view of their virtual ($\gamma^\ominus$) monopolarity, it is not surprising that most water-soluble polysaccharides can play a role in aqueous phase separation, or coacervation, when mixed with ethanol or propanol, or with PEO, or gelatin, or polyvinyl alcohol solutions (Bungenberg de Jong, 1949; Albertsson, 1986; see also van Oss, J. Disp. Sci. Tech., 1988; see also Chapter XXI).

TABLE XVII-8A
Surface Tension Components and Parameters of Carbohydrates, at 20°C, in mJ/m^2; all Measured in the Dried State

Carbohydrate	γ	γ^{LW}	γ^{AB}	$\gamma^\oplus$	$\gamma^\ominus$
Cellulose	54.5	44.0[a]	10.5	1.6[a,b]	17.2[a,b]
Cellulose acetate (film on glass, evaporated from MEK soln.)	52.6	44.9[c]	7.7	0.8[c,d]	18.5[c,d]
Cellulose nitrate	45.1	44.7[e]	0.4	0.003[e]	13.9[e]
Agarose	41	41[f]	0	0	26.9[f]
Ficoll 400 (poly-sucrose)	41.4	41.4[g]	0	0	57.9[g,h]
Dextran T-70 (poly-α (1 $\rightarrow$ 6) glucose)	55.5	41.8[i]	13.7	1.0[i]	47.2[i]
Dextran T-150	42.0	42.0[i,k]	0	0	55.0[i,l]
Arabino-galactan (MW $\approx$ 80,000; from Larex Intl. Co., Tacoma, WA)	50.2	37.6[m]	12.6	0.75[n]	53.1[n]
Inulin (chicory root)	54.0	45.0[o]	9.0	0.37[p]	55.1[p]
Inulin (dahlia tubers)	56.3	45.3[q]	11.0	0.59[r]	51.7[r]
Na-alginate[s] (low viscosity)	50.2	45.6[t]	4.6	0.15[u]	35.3[u]
Amylopectin (from corn)	48.4	40.1[v]	8.3	0.74[w]	23.4[w]
Amylose (from potato)	52.6	42.3[x]	10.3	0.64[y]	41.4[y]
Hydroxyethyl starch	41.2	39.3[z]	0.9	0.007[aa]	29.85[aa]
Sucrose[bb]	41.6	41.6	0	0	36.1
Lactose	41.1	41.1	0	0	26.8
Cellobiose	42.3	42.3	0	0	26.1
D-Mannose	40.6	40.6	0	0	36.2
Galactose	41.1	41.1	0	0	36.0
Glucose[bb]	42.2	42.2	0	0	34.4
α-Cyclodextrin	43.4	43.4	0	0	20.2
β-Cyclodextrin	45.2	45.2	0	0	11.8
γ-Cyclodextrin	45.5	45.5	0	0	25.2

[a] Unpublished results; from θ (DIM) = 29°, θ (αBrN) = 13°.
[b] From θ (H$_2$O) = 53°, θ (Gly) = 50.25°, θ (Fo) = 16°.
[c] Unpublished results; from θ (DIM) = 24°, θ (αBrN) = 13°.
[d] From θ (H$_2$O) = 54.3°, θ (Gly) = 45°, θ (Fo) = 30.5°.
[e] From contact angles in van Oss *et al.*, J. Chromatog., 1987.

[f] van Oss, Good and Busscher, J. Dispersion Sci. Tech., 1990.

[g] Unpublished results; from θ (DIM) = 34.7°, θ (αBrN) = 24.3°.

[h] From θ (H$_2$O) = 31°, θ (Gly) = 56°, θ (Fo) = 42.3°.

[i] van Oss and Good, J. Macromolec. Sci. Chem., 1989

[j] From van Oss et al., 1987a.

[k] Unpublished results; from θ (DIM) = 29.2°, θ (αBrN) = 27.5°.

[l] From θ (H$_2$O) = 4°, θ (Gly) = 26.8°.

[m] Unpublished results; from θ (DIM) = 44.5°, θ (αBrN) = 32°.

[n] From θ (H$_2$O) = 23.5°, θ (Gly) = 47.2°, θ (Fo) = 12.5°.

[o] From θ (DIM) = 23.25°; θ (αBrN) = 13°.

[p] From θ (H$_2$O) = 13°; θ (GLY) = 36°; θ (FO) = 12°.

[q] From θ (DIM) = 21.5°; θ (αBrN) = 13.5°.

[r] From θ (H$_2$O) = 14.6°; θ (GLY) = 29.5°; θ (FO) = 13°.

[s] Courtesy of Prof. E. Dellacherie, Laboratoire de Chimie Physique Macromoléculaire, CNRS, Nancy, France.

[t] From θ (DIM) = 53.1°; θ (αBrN) = 31.5°.

[u] From θ (H$_2$O) = 41.2°; θ (Gly) = 57°; θ (FO) = 38°; θ (EG) = 37.5°.

[v] From θ (DIM) = 38°; θ (αBrN) = 27.5°.

[w] From θ (H$_2$O) = 52.9°; θ (Gly) = 55.5°; θ (FO) = 27.5°.

[x] From θ (DIM) = 26.6°; θ (αBrN) = 22.5°.

[y] From θ (H$_2$O) = 32°; θ (Gly) = 40°; θ (FO) = 19°.

[z] From θ (DIM) = 38°; θ (αBrN) = 32°.

[aa] From θ (H$_2$O) = 55°; θ (Gly) = 60°.

[bb] See Chapter XXII.

TABLE XVII-8B

Surface Tension Components and Parameters of Carbohydrates, at 20°C, in mJ/m^2-monomers and Oligomers (see Table XVII-8C for contact angle data); All Measured in the Dried State

Carbohydrate	γ	γ^{LW}	γ^{AB}	$\gamma^{\oplus}$	$\gamma^{\ominus}$
Glucose	42.2	42.2	0	0	51.1[a]
Fructose	42.5	42.5	0	0	56.1
Sucrose	41.6	41.6	0	0	59.5[a]
Lactose	41.1	41.1	0	0	62.4
Galactose	41.1	41.1	0	0	53.4
Fucose	40.7	40.7	0	0	48.7
Maltose	44.9	41.3	3.6	0.05	63.7
Maltotriose	52.0	43.2	8.8	0.35	55.0
Maltotetraose	53.1	43.0	10.1	0.49	52.3
Isomaltose	42.5	41.8	0.7	0.002	63.4
Isomaltotriose	49.0	38.7	10.3	0.48	55.5
L-Mannose	40.1	40.1	0	0	57.0
D-Mannose	40.6	40.6	0	0	55.2
Cellobiose	42.3	42.3	0	0	62.9
α-Cyclodextrin	53.4	43.4	10.0	0.54	46.0
γ-Cyclodextrin	55.9	45.5	10.4	0.98	27.6

[a] For data on glucose and sucrose in the dissolved state, see Docoslis et al. (2000).

TABLE XVII-8C

Contact Angles (in degrees) on Dried[a] Layers of Carbohydrates: Monomers and Oligomers; All Measured in the Dried State

	Contact angle liquid				
Carbohydrate	Diiomethane	α-Bromo-naphthalene	Water	Glycerol	Formamide
Glucose	31.75°	22.75°	20.75°	57.0°	37.0°
Fructose	30.25°	23.0°	22.0°	52.25°	36.1°
Sucrose	33.5°	24.5°	19.0°	52.5°	34.5°
Lactose	36.5°	23.5°	22.0°	46.0°	40.0°
Galactose	35.5°	24.75°	15.0°	59.0°	36.75°
Fucose	36.0°	26.5°	15.0°	59.0°	42.5°
Maltose	34.0°	25.5°	15.75°	44.5°	31.5°
Maltotriose	28.5°	20.5°	17.6°	36.0°	23.5°
Maltotetraose	29.5°	21.5°	19.75°	35.0°	21.75°
Isomaltose	32.5°	25.0°	20.0°	49.0°	33.0°
Isomaltotriose	42.0°	29.5°	22.0°	42.0°	27.0°
L-Mannose	38.5°	26.5°	13.0°	59.5°	36.0°
D-Mannose	36.0°	27.0°	13.8°	60.0°	35.5°
Cellobiose	31.5°	23.0°	15.0°	41.0°	33.0°
α-Cyclodextrin	28.0°	20.0°	27.5°	34.0°	26.0°
γ-Cyclodextrin	19.0°	15.0°	43.0°	40.0°	17.0°

[a] A 5% solution of the sugar in water was deposited on glass microscope slides and allowed to air-dry overnight. The dried slides were then kept for three days at a constant temperature of 105°C. After this the slides were put into a vacuum desiccator to cool down to 20°C, after which the contact angle measurements were done as soon as possible.

NUCLEIC ACIDS

At first sight the difference between RNA and DNA (Table XVII-9) seems slight, but comparing the values of the dry nucleic acids, DNA is clearly more dipolar and more hydrophobic than RNA. This allows DNA to bind strongly to cellulose esters (in the form of membranes: such DNA binding then is essential to "Southern Blotting"), while RNA will not bind to cellulose esters at all; see Chapter XXIV, and van Oss *et al.* (J. Chromatog., 1987).

CONTACT ANGLE LIQUIDS

In Table XVII-10 the surface tension components and parameters, as well as the viscosities, are given of the liquids most commonly used in contact angle measurements and in wicking.

TABLE XVII-9
Surface Tension Components and Parameters of Nucleic Acids, at 20°C, in mJ/m^2

Nucleic Acid	γ	γ^{LW}	γ^{AB}	$\gamma^\oplus$	$\gamma^\ominus$
RNA, dry (yeast)	35.9	35.9[a]	0	0	73.6[b]
RNA, hydrated (yeast)	62.9	32.0[a]	30.9	4.9[a]	48.7[a]
DNA, dry (calf thymus)	47.15	40.15[a]	7.0	0.62[a]	18.8[a]
DNA hydrated (calf thymus)	63.1	≈36.8[a]	26.3	3.9[a]	44.4[a]

[a] From θ values from van Oss et al., J. Chromatog., 1987. The γ and γ values given in that publication should not be used; the method of their derivation from the various contact angles was still approximative at the time that paper was published. The values given above, which are based on the published contact angles, are the more accurate ones.
[b] From assumption of monopolarity and θ (H$_2$O) = 16.5°; see van Oss et al., J. Chromatog., 1987.

TABLE XVII-10
Surface Tension Components and Parameters of a Number of Liquids Used in Direct Contact Angle Determination, or in Wicking, at 20°, in mJ/m^2, as Well as Their Viscosities, η at 20°C in Poises

Liquid	γ	γ^{LW}	γ^{AB}	$\gamma^\oplus$	$\gamma^\ominus$	η^a
APOLAR						
Decane	23.83[b]	23.83	0	0	0	0.0092
Tetradecane	26.6[b]	26.6	0	0	0	0.0218
Pentadecane	27.07[b]	27.07	0	0	0	0.029
cis-Decalin	32.2	32.2	0	0	0	0.0338
α-Bromonaphthalene	44.4[c]	44.4	≈0	≈0	≈0	0.0489
Diiodomethane	50.8[c]	50.8	0	≈0	0	0.028
POLAR						
Water	72.8[c]	21.8[c]	51.0	25.5	25.5[c]	0.01
Glycerol	64[c]	34[c]	30	3.92[c]	57.4[c]	14.9
Formamide	58[c]	39[c]	19	2.28[c]	39.6[c]	0.0375[d]
Ethylene glycol	48[c]	29[c]	19	3.0[c]	30.1[c]	0.199

[a] From CRC Handbook of Chemistry and Physics.
[b] See Table XVII-1.
[c] See Table XVII-4.
[d] From Merck Index, 11th ed. (1989).

SURFACE TENSION PROPERTIES OF CLAYS AND OTHER MINERALS

The surface tension properties of clay particles and other minerals are treated in Giese et al. (1996) and, more extensively, in Giese and van Oss (2002).

CODA

Tables are given of the γ, γ^{LW}, γ^{AB}, $\gamma^\oplus$ and $\gamma^\ominus$ values of alkanes, other water-immiscible liquids (apolar, monopolar and dipolar), water-miscible liquids, synthetic polymers, plasma proteins, polysaccharides, other carbohydrates, nucleic acids. The chapter includes a table of the γ-values and the viscosities of liquids most used in contact angle measurements. All values are given for 20°C, except when stated otherwise.

XVIII

Attractive LW- and AB-Forces: Hydrophobic Interactions

The molecular mechanism of the strong long-range "hydrophobic" interactions between mainly apolar macromolecules or particles immersed in aqueous media does not yet seem to be established with convincing clarity, or unanimity of pinion (Israelachvili, 1985; Claesson, 1986). Israelachvili (1985, p. 105) states that "on the theoretical side the problem is horrendously difficult, and there are no simple theories of the hydrophobic interaction, though a number of promising approaches have been proposed"; see also Israelachvili (1991, p. 132). Claesson (1986, p. 82) concludes that "several years of experimental and theoretical research work concerning repulsive and attractive hydration forces [must be] anticipated before a satisfactory understanding of the origin[s], properties and implications of these forces are obtained." We showed earlier (van Oss *et al.*, J. Colloid Interface Sci., 1986) that hydrophobic interactions arise out of a combination of Lifshitz–van der Waals forces and hydrogen bonds, where the contribution of the latter forces is the preponderant one by far. We proposed at the time that the attractive part of the term *interfacial forces*, or *interfacial interactions*, would more accurately describe hydrophobic interactions (van Oss and Good, J. dispersion Sci. Tech., 1988). It should be realized that, even teflon or hydrocarbons strongly *attract* water (see also Hildebrand, 1979; Tanford, 1979 and Tanford, 1980, p.3), with a free energy $\Delta G_{iw} \approx -40$ to -50 mJ/m^2. Thus the term "hydrophobic" used to characterize such materials is a misnomer. It may even be suggested that the epithet "hydrophobic" has significantly retarded a more general understanding of the phenomena discussed in this chapter (see also van Oss and Good, 1991b, as well as Chapter IX, which treats hydrophobic interactions from the point of view of the properties or water).

INTERACTION BETWEEN TWO IDENTICAL ORGANIC MOLECULES, IMMERSED IN WATER

Systems in which an organic compound exhibits low solubility in water tend to engage in hydrophobic interactions in aqueous media (Tanford, 1973, 1980; Franks, 1975). It has been said (Tanford 1973, 1980) that water "resists the intrusion of nonpolar solute molecules," or, as Israelachvili puts it (1985, p. 105): "water simply loves itself too much to let some substances get in its way." It is widely held that such interactions, in which there is an apparent strong mutual attraction between apolar molecules immersed in water, have an entropic origin (Israelachvili et al., 1989; Hiemenz, 1986, p. 448). Hydrophobic interactions are believed to be closely related to the "hydrophobic effect" (Israelachvili, 1985), which is described by Tanford (1973, 1980).

The notion that entropic interactions are a salient attribute of the "hydrophobic effect" originates from the marked increase in entropy accompanying the transfer of *hydrocarbons* from an apolar solvent to water (Tanford, 1973, 1980). However, organic compounds other than hydrocarbons also are subject to "hydrophobic interactions," but these can give rise to entropy changes that are quite different from those reported for hydrocarbons.

A more general approach to understanding the thermodynamics of interaction between apolar or polar molecules or particles, immersed in water is to determine the free energy of interaction between molecules or particles, once they are immersed in water (cf. Tanford, 1979), and to do this not only with hydrocarbons, but also with other hydrophobic and even hydrophilic compounds.

An important thermodynamic property in a binary system comprising, e.g., an organic compound (i) and water (w), may be expressed as ΔG_{iwi} (analogous to ΔG_{131} used, e.g., in Chapter IV). This is the free energy change per unit area for the process in which, say, organic molecules of phase (i) are initially present in the aqueous phase (w), with an effectively infinite layer of phase (w) separating two surfaces of phase (i), followed by a coalescence of the organic molecules into one phase (i). If this occurs spontaneously, the organic molecules are designated as hydrophobic; if the reverse process occurs spontaneously, the organic molecules are hydrophilic. In the process, two surfaces of (i) are brought together seamlessly. It has been shown (van Oss and Good, 1991a,b) that the free energy of interaction, ΔG_{iwi}, is directly related to the interfacial tension, γ_{iw}, between compound (i) and water (see eq. [IV-17]):

$$\Delta G_{iwi} = -2\gamma_{iw} \qquad \text{[XVIII-1]}^*$$

To determine the surface thermodynamic functions (in terms of enthalpic and entropic contributions) of attractive interfacial (i.e., "hydrophobic") interactions, [as well as of repulsive interfacial (i.e., hydrophilic) interactions; see Chapter XV], it is necessary to ascertain the interfacial free energy ΔG_{iwi} between the apolar (or the polar) molecules or surfaces, immersed in water, at different temperatures. The theory of interfacial interactions in aqueous media has been given above, in Chapter III; see also van Oss et al. (Adv. Colloid Interface Sci., 1987; Chem. Rev., 1988).

---

* A list of symbols can be found on pages 399–406.

ENTHALPY AND ENTROPY OF HYDROPHOBIC
INTERACTIONS

We found earlier (van Oss and Good, 1991b), basing ourselves on data produced by Harkins (1952) which were obtained by drop-shape/drop-weight methods (see Girifalco and Good, 1957) that interaction energies between various organic liquids and water could be mainly enthalpic, or mainly entropic, or a mixture of the two in almost any proportion (van Oss, 1994). However, the results obtained with drop-shape/drop-weight methods upon which the above conclusions had been based, have since been found unreliable for all organic liquid/water systems, with the exception of alkane/water systems (see Table XVII-1; see also van Oss and Good, 1996; van Oss, Giese and Good, 2002; van Oss and Giese, 2004; 2005; see also Chapter XIII).

It thus became desirable to re-determine the contributions of enthalpic and entropic components of the interaction energies of a number of compounds of different hydrophobicities, when immersed in water. This was done by deriving the ΔG_{iwi} values (i.e., the free energies of interaction between molecules of organic compound, i, when interacting in water, w) from their aqueous solubilities which, contrary to drop-shapes or drop-weights, are not affected by polar/apolar anisotropies. The connection between ΔG_{iwi} and and the aqueous solubilities of organic compounds is expressed as (see also eqs. XIII-2A and XIII-3) as:

$$\Delta G_{iwi} = (kT.\ln s)/S_c \qquad \text{[XVIII-2]}$$

(where k is Boltzmann's constant, T the absolute temperature, s the aqueous solubility in mole fractions and S_c the contactable surface area between two molecules of i, immersed in water). Then, from the ΔG_{iwi} values obtained at different temperatures from the aqueous solubilities of i, at different temperatures, using:

$$\Delta G = \Delta H - T\Delta S \qquad \text{[XVIII-3]}$$

and:

$$\Delta H = (d\Delta G/d\Delta T)T \qquad \text{[XVIII-4]}$$

both the ΔH and $T\Delta S$ components of ΔG could be derived. This holds for the free energy of interaction between two molecules, i, immersed in water, w (i.e., ΔG_{iwi}), as well as for the free energy of hydration of i (ΔG_{iw}). For the latter one uses the Dupré equation (see also eqs. II-36 and IV-6):

$$\Delta G_{iw} = \gamma_{iw} - \gamma_i - \gamma_w \qquad \text{[XVIII-5]}$$

where (see also eqs. XIII-2A and XVIII-1):

$$\gamma_{iw} = -0.5 \, \Delta G_{iwi} \qquad \text{[XVIII-1A]}$$

Thus, for the determination of ΔG_{iw} and its ΔH_{iw} and $T\Delta S_{iw}$ components, ΔG_{iwi} must be known, as well as γ_i and γ_w, at two (or more) temperatures. For γ_i and γ_w values at different temperatures, one can consult Jasper (1972) for a large collection of organic liquids, as well as for water.

New results are given in Table XVIII-1 for a number of liquid solutes and one solid one, from hexane, which is completely hydrophobic (i.e., apolar) ($\Delta G_{iwi} = -102.4$ mJ/m^2) down to sucrose which is closest to being hydrophilic ($\Delta G_{iwi} = -11.3$ mJ/m^2), although technically still slightly hydrophobic, as is also indicated by its finite aqueous solubility (cf. Chapter XIII, and van Oss and Giese, 1995). Completely hydrophilic solutes could not be chosen, because when $\Delta G_{iwi} > 0$, there no longer exists an aqueous solubility that can be quantitatively measured with any accuracy (see Chapter XXII), which makes it impossible to determine its precise ΔG_{iwi} values from solubility data. Within the range of formally hydrophobic solutes (van Oss and Giese, 1995), from completely non-polar hexane to the almost totally polar sucrose, a number of interesting conclusions can be reached, see Table XVIII-1:

1. Hydrophobic interactions are not necessarily mainly entropic. Whilst ΔG_{iwi} for CCl$_4$, benzene and aniline is for more than 50% entropic, for the completely apolar hexane molecules*), ΔG_{iwi} is largely enthalpic and is even associated with a modest but non-neglible negative entropy. On the other hand, for the least hydrophobic solute, sucrose, ΔG_{iwi} is for 80% entropic, a percentage which is nearly as high as the percentage of the entropic contribution found for CCl$_4$.

   There thus appears to be no general rule at all about the percentage of entropic contribution to the free energy of hydrophobic interactions. This should perhaps not be unduly surprising, as in many related systems the proportion of enthalpic to entropic contributions in different systems varies considerably with T, where usually the (negative) value of $T\Delta S$ increases and that of ΔH decreases with an increase in T, a phenomenon called entropy/enthalpy compensation (Mukkur, 1980; 1984).

2. Concerning the free energy of hydration, there does appear to be a general tendency for ΔG_{iw} to manifest a negative entropy value (ΔS_{iw}) in all cases shown in Table XVIII-1, which is entirely plausible, as water of hydration, even of hydrophobic surfaces, may be expected to be more structured than bulk water; see also Tanford (1973; 1980) and Hvidt (1983).

3. The penultimate right hand column of Table XVIII-1, reading from top to bottom, shows a rather regular decrease in the (negative) ΔG_{iwi} values, with the exception of CCl$_4$ and benzene, for which these values are about the

---

* Although our ΔG_{iwi} and ΔG_{iw} values (in mJ/m^2) for hexane, (derived from Harkins' data of 1952) agree with those given by Hildebrand (1979), our ΔH_{iwi} and $T\Delta S_{iwi}$ values, in kcal/M, which we obtained from solubility data, differ from those quoted by Tanford (1980), who used data from Gill et al. (1976), which they obtained by calorimetry, involving the "transfer of hydrocarbons from solutions in organic solvents to water." The difference in experimental approaches most likely account for the variance in ΔH_{iwi} and $T\Delta S_{iwi}$ values found with the two dissimilar systems. Probably for the same reason, our results for benzene also differ from those given by Gill et al. (1975).

same. This decrease in ΔG_{iwi} simply portrays the decrease in hydrophobicity (see van Oss and Giese, 1995). The far right hand column, also reading from top to bottom, shows a strong increase in the (negative) values of the free energy of hydration, ΔG_{iw}, concomitant with the decrease in hydrophobicity. In the cases of CCl_4 and benzene, the ΔG_{iw} values are also about the same, like the ΔG_{iwi} values.

4. It is important to note however, that even the completely "hydrophobic" apolar compound, hexane, still shows a significant free energy of hydration, of -37.5 mJ/m2, thus demonstrating that even the most "hydrophobic" compounds do not really "fear water." This confirms that the word "hydrophobic" is definitely a misnomer, as already pointed out earlier by Hildebrand (1979). In Chapter XI it is shown that the only surface known to be genuinely totally "hydrophobic," i.e., a surface whose ΔG_{iw} is really zero, is the air side of the water-air interface.

5. Finally whilst, by definition, the (negative) value of ΔG_{iwi} increases with an increase in hydrophobicity, the free energy of hydration, $|\Delta G_{iw}|$ decreases with an increase in hydrophobicity. This could be predicted because, as mentioned above, the polar (AB) aspect of the hydration of molecule, i, is linked to the value of $\gamma_i^{\ominus}$ in term V of eq. IV-2 (see Table IV-2), where $\gamma_i^{\ominus}$ (and thus term V) veers to zero the more i tends toward complete hydrophobicity.

APOLAR (LW) AND POLAR (AB) CONTRIBUTIONS TO HYDROPHOBIC INTERACTIONS

HYDROPHOBIC ATTRACTIONS (ΔG_{iwi})

Apolar, i.e., Lifshitz–van der Waals (LW) contributions to hydrophobic interactions in water (w) between apolar molecules of particles (i), are extremely small, as can be seen by comparing the values of ΔG_{iwi}^{LW} and ΔG_{iwi}^{AB}, shown in columns 1 and 2 of Table XVIII-2. For ΔG_{iwi} of hexane, the LW contribution is only 0.8% of the total free energy of interaction (shown in column 3), which means that in purely hydrophobic interactions virtually all the work is done by polar, AB forces. Then, for the slightly more polar compounds as well as for the almost totally and the really totally polar compounds, sucrose and dextran, their ΔG_{iwi}^{LW} values vary only between 6.3 and 6.6 mJ/m2. The ΔG_{iwi}^{LW} values of the more apolar compounds have little or nothing to do with the polarity or otherwise of these molecules, but only with the fact that the most apolar molecules also tend to have the lowest Hamaker constants and thus (see Chapter II) also the lowest γ^{LW} values (see column 1 of Table XVIII-2) which are therefore closest to the γ_w^{LW} of water of 21.8 mJ/m2. Thus their γ_{iw}^{LW} (see eq. II-35) and therefore also their ΔG_{iwi}^{LW} values (see eq. II-33A) remain fairly small.

TABLE XVIII-1

Values of ΔG_{iwi}, ΔH_{iwi}, $T\Delta S_{iwi}$, and ΔG_{iw}, ΔH_{iw}, $T\Delta S_{iw}$; ΔS; Solubilities (s) (in %, w/v). Contactable Surface Areas, S_c; and ΔG_{iwi}; and ΔG_{iw}.

Solute	$\Delta G =$ (kcal/M)	ΔH^d (kcal/M)	$-T\Delta S^f$ (kcal/M)	%$T\Delta S$	ΔS (cal/M/°K)	s^g (%) (at 20°C)	S_c^g (nm²)	ΔG_{iwi}^i (mJ/m²)	ΔG_{iw}^j (mJ/m²)
Hexane	$\Delta G_{iwi}^{a,b} = -6.24$	-7.35	$+1.11$	(neg.) (18%)	-3.8	0.039	0.425	-102.4	-37.5
	$\Delta G_{iw}^c = -2.44$	-4.82	$+2.38$		-8.1				
CCl$_4$	$\Delta G_{iwi}^b = -5.40$	-0.57	-4.83	89%	16.8	0.077	0.453	-83.0	-60.4
	$\Delta G_{iw}^c = -4.00$	-10.98	$+6.98$		-24.2				
Benzene	$\Delta G_{iwi}^b = -4.50$	-1.44	-3.06	68%	10.3	0.175	0.38	-83.2	-60.1
	$\Delta G_{iw}^c = -3.17$	-8.77	$+5.60$		-18.8				
Aniline	$\Delta G_{iwi}^b = -2.87$	-1.11	-1.76	61%	5.8	3.68	0.44	-46.9	-91.0
	$\Delta G_{iw}^c = -5.56$	-8.61	$+3.05$		-10.1				
Sucrose^k	$\Delta G_{iwi}^{b,e} = -1.49$	-0.38	-1.56	80%	5.3	67.1	1.20	-11.3	-208.9
	$\Delta G_{iw}^c = -36.0$	-49.3	$+13.3$		-45.5				

a From Harkins (1952), who used drop-shape/drop-weight analysis for the hexane-water system. Harkins' value of $\Delta G_{iwi} = -102.4$mJ/m² corresponds closely to the value of -102.9mJ/m² found from the aqueous solubility of hexane (interpolated between the solubilities of pentane and heptane; see van Oss and Good, 1996)

b ΔG_{iwi} from the aqueous solubilities of i; see eq. XVIII-2; it should be noted that 1kT = 0.58 kcal/mole, at 20°C.

c ΔG_{iw} from the Dupré equation: $\Delta G_{iw} = \gamma_{iw} - \gamma_i - \gamma_w$ (of eqs. II-37 and IV-6), where $\gamma_{iw} = -0.5\Delta G_{iwi}$.

d See eq. XVIII-3. The two T-values were as follows (in degrees Kelvin): Hexane (293°, 313°); CCl$_4$ (228°, 303°); Benzene (283°, 298°); Aniline (303°, 323°); Sucrose (293°, 303°).

e See also Docoslis et al. (2002).

f $T\Delta S$ from eq. XVIII-3.

g Aqueous solubilities, s, from Stephen and Stephen (1963); in eq. XVIII-3 s must be expressed in mole fractions.

h The contactable surface area, S_c: Hexane, by interpolation between the S_c values of pentane and heptane, see van Oss and Good (1996); CCl$_4$, from its molecular dimensions; see, e.g., Pauling and Hayward (1964); Benzene from van Oss, Giese and Good (2002); Aniline by analogy with toluene, see van Oss, Giese and Good (2002); Sucrose, from Docoslis et al. (2002).

i ΔG_{iwi} in mJ/m², from the compound's aqueous solubility and S_c; see eq. XVIII-2. ΔG_{iwi} (expressed in mJ/m²) also represents the quantitative measure of hydrophobicity (when $\Delta G_{iwi} < 0$) and of hydrophilicity (when $\Delta G_{iwi} > 0$), see van Oss and Giese (1995).

j ΔG_{iw} in mJ/m², i.e. the free energy of hydration of compound, i. ΔG_{iw} can also be used as a more approximative measure of hydrophobicity ($\Delta G_{iw} < |-113$mJ/m²$|$) and of hydrophilicity ($\Delta G_{iw} > |-113$mJ/m²$|$), see Table XIX-1.

k In aqueous solutions; see Docoslis et al. (2000).

TABLE XVIII-2

ΔG^{LW} and ΔG^{AB} Contributions to ΔG_{iwi} and ΔG_{iw}, for the Organic Compounds also Shown in Table XVIII-1, as well as for Dextran. Also Given are the γ^{LW}, γ^{AB}, γ^{TOT}, $\gamma^{\oplus}$ and $\gamma^{\ominus}$ Values for These Compounds. All Data are in mJ/m^2, at 20°C, Except Where Stated Otherwise. For the ΔG^{LW} and ΔG^{AB} Values the Dominant Contributions (either LW or AB) are Also Given, in % (in parentheses) of the Total.

Solute		1 LW Contribution	2 AB Contribution	Total ΔG^c	γ^{LW}	γ^{AB}	γ^{TOT}	$\gamma^{\oplus}$	$\gamma^{\ominus}$
Hexane	ΔG_{iwi}	−0.8	−102.0 (99.2%)	−102.4	16.05	0	16.05	0	0
	ΔG_{iw}	−37.45 (100%)	0	−37.45					
CCL$_4$	ΔG_{iwi}	−0.6	−83.0 (99.3%)	−83.6	27.0	0	27.0	0.88	0
	ΔG_{iw}	−48.5 (83.7%)	−9.5	−58.0					
Benzene	ΔG_{iwi}	−1.0	−82.2 (98.8%)	−83.2	28.9	0	28.9	0	0.96
	ΔG_{iw}	−50.2 (85.5%)	−9.9	−60.1					
Aniline[a]	ΔG_{iwi}	−7.8	−39.1 (83.4%)	−46.9	43.2	≈0	43.2	≈0[b]	9.45
	ΔG_{iw}	−60.2 (66.2%)	−30.8	−91.0					
Sucrose[d]	ΔG_{iwi}	−6.3 (55.7%)	−5	−11.3	41.6	100.2	141.8	28.5	88.0
	ΔG_{iw}	−60.2	−148.7 (71.2%)	−208.9					
Dextran[d]	ΔG_{iwi}	−6.6	+36.4	+29.8	42.0	21.4	63.4	2.0	57.0
	ΔG_{iw}	−60.3	−90.5 (60.0%)	−150.8					

[a] All data on aniline are at 30°C.
[b] There may be a small $\gamma^{\oplus}$ value for aniline, but this has not been determined to date.
[c] To obtain γ^0_{iw}, see eq. XVIII-6.
[d] In aqueous solution; data from Docoslis *et al.* (2000).

HYDROPHOBIC HYDRATION (ΔG_{iw})

Contrary to the mutual hydrophobic attraction between two apolar molecules, immersed in water (ΔG_{iwi}), which is virtually completely Lewis acid-base (AB) driven as shown in Table XVIII-2 for the case of hexane, the free energy of attraction between such apolar molecules and water (ΔG_{iw}), i.e., their free energy of hydration, is exclusively Lifshitz–van der Waals driven; see hexane, where ΔG_{iw} is for 100% LW (Table XVIII-2, column 1). Going down columns 1 and 2, as solutes become less hydrophobic, the % ΔG_{iw}^{LW} diminishes gradually from the 100% for hexane, until ΔG_{iw} is seen to be for 71.2% AB driven with sucrose and for 60% in the case of dextran. The higher ΔG_{iw}^{AB} figure for sucrose, compared with dextran, is due to the much higher $\gamma_i^{\oplus}$ and $\gamma_i^{\ominus}$ values for sucrose than for dextran (see Docoslis *et al.* 2000) which causes higher values of both terms IV and V (eq. IV-2; see Table IV-2) for sucrose, where terms IV and V, together, are proportional to ΔG_{iw}.

From Tables XIII-1 and XIII-2 it can be seen that the most hydrophobic (apolar) compounds are the least soluble in water. (The degree of hydrophobicity is proportional to the positive value of γ_{iw}^o, cf. eq. XVIII-1 and Chapter XIX). Many apolar compounds become less soluble in water as the temperature increases, but with partly polar compounds it usually is the other way around (see, e.g., Stephen and Stephen, 1963). It is also clear, from Tables XVIII-1 and XVIII-2 that the most hydrophobic compounds also have the lowest hydration energies ($|\Delta G_{iw}|$), but the highest free energies of interaction between two molecules of the compound, when immersed in water ($|\Delta G_{iwi}|$). However, otherwise these two types of free energies are not directly connected. Thus Muller (1990) was right in stating that "it is not necessary to accept the paradoxical assertion that hydrophobic hydration increases the solubility of nonpolar species at room temperature," an opinion held, *inter alii*, by Privalov and Gill (1988). As the hydration energy of the most hydrophobic compounds is predominantly apolar, i.e., due to Lifshitz–van der Waals attractions, it is a misapprehension to believe that "van der Waals and hydration effects are contributing to the hydrophobic effect with opposite signs" (Privalov and Gill 1989). This manner of reasoning also leads these authors to believe that "the hydrophobic interaction should be attractive at short distances and repulsive at long distances (exceeding the size of a water molecule)," in spite of much evidence to the contrary (Israelachvili, 1985; Pashley *et al.*, 1985; Claesson *et al.*, 1986; Christenson and Claesson, 1988; Rabinovich and Derjaguin, 1988; see also Chapter XVI). In reality, hydration effects are not *contributing* anything to hydrophobic attractions (ΔG_{iwi}) between *apolar* surfaces, immersed in water, but they can *counteract* the always-present hydrophobic attraction among *polar* surfaces, when immersed in water, see Table IV-2, and Chapter XIX.

CONCLUSIONS

Thus the free energy of purely *hydrophobic interactions* (ΔG_{iwi}) is solely AB driven, whilst the free energy of *hydration* of completely hydrophobic entities (ΔG_{iw}) is solely LW-driven. With more polar compounds the LW component of ΔG_{iwi} increases slightly, but remains small, even with the most polar compounds, whilst the free energy of hydration (ΔG_{iw}) of the more polar compounds can reach 71.2% AB of the total hydration energy, as in the case of sucrose, with 60% AB in the case of dextran, see also Table XVIII-3.

TABLE XVIII-3
Summary of the Involvement of the Physical Forces and the Enthalpy and Entropy
Components of the Free Energies of Hydrophobic and Hydrophilic Interaction
and Hydrophobic and Hydrophilic Hydration; see Tables XVIII-1 and 2

			Nature of force		Free energy component	
			LW (apolar)	AB (polar)	ΔH	ΔS
"Hydrophobic" interactions	Attraction	ΔG_{1w1}	≈ 0	$++++$	$++$	and/or $+++$
Hydrophilic interactions	Repulsion	ΔG_{2w2}	≈ 0	$++++$	$+$	and/or $++$
"Hydrophobic" hydration	Attraction	ΔG_{1w}	$+++$	$+$	$+++$	$++$ (neg.)
Hydrophilic hydration	Attraction	ΔG_{2w}	$++$	$+++$	$+++$	$+++$ (neg.)

Note: Subscript 1 = "hydrophobic" compound; subscript 2 = hydrophilic compound; subscript w = water).

PROPAGATION AT A DISTANCE OF ATTRACTIVE HYDROPHOBIC INTERACTIONS

Attractive H-bonding forces, or hydrophobic interaction forces, make themselves felt at a distance (Pashley *et al.*, 1985; Christenson, 1988). The LW force of attraction between hydrophobic (apolar) molecules or particles and the nearest layers of water molecules, whose free energy of attraction is expressed as ΔG_{iw}, causes local reorganization and an increase in the ordering, especially in the nearest water layer. As this attraction is of purely LW origin in the case of the most hydrophobic surfaces (Table XVIII-2) and thus is largely indifferent to preferential orientation effects, it is bound to give rise to a certain degree of compression in the layers of water molecules closest to the hydrophobic surface. According to Derjaguin's model (Derjaguin *et al.*, 1987; Derjaguin, 1989), a *densification* of a liquid such as water, near the interface, will result in a mutual *attraction* between the hydrophobic moieties that border on the distortion layer(s) of the surrounding water. This local distortion furnishes the mechanism by which attractive hydrophobic interactions are propagated at some distance beyond the initiating interfaces.

Furthermore, it is reasonable to assume that for water the decay length, λ, is the same in cases of apolar, hydrophobic attractions, as in those of polar, hydrophilic repulsions (for which see Chapter XIX, below). However, much larger values for λ have been reported in cases of hydrophobic attraction (Christenson and Claesson, 1988; Rabinovich and Derjaguin, 1988), which more recently turned out to have been based on experimental artefacts, mainly occurring in force-balance types of measurements (see Chapter XIV). [Here forces are frequently measured between two crossed cylinders (see Fig. XIV-1), in a mode where they are approaching one another, and also when they are pulled apart again. Now the crossed cylinders' surfaces are normally exceedingly hydrophilic, as they consist of a thin layer of mica (i.e., Muscovite), whose ΔG_{iwi} value is $+ 25.7$ mJ/m^2 (Giese and van Oss, 2002, p. 244; see also Giese *et al.*, 1996). (Mica is chosen for the half-cylindrical surfaces of most force balances, for its unique, atomic-scale smoothness). Thus, much of the work with hydrophobic surfaces had to be done after coating the hydrophilic mica surfaces, usually electrostatically with an alkyl group, via an amine or a quaternary ammonia moiety. Such attachments were often fairly loose and detached relatively easily with the force balance, especially when used in the separation (pulling-apart) mode. However, when hydrophobization is done by smoothly coating the mica in a firm and robust manner (Wood and Sharma, 1995), the extravagantly long-range effects of hydrophobic interactions alluded to above, revert back to normal].

The *densification* of water layers of hydrophobic hydration also fits well with the thin, dense networks of hydrophobic hydration one encounters in clathrate formation (see Chapter IX), but is not in agreement with a conjecture about a *decrease* in the density of such water of hydration, by Besseling and Lyklema (1997).

Hydrophobic attraction at a distance (in water), while following Derjaguin's model, may be visualized in more detail by realizing that the first layer of water of hydrophobic hydration, which is thinner and thus also somewhat denser than bulk water (and thus has one of the most hydrophobic configurations of which water is capable) also hydrophobically attracts a second layer of water of hydration, albeit more weakly, and so on, as the interaction distance becomes farther and farther removed from the initial hydrophobic molecules or particles. Thus, the hydrophobic attraction energies decrease exponentially with distance, ℓ, in such a manner that the free energy of hydrophobic attraction decays, for two spherical hydrophobic molecules or particles of radius, R, according to:

$$\Delta G_{iwi}^{AB}(\ell) = \pi \lambda R \, \Delta G_{iwi}^{AB''} (\ell = \ell_o) \exp[(\ell - \ell_o)/\lambda] \qquad \text{[XVIII-6]}$$

where $\ell_o = 0.157$ nm, which is the general minimum equilibrium distance of closest approach between two non-covalently interacting surfaces [see Chapter III and van Oss, Chaudury and Good (1988)]; λ is the characteristic (decay) length of water: $\lambda \approx 1.0$ nm (at 20°C) and $\Delta G_{iwi}^{AB''}$ ($\ell = \ell_o$) is the AB free energy of interaction between two parallel flat surfaces, i, in water, at closest approach, such that:

$$\Delta G_{iwi}^{AB''} = -2\gamma_{iw}^{AB} \qquad \text{[XVIII-7]}$$

(see also Table VII-2). The mode of decay of hydrophobic (AB) attractions in water is similar to the mode of decay of hydrophilic (AB) repulsions in water, insofar that they are both exponential, with the same decay length, λ, for water, at 20°C; see Chapter XIX, below.

CAVITATION IS AN EFFECT, NOT A CAUSE OF HYDROPHOBIC INTERACTIONS; WHEN W_{IW} IS SMALLER THAN W_{IWI}, CAVITATION IS FAVORED NEAR THE INTERFACE

When the work of attachment of water to a low-energy surface, i.e., the work of hydration, W_{iw} (= $-\Delta G_{iw}$) is smaller than the work of interfacial ("hydrophobic") attraction, W_{iwi} ($= -\Delta G_{iwi}$), it is easier to detach the water from that low-energy surface, than to break the cohesive water bonds which cause the interfacial attachment between two such low-energy surfaces. It is easily shown (cf. eqs. [IV-17] and [IV-18]) that $W_{iwi} > W_{iw}$ when $(\gamma_i + \gamma_w) < 3\,\gamma_{iw}$. The low-energy surfaces with which this can occur are mainly the alkanes (Table XVII-1) and compounds such as carbon tetrachloride (Tables XVIII-1 and 2), in addition to the more apolar polymers such as Teflon, polyisobutylene, pure propylene and polyethylene (Table XVII-5). Thus, when attempting to separate two low-energy surfaces, attached together by interfacial (hydrophobic) attraction in water, cavitations tend to ensue at the low-energy surface/water interfaces; cf. Yaminsky *et al.* (1983) and Yushchenko *et al.* (1983). Christenson and Claesson (1988) reported on such cavitation effects, concomitant with the attraction between two fluorocarbon (perfluorodecyl) and between two hydrocarbon (octadecyl) surfaces, in water, measured with Israelachvili's surface force balance (cf. Chapter XVI). The observed cavitation was stronger with the fluorocarbon than with the hydrocarbon surfaces, which agrees with the fact that $W_{iwi} - W_{iw}$ has a higher value with the former than with the latter. [With a complete and perfect coating of fluorocarbon, or hydrocarbon, the values of $(W_{iwi} - W_{iw})$ would be about 63 and 53 mJ/m². respectively.] Whilst cavitation phenomena clearly can arise as a result of a hydrophobic attraction, they are not a contributing *cause* of that attraction (Ruckenstein, 1992; see also Ruckenstein and Churaev, 1991).

ATTRACTIVE INTERFACIAL INTERACTIONS
IN NON-AQUEOUS MEDIA

In polar media other than water, analogous interfacial attractions among low-energy materials may occur. However, like water, such polar media must be markedly bipolar (van Oss *et al.*, Adv. Colloid Interface Sci., 1987; van Oss, Ju *et al.*, 1989), i.e., they must have strong electron-donor as well as electron-acceptor parameters. The principal bipolar liquids in which such interfacial attractions can be observed are: glycerol, formamide, ethylene glycol (see Chapter XVII, Tables XVII-4 and XVII-10). One could designate such attractive interactions as "solvophobic."

FURTHER TO THE MECHANISM OF HYDROPHOBIC
INTERACTIONS

It is easily seen that, in interactions of the type described in eq. [XIV-1] the roles of polar surfaces (i) on the one hand and of water molecules (w) on the other are mathematically interchangeable (van Oss and Good, J. Dispersion Sci., Tech., 1988), i.e.,

$$\Delta G_{iwi} = \Delta G_{wiw} \qquad \text{[XVIII-8]}$$

In conjunction with eq. [XVIII-1], it can be seen that:

$$\Delta G_{iwi} = \Delta G_{wiw} = -2\gamma_{iw} \qquad \text{[XVIII-9]}$$

which shows that the subscripts iwi and wiw are mathematically interchangeable. Experimental support for the validity of eq. [XVIII-9] may be found in the fact that the (low) solubility of various apolar or mainly apolar organic liquids in water, in moles/liter, is approximately the same as the solubility of water in the corresponding organic liquids; e.g., benzene, chloroform, toluene, xylene and cyclohexane (Stephen and Stephen, 1963; van Oss, Giese and Good, 2002).

Thus, a very strong interaction between molecules of one of the two components, in a two-component system, gives rise to an equally strong effective interaction, of the same sign, between the molecules or particles of the other component. One may therefore state that the strong (polar) attraction between water molecules "pushes" or squeezes the other molecules or particles together, with the same energy.

In the second edition of his book, Tanford (1980, p. 3) also observes that hydrophobic interactions are principally governed by the "dominant role of water self-attraction in the hydrocarbon-water system"; see also Tanford (1979). The apolar (LW) component of the hydrophobic attraction is usually very slight, in water, because the values for the γ^{LW} of water and the γ^{LW} of most low-energy materials generally are rather close; see eq. [II-35]. *The polar component of that attraction is therefore the dominant one* (see van Oss and Good, J. Dispersion Sci. Tech., 1988), see also Chapter XVI.

OCCURRENCE OF ATTRACTIVE INTERFACIAL ("HYDROPHOBIC") INTERACTIONS BETWEEN SIMILAR SITES

"Hydrophobic" (AB) interactions are pervasive in all interactions among low- or medium-low-energy compounds immersed in water. In most such cases $|\Delta G_{iwi}^{AB}| \gg |\Delta G_{iwi}^{LW}|$ and usually, $|\Delta G_{iwi}^{AB}| > |\Delta G_{iwi}^{EL}|$. Some examples are:

Insolubility of organic compounds in water (Chapter XXII);
Flocculation of particles and cells in aqueous media (Chapters XXIII and XXIV);
Micelle formation (Chapter XXII); see also Tanford (1973, 1980);
Organization of the phospholipid bilayer; see also Israelachvili (1991).

INTERACTIONS BETWEEN TWO DIFFERENT ORGANIC COMPOUNDS IMMERSED IN WATER

Attractive interfacial (or "hydrophobic") interactions not only take place between identical organic molecules immersed in water, but also between two different low-energy moieties (1 and 2) in water (w). According to the Dupré equation in the form of:

$$\Delta G_{1w2} = \gamma_{12} - \gamma_{1w} - \gamma_{2w} \qquad \text{[XVIII-9]}$$

and the equation for the interfacial tension:

$$\gamma_{ij} = (\sqrt{\gamma_i^{LW}} - \sqrt{\gamma_j^{LW}})^2 + 2(\sqrt{\gamma_i^\oplus \gamma_i^\ominus}$$
$$-\sqrt{\gamma_j^\oplus \gamma_j^\ominus} - \sqrt{\gamma_i^\oplus \gamma_j^\ominus} - \sqrt{\gamma_i^\ominus \gamma_j^\oplus}$$

[IV-9A]

The total interaction can be expressed as (van Oss *et al.*, 1988a); eq. IV-16:

$$\Delta G_{1w2} = (\sqrt{\gamma_1^{LW}} - \sqrt{\gamma_2^{LW}})^2 - (\sqrt{\gamma_1^{LW}} - \sqrt{\gamma_W^{LW}})^2$$
$$- (\sqrt{\gamma_2^{LW}} - \sqrt{\gamma_W^{LW}})^2 + 2[\sqrt{\gamma_W^\oplus}(\sqrt{\gamma_1^\ominus} + \sqrt{\gamma_2^\ominus} - \sqrt{\gamma_W^\ominus}) \quad \text{[XVIII-10]}$$
$$+ \sqrt{\gamma_W^\ominus}(\sqrt{\gamma_1^\oplus} + \sqrt{\gamma_2^\oplus} - \sqrt{\gamma_W^\oplus}) - \sqrt{\gamma_1^\ominus \gamma_2^\oplus} - \sqrt{\gamma_1^\oplus \gamma_2^\ominus}]$$

Eqs. [XVIII-9 and 10] describe interfacial (hydrophobic) attractions when $\Delta G_{1w2} < 0$, and interfacial (hydrophilic) repulsions when $\Delta G_{1w2} > 0$.

It is easy to understand that when $\Delta G_{1w2} < 0$, and compounds (1) and (2) are both low-energy compounds (i.e., non-polar surfaces, or surfaces of low polarity), an interfacial, or "hydrophobic" attraction prevails between (1) and (2) when immersed in water, in the same manner as when $\Delta G_{1W1} < 0$. It still is less generally realized that $\Delta G_{1w2} < 0$ can also occur in the case of the interaction between a low-energy ("hydrophobic") compound (1) and a hydrophilic compound (2) (van Oss *et al.*, 1986). In 1989 van de Ven (1989, p. 49) stated: "It is not known whether or not any short-range force acts between a hydrophobic surface and a hydrophilic surface." Well, it does! This variety of interaction, which is typified by the adsorption of, e.g., (hydrophilic) proteins onto relatively low-energy surfaces in water, has been used quite consciously for about thirty years, for protein fractionation by "hydrophobic interaction chromatography" (Rosengren *et al.*, 1975; Hofstee, 1973, 1976). "Reversed-phase liquid chromatography" (RPLC) is also based upon the attachment of hydrophilic proteins or peptides to low-energy surfaces in aqueous media, followed by their detachment from these low-energy surfaces by the admixture of water-miscible organic solvents. This technique is based on work by Howard and Martin (1950), and has also been in extensive use for over twenty-five years; see, e.g., Krstulovic and Brown (1982); see also Chapter XXIV. We analyzed the isolation of human immunoglobulin G (IgG) by RPLC on a phenyl-sepharose column (van Oss, Israel J. Chem., 1990), based upon data from van Oss, Absolom and Neumann (1979). It could be shown that upon the gradual addition of ethylene glycol, IgG started to elute from the phenyl-sepharose beads, close to the point where ΔG_{132} changed from a negative to a positive value. Van Oss *et al.* (J. Colloid Interface Sci., 1986) compared the adsorption isotherms of two major serum proteins [human serum albumin (HSA) and IgG] onto poly(tetrafluorethylene) and polystyrene surfaces, with ΔG_{1w2} data obtained from contact angle measurements. It was found that the ΔG_{1w2} values thus obtained correlated reasonably well with the values derived from the adsorption isotherms, but it should be kept in mind that the methodology including the $\gamma^\oplus$ and $\gamma^\ominus$ values into ΔG_{1w2} (van Oss *et al.*, Chem. Rev., 1988) was not yet utilized in van Oss *et al.* (J. Colloid Interface Sci., 1986). A comparison between the relative hydrophobicity of serum proteins and the adsorption isotherms of the same proteins on poly(tetrafluoroethylene) shows good qualitative agreement. Van Oss, Absolom and Neumann (1979) showed by RPLC (then still called "hydrophobic chromatography") that serum proteins, in the order of increasing hydrophobicity (as judged by the order of their elution by increasing ethylene glycol concentration and/or by contact angle determination), are: α_2 macroglobulin (α_2M), HSA, immunoglobulin M (IgM) and IgG. Subsequently, adsorption isotherms were given of these four serum proteins on poly(tetrafluorethylene) (van Oss *et al.*, 1981); the slopes of the Langmuir isotherms (at low protein concentration) also increased in the order: α_2M, HSA, IgM, IgG.

OCCURRENCE OF ATTRACTIVE INTERFACIAL ("HYDROPHOBIC") INTERACTIONS BETWEEN TWO DISSIMILAR SITES

Hydrophobic (AB) interactions between dissimilar sites are also quite prevalent in biological and other aqueous systems. Here also, usually: $|\Delta G_{1w2}^{AB}| \gg |\Delta G_{1w2}^{LW}|$ and often: $|\Delta G_{1w2}^{AB}| > |\Delta G_{1w2}^{EL}|$ Some examples are:

Specific interactions, such as antigen-antibody, lectin-carbohydrate, enzyme substrate, generally: ligand-receptor interactions (Chapter XXIV); protein-folding; protein-lipid interactions; intra-cellular organization; see, e.g., Tanford (1973, 1980).

Protein adsorption (see above): other types of adsorption in aqueous media; chromatography: HPLC in general; reversed-phase liquid chromatography; hydrophobic interaction chromatography; thin layer chromatography; side effects in pore exclusion and ion exchange chromatography; various other separation methods (van Oss *et al.*, Separ. Sci. Tech., 1987).

CODA

Hydrophobic interaction is a misnomer because all hydrophobic surfaces attract water to a non-negligible degree. It is shown that the attraction between hydrophobic moieties immersed in water is a consequence of the hydrogen-bonding energy of cohesion of the surrounding water molecules. Cavitation is shown to be an effect of attractive hydrophobic interactions and not their cause. By comparing the hydrophobic energies of interaction between various low-energy compounds, i, immersed in water, w: ΔG_{iwi}, at different temperatures, values for both the enthalpic (ΔH) and entropic ($T\Delta S$) contributions to these interactions could be obtained. It followed that the hydrophobic interactions are mainly entropic in some cases and mainly enthalpic in others. Whilst hydrophobic energies of attraction, ΔG_{iwi}, are of polar (AB) origin, the free energy of hydration of low-energy entities, ΔG_{iw} is mainly due to Lifshitz–van der Waals (LW) forces. In all cases studied, the free energy of hydration, ΔG_{iw} shows a negative entropy (ΔS) component which is presumably indicative of an increase in organization among water molecules of hydration.

It is hydrophobic attraction which makes low-energy compounds so insoluble in water. Hydrophobic interactions are relatively short-range in nature, with a decay length, λ, for water, of the order of 1.0 nm.

It is stressed that hydrophobic attractions not only occur between two similar low-energy compounds immersed in water ($\Delta G_{iwi} < 0$), but they also quite readily obtain between one low-energy (hydrophobic) entity (i) and one high-energy (hydrophilic) compound (j) immersed (or as far as the latter is concerned, dissolved) in water ($\Delta G_{iwj} < 0$). Examples of the latter phenomenon are, e.g.: adsorption of proteins and other biopolymers; cell adhesion; various modes of liquid chromatography such as reversed-phase, hydrophobic interaction; and specific interactions, such as antigen-antibody binding and affinity chromatography.

Attractive interfacial solvophobic interactions also occur in non-aqueous bipolar, organic liquids. They are analogous to, but not as strong as, hydrophobic interactions occurring in water.

XIX Repulsive AB-Forces: Hydrophilic Interactions— Osmotic Pressures of PEO Solutions

NEGATIVE INTERFACIAL TENSIONS AND POLAR REPULSION

As has been shown in Chapter IV, the polar component of the interfacial tension:

$$\gamma_{12}^{AB} = 2(\sqrt{\gamma_1^{\oplus}\gamma_1^{\ominus}} + \sqrt{\gamma_2^{\oplus}\gamma_2^{\ominus}} - \sqrt{\gamma_1^{\oplus}\gamma_2^{\ominus}} - \sqrt{\gamma_1^{\ominus}\gamma_2^{\oplus}})$$ [IV-7]*

can readily assume a negative value when

$$\gamma_1^{\oplus} < \gamma_2^{\oplus} \ \text{and} \ \gamma_1^{\ominus} > \gamma_2^{\ominus}$$ [IV-8A]

or when

$$\gamma_1^{\oplus} > \gamma_2^{\oplus} \ \text{and} \ \gamma_1^{\ominus} < \gamma_2^{\ominus}$$ [IV-8B]

And when $|\gamma_{12}^{AB}| > \gamma_{12}^{LW}$, *and* $\gamma_{12}^{AB} < 0$, the total interfacial tension:

$$\gamma_{12} = (\sqrt{\gamma_1^{LW}} - \sqrt{\gamma_2^{LW}})^2 + 2(\sqrt{\gamma_1^{\oplus}\gamma_1^{\ominus}} + \sqrt{\gamma_2^{\oplus}\gamma_2^{\ominus}}$$
$$- \sqrt{\gamma_1^{\oplus}\gamma_2^{\ominus}} - \sqrt{\gamma_1^{\ominus}\gamma_2^{\oplus}})$$ [IV-9]

will have a negative value. In practice this occurs quite readily in biological and other aqueous systems.

Using the Dupré equation (cf. eqs. [II-37], [IV-15]):

$$\Delta G_{132} = \gamma_{12} - \gamma_{13} - \gamma_{23}$$ [XIX-1]

and for compound (1) immersed in liquid (2), cf. eq. [II-33B]:

---

* A list of symbols can be found on pages 399–406.

$$\Delta G_{121} = -2\gamma_{12} \qquad\qquad\qquad [XIX\text{-}2]$$

or, in aqueous media (eq. [XVIII-7]):

$$\Delta G_{iwi} = -2\gamma_{iw} \qquad\qquad\qquad [XIX\text{-}3]$$

ΔG_{iwi} is the free energy change in bringing molecules of compound (i), immersed in water (w) from infinity to the minimum equilibrium distance, in which a very thin sliver of (w) separates two macroscopic bodies of (i); see van Oss and Good (1984). During the process of bringing the two materials together at the equilibrium separation distance, the work may either be done *by* the system (i.e., when the operating forces are attractive) or work needs to be done *on* the system (i.e., when the operating forces are repulsive). It can therefore be seen that attractive forces will contribute to a positive interfacial tension (free energy per unit area) whereas repulsive forces will give rise to a negative interfacial tension. Since the net interfacial tension (γ_{12}, or in aqueous systems, γ_{iw}) also comprises the apolar (LW) as well as the polar (AB) component, where the LW component is always attractive, the total interfacial tension can consist of contributions by both attractive and repulsive forces simultaneously, so that the resultant γ_{12}, in general, can be either positive or negative; see also Chapter XIII.

A net negative interfacial tension normally gives rise to a thermodynamically unstable interface, so that the system will spontaneously relax by either abolishing the interface or by reverting to a more stable geometric configuration, ultimately resulting in $\gamma_{12} \to 0$. An example of such a process may be found in the formation of microemulsions. Prince (1967), while discussing the formation of micro- and macro-emulsions, suggested that a small but positive interfacial tension at an oil/water interface with adsorbed surfactant could favor the formation of macroemulsions, whereas a negative interfacial tension would cause the onset of microemulsion formation (see also van Oss *et al.*, Advan. Colloid Interface Sci., 1987; and Chapters XXI and XXII). Other examples of the effects of the decay of negative interfacial tensions are: "anomalous" high osmotic pressure of solutions of monopolar compounds in water (see below); phase separation in aqueous polymer solutions (Chapter XXI); pronounced solubility of biopolymers and other polar compounds in water or in other hydrogen-bonded polar liquids (Chapter XXII); "steric" stabilization of hydrophobic particles in aqueous media; "depletion" interactions (Chapter XX).

There are, however, cases where a net negative interfacial tension does not find scope to relax and tend to $\gamma_{12} \to 0$. This occurs when polar materials, due to crystallization, or strong covalent crosslinking, notwithstanding considerable negative values of γ_{12}, cannot dissipate or dissolve in an aqueous or other polar liquid medium with which they are in contact. For instance, mineral solids, all displaying considerable negative γ_{iw} values, nevertheless do not give rise to a significant degree of solubilization in water (Giese and van Oss, 2002).

MONOPOLAR SURFACES

Inspection of eqs. [IV-7] and [IV-9], and the conditions for $\gamma_{12} < 0$, given in eqs. [IV-8A and B] will elucidate that when, e.g., $\gamma_1^\oplus \approx 0$ and $\gamma_1^\ominus > \gamma_2^\ominus$, a negative polar interfacial tension can ensue. Upon contact angle measurement, using apolar, as well as a number of well-characterized polar liquids (i.e., water, glycerol, formamide), it became clear that, especially among biopolymers, the occurrence of surfaces manifesting a sizeable value for $\gamma_1^\ominus$ and a zero or negligible value for $\gamma_1^\oplus$ is quite common (van Oss *et al.*, Advan. Colloid Interface Sci., 1987; see also Tables XVII-5 to XVII-8). Such surfaces may be designated as "monopolar surfaces," and the compounds manifesting these as "monopolar compounds." Compounds in which cohesive hydrogen bonds (or in more general terms, cohesive Lewis acid-base interactions) occur are designated as "dipolar compounds." It became apparent (van Oss *et al.*, Advan. Colloid Interface Sci., 1987) that monopolar compounds with a high $\gamma_1^\ominus$ value and a zero or negligible value for $\gamma_1^\oplus$ are quite prevalent in nature (see also Chapter XVII), whilst monopolar compounds with a sizeable $\gamma_1^\oplus$ and zero $\gamma_1^\ominus$ are quite rare. (Chloroform and diiodomethane are examples of the latter variety; see Table XVII-2.)

From eq. [IV-5], it is clear that for a true monopolar compound, $\gamma_i^{AB} = 0$, so that $\gamma_i = \gamma_i^{LW}$. Thus, once it has been determined for a polar compound (1) that, e.g., $\gamma_1^\oplus = 0$, using eq. [IV-13A], proving compound (i) to be monopolar, then its interfacial tension with water can be determined directly from its contact angle with water, using the following version of Young's equation:

$$\gamma_{iw} = \gamma_i^{LW} - \gamma_w \cos \theta_w \qquad [XIX-4]$$

It is easily shown that for $\gamma^\ominus$ monopolar compounds, with a γ^{LW} value of about 40 mJ/m² (which is a rather commonly occurring value for such compounds, see Chapter XVII), $\gamma_i^\ominus$ must be of the order of 28.3 mJ/m² or higher for γ_{iw} to be negative. It should be noted that the underlying "hydrophobic" attraction always persists, even when a net hydrophilic repulsion prevails. Reiterating eq. [IV-17], for water:

$$\Delta G_{1W1} = -2(\sqrt{\gamma_1^{LW}} - \sqrt{\gamma_W^{LW}})^2 - 4(\sqrt{\gamma_1^\oplus \gamma_1^\ominus}$$
$$+ \sqrt{\gamma_W^\oplus \gamma_W^\ominus} - \sqrt{\gamma_1^\oplus \gamma_W^\ominus} - \sqrt{\gamma_1^\ominus \gamma_W^\oplus}) \qquad [IV-17A]$$

it is obvious that the term $-4\sqrt{\gamma_W^\oplus \gamma_W^\ominus}$ always has the finite value of -102 mJ/m² at 20°C and represents the basic "hydrophobic" attraction contribution due to the free energy of cohesion of water. The role of monopolarity in hydrophilicity now becomes apparent, because with $(\gamma_1^\ominus)$ monopolarity, $\sqrt{\gamma_1^\oplus \gamma_1^\ominus} = 0$ and $\sqrt{\gamma_1^\oplus \gamma_W^\ominus} = 0$, leaving the necessity for $4\sqrt{\gamma_1^\ominus \gamma_W^\oplus}$ to be larger than 102 mJ/m², or for $\gamma_1^\ominus$ to be larger than 25.5 mJ/m² (given that $\gamma_W^\oplus = \gamma_W^\ominus = 25.5$ mJ/m²). In addition, the relatively slight LW attraction (of about $\Delta G_{1W2}^{LW} \approx -5.5$ mJ/m²) has to be overcome; see eq. IV-17A and Table IV-2. Thus, using eq. [XVIII-1], it becomes clear that in the case of $\gamma^\ominus$ monopolar compounds with $\gamma_1^\ominus \geq 28.3$ mJ/m², so that $\gamma_{iw} < 0$ and $\Delta G_{iwi} > 0$, the molecules of such monopolar compounds, immersed in water, *will repel each other.* Monopolar compounds of this class include, see Chapter XVII: polyvinylpyrrolidone, polyethylene oxide, co-poly(ethylene glycol, propylene glycol), human serum albumin (hydrated), human immunoglobulin-G (hydrated), fibrinogen (dry as well as hydrated), human low density lipoprotein, ficoll, dextran, arabino-galactan, inulin, Na-alginate (only compounds with $\gamma^\oplus$ between 0 and 1 mJ/m^2 and $\gamma^\ominus > 28.3$ mJ/m^2 are listed here).

QUANTITATIVE EXPRESSION OF HYDROPHILICITY AND HYDROPHOBICITY

In the condensed state all compounds attract water molecules to a considerable degree, i.e., with a free energy varying from a value of –40 mJ/m^2 for the most apolar compounds to –140 mJ/m^2 for the most strongly monopolar hydrophilic materials; see Tables XVIII-1 and XIX-1. Thus, as already stated in the preceding Chapter, to designate apolar compounds as "hydrophobic" is less than appropriate. However, as the use of that adjective in this context has become too wide spread to hope for its speedy eradication, we shall continue to use it for the sake of clarity.

Table XIX-1 shows a list of compounds, varying from totally apolar to very polar, or if one wishes, from quite hydrophobic to extremely hydrophilic, showing: (1) their electron-donor surface tension parameters; (2) their free energies of hydration; and (3) their free energies of interfacial interaction.

1. *The electron-donor surface tension parameter* ($\gamma^\ominus$) is a fairly good semi-quantitative indicator of the degree of hydrophilicity of a compound, but hydrophilicity also remains linked to the values of the γ^{LW} component and the $\gamma^\oplus$ parameter (if any) of the surface tension. For instance, when $\gamma_i^{LW} = 40$ mJ/m^2, and $\gamma_i^\oplus = 0$, it can be stated that with a $\gamma^\ominus \geq 28.3$ mJ/m^2, a compound (i) is genuinely hydrophilic, i.e., its molecules, when immersed in water, will tend to repel one another, a condition which normally leads to solubility (see Chapter XIX). However, when there is a finite $\gamma_i^\oplus$, or when $\gamma_i^{LW} \neq 40$ mJ/m^2, the cut-off for hydrophilicity occurs for $\gamma_i^\ominus$ values that can be lower or higher than 28.3 mJ/m^2. Thus, in many cases, judging hydrophilicity of a compound solely by its $\gamma^\ominus$ value may be somewhat imprecise.

2. *The free energy of hydration* (ΔG_{iw}) would, by its very meaning, appear to be an ideal way to characterize hydrophilicity (or its opposite), noting that according to the Dupré equation (cf. eq. [IV-6]):

$$\Delta G_{iw} = \gamma_{iw} + \gamma_i + \gamma_w \qquad\qquad [XIX-5]$$

For protein adsorption and cell adhesion onto low-energy surfaces (see Chapter XXIV) the correlation is quite good, albeit not perfect; see also Table XIX-2. And, whilst ΔG_{iw} follows the series shown in Table XIX-1 fairly closely, there also are some discrepancies (cf. benzene and octanol).

TABLE XIX-1

Parameters in mJm² Relating to the Hydrophilicity or "Hydrophobicity" of Various Compounds, in the Order of Increasing ΔG_{iwi}, at 20°C.

Compound	Free energy for interfacial interaction in water, ΔG_{iwi}	Free energy of hydration, ΔG_{iw} [c]	Electron-donor parameter, $\gamma_i^{\ominus}$ [d]	Electron-acceptor parameter, $\gamma_i^{\oplus}$ [d]	Polar surface tension component, γ_i^{AB} [d]	Apolar surface tension component, γ_i^{LW} [d]
Diiodomethane	-112.2[a]	-67.5	0	≈0.01	0	50.8
Hexane	-102.3[a]	-40.1	0	0	0	18.4[e]
CCl₄	-83.6[a]	-58.0	0	0.9	0	27.0[e]
Benzene	-83.2[a]	-60.1	1.0	0	0	28.85[e]
Chloroform	-77.6[a]	-61.2	0	1.5	0	27.15[e]
Octanol	-61.6[a]	-69.5	4.1	0	0	27.5[e]
Ethyl ether	-42.0[a]	-68.8	9.0	0	0	17.0[e]
PMMA*	-37.8[b]	-94.5	12.0	0	0	40.6
Fibrin (Bovine)*	-34.2[b]	-99.7	12.0	0.3	3.8	40.2
Zein*	-20.9[b]	-105.2	18.4	0.04	1.7	41.1
Gelatin*	-19.4[b]	-100.7	18.5	0	0	37.6
Agarose*	-3.2[b]	-112.2	26.9	0	0	41.0
PVP*	+0.72[b]	-116.6	29.7	0	0	43.4
Glucose*	+35.7[b]	-132.9	51.1	0	0	42.2
Dextran*	+41.2[b]	-135.4	55.0	0	0	42.0
Fibrinogen (human)*	+41.8[b]	-134.3	54.9	0	0	40.6
PEO*	+52.5[b]	-142.0	64.0	0	0	43.0

Molecules which attract each other in water, i.e. Hydrophobic

Molecules which repel each other in water, i.e. Hydrophilic

* Dried

[a] From its aqueous solubility, cf. eqs. XIII-2A; XIII-3; see also Table XIII-2

[b] Data from Chapter XVII, cf. eq. IV-17

[c] Data from Chapter XVII, cf. eq. IV-18

[d] Data from Chapter XVII

[e] See also Jasper (1972)

However, the free energy of hydration (in which one might identify a cut-off of $\Delta G_{iw} \approx -113$ mJ/m^2, where more negative values would indicate genuine hydrophilicity) is largely of an apolar (LW) nature, especially for the more "hydrophobic" compounds, see Chapter XVIII, Table XVIII-2.

3. *The free energy of interfacial interaction* between molecules of a compound (i) immersed in water (ΔG_{iwi}) clearly is the most appropriate measure of hydrophilicity, e.g., *with respect to the solubility of compound (i) in water* (see Chapter XXII). When the net free energy of interaction between molecules of compound (i) immersed in water is attractive (i.e., ΔG_{iwi} has a negative value), the molecules of compound (i) have less affinity for water than for themselves; they thus are hydrophobic, i.e., they are more autophilic than hydrophilic. When the net free energy of interaction between molecules of compound (i) immersed in water is repulsive (i.e., ΔG_{iwi} has a positive value), the molecules of compound (i) have more affinity for water than for themselves, in which case they are genuinely hydrophilic. The more negative ΔG_{iwi}, the more hydrophobic compound (i) is; the more positive ΔG_{iwi}, the more hydrophilic compound (i); see Table XIX-1. In general (and in contrast with ΔG_{iw}), ΔG_{iwi} is for 95% or more due to polar (AB) forces (see Table XVIII-2).

For purposes of comparison of the degree of hydrophilicity between different surfaces, it is appropriate to express ΔG_{iwi} in terms of free energy per unit surface area (i.e., in S.I. units of, e.g., mJ/m^2). However, in special cases, one may wish to

---

TABLE XIX-2
Free Energies (in mJ/m^2) of Interfacial Interaction (ΔG_{iwi}), in Water, and of Hydration (ΔG_{iw}), of a Number of Hydrophilic Entities, Classified in the Order (from top to bottom, in each category, see arrows) of Their Increasing Adsorbability onto Low-Energy Surfaces

	ΔG_{iw}	ΔG_{iwi}
Cells[a]		
S. epidermidis (caps)	−141.6	+18.5
HELA cells	−138.3	+6.5
S. epidermidis (non-caps)	−126.7	+37.7
Fibroblasts	−119.7	−6.6
Proteins[a]		
HSA (hydrated)	−145.7	+20.4
Fibronectin	−144.4	+24.2
IgG (hydrated)	−138.0	+27.7
Fibrinogen	−134.3	+41.7
Fibrin	−99.7	−34.1

[a] The arrows point in the direction of the highest degree of adsorption onto low-energy surfaces (van Oss, 1991; see also Chapter XXI).

express hydrophilicity on a molecular scale. In such cases ΔG_{iwi} is best expressed in units of kT, so that $\Delta G_{iwi} = -\chi_{iw}$ (eqs. [VI-7] and [VI-8]). The molecular degree of hydrophilicity then relates directly to the material's solubility (Chapter XXII). Thus, dry dextran ($\Delta G_{iwi} = +41.2$ mJ/m^2) is somewhat less hydrophilic than dry polyethylene oxide (PEO) ($\Delta G_{iwi} = +52.2$ mJ/m^2) (Table XIX-1), if one compares global surface properties. However, the contactable surface area (S_c) between dextran (Mw $\approx$ 180,000) chains $\approx$ 34 Å^2, while for PEO (Mw $\approx$ 6,000), $S_c \approx 21.2$ Å^2 (see this chapter, below, under Osmotic Pressures of PEO-water solutions). Then, expressed in units of kT, ΔG_{iwi} for dextran is +3.5 kT, while ΔG_{iwi} for PEO is +2.77 kT. Thus on a molecular level, in aqueous solution, dextran is somewhat more hydrophilic than PEO, which agrees well with the behavior of these polymers observed in aqueous phase separation by Albertsson (1971, p. 29; 1986, p. 17).

Conclusion: ΔG_{iwi} is the ideal standard quantitative measure of the degree of hydrophobicity/hdyrophilicity (van Oss and Giese, 1995).

Thus, when $\Delta G_{iwi} < 0$, i is formally hydrophobic and when $\Delta G_{iwi} > 0$, i is hydrophilic (van Oss and Giese, 1995), where ΔG_{iwi} is expressed in mJ/m^2. This is why the ΔG_{iwi} column in Table XIX-1 is arranged, from top to bottom, in the order of decreasing hydrophobicity, followed (after ΔG_{iwi} turns from negative to positive) by ΔG_{iwi} values, arranged from top to bottom, in the order of increasing hydrophilicity.

PROPAGATION AT A DISTANCE OF REPULSIVE HYDROPHILIC INTERACTIONS

Like attractive hydrophobic interactions, repulsive hydrophilic interactions (i.e., hydration forces) make themselves felt at a distance (see, e.g., Claesson, 1986). Here also the properties of successive layers of water of (in this case, hydrophilic) hydration play a role, like the layers of water of hydrophobic hydration (see Chapter XVIII). However where the first layer(s) of water of hydrophobic hydration are flatter and thus more dense than bulk water, in hydrophilic hydration the layers of hydration water are oriented perpendicular to the surface of the hydrophilic material and thus less dense than bulk water (see Fig. X-1). Here, also in accordance with Derjaguin's model (Derjaguin *et al.*, 1987; Derjaguin, 1989; see also Chapter XVIII), a local *decrease* in the density of water of hydration (nearest a hydrophilic surface surface immersed in water) gives rise to a *repulsion* between two hydrophilically hydrated surfaces (which are hydrophilic as a consequence of their pronounced surface electron-donicity) as they attract the water molecules of hydration by their H-atoms, i.e., by the water molecules' electron-acceptors (see Fig. X-1; see also Parsegian *et al.*, 1985). Thus, hydrophilic repulsion at a distance (in water), while obeying Derjaguin's model, may be visualized in more detail by realizing that the first layer of water of hydrophilic hydration where most, if not all, aqueous O-atoms are oriented distally with respect to the hydrophilic surface and point away into the bulk water, thus presenting a strongly electron-donating layer of water of hydration to the next layer of water. This layer then orients itself similarly (but to a somewhat lesser degree) toward the next layer of water, and so on. The hydrophilic repulsion energies then decrease exponentially with distance in the case of two spherical molecules or particles of radius, R when immersed in water, according to eq. XVIII-6 (see Chapter XVIII) as a function of distance, ℓ; see Chapter XVIII and see also Table VII-2. Analogous to the decrease of hydrophobic attraction at a distance, the exponential decrease in hydrophilic repulsion at a distance is also linked to a decay length of water, at 20°C, of $\lambda \approx 1.0$ nm.

LINKAGE BETWEEN EL AND AB FORCES

One frequently encounters a linkage between EL and AB forces. In aqueous media, surfaces with a sizeable ζ-potential also usually have a high $\gamma^\ominus$ value. There are, however, many hydrophilic materials with a high $\gamma^\ominus$ value with a negligible or a zero ζ-potential (e.g., dextran and other neutral polysaccharides). Among hydrophilic materials with a non-negligible ζ-potential the following properties can be noted (see Schematic Chart XIX-1):

SCHEMATIC CHART XIX-1   Linkage between EL and AB forces

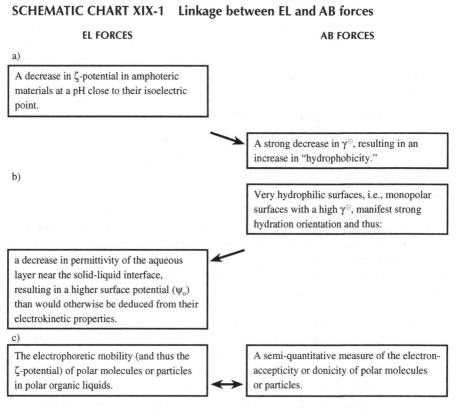

a Amphoteric molecules tend to become very "*hydrophobic*" under conditions of pH close to the isoelectric point.

b Very hydrophilic surfaces (i.e., monopolar surfaces with a high $\gamma^\ominus$ value) display a strong degree of hydration orientation, which gives rise to a decrease in the dielectric constant of the liquid medium.

c There is a correlation between the electrophoretic mobility of polar molecules or particles in polar organic media and their electron-donicity or accepticity; see the last section of Chapter XV before the Coda.

a. Holmes-Farley *et al.* (1985) showed that while contact angles measured with buffered water of different pH's on amphoteric surfaces were quite low at pH values fairly far removed from the isoelectric point of the amphotere, they were very high in the pH range close to or at the isoelectric point. As indicated above, serum albumin, even though still somewhat hydrated at the isoelectric pH of 4.9, becomes significantly hydrophobic at that pH (van Oss and Good, 1988). It is well known that all proteins are least soluble in water at their isoelectric point. It can be argued that the decreased solubility of, e.g., albumin at pH 4.9, is not directly caused by the elctroneutrality of the protein at its isoelectric pH (as the phenomenon is not dependent on the dielectric constant of the medium), but rather by the increased hydrophobicity of the protein at that pH (van Oss, J. Protein Chem., 1989). In aqueous media, on a macroscopic scale, a net equality of positive and negative charges on an amphoteric surface thus does not show an equal degree of electron donicity and accepticity but, on the contrary, appears to result in the complete disappearance of both $\gamma^{\oplus}$ and $\gamma^{\ominus}$. Also, neutralization of the charge of negatively charges surfaces, cells, or particles, by the admixture of Ca^{++} ions, not only decreases their ζ-potential, but in addition, through the action of the Ca^{++} ions as electron-acceptors, significantly decreases their electron-donicity parameter ($\gamma^{\ominus}$), and thus renders them strongly hydrophobic (Ohki, 1982; Gallez *et al.*, 1984; van Oss *et al.*, Advan. Colloid Interface Sci., 1987; van Oss *et al.*, in Molecular Mechanisms of Membrane Fusion, 1988; van Oss, Giese and Wu, 1993; Mirza *et al.*, 1998).

b. As shown in Table XIX-4 and as illustrated in Fig. XIX-1, strongly hydrophilic surfaces (i.e. monopolar surfaces with a high $\gamma^{\ominus}$ value) give rise to a strong orientation of the water molecules in the first layer adjoining the interface and, to a diminishing degree, in the second and subsequent layers. This orientation of water molecules not only causes "hydration pressure" to be measurable some distance away from the interface, but it also locally decreases the permittivity (ε) of water to ε'. For a given electrophoretic mobility, this effect van cause an increase in the real surface potential (ψ_o) by, roughly, $\sqrt{\varepsilon/\varepsilon'}$ (where ψ_o was calculated with the assumption that ε = 80). It thus can be estimated that, with a degree of orientation of 73.5% in the first hydration layer of albumin (see Table XIX-4), ε' would be of the order of ≈ 20 in that layer. However, the influence of this phenomenon on the ζ-potential, which is the potential (at the slipping plane) that can actually be derived from the electrophoretic mobility, the error is less pronounced. For an orientation of the order of 31%, occurring in the second layer of hydration of albumin (Table XIX-4), a rough estimation of ε' would be about 55, which rapidly grows to 80 with increasing distance, ℓ. The ζ-potential, derived from the electrophoretic mobility, thus would only be about 17% higher than was at first supposed. At the same time, however, the thickness of the diffuse ionic double layer, $1/\kappa$ (eq. [IV-2]), would be considerably less than at first assumed (Vaidhyanathan,

1988), so that at higher ionic strengths the decay of ΔG^{EL} as a function of ℓ will be significantly steeper. As a result, the permittivity gradient in the immediate vicinity of hydrophilic monopolar surfaces gives rise to a considerably higher ψ_o value than the one obtained when adhering to a constant value of $\varepsilon = 80$ for water. However, the ζ-potential, taken at the slipping plane, which is 1 to 2 layers of water molecules farther out, fares somewhat better and usually will only be underestimated by about 17%. Estimation of an *operative* ψ_o value, from ζ, *via* (eq. V-1] therefore will not usually give rise to large errors. The Debye length $(1/\kappa)$ will be considerably smaller than is usually calculated for high ionic strengths, but at ionic strengths of 0.01 or less, $1/\kappa$ is not significantly affected. (For considerations on the dielectric profile near biological systems, due to ion-ion interactions, see Vaidhyanathan, 1982, 1985, 1986, 1988.)

c. A correlation has been observed between electrophoretic mobility in polar organic media and electron-donicity and accepticity. The degree of electron-donicity of a macromolecule or particle can be estimated by its electrophoretic mobility in Lewis acid organic liquids and the degree of electron-accepticity can be obtained via electrophoresis in Lewis base organic liquids (Fowkes *et al.*, 1982; Labib and Williams, 1987; Labib, 1988). It is therefore not surprising that amphoteric entities which are immersed in water at a pH close to their isoelectric point, and consequently have no electrophoretic mobility, also have zero electron-donicity as well as zero electro-accepticity, which then corresponds to apolarity, or "hydrophobicity." For further observations on the correlation between the donicity scale of a number of inorganic oxides and their electrokinetic potentials in water, as a function of pH, see Labib and Williams (1986); see also the end of Chapter XV, for an indication as to how that qualitative correlation may be molded into a quantitative correspondence.

The role of the linkage between AB and EL forces in particule flocculation, is further treated in Chapter XXIII.

OSMOTIC PRESSURE EFFECTS

As shown by van Oss, Arnold *et al.* (1990) (see also Chapter VI) the strong monopolar repulsion between polar polymer molecules in aqueous solution at high concentrations gives rise to a strongly negative Flory-Huggins χ_{12} parameter, which can be expressed as

$$\chi_{12} = -S_c \Delta G_{121}/kT, \text{ or } \chi_{12} = 2 \, S_c \gamma_{12}/kT \qquad [\text{VI-8A}]$$

where S_c is the contactable surface area between two polymer molecules (van Oss and Good, 1989). The effect of this is that the second virial coefficient B_2 (eq. [VI-3]) of the osmotic pressure equation (eq. [VI-2]) becomes strongly positive; this results in unusually high osmotic pressures, especially at high concentrations of strongly monopolar polymers dissolved in water. In addition, at high concentrations, these osmotic pressures are virtually independent of the molecular weight of the polymer (at Mw ≥ 1,000) (van Oss, Arnold et al., 1990). These considerations relate to the measurements done on the osmotic pressures of polyethylene glycols, or polyethylene oxides (PEO) of various molecular weights by Gawrisch (1986) (see also Gawrisch et al., 1988; Arnold et al., 1988), and analyzed by van Oss, Arnold et al. (1990), in the context of the theory discussed in Chapter VI; see below.

OSMOTIC PRESSURES OF PEO-WATER SOLUTIONS

Osmotic pressures of polyethylene oxide solutions in water have been reported by several authors. Malcolm and Rowlinson (1957) measured water activities of PEO water solutions via the water vapor pressure for PEO of molecular weights 300, 3,000 and 5,000. The measurements were done in the region of higher PEO concentrations (above 50 wt.%) and mainly at higher temperatures (typically at 50° to 65°C; i.e., at or above the θ. Rogers and Tam (1977) measured water activities for PEO 600, 1,001, 1,513, 3,035 and 7,980 solutions in the region of moderately dilute solutions (less than 20 wt.% PEO) at the temperatures, 35, 45 and 69.5°C. Chirife and Fontan (1980) measured water activities for PEO 200, 400 and 600 for PEO concentrations between 0 and 65 wt.% at 25°C.

For the data analysis, the osmotic pressures of aqueous solutions of polyethylene oxides with molecular weights from 150 to 20,000 and concentrations from 0–60 wt.% were measured at 25°C. The data were calculated from the water vapor pressure over the solution, relative to the vapor pressure over pure water at the same temperature (Gawrisch, 1986; Arnold et al., 1988). The experimental data of the osmotic pressures were fitted to an analytical expression proposed by Norrish (1966); these are shown in Fig. XIX-1. The deviations of the single experimental points from these curves are smaller than ± 0.5 MPa. The osmotic pressure data for PEO 6,000 are in good agreement with the data of Michel and Kaufman (1973) (0–30 wt.% PEO). The PEO 20,000 data agree reasonably with data published by Parsegian et al. (1986). The PEO 400 data are nearly identical with the values published by Chirife and Fontan (1980).

With polymers that act as strong electron donors when they take part in hydrogen bonding, γ_{12} has been determined (van Oss et al., Advan. Coll. Interface Sci., 1987) on the dry solid, and on solids swelled by amounts of water equivalent to one or two layers of hydration per molecule (see above). It was found that the exponential decay of the apparent value of γ_{12} sets in strongly with the second layer of hydration. For 60% PEO/40% water (which is close to the limit of feasible solubility of PEO), the first layer of hydration is barely completed.

So it may be assumed that the value of γ_{12} found with water on dry PEO gives a fair approximation to the value of PEO in 60% solution. This value was found to be about –26 mJ/m², from the data given in van Oss et al. (Advan. Coll. Interface Sci., 1987); see also Chapter XVII.

If a plot of log ΔG^{AB} vs. ℓ may be assumed to be a straight line, and assuming that at 60% PEO, hydration may be ignored, this makes it possible to establish one point

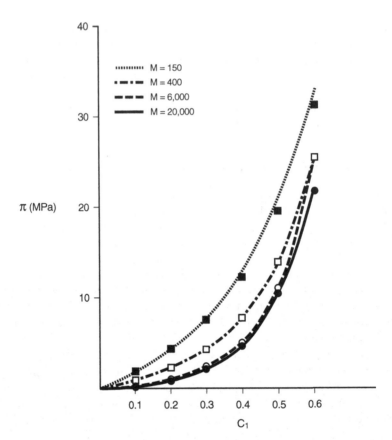

FIGURE XIX-1 Plot of the osmotic pressure, π, vs. the PEO concentration, c1, for four different molecular weight sof PEO. The *curves* are based upon the experimental results obtained by Arnold *et al.* (1988). The *points* are calculated *via* eq. [VI-2]: M = 150; M = 400; M = 6,000; M = 20,000, using the χ_{12} values from Table XIX-5. (From van Oss, Arnold *et al.*, 1990b.)

on that straight line, i.e., $\Delta G^{AB} = 2 \times 26.25$ or 52.5 mJ/m² at 60% PEO. The thickness of the PEO strands is about 4.6 Å, being about the same as that of linear polyethylene (Bunn, 1939), so an estimate of the contactable surface area is $S_c = 4.6^2 = 21.2 Å^2 = 21.2 \times 10^{-20}$ m². Thus $\chi_{12} = 2S_c\gamma_{12} = -1085 \times 10^{-23}$ J. At room temperature, kT ≈ 4 × 10^{-21}J, so $\chi_{12} = -2.78$, for 60% PEO.

The slope of ΔG^{AB} vs. ℓ is not known directly, but one can estimate another point on the line, because it is known (Albertsson, 1971, p. 38) that incipient phase separation takes place with 4% PEO 6,000 and 4% dextran (M ≈ 180,000), in water. At this point one might take $\Delta G_{121}^{AB} = + 1.5$ kT, or + 1.0 kT, or +0.5 kT. The term 1.5 kT (which comes from the kinetic theory of gases) pertains to the interaction between spherical (or point) particles with three translational degrees of freedom. For mutually repelling polymer chains, the motion in which contact is established or broken is predominantly along the line perpendicular to the axes of the two chains, i.e., it is one-dimensional. Hence, only one degree of freedom is involved, and 0.5 kT is thus the appropriate value of ΔG_{121}^{AB}. The strand width of dextran is about 7.5 Å, so that $S_c = 4.6 \times 7.5 \approx 34$ Å²; see Table XIX-3 (which also shows the densities of the various PEO solutions used). Substituting another 4% PEO 6,000 for the 4% dextran, we find for just 8% PEO 6,000 in water:

$$\chi_{12}^{(8\%)} = -\frac{52.5}{46.8} \cdot \frac{(4.6)^2}{4.6 \times 7.5} \times 0.5 = -0.34$$

The quotient, 52.5/46.8, is the correction needed to convert from the interaction energy between PEO and dextran in water $\Delta G_{132} = +46.8$ mJ/m² , using eq. [XVIII-9] and the data given in Tables XVII-5 and XVII-8) to the interaction energy between two PEO molecules in water $\Delta G_{121} = +52.5$ mJ/m²). Data derived from $\Delta G = 0.5$ kT (for incipient phase separation of 4% PEO and 4% dextran) are used here for the calculation of χ_{12} values for PEO at concentrations c_1 from 0.1 to 0.5. The results are shown in Table XIX-4.

Table XIX-5 shows the values of osmotic pressure calculated using eq. [VI-2], for the four different molecular weights and six different concentrations. The concentration-dependent values of χ_{12} were employed, as were the solute densities from Table XIX-3. The contributions to II from each of the first three virial coefficients are shown, together with the total osmotic pressure.

Figure XIX-1 shows the osmotic pressures calculated for PEO of MW 150, 400, 6,000, and 20,000 via eq. [VI-2], superimposed on the values found experimentally

---

TABLE XIX-3

Properties of Dextran (M = 180,000) and Polyethylene Oxide (PEO) Used in Calculations of χ_{12} and Osmotic Pressure

	Concentration c_1 in aqueous solution (w/v)	Density d_1 (with respect to water)	Thickness (Å)
Dextran (180,000)			7.5[a]
PEO (unhydrated)	0.6	1.15[b]	4.6[c]
PEO (hydrated)	0.5	1.124[d]	
PEO (hydrated)	0.4	1.1098[d]	
PEO (hydrated)	0.3	1.073[d]	
PEO (hydrated)	0.2	1.048[d]	
PEO (hydrated)	0.1	1.023[d]	

[a] Estimated using the analogy of the crystal structure of cellulose (Hui *et al.*, 1985).
[b] *Merck Index.*
[c] Estimated from the crystal structure of linear polyethylene (Bunn, 1939).
[d] Interpolated on a straight line in a semilogarithmic plot, between $d_1 = 1.15$ at $c_1 = 0.6$, and $d_1 = 1.00$ at $c_1 = 0$.

TABLE XIX-4

Values of χ_{12} for the PEO-Water System

c_1 fractional concentration (weight/volume)	χ_{12}
0.08	−0.34[a]
0.1	−0.37[b]
0.2	−0.55[b]
0.3	−0.83[b]
0.4	−0.24[b]
0.5	−1.86[b]
0.6	−2.78[c]

[a] Based upon the assumption that incipient phase separation between PEO 6,000 and dextran 180,000 occurs at $\Delta G_{121} = \tfrac{1}{2}\,kT$; see text.
[b] Interpolated values on a semilog plot; see Appendix.
[c] From $\Delta G121 = +52.5$ mJ/m2 and $Sc = 21.2$ Å2; see eq. [VI-8].

by Arnold *et al.* (1988) with aqueous solutions of the same polymers. The agreement is quite close.

Thus, in cases where a strong polar repulsion exists between polymer moieties dissolved in water (as occurs with polyethylene oxides to a very pronounced degree), $S_c\Delta G_{121}$ is strongly positive, making χ_{12} strongly negative (eqs. [VI-7] and [VI-8]). A strongly negative χ_{12} in the second virial coefficient (eq. [VI-9]) indicates that the osmotic pressure of such mutually repulsive polymers is extremely high, especially at higher concentrations.

Inspection of Fig. XIX-1 and Table XIX-5 shows that the second virial coefficient is the major cause for the striking lack of dependence of Π on the molecular weight of the polymer. The major factor in this term in eq. [VI-3] is the square of the (inverse) molecular weight of the polymers' *subunits*; and this molecular weight ($M/n = 44$ for PEO) is constant for all linear polymers of the same monomer. But this term can only attain significant influence when the value for χ_{12} is negative and of the order of −1 or −2 or so, as is indeed the case with PEG. This is also the case with dextrans, although to a lesser extent. For dextrans, $M/n = 180$, which makes for a much lower value $(1/180)^2$, as compared to $(1/44)^2$ for PEO (i.e., 16.7 times lower). However, the S_c value for dextran is 2.66 times greater than for PEO. Thus, in general the magnitude of the value of the second virial coefficient (eq. [VI-2]) for dextran would be only about 6.3 times smaller than that for PEO. Hence, for dextran also, Π still is strongly dependent on the second virial coefficient, and thus, also, largely independent of molecular weight (see, e.g., Evans and Needham, 1988), especially for M_1 1000.

In Fig. XIX-1, it can be seen that by using the χ_{12} values obtained for PEO, in calculating the second viral coefficient, Π values are found which correspond closely to the experimental results for PEO 150, 400, 6,000 and 20,000. The third virial coefficient was also taken into account in the computation, as it can represent up to 6.5% of the total value of Π for aqueous PEO solutions. For molecular weights of

TABLE XIX-5
Osmotic Pressures, in MPa, of Aqueous Solutions of PEO of Various Molecular Weights Calculated, Broken Down into the First Three Virial Coefficient Terms

Concentrations c_1	MW	$(\tfrac{1}{2} - \chi)$	Π1	+	Π2	+	Π3	=	Π total
0.6	150	3.28	9.76[a]	+	20.25[a]	+	1.07[a]	=	31.09[a]
	400		3.66	+	20.25	+	1.07	=	24.99
	6,000		0.34	+	20.25	+	1.07	=	21.57
	20,000		0.07	+	20.25	+	1.07	=	21.40
0.5	150	2.36	8.13	+	10.59	+	0.66	=	19.37
	400		3.05	+	10.59	+	0.66	=	14.30
	6,000		0.20	+	10.59	+	0.66	=	11.44
	20,000		0.06	+	10.59	+	0.66	=	11.31
0.4	150	1.74	6.51	+	5.25	+	0.37	=	12.13
	400		2.44	+	5.25	+	0.37	=	8.05
	6,000		0.17	+	5.25	+	0.37	=	5.78
	20,000		0.05	+	5.25	+	0.37	=	5.66
0.3	150	1.33	4.88	+	2.37	+	0.17	=	7.42
	400		1.83	+	2.37	+	0.17	=	4.37
	6,000		0.12	+	2.37	+	0.17	=	2.66
	20,000		0.04	+	2.37	+	0.17	=	2.57
0.2	150	1.05	3.25	+	0.87	+	0.05	=	4.16
	400		1.22	+	0.87	+	0.05	=	2.14
	6,000		0.08	+	0.87	+	0.05	=	1.00
	20,000		0.02	+	0.87	+	0.05	=	0.94
0.1	150	0.87	1.63	+	0.19	+	0.006	=	1.83
	400		0.60	+	0.19	+	0.006	=	0.80
	6,000		0.04	+	0.19	+	0.006	=	0.24
	20,000		0.01	+	0.19	+	0.006	=	0.21

[a] All figures are rounded off to the nearest 10 KPa.

PEO of 6,000 and higher, the first term of Π (RTc_1/M_1) accounts for less than 10% of the total at $c_1 > 0.2$. Thus, Table XIX-5 reveals the fundamental reasons for the molecular weight independence of Π at the higher concentrations, in this system.

Thus, for linear polymers which have a negative interfacial tension with the solvent, and which have subunits of a small molecular weight (such as PEO), the osmotic pressure, Π, with the higher molecular weights, mainly depends upon the value of the second virial coefficient (see eqs. [VI-2] and [VI-3]) and this coefficient is governed by the value of χ_{12}. Clearly, strongly negative χ_{12} values will give rise to more highly positive values of the second virial coefficient. Therefore, for the proper estimation of the polymer/liquid interaction parameter, χ_{12}, it is imperative to determine that value *at the highest possible polymer concentration* (cf. Tables XIX-4 and XIX-5), and not as was customary to date, at the greatest possible dilution, because the latter practice yields the rather uninformative χ_{12} values which only vary between +0.40 and 0.49* (see, e.g., Barton, 1983). The latter values have caused some of the remarkable properties of polar polymers to remain unsuspected until fairly recently.

While it has been recognized in the past the negative χ_{12} values can occur, they are not encountered with great frequency in the literature. One case, given by Hermans (1949), shows a value of $\chi_{12} = -1.8$ for cellulose acetate in tetrachloroethane, which is a weak electron acceptor. That value is now entirely understandable, as cellulose acetate is a fairly strong electron donor (van Oss *et al.*, Advan. Colloid Interface Sci., 1987); see also Table XVII-8. Its ΔG_{121} value (see eq. [VI-8]) in tetrachloroethane should be positive, which would give rise to a negative χ_{12} of about the order of magnitude found by Hermans (1949). With most other organic solvents, such situations would, however, be rather unlikely. However, with solvents such as chloroform [which is an electron-acceptor (van Oss and Good, 1989), and water, formamide, or glycerol (which are self-hydrogen bonding, and thus bipolar], negative χ_{12} values can occur with polymers that have electron donor properties. It should be noted that the positive χ_{12} values reported for PEO of various Mw's in water by Molyneux (1983) all pertain to conditions above the θ-point (i.e., from 55°C to 65°C), at which water becomes a less than ideal solvent for PEO (Fowkes, 1987).

In eq. [VI-2], the activity coefficients of PEO are not taken into account. However, it should be understood that eq. [VI-2] serves to *determine* the activity of PEO at various concentrations, and relate it to the osmotic pressure, Π. So to include activity coefficients in eq. [VI-2] would be redundant. It should also be noted that, in the case of aqueous solutions of PEO, Π/c_1 increases with the concentration, c_1; this is a phenomenon which has been recognized as peculiar to polymers dissolved in "good" solvents (Napper, 1983). Thus, in the case of PEO, its activity increases with concentration. This is largely due to the fact that, with increasing concentration, PEO becomes less hydrated (because of the lack of a surplus of water molecules) which leads to an increase in the positive value of ΔG_{121}^{AB}, i.e., to an increase in mutual repulsion between PEO molecules, and a tendency to draw more water into the

---

* Various approaches to the interpolation of values for χ_{12} are possible, for polymer concentration c_1 between 0.08 and 0.60; their merits and drawbacks are discussed in the Appendix to this chapter.

polymer—in other words, high osmotic pressure. Whilst oriented hydration is, of course, the mechanism by which polar repulsion is propagated some appreciable distance beyond the very close range of direct hydrogen bonds (see above), the hydration orientation decreases with distance, and the result is the decay of polar repulsion (see Fig. XIX-1). Thus, the greater the dilution, the weaker the polar repulsion. The strong positive value of ΔG_{121} for PEO in water, and thus its strongly negative χ_{12} value, therefore not only cause the unusually high (and largely molecular weight-independent) osmotic pressure, but it also is the real origin of the strong stabilizing power of PEO for aqueous suspensions of hydrophobic particles (see, e.g., Napper, 1983), as well as of its flocculating power for various proteins (van Oss, 1988), and of its capacity to cause and/or to facilitate cell or liposome fusion (Arnold *et al.*, 1988; van Oss *et al.*, *Molecular Mechanisms of Membrane Fusion*, 1988).

MONOPOLAR REPULSION BETWEEN DISSIMILAR POLAR ENTITIES

In the interfacial interaction between two dissimilar "hydrophobic" entities immersed in water, or between one "hydrophobic" entity and one hydrophilic entity in water, ΔG_{1w2} (see eq. [XVIII-9]) usually has a negative value. However, a positive value for ΔG_{1w1} is also possible. In the latter case a repulsion ensues, i.e., generally when the entities (1) and (2) are hydrophilic, i.e., if they are electron-donor monopolar compounds, and the average value of their $\gamma^{\ominus}$ parameters is greater than 28.3 mJ/m²i.

There are some important differences between the behavior of similar and that of dissimilar entities, in the outcome of the repulsive interactions which they undergo, when immersed or dissolved in water. A positive value of ΔG_{1w1} gives rise to a pronounced solubility of compound (1) in water (see Chapter XXII). A positive value of ΔG_{1w2} [where both compounds (1) and (2) are dissolved in water] tends to result in phase separation (see Chapter XXI). When ΔG_{1w2} has a negative value, and when, e.g., entity (1) is moderately "hydrophobic" and entity (2) hydrophilic (so that, in *water*, an attraction occurs), that attraction between (1) and (2) can be changed into a repulsion by the admixture of a water miscible *organic solvent*; see below.

MONOPOLAR REPULSION IN NON-AQUEOUS POLAR MEDIA

Due to the much smaller polar energy of cohesion of non-aqueous polar liquids than of water, the net interfacial ("solvophobic") attraction of low energy solutes in such non-aqueous polar liquids is much smaller than their "hydrophobic" attraction in water. Thus, in general, in the case of an attraction between low-energy surfaces and hydrophilic compounds dissolved and/or immersed in water ($\Delta G_{132} < 0$, where the subscript, 3, denotes the liquid), by admixture of a water-miscible polar solvent to the water, the attraction between 1 and 2 can be turned into a repulsion ($\Delta G_{132} > 0$). This phenomenon is the basis of "reversed-phase liquid chromatography" (RPLC), where hydrophilic compounds (e.g., proteins, peptides, etc.), dissolved in water, are adsorbed onto chromatographic carriers with a low-energy ("hydrophobic") surface. The adsorbed hydrophilic compound can then be desorbed, or "eluted" from the chromatographic column by the admixture of increasing amounts of a water-miscible organic solvent [e.g., ethylene glycol (van Oss, Israel J. Chem., 1990), dimethylsulfoxide, acetonitrile, etc.]; see Chapter XXIV. In principle the same phenomenon also underlies the mechanism of removing low-energy material (e.g., "dirt") from hydrophilic tissue by means of an aqueous surfactant solution, that is, "washing" (van Oss and Costanzo, 1992); see also Chapter XXII.

PERSISTENCE OF ΔG_{iwi} INTERACTIONS AT A DISTANCE

The mechanism of propagation at a distance, of the *repulsion* between strong electron-donor molecules, immersed in water, can be readily understood, when one takes into account the orientation of the water molecules of hydration. This hydration orientation [first noticed as "hydration pressure" (Parsegian *et al.*, 1985)] decreases exponentially with the distance, ℓ, and a characteristic property of the liquid medium, i.e., its "decay length," λ:

$$\Delta G_\ell^{AB} = \Delta G_{\ell_0}^{AB} \exp\left(\frac{\ell_0 - \ell}{\lambda}\right)$$ [VII-4]

where ℓ_0 is the minimum equilibrium distance to which atoms and molecules can approach each other; $\ell_0 \approx 0.16$ nm (see Chapter VII). ΔG_{iwi} (eq. [XVIII-1]) and ΔG_{iw} (eqs. [II-36] and [IV-6]) are to be identified with ΔG_{ℓ_0}, in eq. [VII-4], and not with ΔG_ℓ.

The value of λ for water is not known with precision. It may be argued that λ is of the order of magnitude of the radius of gyration (Rg) of single water molecules, i.e., $\lambda \approx 0.2$ nm (Chan *et al.*, 1979). However, given that at any given moment, at 20°C, a significant proportion of the molecules of liquid water are hydrogen-bonded to one or several other water molecules (Eisenberg and Kauzmann, 1969, p. 178) (at 20°C the cluster size of water is about 4.5 molecules per cluster; see Chapter VIII), their effective Rg value must be greater than 0.2 nm, although it is unlikely to surpass 1.0 nm. A rough estimate for the mean value of λ for water appears to be close to 0.6 nm (see Chapter XXIII; also, van Oss, *Biophysics of the Cell Surface*, 1990). Within this distance from the non-aqueous molecule, the water molecules are the most highly oriented. As layers of water molecules which are all oriented the same way have a somewhat lower density than randomly oriented bulk water, such boundary layers of decreased density, which surround polar molecules or particles, will mutually repel each other, according to Derjaguin's model (Derjaguin *et al.*, 1987, pp. 25–52; Derjaguin, 1989, pp. 23–31).

Thus, the propagation at a distance of repulsive H-bonding forces, under the influence of a local distortion or gradient of a given property of the liquid medium, is analogous to the repulsion at a distance between molecules or particles surrounded by a diffuse ionic double layer, where the decay length is the thickness of that double layer, i.e., the Debye length, which is inversely proportional to the square root of the ionic strength (Chapter V).

CODA

Hydrophilic (AB) repulsions occur among particles and molecules with a $\gamma^\oplus$ parameter which is negligible, or zero, and a $\gamma^\ominus$ parameter which is greater than 28.3 mJ/m². Such entities are ($\gamma^\ominus$) monopolar. The monopolarity *and* the high value of $\gamma^\ominus$ are necessary for a repulsion to occur in aqueous media, because of the high value of the underlying hydrophobic attraction energy ($\Delta G^{AB} = -102$ mJ/m²) *which is always present* and must be overcome before repulsion can prevail. Hydrophilicity can be quantitatively defined in three different ways: (1) as the free energy of interfacial interaction in water, ΔG_{iwi} (which is mainly AB), (2) as the free energy of hydration, ΔG_{iw} (which is part LW and part AB), and (3) as the electron-donor parameter, $\gamma_i^\ominus$. No. 1 relates most closely to the aqueous solubility of i; when $\Delta G_{iwi} > 0$, i is very soluble in water. No. 2 generally indicates hydrophilicity when $\Delta G_{iw} -113$ mJ/m². No. 3 usually indicates hydrophilicity when $\gamma_i^\ominus > 28.3$ mJ/m². All three hydrophilicity criteria are tabulated for a number of hydrophobic and hydrophilic compounds in Table XIX-1. Of these, the only *absolute* quantitative measure of hydrophobicity/hydrophilicity for condensed-phase materials, i, is ΔG_{iwi}. When $\Delta G_{iwi} < 0$, i is hydrophobic and when $\Delta G_{iwi} > 0$, i is hydrophilic, whilst the quantitative value of ΔG_{iwi} denotes the *degree* of hydrophobicity (when negative) and of hydrophilicity (when positive); see van Oss and Giese (1995).

The long-range nature of hydrophilic repulsion is a consequence of the orientation of the water molecules of hydration of hydrophilic surfaces, which continues in an exponentially decreasing fashion into the more distal layers of neighboring water molecules, thus giving rise to "hydration pressure"; see Chapter X. The decay length of water is ≈ 1.0 nm; see Chapter VII.

Whilst AB repulsions can occur among electrically totally neutral hydrophilic entities, when an electric surface potential does occur, it is always accompanied by a concomitant repulsive AB interaction. There thus is a linkage between EL and AB interactions (but not necessarily between AB and EL interactions). When electrically charged hydrophilic surfaces are neutralized, either by means of a change in pH, or through the admixture of plurivalent counterions, such surfaces become "hydrophobic."

It can be shown that the osmotic pressure of concentrated hydrophilic polymer solutions in water (of, e.g, polyethylene oxide, PEO) is mainly due to the AB repulsion engendered by the $\gamma^\ominus$-monopolar monomeric sections of the polymers. The osmotic pressure of concentrated aqueous PEO solutions thus is largely molecular weight independent. The Flory-Huggins solubility parameter, χ_{iw}, is equal to the (mainly AB) work of repulsion, expressed in units of kT. It is important to determine χ_{iw} at the highest possible polymer concentration, and not at the greatest dilution. PEO osmotic pressures thus calculated correspond precisely to experimentally measured osmotic pressures; see Fig. XIX-1, above.

Monopolar repulsions also occur in various non-aqueous polar solvents; in a sense they even occur more readily than in water, because their γ^{AB} (and thus the origin of "solvophobic" attraction energies tends to be lower than that of water. This effect is especially important in mixtures of polar solvents with water, which are used (e.g., in reversed-phase liquid chromatography) to elute relatively hydrophilic solutes from "hydrophobic" attraction into a "solvophilic" repulsion through the decrease the non-aqueous solvents bring about in the γ^{AB} of the liquid mixture.

APPENDIX: ON THE INTERPOLATION OF VALUES OF γ_{12} VERSUS PEO CONCENTRATION

The exponential decay relation, eq. ([VII-4]), was developed to describe the interaction between two parallel, semi-infinite slabs. However, for a pair of real, non-parallel polymer chains, the decay cannot be given exactly, by eq. [VII-4]). The change in geometry, from parallel slabs to crossed cylinders, should lead to a change in the decay function. There is, in addition, the important question of how the first hydration layer affects the mechanics of attraction of the crossed cylinders: the effective contactable area will be larger if the hydration layer is so tightly bound that, in the attraction or repulsion, it acts as part of the core of the chain. This increase in S_c will tend to counterbalance the decrease in ΔG that accompanies dilution and the increase in ℓ.

For the purpose of interpolating χ_{12} as a function of concentration, between the values at $c_1 = 0.08$ and $c_1 = 0.6$, eq. [VII-4] points to the plotting of $\log \chi_{12}$ vs. $(c_1)^{-1/3}$, for spherical particles, or $\log \chi_{12}$ vs. for parallel, cylindrical particles. On the other hand, following Napper (1983), one may (at least at the lower concentrations) expect a reasonable correlation when plotting ΔG directly vs. c_1. However, especially if one wishes to include the higher concentrations, a more reasonable connection between the virial coefficients and the concentration (by analogy with Napper, 1983), leads to a correlation between ΔG (and thus χ_{12}) and c_1^2; cf. eq. [VI-2].

The order of decreasing correlation between the observed values for the osmotic pressure, Π, and the values calculated by means of eq. [VI-2], obtained with χ_{12} for $c_1 = 0.1$ to 0.5, for the various methods for interpolating χ_{12} vs. c_1 between the known values for χ_{12} for $c_1 = 0.08$ and $c_1 = 0.06$, was found to be:

$$\log \chi_{12} \text{ vs. } c_1 \qquad\qquad\qquad\qquad [A]$$

$$\chi_{12} \text{ vs. } c_1^2 \qquad\qquad\qquad\qquad [B]$$

$$\log \chi_{12} \text{ vs. } (c_1)^{1/3} \qquad\qquad\qquad\qquad [C]$$

$$\chi_{12} \text{ vs. } c_1 \qquad\qquad\qquad\qquad [D]$$

$$\log \chi_{12} \text{ vs. } (c_1)^{1/2} \qquad\qquad\qquad\qquad [E]$$

Figure XIX-1 was obtained by using approach [A]. [B] yields very similar results. The others gave Π values which, from [C] to [E], increasingly diverge from the observed osmotic pressures; see van Oss, Arnold et al. (1990).

XX The Primary and Secondary Interactions

This seems to be an appropriate junction to reflect on which of the covalent forces discussed so for are genuine primary driving forces, and which are secondary interactions, caused by the primary interactions. In past years one has tended to be confronted with a growing number of different forces, which are commonly believed to play a role in the interactions between biological and other polar entities, such as cells, biopolymers and other polar polymers or particles, when immersed or dissolved in water. This expanding catalogue of (non-covalent) interaction forces comprises (van Oss, J. Dispersion Sci. Tech./Biocoll. Biosurf., 1991):

1. London-van der Waals, or dispersion forces;
2. Debye-van der Waals, or induction forces;
3. Keesom-van der Waals, or orientation forces;
4. Electrostatic, or Coulombic forces;
5. Hydrogen-bonding forces;
6. Hydrophobic interactions;
7. Hydration forces;
8. Brownian movement forces;
9. Osmotic pressure;
10. Disjoining pressure;
11. Structural forces;
12. Steric interactions;
13. Depletion interactions;
14. Entropy-driven interactions;
15. Enthalpy-driven interactions;
16. Cross-binding interactions;
17. Specific interactions.

For linguistic reasons, the terms: "force," "pressure," and "interaction," are here used interchangeably. Where these interactions are discussed specifically, they are treated in units of *energy*, or in units of *energy per surface area*, except when explicitly stated otherwise.

Whilst some are more felicitously named than others, all these forces, interactions and pressures indubitably exist and contribute, to various degrees, to the sum total of the interaction between biological and other polar molecules and/or particles, immersed in water. The question is: which ones of these are primary and separate physical-chemical forces, and which ones are secondary manifestations of one or the other of the more fundamental forces?

Apart from the undoubted advantage of simplicity, and the satisfaction of complying with philosophical orthodoxy,* there are a few even more cogent reasons for answering this question:

1. If any one of these interactions is just a manifestation, or a result, of one or several of the more fundamental physical forces, then, by not taking that fact into account, one would erroneously include the same force more than once into the sum total of the interactions pertaining to a given system, as has often been done in the past.
2. Different types of forces tend to decay with distance according to different regimes.
3. Similar forces obey the same combining rules, forces of a different nature obey entirely different rules.
4. Various non-covalent interactions can be caused by a combination of two or more different forces.
5. Seemingly different interactions can be caused by the same primary force.

VAN DER WAALS INTERACTIONS (PRIMARY)

In macroscopic systems, in the condensed state, according to Lifshitz's treatment (1955; Dzyalochinskii *et al.*, 1965; see also Chaudhury, 1984) all three van der Waals interactions obey the same rules and, in the non-retarded regime (i.e., at distances smaller than 5 to 10 nm), decay with distance at the same rate, i.e., for energies per unit surface area and for semi-infinite parallel flat slabs with a distance (ℓ) between them, they decay as ℓ^{-2}. London-van der Waals (dispersion) forces undergo retardation at $\ell > 5$ to 10 nm, after which they decay as ℓ^{-3} (see Chapters II, III and XVI). Debye-van der Waals and Keesom-van der Waals forces on the other hand are not subject to retardation. However, in water, London-van der Waals (dispersion) interactions usually represent more than 95% of the total Lifshitz–van der Waals interaction, so that retardation effects cannot always be neglected. However, as all three varieties of van der Waals forces should otherwise be treated in the same manner (as far as interactions in liquids between macromolecules or particles are concerned), it is best to designate them collectively as Lifshitz–van der Waals forces, using the superscript LW (Chapter II). It should be remembered that LW interactions between two *different* materials, in a liquid, can be either repulsive or attractive, depending on the properties of the materials and the liquid (Chapter III).

---

* "*Pluralitas non est ponenda sine necessitate*" [Ockham, *ca.* 1320]

ELECTROSTATIC, OR COULOMBIC INTERACTIONS (PRIMARY)

Electrostatic interactions between charged macromolecules and/or particles, immersed in water, are generally the outcome of the repulsion between two entities with the same sign of charge. The repulsive interaction energy is a function of the surface potential of the molecules or particles, the value of which is relatively easily determined by an electrokinetic method; usually electrophoresis. The interaction energy decays exponentially with distance (ℓ), as $\exp(-\kappa\ell)$, where $1/\kappa$ is the thickness of the Gouy-Chapman electrical double layer (or Debye length). $1/\kappa$ varies with the ionic strength (Chapter IV).

The classic DLVO (for Derjaguin-Landau-Verwey-Overbeek) theory of stability of particle suspensions (Chapter XXIII), which was the accepted approach for at least a generation (roughly from 1948 to 1990), is based on the balance between the van der Waals-London (dispersion) attraction and the electrostatic (EL) repulsion between particles, i.e.:

$$\Delta G^{DLVO} = \Delta G^{LW} + \Delta G^{EL} \qquad\qquad [XX\text{-}1]^*$$

Like LW forces in liquids, EL forces between different particles can be repulsive, or attractive, depending on the sign and value of their surface potential. The interaction between two charged particles (of the same sign of charge) can also be *attractive*, when: a) the values of their surface potentials differ widely (Derjaguin, 1954; Overbeek, 1988), and b) when cross-bound by means of plurivalent counterions, such as Ca^{++}, in the case of negatively charged particles (Chapter XXIV).

In polar media, e.g., in water, an extended DLVO (XDLVO) approach has to be utilized, such that (cf. Chapters VII and XXIII):

$$\Delta G^{XDLVO} = \Delta G^{LW} + \Delta G^{EL} + \Delta G^{AB} \qquad\qquad [XX\text{-}2]$$

LEWIS ACID-BASE INTERACTIONS (PRIMARY)

As it has been shown that *hydrophobic interactions* represent the attractive mode, and *hydration pressure* is a manifestation of the repulsive mode of hydrogen-bonding interactions (Chapters XVIII and XIX), these two types of interactions (Nos. 6 and 7 from the interactions listed above) are also treated here under the general heading of *Lewis acid-base forces*. Contrary to LW interactions, but similar to EL interactions, Lewis acid-base interactions between *identical* materials, in water, can be repulsive with very hydrophilic materials (Chapter XIX). Lewis acid-base interactions between more hydrophobic materials, immersed in water, are strongly attractive (Chapter XVIII). In water, the interaction energies between strongly or weakly polar molecules

---

* A list of symbols can be found on pages 399–406.

or particles, whether they are attractive or repulsive, tend to be quantitatively much stronger than either LW or EL interactions. Thus, *in water*, energy balances of the strict DLVO-type, which take only LW and EL interactions into account, do not accurately reflect the total interaction energies (Chatper XXIII).

Hydrogen-bonding interactions, i.e., hydrogen-donor/hydrogen-acceptor interactions or Brønsted acid-base interactions, are a sub-set, albeit an important subset, of the more general Lewis acid-base, or electron-acceptor/electron-donor interactions. We have therefore designated hydrogen-bonding interactions, together with other polar interactions, as (Lewis) acid-base, or AB interactions (Chapter IV). AB interactions, somewhat like EL interactions, decay exponentially with distance (ℓ), as $\exp{(\ell_o - \ell)/\lambda}$ where λ is the decay length characteristic of the liquid medium. In water, λ equals about 1.0 nm, at 20°C, at $\ell \geq 1.0$ nm, for AB interactions.

It should be stressed again that repulsive Lewis acid-base interactions or *hydration forces* are not only the initiating forces of the strongly repulsive interaction between two very hydrophilic (usually exclusively hydrogen-acceptor) surfaces at very short range, when immersed in water. They are, in addition, the result of the interaction at a distance between opposing layers of oriented water molecules of hydration (Chapter XIX), which causes the net AB repulsion. The orientation of these water molecules is itself caused by the high density of opposing electron-donors, and the orientation of water molecules extends a sizeable distance away from the hydrophilic (electron-donating) surfaces, which are the real initiating cause of the repulsive interaction. It is the decrease in hydration orientation with distance (ℓ), which declines exponentially and which causes the effect of AB-repulsion to decrease as $\exp{(\ell_o - \ell)/\lambda}$. Some authors designate these longer-range effects, caused by the orientation of molecules of solvation as "structural forces" (Israelachvili, 1985; Derjaguin *et al.*, 1987; Derjaguin, 1989).

BROWNIAN MOVEMENT INTERACTIONS

It should be realized that every single detached molecule or particle, immersed in a liquid medium, is endowed with a Brownian (BR) energy of ≈ 1 kT. This energy keeps it in solution or in suspension, provided the energy of attraction between similar molecules (or particles) immersed in that liquid is less than 1 kT per pair of molecules or particles. But while a very small molecule has a Brownian energy of ≈ 1 kT, a large particle also has a Brownian energy of only ≈ 1 kT. Thus, micron sized particles, each pair of which would typically have a contactable surface area (S_c) of the order of 10^4 nm^2, will overcome the repulsive forces of Brownian motion and become destabilized even if their energy of mutual attraction, in a given liquid at close range, is as small as 10^{-3} mJ/m^2. Whilst the energies of thermal motion, or diffusion, are relatively small, they are not necessarily negligible; in apolar media they can be the major contribution to solubility or stability. Thus, as all single entities immersed in a liquid are endowed with $\Delta G^{BR} \approx + \frac{1}{2}$ kT per degree of freedom, this contribution clearly plays the greatest role in the case of the smallest molecules or particles, as these have the smallest surfaces of mutual contact (Chapter VI).

PRIMARY FORCES AND SECONDARY PHENOMENA

The interaction forces described under Nos. 1–8, in the list given above, may be considered as primary forces, the others (ns. 9–17) are secondary phenomena, discussed in the following Section.

The few really primary non-covalent interaction forces, operative in condensed media, are:

1. Electrodynamic, or Lifshitz–van der Waals (LW) interactions.
2. Electrical double layer (EL), or Coulombic interactions.
3. Polar, or electron-acceptor-electron-donor [or Lewis acid-base (AB)] interactions, of which hydrogen-donor/hydrogen-acceptor (Brønsted acid-base) interactions, or hydrogen-bonding interactions, are a subset.

In addition, there is a fourth primary interaction which is mainly important among molecules and particles in the 1 to 100 nm range.

4. The Brownian movement (BR) or diffusion interaction, which for purposes of *binding* first has to be overcome, but which for purposes of *repulsive stabilization*, can be made use of with some advantage, especially when contributing to "steric stabilization" of particles by means of adsorbed polymer molecules in apolar media.

OSMOTIC PRESSURE

Osmotic (OS) pressure interactions do not comprise any components that have not already been accounted for above. The van't Hoff part of the osmotic pressure (Π) equation, $\Pi = RTc/M_n$, covers just the Brownian movement–driven interactions (cf. eq. [VI-1]). Van't Hoff's equation is valid for solute molecules, which neither attract nor repel each other, in the liquid medium in which they are dissolved. As soon as such solute molecules do either attract or repel each other, a second term containing the second virial coefficient, B, needs to be introduced (cf. eq. [VI-9]):

$$\Pi = RTc/M_n + BRTc^2 \qquad [\text{XX-3}]$$

where the value of B is a measure of the non-Brownian interactions. When these intermolecular interactions are attractive, $B < O$, and when they are repulsive, $B > O$. The possible non-Brownian interactions comprise the Lifshitz–van der Waals, the electrostatic and the polar (Lewis) acid-base interactions, which include the hydrogen-bonding interactions (27). Thus:

$$\Delta G^{OS} = \Delta G^{BR} + \Delta G^{LW} + \Delta G^{EL} + \Delta G^{AB} \qquad [\text{XX-4}]$$

But it may also be stated that:

$$\Delta G^{TOT} = \Delta G^{BR} + \Delta G^{LW} + \Delta G^{EL} + \Delta G^{AB} \qquad [\text{XX-5}]$$

so that it is clear that ΔG^{OS} should not be further included in ΔG^{TOT}, provided the latter comprises all the applicable terms outlined in eq. [XX-5].

DISJOINING PRESSURE

Disjoining pressure alludes to, e.g., the interaction between the two thin liquid layers which border immediately on two opposing solid surfaces. Initially it alluded to repulsive interactions: when two such thin liquid layers (the molecules of which have been changed in orientation and/or in density of packing, due to their vicinity to the solid surfaces) overlap, they give rise to a *disjoining pressure* (Reviews by Derjaguin *et al.*, 1987; Derjaguin, 1989). In more recent usage disjoining pressure may be either repulsive or attractive. For instance, in terms of Derjaguin *et al.* (1987), hydration pressure is a positive disjoining pressure, whilst an attractive Lifshitz–van der Waals interaction may be called a negative disjoining pressure (see, e.g., Derjaguin, 1989, p. 67). Thus the term disjoining pressure, in the larger sense which it seems to have acquired, now embraces the longer-range manifestations of all the forces listed in the first page of this chapter, from 1–7, i.e., of the LW, EL and AB forces. Disjoining pressure, in its larger sense is therefore not a novel force or interaction and there seems to be no compelling advantage in using the term. It is, however, useful to understand its variegated meaning when consulting the Russian literature in colloid and surface science.

STRUCTURAL FORCES

Structural forces allude to the interactions engendered by either the orientation of the molecules of solvation (in water: hydration) (Israelachvili, 1985; Derjaguin *et al.*, 1987; Derjaguin, 1989) or by the finite and discrete size and shape of the liquid molecules in the vicinity of interacting surfaces (Israelachvili, 1985, 1991; see also Chapter XVI).

The effects of the orientation of molecules of hydration have already been discussed above, under Lewis Acid-Base Interactions. A decrease in the orientation of water molecules in the vicinity of hydrophobic surfaces has been proposed as a rather interesting explanation of the attractive effect of hydrophobic interactions (Israelachvili, 1985, 1991, p. 198); see also Derjaguin *et al.* (1987, pp. 25–28) and Derjaguin (1989, pp. 23–27), and Chapter XVIII, above.

On the other hand, the oscillating phenomena caused by the size and shape of the liquid molecules, which can give rise to an alternative attraction and repulsion, as a function of distance (ℓ) at fairly close range are an entirely different matter. However, apart from measurements made with the help of a force balance (Israelachvili, 1985, 1991; see Figure XVI-1) on the interaction between hard surfaces, in liquids consisting of rather rigid and fairly large molecules [such as benzene, cyclohexane, or octamethyl cyclotetrasiloxane (Israelachvili, 1985, 1991; Horn and Israelachvili, 1981)], the effects of the second type of structural forces, while real, are more a curiosity than a phenomenon which is frequently encountered. Thus, in aqueous media, structural forces of the first kind are comprised in the hydrogen-bonding interactions (see above), and structural forces of the second kind can generally be neglected; see also Israelachvili *et al.* (1989).

STERIC INTERACTIONS

"Steric stabilization" is a term that was coined to explain the phenomenon of stabilization of hydrophobic particles in water by means of non-ionic surfactants, at a time when the strength of the polar (AB) repulsion between the surfactant molecules was not yet universally realized, although its importance more recently has become increasingly accredited (Napper, 1987). However, once one does realize the quantitative significance of the polar repulsion between, e.g., polyethylene oxide (PEO) molecules, it becomes obvious that the stabilization of *aqueous suspensions* of hydrophobic particles, upon adsorption (or covalent attachment) of PEO molecules, is preponderantly due to the polar repulsion between these molecules. Thus, prior to the realization of the importance of the role of the (non-electrostatic) polar repulsion between non-ionic surfactants, other interaction varieties, which also to various degrees can contribute to this type of stabilization, tended to be stressed more strongly, or were even held to be solely instrumental in bringing about the entire phenomenon. Some of the types of repulsive interaction which have been invoked are: steric hindrance, changes in configuration entropy, excluded volume effects.

There now is a sizable body of literature on the subject of steric stabilization, of which the most thorough and extensive are the monographs by Napper (1983) and by Sato and Ruch (1980). A more compact but very lucid compendium, giving an excellent overview of the phenomenon and its underlying forces, is a chapter written by Ottewill (1967). Adopting Ottewill's nomenclature:

$$\Delta G_{steric} = \Delta G_{surface} + \Delta G_{osmotic} + \Delta G_{chain\ elasticity} \qquad [XX\text{-}6]$$

The three components can be treated as follows: $\Delta G_{surface}$ represents Mackor's configurational entropic repulsion (Mackor, 1951; Mackor and van der Waals, 1952); this corresponds to our ΔG^{BR} (see also Chapter VI). As Ottewill already observed (1967), these interaction forces are fairly feeble. Thus, $\Delta G_{surface}$, or ΔG^{BR}, only becomes important when overcoming the usually rather small Lifshitz–van der Waals attraction energies between particles or macromolecules, when immersed in apolar media. In water, however, the contribution of Brownian movement forces to stability is rather slight; see above. $\Delta G_{osmotic}$, in the sense discussed by Ottewill (1967), corresponds to the second virial coefficient of the osmotic pressure equation, i.e., to the last term in eq. [XX-3]. In the case of PEO, dissolved in water, B is strongly positive, in conformation with its exceptionally high largely molecular-weight-independent osmotic pressure at high concentrations (see Chapter XIX). Thus, in the case of PEO, $\Delta G_{osmotic}$ is principally identifiable with ΔG^{AB}.

Whilst $\Delta G_{surface}$ and $\Delta G_{osmotic}$ thus belong to interaction types that have already been identified above, $\Delta G_{chain\ elasticity}$ is a new entity. It involves estimating the deformation energy of compressible adsorbed polymer layers, as a function of the distance of overlap of two such layers, during the elastic collision between two particles. $\Delta G_{\text{chain elasticity}}$ can be calculated, provided that a value can be found for the elastic modulus (Q) of the adsorbed layer (Q α $\Delta G_{\text{chain elasticity}}$). The difficulty lies in determining the value of Q (Ottewill, 1967); see also Napper (1983), and Sato and Ruch (1980). It would of course be useful if one could determine Q, involved in the interaction between two cells *via* their glycocalices, but this still is hard to do, as it remains based on a number of assumptions which, while probably reasonable, are not easily experimentally verifiable. Nevertheless, it is now possible to measure all the other interaction energies and thus to determine the total interaction energy between the interacting polymer or surfactant molecules that are attached to the particles, and one must content oneself for the time being to achieve only a rough estimate of the short-range elastic interactions to which the cells or particles are subjected upon the close approach of the stabilizing polymer of surfactant molecules.

One thing, however, should be stressed more emphatically than is usually done in disquisitions on steric stabilization. This is that the *conditio sine qua non* for steric stabilization is that the *stabilizing polymer or surfactant must be soluble in the liquid medium* (see also Chapter XXIII). Questions of feasibility of steric stabilization thus can always be reduced to the matter of solubility of the stabilizing agent in the chosen liquid. And polymer solubility can always be predicted to occur when, for apolar systems, $-1kT < \Delta G^{\text{TOT}}$. Complete solubility, or miscibility, should prevail when $\Delta G^{\text{TOT}} > 0$ (van Oss and Good, 1989; see also Chapter XXII). In all these cases ΔG^{TOT} signifies $\Delta G_{121}^{\text{TOT}}$, where 1 is the solute, and 2 the solvent.

DEPLETION INTERACTIONS

Depletion flocculation is the aggregation of particles suspended in a liquid, due to the presence of dissolved polymer molecules that remain unattached to the particles. Parts of the polymer molecules may attach themselves to the particles, but the flocculation is nevertheless due to the action of the polymer molecules that remain in free solution outside the particles (in the absence of cross-binding interactions, see below). And depletion stabilization is in a way its inverse; here stability rather tends to be imparted to a suspension by adsorbed macromolecules than by those which are free in solution. Both depletion flocculation and depletion stabilization probably are best considered from the viewpoint of phase separation (see Chapter XXI). In aqueous phase separation dissolved molecules or suspended particles that have strong (monopolar) electron-donor properties repel one another (Chapter XIX). When two different types of molecules or particles present in an aqueous medium repel each other with different repulsion energies, and when they are present at a sufficiently high concentration, phase separation ensues (Chapter XXI). Even monopolar molecules of the *same kind*, when present at different concentrations in different guises, even one species of polar polymer, can give rise to phase separation, e.g., in one case in free solution, and in the other adsorbed onto suspended particles, as their different surfaces then tend to have different energies of repulsion (van Oss, Arnold and Coakley, 1990).

Now, one may ask, what does this have to do with depletion? Depletion itself has little to do with the actual cause of aqueous phase separation or coacervation. But the existence of a depletion layer between the particle surface and the bulk polymer solution has been postulated by different workers (Napper, 1987). In a number of cases it could be demonstrated that a depletion layer indeed exists. For instance, Arnold *et al.* (1988) showed the existence of an exclusion layer between the naked surface of liposomes and a bulk PEO solution. And Baümler and Donath (1987) and Arnold *et al.* (1988) showed that a depletion layer occurs between a dextran layer, adsorbed onto erythrocytes, and the bulk dextran solution surrounding these erythrocytes. The mutual (AB) repulsion between free and adsorbed dextran gives rise to a depletion layer of a thickness that is approximately equal to the radius of gyration of the dextran molecules (Baümler and Donath, 1987; see also Arnold *et al.*, 1990). Thus, depletion does tend to accompany aqueous phase separation, even if it does not cause it; see also Chapter XXIII, and Chapter V, under Brooks Effect.

It is easy to see how the repulsion between dissolved polymer molecules and particles in suspension can destabilize the particles and thus cause their flocculation. But it is at first sight unclear how an even stronger repulsion of this type can give rise to *stabilization* (depletion stabilization typically occurs at higher free polymer concentrations than depletion flocculation). Various theories endeavoring to explain this phenomenon have been proposed by a number of workers, including, e.g., scaling theories (de Gennes, 1982), Monte Carlo approaches (Feigin and Napper, 1980), statistical thermodynamics (Scheutjens and Fleer, 1982), etc., all of which have been extensively discussed by Napper (1983). However, the few observed phenomena that one could be tempted to designate as depletion stabilization are open to a simpler and more plausible explanation.

From observations on the adsorption of dextrans on red cells (van Oss and Coakley, 1988), it became apparent that whilst at fairly low free solute concentrations (c_f) (e.g., below 1.5%), the concentration of adsorbed dextran (c_a) is somewhat higher than c_f, at higher free solute concentrations (e.g., between 2 and 7%), $c_f > c_a$, upon which flocculation is favored in the phase separation system that ensues. However, at even higher free solute concentrations (e.g., higher than 10%), when once more $c_a > c_f$, stability again prevails (van Oss, Arnold and Coakley, 1990).

Few data appear to be available on the concentration of PEO adsorbed onto polystyrene latex particles, as a function of free PEO present in the bulk solution, in the experiments reported by Li-In-On *et al.* (1975) and Cowell *et al.* (1978), but it is reasonable to assume that in this system also, at lower PEO concentrations in solution, $c_f > c_a$, while at the higher concentration range a regime occurs where $c_a > c_f$. Then also, in the first case flocculation should occur (as it does) and in the second case, stability should begin to re-establish itself (as is the case).

Thus, flocculation of particles (in a non-charged system) in the presence of dissolved polymer molecules occurs when $\Delta G^{AB}_{polymer} > \Delta G^{AB}_{particle}$ and stabilization prevails when $\Delta G^{AB}_{polymer} < \Delta G^{AB}_{particle}$ (both terms being positive). A depletion layer does exist in both cases, but it is a secondary, accompanying phenomenon of the polar (AB) repulsion-driven phase separation system which establishes itself between the bulk polymer solution and the particle-with-adsorbed-polymer suspension. The adjective "depletion," applied to either the flocculation or the stability mode of such systems, is therefore more misleading than explanatory. In depletion stabilization, the stabilization is not caused by interaction with the free polymer, but occurs notwithstanding the presence of free polymer. Here, the only contribution to stabilization by free polymer is in the increasing proportion of free polymer that becomes adsorbed, upon which it clearly is no longer free.

In those cases where no polymer molecules adsorb onto the particles (see, e.g., Arnold *et al.*, 1988), the same mechanism applies as far as depletion flocculation is concerned. However, in such cases, depletion stabilization of the type described here cannot take place.

ENTROPY-DRIVEN AND ENTHALPY-DRIVEN INTERACTIONS

Napper describes a number of conditions under which an enthalpic-type, an entropic-type, or an (enthalpic-entropic)-type of stabilization occurs (1987). Generally, however, these terms tend to be used more loosely. Stabilizing interactions in apolar media, which usually are mainly due to Brownian motion (i.e., ΔG^{BR}, or $\Delta G_{surface}$; see above), may also be alluded to as entropic stabilization, whilst stabilization of hydrophobic particles by PEO or aqueous media is largely enthalpic. But on the whole it is probably wisest to avoid this type of terminology altogether, unless specific measurements have been done by microcalorimetry, or at least at a number of different temperatures, allowing the determination of ΔG, as well as the precise evaluation of its constituent entities, ΔH and $T\Delta S$; see also Chapter XVIII.

CROSS-BINDING INTERACTIONS

Cross-binding of particles or cells by polymer molecules which can bind two (or more) particles together (also called polymer-binding, or bridging) occurs fairly readily under appropriate circumstances. A typical example is the cross-binding of red cells by fibrinogen, or by dextran molecules (at a dextran concentration $\approx 1\%$, $M_w \approx 500,000$), giving rise to the formation of stacks of cells, or "rouleaux" (van Oss and Coakley, 1988). In the binding of dextran to the glycoprotein strands (Chien *et al.*, 1977) of the cells' glycocalices, only LW and AB forces are involved. The aligning of the discoid erythrocytes in cylindrical stacks (rouleaux) is due to the fact that this is energetically the most favored configuration of attraction between these cells (van Oss and Absolom, 1987), at the secondary minimum of attraction (involving LW, EL and AB interactions), at a distance of approximately 5 nm between the distal ends of opposing glycocalyx strands (van Oss, *Biophysics of the Cell Surface*, 1990). However, in the absence of the stabilizing influence of cross-binding polymer molecules, no rouleau-formation occurs, as the attraction between cells or particles at the secondary minimum is too unstable to give rise to an attraction which is sufficiently long-lived to be observable. At higher polymer concentrations, cross-binding ceases to dominate and phase separation phenomena prevail (van Oss and Coakley, 1988).

SPECIFIC INTERACTIONS

By specific interactions, we understand the non-covalent interactions occurring between antigens and antibodies, carbohydrates and lectins, ligands and their specific receptors, and in many cases, between enzymes and their substrates. One of the reasons why it seems useful to include these specific interactions in this catalog is that one still frequently observes, e.g., at meetings where various aspects of cellular or bacterial adhesion are discussed, that a number of the research workers in these fields still are of the opinion that the specific forces involved in these biological interactions are somehow not governed by the laws of physical chemistry. In actual fact, all these specific, non-covalent interactions involve only LW, EL and AB forces and must, if specific binding is to be achieved, overcome the BR repulsion. Only enzyme-substrate interactions may, in certain instances, involve covalent (e.g., disulfide) bonds, in addition to LW, EL and/or AB bonds. What sets these specific interactions somewhat apart from, e.g., simple aspecific adsorption, is the relatively small surface area of the actual interaction sites, compared to the total surface area of the interacting bodies or macromolecules. The actual binding energies are nevertheless fairly strong, thanks to the fact that due to the excellent "fit" between the opposing binding sites, the distance (ℓ) between two such binding surfaces becomes so small as to maximize their interaction forces (see Chapter XXIV).

CODA

Mention of at least 17 different varieties of non-covalent interactions in polar media can be encountered in the recent literature, but it can be shown that most of these are not separate primary physical-chemical forces, but secondary manifestations of one or several such primary forces. The underlying primary forces which play a role in aspecific or specific noncovalent interactions between colloidal entities in aqueous media are limited to: Lifshitz–van der Waals forces, electrostatic forces, hydrogen-bonding interactions (or more generally, polar, or Lewis acid-base interactions) and, in some cases, Brownian motion-induced interactions. The secondary interaction types are here discussed in somewhat more detail, including their place in the hierarchy with respect to the primary interaction classes mentioned in the first part of this chapter. For further discussions on "steric" and "depletion" stabilization (and flocculation), see also Chapter XXIII.

XXI Phase Separation in Polymer Solutions; Coacervation and Complex Coacervation

PHASE SEPARATION OF POLYMERS IN ORGANIC MEDIA

When two low molecular weight solutes are present in the same solution, i.e., in a common solvent, it is rare for the solution to break up spontaneously into two phases. However, when two solutions of different polymers in a common solvent are mixed, it is often found that two phases are formed, in organic (Dobry and Boyer-Kawenoki, 1947) as well as in aqueous media (Albertsson, 1986).

Many studies have been done on phase separation of polymer solutions. According to Vrij (1968), solutions of two different polymers in a common solvent will separate into two homogeneous phases when the entropy of mixing is small and there is a slight positive enthalpy of mixing, which gives rise to a positive Gibbs free energy of mixing. Most studies have been based on the determination of the enthalpy and entropy of mixing, using the Flory-Huggins formulation of the combinatorial entropy (Flory, 1953). Dobry and Boyer-Kawenoki (1947) observed that some polymer pairs are compatible in some solvents but incompatible in others. Bank *et al.* (1971) observed that polystyrene and poly(vinyl methyl ether) are compatible in toluene, benzene, or perchloroethylene, but incompatible in chloroform, methylene chloride, or trichorethylene. Thus, the solvent plays a crucial role in phase separation of polymer solutions. The incompatibility of polymers in a given solvent increases rapidly with an increase in molecular weight. From a thermodynamic point of view, if two dilute solutions of low molecular weight solutes in the same solvent are mixed, the system will form a single phase since the gain in entropy then usually outweighs the enthalpy of mixing. But if the solutions contain large molecules, the entropy of mixing becomes negligible. The solutions may then resist mixing and separate into two phases for any slightly positive enthalpy of mixing (Vrij, 1968). The Flory-Huggins approach to the determination of the thermodynamic properties of polymer solutions is usually considered applicable to the study of the compatibility or incompatibility of polymer systems (Flory, 1953). The Flory interaction parameters

275

(χ) between each polymer and the solvent and between the two polymers have been held to characterize intermolecular interactions completely. It can be shown (Chapters VI and XIX) that χ and ΔG are related in a simple manner.

In the final analysis, when the total free energy of interaction (ΔG_{132}) of a system (comprising polymers 1 and 2 dissolved in a common solvent 3) is negative, polymers 1 and 2 will attract each other, so that miscibility is favored. And when ΔG_{132} has a positive value, polymers 1 and 2 repel each other, which causes a phase separation (van Oss *et al.*, Separ. Sci. Tech., 1989). In *apolar* systems this approach had already been applied successfully in 1979 (van Oss, Omenyi and Neumann). Here a positive value for ΔG_{132} (or a negative value for the Hamaker constant A_{132} of the system; see also Chapters II and III) implies a repulsion between polymers 1 and 2 dissolved in solvent 3, and indeed, whenever $\Delta G_{132} > 0$, in apolar systems, separation occurs, and in apolar systems for which $\Delta G_{132} < 0$, miscibility is observed (van Oss, Omenyi and Neumann, 1979). The Flory-Huggins approach applied to such apolar systems will yield the same correlation. However, recent studies on the interaction forces between polar materials have focused on certain unique attributes of these materials that place them outside the scope of the classical Flory-Huggins approach. Among these is the pronounced asymmetry in the respective strengths of the electron-donor and electron-acceptor ($\gamma^{\ominus}$ and $\gamma^{\oplus}$) parameters, peculiar to many polar and especially to many hydrogen-bonding compounds, which makes it possible for two different polar polymers to interact in much the same manner with a given polar solvent, while remaining capable of reacting very differently with each other, either through a strong mutual attraction (Molyneux, 1983) or by means of a strong repulsion (van Oss *et al.*, Advan. Coll. Interface Sci., 1987), both for purely polar reasons.

More than ten years ago a first attempt was made (van Oss, Omenyi and Neumann, 1979) to establish a correlation between the phase behavior of such polymer solutions and the sign of the Hamaker coefficient (Chapters II and III) which describes the three-component interaction in the system. The Hamaker coefficient (A) was taken to be directly proportional to the free energy of interaction, according to eq. [III-4]. For polymers 1 and 2, dissolved in solvent 3 (cf. eqs. [II-5] and [II-32]):

$$\Delta G_{132}^{LW} = \frac{-A_{132}}{12\pi\ell_o^2} \qquad\qquad \text{[XXI-1]}^*$$

where in accordance with the three-condensed phase analogue of the Dupré equation:

$$\Delta_{132}^{LW} = \gamma_{12}^{LW} - \gamma_{13}^{LW} - \gamma_{23}^{LW} \qquad\qquad \text{[IV-15]}$$

it could be observed (van Oss, Omenyi and Neumann, 1979) in all but five of 31 pairs of polymers dissolved in various organic solvents, that when ΔG_{132}^{LW} was positive, two phases were formed, one containing mainly polymer 1, the other containing mainly polymer 2. When ΔG_{132}^{LW} was negative, only one phase resulted.

---

* A list of symbols can be found on pages 399–406.

In the light of more recent developments, however, the concept of a single coefficient that would permit the characterization of both apolar and polar interactions with one factor now is known to be a serious oversimplification; see Chapters IV, XII and XIV. Thus, these older experiments were reconsidered and a number of determinations were redone (van Oss et al., Separ. Sci. Tech., 1989), while systems comprising polystyrene, polymethylmethacrylate and polyethyleneoxide dissolved in chloroform now were also included. As in the cases already discussed by van Oss, Omenyi and Neumann (1979), phase separation, either visible with the naked eye or microscopically, would occur within 4 days at polymer concentrations of up to 6% (w/v). When after that lapse of time no phase separation could be observed in a given system either macroscopically or microscopically, compatibility (or miscibility) was assumed to exist. A few data from van Oss, Omenyi and Neumann (1979) were not reconsidered; i.e., polymers dissolved in dichlorobenzene and in cyclohexanone, because of an absence of data on the polar parameters of the surface tension of these liquids. The polar parameters of the surface tension of methylethylketone and of tetrahydrofuran were also unknown, but could be estimated rather closely from the phase separation results themselves (see below). The surface tension parameters of the organic liquids used are given in Table XXI-1 and the surface tension parameters of the polymers are shown in Table XXI-2. For the various reasons mentioned in Chapter XIII, the surface tension parameters for CTC, CFO, TOL and BNZ, listed in Table XXI-1, below (which are derived from these liquids' interfacial tensions with water) have been updated for the present edition of this book.

TABLE XXI-1
Surface Tension Parameters of Organic Liquids Used (in mJ/m²)

Liquid[a]	γ^b	γ^{LW}	γ^{AB}	$\gamma^{\oplus}$	$\gamma^{\ominus}$
MEK	24.6	24.6			24[c]
CHX	25.24	25.24			
CTC	26.8	26.8		0.95[h]	
CFO	27.3	27.3		1.5[d]	
THF	27.4	27.4			15.0[e]
TOL	28.3	28.3			0.72[d]
BNZ	28.9	28.9			0.96[d]
CBNZ	33.6	32.1	1.5[f]	ND[g]	ND[g]

[a] Abbreviations: methylethylketone, MEK; cyclohexane, CHX; carbon tetrachloride, CTC; chloroform, CFO; tetrahydrofuran, THF; toluene, TOL; benzene, BNZ; chlorobenzene, CBNZ.

[b] From Jasper, 1972.

[c] Average of the maximum value of 27.9 and the minimum value of 20.4 mJ/m2; see text and Table XXI-5.

[d] See Table XVII-2.

[e] Average of the maximum value of 19.4 and the minimum value of 10.5 mJ/m2; see text and Table XXI-5.

[f] From contact angle determinations on Teflon; i.e., 62° for CBNZ and 75° for NBNZ.

[g] ND = Not yet determined.

[h] New data point, from solubility, using eqs. [XIII-3] and [IV-9]; see van Oss, Wu et al. (2001).

From van Oss et al., Separ. Sci. Tech., (1989).

TABLE XXI-2
Surface Tension Parameters of the Polymers Used (in mJ/m²)

Polymer[a]	γ	γ^{LW}	γ^{AB}	$\gamma^{\oplus}$	$\gamma^{\ominus}$
PIB	25[b]	25[b]			
PPL	25.7[c]	25.7[c]			
CLA	43	38[d]	5.2	0.3[d]	22.7[d]
PMMA	40.6[e]	40.6[e]			12.0[e]
PST	42[f]	42[f]			1.1[g]
PVC	43.8	43[h]	0.75	0.04[i]	3.5[i]
PEO	43	43[j]			64[j]

[a] Abbreviations: polyisobutylene, PIB; polypropylene, PPL; cellulose acetate, CLA; polymethyl methacrylate, PMMA; polystyrene, PST; polyvinyl chloride, PVC; polyethyleneoxide, PEO.
[b] From $\theta = 60°$ with α-bromonaphthalene and assumption of apolarity.
[c] From $\theta = 58.5°$ with α-bromonaphthalene and assumption of apolarity.
[d] From $\theta = 58°$ with water, $\theta = 56.5°$ with glycerol, and $\theta = 20°$ with dimethylsulfoxide.
[e] From Chaudhury, 1984.
[f] From Good and Kotsidas, 1979.
[g] From $\theta = 91.4°$ with water; see van Oss et al., J. Coll. Interf. Sci., 1986.
[h] From $\theta = 15°$ with α-bromonaphthalene.
[i] From $\theta = 82.5°$ with water and $\theta = 68°$ with glycerol.
[j] From van Oss et al., Advan. Coll. Interface Sci., (1987).
Source: van Oss et al., Separ. Sci. Tech., (1989).

Table XXI-3 lists the ΔG_{132}^{TOT} values and the miscibility *vs.* separation characteristics of apolar systems, and in Table XXI-4 those of a few polar systems are shown. It should be noted that most of the polymers as well as most of the solvents of the systems treated as apolar do have some polar characteristics, but these are all of a monopolar nature, involving monopoles of the same sign, such that the entire polar part of the right-hand terms of eq. [IV-16] remains zero, leaving only the three first (LW) right-hand terms. Table XXI-4 also comprises a number of monopolar compounds (solid and liquid) but these involve interactions between polarities of opposite signs, which strongly impact on the right-hand side of eq. [IV-16], which is obtained from eq. [IV-15] by expressing all three interfacial tensions in terms of their constituent sub-components and parameters, according to:

$$\Delta G_{132} = (\sqrt{\gamma_1^{LW}} - \sqrt{\gamma_2^{LW}})^2 - (\sqrt{\gamma_1^{LW}} - \sqrt{\gamma_3^{LW}})^2 - (\sqrt{\gamma_2^{LW}} - \sqrt{\gamma_3^{LW}})^2$$
$$+ 2[\sqrt{\gamma_3^{\oplus}}(\sqrt{\gamma_1^{\ominus}} + \sqrt{\gamma_2^{\ominus}} - \sqrt{\gamma_3^{\ominus}})$$
$$+ \sqrt{\gamma_3^{\ominus}}(\sqrt{\gamma_1^{\oplus}} + \sqrt{\gamma_2^{\oplus}} - \sqrt{\gamma_3^{\oplus}})$$
$$- \sqrt{\gamma_1^{\oplus}\gamma_2^{\ominus}} - \sqrt{\gamma_1^{\ominus}\gamma_2^{\oplus}}]$$

[IV-16A]

While the surface tensions of MEK and THF are known with precision (Jasper, 1972), their $\gamma^{\ominus}$ parameters are not. From contact angle measurements on Teflon it could be ascertained that for both liquids, $\gamma \approx \gamma^{LW}$, i.e., their $\gamma^{AB} \approx 0$. They nevertheless

TABLE XXI-3
ΔG_{132}^{TOT} Values and Miscibility versus Separation in Apolar Systems

System	ΔG_{132}^{TOT} (in mJ/m2)	Observation
1. PMMA and PST in MEK	−4.3	Miscible
2. PMMA and PST in THF	−2.8	Miscible
3. PMMA and PST in BNZ	−2.2	Miscible
4. PIB and PST in BNZ	+0.8	Separation
5. PIB and PMMA in BNZ	+0.8	Separation
6. PIB and PST in TOL	+0.7	Separation
7. PIB and PST in THF	+0.6	Separation
8. PIB and PMMA in THF	+0.5	Separation
9. PIB and PST in CTC	+2.5	Separation

From van Oss *et al.*, Separ. Sci. Tech., 1989.

TABLE XXI-4
ΔG_{132}^{TOT} Values and Miscibility versus Separation in Polar Systems, in mJ/m²

	ΔG_{132}^{LW}	+	ΔG_{132}^{AB}	=	ΔG_{132}^{TOT}	Observation
10. PIB and PPL in CBNZ	−1.2	−	0	=	−1.2	Miscible
11. PMMA and PST in CFO[a]	−2.9	+	11.1	=	+9.8[b]	Separation
12. PMMA and PEO in CFO[a]	−3.1	+	22.9	=	+19.8[b]	Separation

[a] Recalculated with new, corrected values for CFO; see Table XIX-1..
[b] If it were not for the AB interactions, miscibility would have prevailed in these cases.
From van Oss *et al.*, Separ. Sci. Tech., 1989.

have a sizable $\gamma^{\ominus}$ parameter, e.g., from their chemical structure, and by analogy with the measured value for dimethylsulfoxide (van Oss, Ju *et al.*, 1989). Measurement of the $\gamma^{\ominus}$ of these liquids by contact angle measurements on solids with a known $\gamma^{\oplus}$ is not feasible because the γ_L of both liquids is significantly smaller than the γ_S of available polar solids, which thus would lead to spreading. Encasing these liquids in a gel, as was done with, e.g., dimethylsulfoxide (van Oss, Ju *et al.*, 1989), while possible, would not lead to a measurable contact angle with other liquids on account of the adverse effects of the volatility of both MEK and THF. Thus the present method of estimating the $\gamma^{\ominus}$ values of these liquids from phase separation data may, for the moment, be the only feasible approach; see Table XXI-5. The importance of taking ΔG_{132}^{AB} into account in predicting phase separation can be seen in Tables XXI-4 and 5, polymer pairs 11, 12, 19, 20 and 21; in all these cases the observed separation could not have been predicted from the ΔG_{132}^{LW} alone.

All nine apolar systems listed in Table XXI-3 show a complete correlation between a negative ΔG_{132} and mixing, and a positive ΔG_{132} and separation. There

TABLE XXI-5

Values and Miscibility versus Separation in Polar Systems with MEK and THF Solvents, in mJ/m²

		ΔG_{132}^{LW}	+	ΔG_{132}^{AB}	=	ΔG_{132}^{TOT}	Observation
13.	PMMA and PVC in THF	−3.0	−	0.2	=	−2.8	Miscible
14.	PST and PVC in MEK	−4.9	+	1.6	=	−3.3	Miscible
15.	PMMA and PVC in MEK	−4.5	+	0.6	=	−3.9	Miscible
16.	PST and PVC in THF	−4.8	+	1.5	=	−3.2	Miscible
17.	CLA and PVC in THF[a]	−2.4	+	1.8	=	−0.6	Miscible
18.	CLA and PVC in MEK[b]	−3.8	+	3.3	=	−0.5	Miscible
19.	CLA and PST in MEK[c]	−3.7	+	4.2	=	+0.5[d]	Separation
20.	CLA and PST in THF[e]	−2.3	+	3.1	=	+0.8[d]	Separation
21.	PIB and CLA in THF	+0.4	+	4.2	=	+4.6	Separation

[a] $\gamma^{\ominus}$ must be less than 19.4 mJ/m² for the sign of ΔG_{132}^{AB} to be in accordance with the observation.
[b] $\gamma^{\ominus}$ must be less than 27.9 mJ/m² for the sign of ΔG_{132}^{AB} to be in accordance with the observation.
[c] $\gamma^{\ominus}$ must be more than 20.4 mJ/m² for the sign of ΔG_{132}^{AB} to be in accordance with the observation.
[d] If it were not for the AB interactions, miscibility would have prevailed in these cases.
[e] Here, $\gamma^{\ominus}$ must be more than 10.5 mJ/m² for the sign of ΔG_{132}^{AB} to be in accordance with the observation.
From van Oss *et al.*, Separ. Sci. Tech., (1989).

is also complete correlation with the polar systems listed in Table XXI-4. Of the systems pertaining to MEK and THF, Nos. 17 to 20 were used to determine the limits for the $\gamma^{\ominus}$ values of MEK and THF (see Table XXI-5). In the three systems shown in Table XXI-4, the systems' surface tension parameters of the polymers as well as of the solvents were determined independently of the outcome of the phase separation (or miscibility) observations. Thus, not only in exclusively apolar systems, but also in polar systems, phase separation of polymer pairs dissolved in the same organic solvent conforms to the surface thermodynamic treatment incorporated in eq. [IV-16].

PHASE SEPARATION OF POLYMERS IN AQUEOUS MEDIA

Following a number of observations by Beijerinck (1896, 1910) on the phase separation or coacervation (see below) between water-soluble polymers dissolved in water, such as gelatin and agar, and gelatin and soluble starch, Albertsson (1971, 1986) pioneered the use of other water-soluble systems, starting in the 1950s. Of these, the dextran-polyethylene oxide system is typical and now also the best known example. Nevertheless, as Albertsson still stated in the most recent edition of his work (1986, p.3): "The mechanism governing [aqueous phase] partition is largely unknown."

We have seen in the preceding section that the phase separation frequently observed in solutions of two different (large apolar) polymers in organic solvents is generally due to a van der Waals repulsion (Table XXI-3). However, with water-soluble polymers the mechanism is not nearly as obvious. The experimental observation is that with dextran and polyethylene glycol in water, phase separation occurs when the two polymers are present at concentrations of above ≈8% (w/v) each or ≈8% total polymer (Albertsson, 1986). Now, a van der Waals repulsion cannot exist between dextran and polyethylene oxide (PEO) dissolved or suspended in water, because the Lifshitz–van der Waals surface tension (γ^{LW}) of water is only 21.8 mJ/m^2 (see Tables XVII-4 and 10), while the γ^{LW} of both dextran and PEO is of the order of 42 to 43 mJ/m^2 (see Tables XIII-8 and 5), so that nonhydrated dextran and PEO would attract each other (in water) with an apolar energy of about 7 mJ/m^2. Second, dextran is one of the biopolymers endowed with a remarkably low surface charge (when dissolved in water): by moving boundary electrophoresis, the electrophoretic mobility of dextran (T-40) at $\mu = 0.15$ was –0.002 μm/s/V/cm (van Oss et al., 1974), corresponding to ζ–potentials of –0.06 and –0.7 mV, respectively, for the lower molecular weight dextrans, and ≈33% less than these values for the higher molecular weight dextrans. Thus in systems containing dextran (DEX) as one of the repelling partners, electrostatic repulsion may also be totally ruled out.

Using eq. [IV-16] and the data from Tables XVII-5 and 8, the total interaction energy ΔG_{132}^{TOT} between DEX and polyethylene oxide (PEO), immersed in water, is found to equal +46.9 mJ/m^2. This, of course, signifies a strong repulsion, which is wholly due to polar (AB) repulsive forces: $\Delta G_{132}^{TOT} = \Delta G_{132}^{LW} + \Delta G_{132}^{AB}$, where $\Delta G_{132}^{LW} = $ –6.8 mJ/m^2 and $\Delta G_{132}^{AB} = $ +53.7 mJ/m^2. However, the repulsion energy expressed by $\Delta G_{132}^{TOT} = 46.9$ mJ/m^2 applies only to the interaction between negligibly hydrated DEX and PEO molecules; these high repulsion energies therefore are only operative at very high DEX and PEO concentrations. When dissolved in water at low or intermediate concentrations, DEX and PEO are surrounded by oriented layers of water of hydration, which beyond the first layer are subject to an exponential decrease in the degree of orientation (see Fig. X-1), which results in a decay in the actual value of ΔG_{132}^{TOT} with distance (ℓ) (eq. [V-4]). Thus, greater dilution leads to a decrease in the effective value of $|\Delta G_{132}^{TOT}|$. For all practical purposes one is sufficiently close to being correct, when assuming that at infinite dilution the value of ΔG_{132}^{TOT} (for $\gamma^{\ominus}$–monopolar polymers dissolved in water) approaches zero. The precise form of the function describing the decay of ΔG_{132}^{TOT}, going from very high to very low solute concentrations, is difficult to assess (see also the section on the osmotic pressure of aqueous PEO-solutions in Chapter XIX). However, the assumption of a decay which is linearly proportional to the decrease in polymer concentration (see van Oss, Chaudhury and Good, Separ. Sci. Tech., 1987) serves rather well to predict the polymer concentration level at which phase separation commences. [It should be noted that for polymers in the hydrated state, the ΔG_{132}^{LW} component becomes quite small. The γ^{LW} of hydrated biopolymers, for example, tends to be of the order of 26 to 27 mJ/m^2 (see, e.g., Table XVII-6) which yields values for ΔG_{132}^{LW} of only –0.37 to –0.56 mJ/m^2, at $\ell = \ell_o$.]

It is reasonable to estimate that phase separation sets in when $\Delta G_{132}^{TOT} \geq +0.5$ kT (see Chapter XIX), where ΔG is converted from energy per unit surface area to energy per minimum contactable surface area (S_c) per molecule pair (see Chapters VI and XIX) and expressed in units of kT (1 kT = 4×10^{-14} ergs = 4×10^{-21} J). Thus the onset of aqueous phase separation* is not only a function of the concentration of each polymer species, but also of their minimum contactable surface area, S_c. And as the dimensions of polymer molecules in aqueous solution are a function of their moleculer weight (cf. Table VI-2 for the dimensions of dextran molecules of different molecular weights in aqueous solution), the onset of aqueous phase separation also depends on the respective molecular weight of each polymer species; the higher the molecular weight of a polymer, the lower its minimum concentration required for the onset of phase separation (see Albertsson, 1971, 1986; Walter *et al.*, 1985). For dilute aqueous polymer solutions it has been attempted to incorporate the concentration of each polymer (as the volume fraction, c_i) as well as the number-average molecular weight (M_n), into the Dupré equation (eq. [IV-15]); see van Oss, Arnold *et al.* (1990):

$$\Delta G_{132}(\text{dil.}) = \gamma_{12}\sqrt{m_1 c_1 m_2 c_2} - \gamma_{13} m_1 c_1 - \gamma_{23} m_2 c_2 \qquad \text{[XXI-2]}$$

where:

$$m_i \approx f_i \sqrt{M_n} \qquad \text{[XXI-3]}$$

(see Napper, 1983, p. 379), in which the factor f_i may be approximated from the binodials determined for polymers of different molecular weights (see, e.g., Albertsson, 1971, Table 2.10). For dextran ($M_w \approx 500,000$) and PEO ($M_w \approx 6,000$), we estimated $f_i \approx 0.005$ (van Oss, Arnold and Coakley, 1990). Nevertheless, it should be realized that both eqs. [XXI-2 and 3] are, at best, rough approximations. For instance, at constant molecular weights of the polymers, the minimum contactable surface area (S_c) increases with decreasing concentration, due to the increase in size of the hydration envelopes surrounding the polymer molecules.

Albertsson (1971, 1983) shows a rather modest influence of temperature on phase separation: at lower temperatures phase separation sets in at slightly lower concentrations, conforming with the somewhat lower value of 0.5 kT at lower temperatures; see also Walter *et al.* (1985).

PHASE SEPARATION OF WATER-SOLUBLE POLYMERS IN POLAR ORGANIC MEDIA

Dextran and polyethylene oxide, dissolved in organic solvents with a sizable electron-acceptor parameter ($\gamma_3^\oplus$), even though considerably smaller than the $\gamma_3^\oplus$ for water ($\gamma_w^\oplus = 25.5$ mJ/m²), should also give rise to phase separation (cf. eq. [IV-16]). For instance, DEX and PEO solutions in formamide (see Table XVII-10) show phase separation after mixing. In formamide, the ΔG_{132} value for DEX and PEO amounts to

---

* The onset of phase separation corresponds to the "critical point" in the phase diagram, as indicated by Albertsson (1971, Fig. 2.5).

+5.6 mJ/m^2, which is only 12% of the ΔG_{132} value for the same compounds in water, but still sufficient for phase separation.

Finally, there is a hybrid situation, where the water-soluble PEO, and water-insoluble PMMA, dissolved in chloroform, give rise to a phase separation; see Table XXI-4. (Dextran is, contrary to what might be expected at first sight, insoluble in chloroform; see Chapter XXII.)

MECHANISM OF POLYMER PHASE SEPARATION IN APOLAR SYSTEMS

In apolar media, or in media which can be considered apolar vis-à-vis the dissolved polymers because they are endowed with monopolarity of the same sign (see Table XXI-3), miscibility or separation depend solely on LW interactions. Thus in these cases miscibility prevails when $\Delta G_{132}^{LW} < 0$ and phase separation occurs when $\Delta G_{132}^{LW} > 0$. Therefore, in apolar systems there are only two possibilities: there will be either one phase, or two phases.

Suppose that one has an LW system with two polymers (1 and 2), dissolved in apolar solvent (3), which separate. This occurs when $\gamma_1^{LW} > \gamma_3^{LW} > \gamma_2^{LW}$, or when $\gamma_1^{LW} < \gamma_3^{LW} < \gamma_2^{LW}$ (see Chapter III and Table III-1A). Assume the first contingency, i.e., $\gamma_1^{LW} > \gamma_3^{LW} > \gamma_2^{LW}$. If one now adds a solution of a polymer (4), in the same solvent, then there are only two possibilities: *either* $\gamma_4^{LW} > \gamma_3^{LW}$, *or* $\gamma_4^{LW} < \gamma_3^{LW}$. In the first case, polymer (4) will join the phase already occupied by polymer (1) and polymers (4) and (1) mix in that phase, while both polymers (4) and (1) repel polymer (2); there still are only two phases. In the second case, polymer (4) will join the phase already occupied by polymer (2) and polymers (4) and (2) mix in that phase, while both polymers (4) and (2) repel polymer (1); there are, again, still only two phases.

Thus when two phases occur in apolar systems, all polymers in phase A repel all polymers in phase B, while all polymers in phase A attract each other, and all polymers in phase B attract each other.

MECHANISM OF POLYMER PHASE SEPARATION IN POLAR AND ESPECIALLY IN AQUEOUS SYSTEMS

Albertsson (1971, 1986) was right in proposing that aqueous phase separation between dissolved polymers results "from a repulsive interaction between unlike molecules." However, all aspects of aqueous phase separation phenomena can only be understood when it is realized that in all these cases *identical* polymer molecules also repel each other. Indeed, in strongly polar systems, e.g., in aqueous phase separation, even relatively similar polymer molecules can repel one another, but the energies of repulsion between polymer molecules (1) and (1) still are not quite the same as the repulsion energies between polymer molecules (2) and (2). That condition tends to give rise to phase separation by the following mechanism: In water (3), polymer molecules (1) repel each other with energy ΔG_{131} and polymer molecules (2) repel each other with energy ΔG_{232}, while polymer molecules (1) and (2) repel each other with energy ΔG_{132}, in such a manner that when, e.g., $\Delta G_{131} > \Delta G_{232}$, then also $\Delta G_{131} > \Delta G_{132} > \Delta G_{232}$. Such a system will only come to equilibrium after sorting-out, i.e., when most polymers (1) have congregated to one phase and most polymers (2) to the other phase, to which effect it also in necessary that $\Delta G_{132} > 0.5$ kT. This often first gives rise to an emulsion-like phase separation of droplets of one phase suspended in the other phase; usually the droplets then gradually coalesce and two distinct phases are formed with typically a markedly small interfacial tension between the two phases of the order of 10^{-4} to 10^{-1} mJ/m². In polar systems the driving force for the phase separation (or for microemulsion formation) is the initial negative interfacial tension, which decays to a value close to zero upon attaining equilibrium, at which value it levels off (van Oss *et al.*, Advan. Colloid Interface Sci., 1987); see below.

Thus in polar, and especially in aqueous phase separation systems and in contrast with apolar systems, it is in principle possible to produce as many phases as the number of different polymers involved (van Oss, Abstr. Partitioning Conf., 1989). Albertsson (1971, 1986, Figure 2.2) gives an illustration of multiple phase systems, comprising various differently substituted dextrans. The right hand part of Albertsson's (1986) Figure 2.2 displays 18 separate phases in one tube.

In earlier work on aqueous phase separation, osmotic interactions were invoked, to account for the observed phase separation; see, e.g., Edmond and Ogston (1968). However, on closer comparison between the results predicted by the osmotic hypothesis and the results obtained by Edmond and Ogston (1968), Molyneux (1983) concluded that an actual repulsion between the dissolved polymers had to exist over and above their osmotic interaction. If, however, the total osmotic interaction had been properly computed, using eqs. [VI-9] and [VI-10], and determining χ (eq. [VI-10]) at high polymer concentrations (where a negative value would have been found for χ), instead of at a very low polymer concentration (where χ invariably has a value which is very close to +0.5), the polar repulsion between polymer molecules should also have shown up *via* the osmotic approach.

THE INTERFACIAL TENSION BETWEEN
AQUEOUS PHASES

The interfacial tension between phases is exceedingly low [e.g., between 5×10^{-4} to 7×10^{-2} mJ/m² (Albertsson, 1971, Table 2.11; 1986, Table 2.7)], and is at its lowest at the critical point, i.e., at the point where the compositions and the volumes of the two phases approach equality. Here, as with microemulsions (see below, and Chapter XIX), it is reasonable to presume that, upon mixing the two components together, negative interfacial tensions occur, which are the driving force for the phases separation. When the phase separation is complete, the initial negative interfacial tension decays to a value, at equilibrium, which from there on remains close to zero; see van Oss *et al.* (Advan. Colloid Interface Sci., 1987).

The extremely low interfacial tension between, e.g., DEX and PEO solutions, has been applied to contact angle determination in two-liquid systems (see Chapter XII), e.g., by Schürch *et al.* (1981).

USE OF AQUEOUS TWO-PHASE SYSTEMS FOR CELL AND BIOPOLYMER SEPARATION

DEX/PEO (and some other) aqueous two-phase systems are widely used for separation and purification of cell particles, proteins and other biopolymers (Albertsson, 1971, 1986). A given biopolymer (subscript k) will have a preference for that phase in which the polymer (i) preferentially resides, to which it is most strongly attracted, or by which it is least strongly repelled, thus, when $\Delta G_{iwk} < \Delta G_{jwk}$. In the most widely used aqueous two-phase system, comprising DEX and PEO, i then stands for either DEX or for PEO.

As an example, human serum albumin (HSA) is preferentially found in the DEX-rich phase in the DEX/PEO system (Albertsson, 1971, Fig. 4.9; 1986, Fig. 5.1b). Taking the surface tension values for HSA from Table XVII-6, those for DEX (T-170) from Table XVII-8, and the values for PEO from Table XVII-5, and usidng eq. [IV-16], one can see that:

$$\Delta G_{HSA \text{ (hydrated)} - w - PEO} = +34 \text{ mJ/m}^2$$

while:

$$\Delta G_{HSA \text{ (hydrated)} - w - DEX} = +31 \text{ mJ/m}^2$$

so that HSA should indeed prefer the DEX-rich phase, as it is more strongly repelled by PEO than by DEX. [There is no need to convert the ΔG values to kT units, because in both cases the operative contactable surface area (cf. eq. [VI-8]) is that of HSA, so that ΔG in mJ/m² remains proportional to ΔG in kT.]

POLYMER PHASE SEPARATION SUMMARIZED

The major condition for a phase separation to occur in a solution containing two different solutes (of which one, in general, is a macromolecule) dissolved in the same solvent is that the free energy of interaction between solutes (1) and (2) while dissolved in solvent (3) must have a positive value, i.e., $\Delta G_{132}^{TOT} > 0$. This condition ensures that the dissolved solutes (1) and (2) repel each other while immersed in solvent (3). However, this condition does not automatically guarantee phase separation, for as long as the thermal motion energy of the dissolved (macro-)molecules is large enough to match the energy of interaction between dissolved molecules (1) and (2), any separation that might occur would be undone by thermal remixing. Separation therefore can only prevail over remixing when $\Delta G_{132}^{TOT} \geq +0.5$ kT, where ΔG_{132}^{TOT} is expressed in terms of free energy of interaction between (macro-)molecules (1) and (2) per unit area of contactable surface. The second condition can only be fulfilled (at a given temperature T) when the solutes have the appropriate molecular weight, to ensure reaching the required contactable surface area (S_c). And finally, it also can only be fulfilled when a critical minimum concentration of solutes (1) and (2) has been reached to ensure: (a) a sufficient likelihood of repulsive encounters (in apolar systems), and/or (b) a sufficient degree of desolvation of the macromolecules to allow for interactions between macromolecules that are not largely masked by excessively thick layers of molecules of solvation (in polar systems). The mutual repulsion of all macromolecular solutes in a polar system makes it possible for such a system to split into multiple phases, i.e., as many phases as there are different macrosolutes. Apolar systems, however, cannot give rise to more than two phases.

COACERVATION AND COMPLEX COACERVATION

Phase separation readily occurs when aqueous solutions of two different charged or neutral polymers are mixed together. The phenomenon was first described by Beijerinck (1896, 1910) almost a century ago and was later intensively investigated by Bungenberg de Jong (1949; Bungenberg de Jong and Kruyt, 1930), who called it "coacervation" (from the Latin *coacervo*, or heaping together), culminating in a number of chapters in the second volume of Kruyt's *Colloid Science* (Bungenberg de Jong, 1949). Notwithstanding the curious colloidal properties of coacervates and the potential uses of coacervation in separation processes, research on coacervation has declined in the last 60 years.

TABLE XXI-6

Distinctions Between and Conditions for Coacervation and Complex Coacervation, Given by Bungenberg de Jong (1949, p. 232), Using a Mixture of Gelatin and Gum Arabic Solutions (sols) as Examples

Coacervation	Complex coacervation
Occurs at pH > I.E.P.[a] of the gelatin (at which both colloids are negative charged). Is only possible on mixing concentrated sols. The coacervation disappears on adding water.	Only occurs at pH < I.E.P.[a] of the gelatin (at which the two colloids have opposite charges). Does not occur in concentrated sols, occurs, however, after dilution with water. Still occurs even on added mixing 0.001% sols.
Indifferent salts do not suppress coacervation, appear rather to promote coacervation in certain cases. The ions arrange themselves in their effectiveness in lyotropic series.	Indifferent salts suppress complex coacervation. The valency of both ions is of primary importance in the matter. The position of the ions in the lyotropic series is of very minor significance.
The drops exhibit no disintegration phenomena in a D.C. electric field.	The coacervate drops exhibit disintegration phenomena in a D.C. electric field.
Both liquid layers are rich in colloid, each layer containing in the main one of the colloids.	One layer, the complex coacervate, is rich in colloids, the second layer, however, is poor in colloid. Both layers contain gelatin and gum arabic in not very different ratios.
Principal Condition	**Principal Condition**
Water deficit in the total system.	Adequate charge opposition between the two colloids.

[a] I.E.P. = isoelectric gelatin is heterogeneous; at the I.E.P. it consists of an equal amount of positive and negative charged proteins, which mixture is also close to being hydrophobic. See also Chapter XVII. From van Oss, J. Dispersion Sci. Tech., 1988.

In Table XXI-6 Bungenberg de Jong's description (1949, p. 232) of simple and complex coacervation is recapitulated (see also van Oss, J. Dispersion Sci. Tech., 1988). In that description some of the differences between simple and complex coacervation are enumerated. The salient item in the description is the principal condition Bungenberg de Jong advances as favoring simple coacervation, i.e., a "water deficit in the total system." Now, a relative deficit may play a role in the volume ratios of the phases occurring in coacervation (we shall omit the adjective "simple," and only use the terms coacervation and complex coacervation for the two different phenomena), but in the light of more recent studies on the polar components of interfacial interactions in aqueous media (see Chapters III, XIX and this chapter, above), some of the older assumptions need to be reexamined.

Taking the example given in Table XXI-6, of a mixture of aqueous gelatin and gum arabic solutions under conditions where both polymers are negatively charged (and thus will not interact by electrostatic attraction), it should be noted that both polymers have fairly strong monopolar electron-donor properties (see Tables XVII-7 and 8) and that their free energy of interaction in water, ΔG_{132}, thus has a positive value, so that these will repel each other. Thus, *coacervation* in the sense of Bungenberg de Jong (1949, p. 232), comprises phase separation in aqueous polymer solutions (see above), and is even synonymous with it in most cases. However, coacervation, i.e., phase separation, can also occur with systems consisting of one polar polymer and one polar microsolute (e.g., ethanol; see Table XXI-8).

ROLE OF LOW MOLECULAR WEIGHT SOLUTES IN COACERVATION

It is frequently held that, "upon the admixture of organic solvents, the permittivity (dielectric constant) of the medium becomes significantly decreased, which favors the complex formation between amphoteric polyelectrolyte molecules through the interaction between a positively charged moiety on one molecule with the negatively charged moiety on another molecule"; see, e.g., van Oss, J. Dispersion Sci. Tech., (1988). On closer scrutiny this reasoning may, in certain cases, be incorrect, e.g., in the case of the interaction of ethanol on proteins in aqueous solution, at fairly low temperatures: at the temperatures employed in the cold ethanol precipitation method of phasma protein fractionation *the dielectric constants of the mixtures used are very close to that of water at 20°C* (Wyman, 1931; see also van Oss, J. Protein Chem., 1989). Upon closer scrutiny of the mechanism involved in these cases, it can be shown that ethanol has a strongly dehydrating influence on dissolved proteins, and thus causes these proteins to become more hydrophobic, which allows them to attract each other, i.e., make them insoluble. However, most proteins are insoluble in ethanol [with the exception of the corn protein, *zein* (see Table XVII-7), which is insoluble in water but soluble in ethanol]. Thus, increasing the ethanol concentration in aqueous protein solutions causes the proteins to precipitate. The main reason for precipitating proteins in *cold* water-ethanol mixtures is that the insolubilization process is less denaturing for protein molecules than water at, e.g., 20°C. Also, at temperatures close to 0°C, water becomes less Lewis acidic and thus is less hydrating for monopolar proteins than at room temperature (see Chapter VIII and Table VIII-1), which enhances the desolubilizing influence of cold ethanol-water mixtures.

Na_2SO_4 (Bungenberg de Jong, 1949a, p. 232; Table XXI-8) [and especially $(NH_4)_2SO_4$] can be used to dehydrate proteins in aqueous solution and as a result, to precipitate these proteins, a process known as "salting-out" (van Oss, Moore et al., 1985; van Oss et al., J. Protein Chem., 1986; see also Chapters IX and XXII). In "salting-out" the salt ion interactions enhance the free energy of cohesion of the aqueous salt solution, which increases the hydrophobizing capacity of that solution and, concomitantly, favors the insolubilization of proteins (see Chapter IX). On the other hand, in coacervation (as opposed to flocculation, see below), the principal effect of Na_2SO_4 (and of ethanol) is due to its strong electron donicity. Recent work by Ananthapadmanabhan and Goddard (1987) clearly illustrates the difference between the effects of "salting-out" and repulsion: plots of aqueous biphase formation and clouding temperatures of polyethylene oxide as a function of salt concentration show a pronounced negative slope ("salting-out"), while plots of aqueous biphase formation temperatures as a function of dextran concentration (coacervation) manifest a strongly positive slope. However, in the coacervation occurring through the interaction between an inorganic salt and a polyelectrolyte that is preponderantly an electron-donor, the place of the salt anion in the lyotropic (or Hofmeister) series, plays an important role (Ataman and Boucher, 1982). Here the anions which are considered the "hardest" Lewis bases (Ho, 1975) are the most effective in inducing coacervation in conjunction with polyethyleneoxide. The influence of the cation on this effect is less strong; here the cations that are the "softer" Lewis acids have a somewhat stronger influence on depressing the theta-temperature (Boucher and Hines, 1976) and on the coacervation of polyethyleneoxide (Boucher and Hines, 1978), which appears reasonable, as cations that are weaker electron acceptors tend to have less effect on the electron donicity of the anion.

COMPLEX COACERVATION

Complex coacervation is an entirely different phenomenon: in many ways it is the opposite of (simple) coacervation, see Tables XXI-6 and 7. It is also useful to compare both phenomena with flocculation: both coacervation and complex coacervation can give rise to flocculation (see Tables XXI-7 and 9), but flocculation differs from both in that it gives rise to an insoluble component (see below).

Table XXI-8 also shows a number of examples of complex coacervation occurring between positively charged and negatively charged polymers, and between positively charged polymers and negatively charged small ions, as well as between negatively charged polymers and positively charged small ions. The mechanism of complex coacervation between oppositely charged polyelectrolytes has been thoroughly analyzed by Voorn (1956).

As Molyneux (1983) rightly remarked, complex coacervation occurs not only as a result of the interaction between positively and negatively charged polymers, but also when the pair of water-soluble polymers consists of both a strong hydrogen bond donor and acceptor. Even though this aspect of complex coacervation has

TABLE XXI-7

Mechanisms of and Conditions for Coacervation, Complex Coacervation and Flocculation, Taking the Polar Interfacial Interaction Approach into Account

Coacervation	Complex coacervation	Flocculation
Polar (AB) and/or apolar (LW) *repulsion* between two solutes, of which one or both must be a (generally asymmetrical) polymer dissolved in the same solvent, resulting in a phase separation, where one solute resides mainly in one phase and the other solute in the other phase. One phase is rich in molecules of one species, the other phase is enriched in molecules of the other species.	Electrostatic (or polar) *attraction* between two (usually asymmetrical) polymers, of opposite signs of charge (or of opposite signs of Lewis acid-base behavior), resulting in a phase separation, where one phase is rich in both polymers and the other polymer-poor. However, the concentration *ratio* of the solutes of both species remains the same in both phases, as it was in the initial solutions.	Electrostatic and/or interfacial *attraction* between two (not necessarily asymmetrical) polymers, of the same or of different species, resulting in insoluble floccules. The same mechanisms operative in either coacervation or in complex coacervation can result in flocculation, when the polymers or the complexes become insoluble.
Occurs in concentrated solutions or sols. Effect disappears at low concentrations. Ionic strength has little influence on the phenomenon. D.C. electric fields have no influence on coacervate drops of the different phases.	Does not occur in concentrated solutions or sols. Effect occurs at low concentrations. Neutral salts suppress complex coacervation; the valency of the ions is of crucial importance. Complex coacervate drops disintegrate in a D.C. field.	In some cases the mechanisms operative in both coacervation and complex coacervation can play a role in flocculation.
The polymer molecules remain in solution in both phases		*The polymer molecules, or their complexes, become insoluble.*

From van Oss, J. Dispersion Sci. Tech., 1988.

TABLE XXI-8
Examples of Coacervation and Complex Coacervation

Coacervation[a]	Complex coacervation[b]
A Polar interactions	**A** Electrostatic interactions
Negatively charged gelatin and gum arabic[1]	Positively charged gelatin and negatively charged gum arabic[1,5]
Agar and ethanol[1]	
Dextran and propanol[2]	Positively charged gelatin and soy bean phosphatide[5]
Gelatin and resorcinol[1]	Positively charged gelatin and nucleic acid[5]
Isoelectric gelatin and Na_2SO_4[1]	Positively charged gelatin and negatively
Dextran and polyethylene glycol[2]	charged soluble starch[5]
Polyvinyl alcohol and dextran[2]	Positively charged gelatin and agar[5]
Polyvinyl alcohol and polyethylene glycol[2]	Positively charged ichthyocoll (fish-glue) and
Many other polymer pairs, dissolved in water[2]	$K_3Fe(CN)_6$[5]
	Positively charged hexolnitrate[c] and gum arabic[5]
	Positively charged hexolnitrate and Na-pectate[5]
	Isoelectric gelatin and isopropanol[d]
	Isoelectric gelatin and phenols[d]
B Apolar Interactions	**B** Polar (Lewis acid-base) interactions
Cellulose acetate and ethanol, in chloroform[3]	Polyacrylic acid and polyvinyl methylether[6]
Polyisobutylene and polystyrene, in benzene, and many other polymer pairs, dissolved in organic solvents[4]	Polyacrylic acid and polyvinyl pyrrolidone[6]
	Polyacrylic acid and polyethyleneoxide[e]
	Polymethacrylic acid and polyethyleneoxide[7]
	Nucleic acid complexes[8]
	C Immune reactions
	Antigens and antibodies[f,9,10,11]

[a] Water is the solvent in all cases, except where mentioned otherwise.

[b] Water is the solvent in all cases.

[c] $[Co[(OH)_2Co(NH_2CH_2\text{-}CH_2NH_3)_2]_3](NO_3)_6$.

[d] Bungenberg de Jong classifies these as examples of coacervation; the real mechanism however lies in the dehydrating power of ethanol, which causes the protein to become more "hydrophobic." There is no significant polar repulsion between isoelectric proteins and monopolar alcohols.

[e] Also called polyethylene glycol.

[f] Complex formation between antigens and antibodies can on rare occasions give rise to the complex coacervation type of complex-formation, and even to gel-formation,[11] usually, however, flocculation occurs; see Table XXI-9.

[1] Bungenberg de Jong, 1949, p. 232.

[2] Albertsson, 1986.

[3] Dobry, 1938, 1939.

[4] See this chapter, above.

[5] Bungenberg de Jong, 1949, p. 335.

[6] Dobry, 1948.

[7] Smith et al., 1959.

[8] Davidson, 1977.

[9] Polson, 1977.

[10] van Oss, 1984.

[11] van Oss and Bronson, unpublished observations

From van Oss, J. Dispersion Sci. Tech., 1988.

TABLE XXI-9
Examples of Flocculation of Different Types

Coacervation type	Complex coacervation type
Polyethylene glycol and human immunoglobulin-G[1a]	Polycations and polyanions[4]
Polyethylene glycol and immune complexes[1a]	Proteins and heavy metals[2]
$(NH_4)_2SO_4$ and various globulins[2,3b] (salting-out)	Proteins and other positively charged compounds (e.g., Rivanol)[2]
Ethanol and plasma proteins[2,5c]	Antigens and antibodies[4]

[a] Dehydration also plays a role here[1,3].
[b] Here dehydration by "salting-out" is the principal mechanism.[5]
[c] Here dehydration is the principal mechanism, see text, above.
[1] Polson, 1977.
[2] Schultze and Heremans, 1966.
[3] van Oss, Moore *et al.*, 1985.
[4] van Oss, 1984, 1988.
[5] van Oss, J. Protein Chem., 1989.

mainly been studied in aqueous systems, Bungenberg de Jong does not appear[*] to have included it among the systems he discussed. It was again Dobry (1948) who pioneered this aspect of polymer interactions, followed by the Union Carbide group (Smith *et al.,* 1959); see also Molyneux (1983). It should not be forgotten that the (specific) interaction between nucleic acids also is essentially a hydrogen-bonding interaction (Davidson, 1977; also, Chapter XXIV).

Finally, a more complicated interaction between polymer molecules, i.e., the specific interaction between antigens (Ag) and antibodies (Ab) can, on occasion, give rise to the phenomenon of complex coacervation, even though the much more frequent outcome of these immunochemical reactions is flocculation; see below. The Ag-Ab bond can be hydrophobic, electrostatic, or more rarely, hydrogen-bonding in nature. Often the first two mechanisms play a simultaneous, or consecutive role in Ag-Ab bond formation; see Chapter XXIV.

Thus, the mechanism and most of the effects of complex coacervation are the opposite of those of coacervation; cf. Tables XXI-7 and 8.

FLOCCULATION

The mechanisms operative in coacervation (and in "salting-out"), as well as in complex coacervation, can readily lead to flocculation, whenever the brink of solubility is transgressed. Curiously, even though coacervation and complex coacervation are in

[*] Bungenberg de Jong's output of publications in this field has been so prodigious that it is impossible to study them all (and many of them are not very accessible in the United States), and it thus is perhaps temerarious to state that he has not included a given concept. But he certainly has not treated hydrogen bonding in this context in his major reviews (1949).

many ways the opposites of each other (see above), either phenomenon can lead to flocculation; see Table XXI-9. To understand this it should be realized that in most cases the major contribution to the insolubilization of dissolved polymer molecules and/or for the destabilization of polymer particles, ultimately is an attractive interfacial (usually hydrophobic) interaction between low-energy solutes or particles in a polar medium (see Chapters XIV, XIX and XX).

The *coacervation approach* to flocculation (see Table XXI-9) is one of dehydration, by concentrating the dissolved polymer molecules (1) into a much smaller volume (or phase), thus changing the positive value of ΔG_{iwi} to a negative value in the phase enriched in polymer 1, as well as increasing the opportunities for interaction. The polyethylene glycol precipitation of immunoglobulins and of immune complexes (Polson, 1977) is the simple coacervation class.

The *complex coacervation approach* to flocculation (see Table XXI-9) works through the interaction between two different polymers and is also due, in part, to the increase in size of the polymer complexes and/or particles. The larger the polymer molecules or particles grow, the larger their contactable surface area becomes, and the more readily the energy of attraction between two such macromolecules or particles significantly surpasses -1 kT, leading to irreversible attraction, followed by the formation of even larger complexes, and insolubilization, i.e., flocculation.

Polycation-polyanion complexes (e.g., polyvinyl benzyltrimethyl ammonium-polystyrene sulfonate) have been much studied (Michaels and Miekka, 1961; Bixler and Michaels, 1969), and have become an important material for the casting of anisotropic ultrafiltration membranes, extensively used in many biological and biomedical applications. Precipitation of euglobulins at low ionic strength and at the isoelectric pH still is a simple and much used method for isolating such proteins (Schultze and Heremans, 1966; van Oss, 1982; van Oss *et al.*, J. Protein Chem., 1986; see also Chapter XXII), even though the flocculation often is only partly reversible. This is a form of complex coacervation which is perhaps best called "auto-complex coacervation", as in this case a negative site of one protein molecule combines with the positive site of another protein molecule *of the same species*. Resolubilization occurs by increasing the ionic strength of the medium ("salting-in"), which allows the small counterions to "mask" the positive and the negative sites on the protein molecules, thus favoring the dissociation of large protein-protein complexes formed through electrostatic interaction; see also Chapter XXII. Along the same principle as protein-protein interaction, protein flocculation can also be effected by protein-heavy metal, or protein–positively charged organic solute interactions (Schultze and Heremans, 1966).

An unusual aspect of the complex coacervation type of flocculation is its non-stoichiometry, which can lead to a number of curious phenomena when flocculation is made to occur by double-diffusion encounters in gels (van Oss, 1984). However, the maximum amount of flocculate always occurs at close to stoichiometric ratios. Finally, a much-studied type of flocculation, also of the complex-coacervation class, is the flocculation of antigen-antibody (Ag-Ab) complexes; see Chapter XXIV.

It is important to note that flocculation *via* coacervation is usually more completely reversible than the complex coacervation type of flocculation.

PHASE FORMATION IN MICROEMULSIONS

We proposed earlier (van Oss *et al.*, Advan. Coll. Interface Sci., 1987) that the extremely low interfacial tensions which have been reported to exist between the major phases which are visible with the naked eye (Ostrovsky and Good, 1984) most likely are the ultimate equilibrium situation resulting from the decay of initially strongly negative interfacial tensions, which are, typically, of the order of -16 mJ/m^2, in a mineral oil-docusate sodium (Aerosol OT)-water system. The initial negative interfacial tension then is the driving force for the formation of the microemulsion.

In principle, in an aqueous multi-component system, it is possible to obtain multiple phases. And microemulsions can be considered to be multiple-component systems, as long as droplets of different sizes can occur. This is because even if the interfacial tensions between, e.g., oil droplets and water are the same, when droplets of different sizes exist, the free energies of repulsion (ΔG_{iwi}) between such droplets of different sizes (expressed as energy per droplet-pair) will differ. Even though only one or at most two phase boundaries are normally visible with the naked eye, it could be expected that with an optical system which visualizes differences in refractive index, such as schlieren optics, more phase boundaries would become, apparent, if multiple phases indeed occur. This was first verified in an analytical ultracentrifuge (Beckman-Spinco, Model E), and multiple phases were visible early on in each run (at low g-forces), and persisted for at least one hour at about $100,000 \times$ g (Good *et al.*, 1986; Broers, 1986; Yang, 1990); see Figs. XXI-1 and XXI-2.

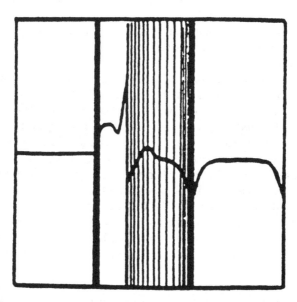

FIGURE XXI-1   Drawn reproduction of schlieren picture of analytical ultracentrifugation (Beckman Model E instrument) of microemulsion (n-octane, brine and docusate sodium); $100,000 \times$ g; 14 minutes from start; 70° schlieren angle. (From Broers, 1986; see also Good *et al.*, 1986.)

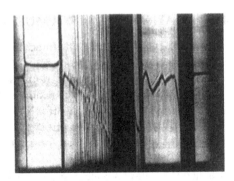

FIGURE XXI-2   Schlieren picture of analytical ultracentrifugation of a microemulsion as above; 100,000 × g; speed just attained. As in Fig. XXI-1; the multiple phases in the middle (microemulsion) phase are clearly visible. From left to right: air phase, hydrocarbon phase, microemulsion phase and brine phase. (From Yang 1990.)

It could not, however, be completely ruled out that the high pressures occurring during centrifugation and ultracentrifugation, and/or other high g-force effects might have contributed to the observed multiple phase separation. Therefore, it was also attempted to use schlieren optics for the study of microemulsions at ambient gravity. To that effect a moving boundary ("Tiselius") electrophoresis apparatus (Beckman-Spinco, Model H) was used, in the presence as well as in the absence of an electric field; see Figs. XXI-3 and XXI-4.

Without an electric field, and at ambient gravity, multiple phases can indeed be observed, usually starting within about 24 hours, and persisting for several days; see Fig. XXI-3 (Yang, 1990).

In the presence of an electric field, the various peaks (phases) were observed traveling toward the anode (Fig. XXI-4) (Yang, 1990), with, however, a few peaks traveling in the opposite direction. Now, the existence of positively charged droplets, in the presence of a much larger number of negatively charged droplets, all freely suspended in a liquid medium, is not as improbable as might be thought *a priori*. For particles of opposite charges to coexist without interacting and neutralizing each other, it suffices for them to have a higher positive value of ΔG_{132}^{AB}, than the negative value of their ΔG_{132}^{EL} (assuming ΔG_{132}^{LW} to be negligible). Taking the highest observed negative ζ potential of –60 mV and the highest estimated positive ζ potential of +7 mV, this would yield an attractive potential $\psi_o \approx 30.6$ mV, which for R = 100 nm, yields $\Delta G_{132}^{EL} \approx -1$ kT, whilst for $\gamma_{12} = -16$ mJ/m^2, $\Delta G_{132}^{AB} \approx +1500$ kT (at close range). In brine ($\approx 3\%$ NaCl), the decay with distance of ΔG^{EL} is steeper than the decay of ΔG^{AB}.

Based on the ($\gamma^\ominus$) monopolarity of the surfactants used in microemulsion formation, and on the observations made on microemulsions with schlieren optics, it may be concluded that in addition to a moderate EL repulsion and to a relatively weak LW attraction, microemulsion formation and maintenance is predominantly due to a strong AB repulsion.

FIGURE XXI-3   Schlieren picture of sedimentation at ambient gravity of microemulsion (as above), taken after 120 hours (Beckman Model H instrument). The multiple phases persisting in the microemulsion (middle) phase are clearly visible. (From Yang, 1990.)

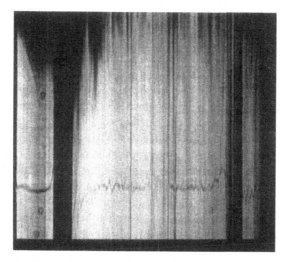

FIGURE XXI-4   Schlieren picture of electrophoresis of microemulsion (as above), at about 10 V/cm; as above, taken after 3 minutes (Beckman Model H Tiselius electrophoresis instrument). Shortly after connecting the DC field to the microemulsion phase, drastic structure changes occurred in the microemulsion phase. Small peaks can be seen to cover the whole microemulsion phase. These small peaks quickly sorted into three major peaks and moved to the positive electrode. Observations made during the run did reveal some peaks moving in the opposite direction; however, this opposite movement cannot be seen clearly in the photographs. This is mainly because the peaks were constantly changing in size and shape, and could not be easily followed. The average zeta potential of particles calculated based on the mobility of the largest peak was −59.3 mV; see text. (From Yang, 1990.)

CODA

Phase separations are treated which occur between different polymers dissolved in various apolar, polar and aqueous media. In apolar systems phase separation can only occur between two *different* polymers, one having a higher and the other a lower Hamaker coefficient, A, than that of the liquid; more than two phases cannot occur in such systems, even if more than two different polymers participate. On the other hand, in polar and especially in aqueous systems, as many phases can occur as there are (different) polymers. This is because, e.g., in water, even similar polymer molecules repel each other, even though they do not form different phases. However, different polymer species not only repel one another within each species, but they also repel each other *between* species, in each case with different energies of repulsion. This gives rise to sorting-out and to phase-formation, if the different polymers also have at least slightly different specific densities. A minimum concentration of each polymer species is required to effect phase separation; this is because the interaction energy between two polymer species needs to be at least equal to +0.5 kT for separation to prevail over mixing. Thus the higher molecular weight of a given polymer, the lower its concentration needs to be for separation to ensue (because higher molecular weight polymers also have a greater contactable surface area, so that for a given repulsive energy per unit surface area, 0.5 kT is reached more readily).

The above phenomena can also be designated as *coacervation*, i.e., the separation into two (or more) phases, each one mainly containing one polymer species, while being then depleted of the other polymer species, as a consequence of a *repulsion* between polymer molecules. In contrast with these phenomena, there is *complex coacervation*, where the molecules of two polymer species *attract* each other, thus creating one polymer-rich phase (comprising polymer molecules of both species, complexed with each other) and one phase which is depleted of polymer molecules. Complex coacervation can occur at much lower polymer concentrations than coacervation.

In microemulsions many dozens of different phases exist, which are invisible with the naked eye, but can be detected by means of schlieren optics (*via* which differences in refractive index can be visualized). The driving force of polymer phase separation (in the sense of coacervation) as well as of microemulsion formation originates in negative interfacial tensions, which at equilibrium decay to 0. (It should be recalled that between polymer phases in solution, as well as between microemulsion phases, the measured interfacial tensions always are ultra-low).

XXII | Solubility of Polymers and Other Solutes

In polymer solubility studies, polar molecular interactions have been treated for a number of years as being comprehensible in the same terms, and with the same formalism, as apolar interactions. This is tacitly assumed in the Scatchard-Hildebrand theory (Hildebrand and Scott, 1950), and, for polymers, by Flory (1953) and Huggins (1941, 1942). [See also Prausnitz *et al.* (1986), Patterson (1969), Hansen (1967, 1969, 1970), Hansen and Beerbower (1971), and Barton (1983).] It also has been recognized for some time that, for polar polymers that form hydrogen bonds, accurate and quantitative predictions of solubility from the molecular properties are not easily achieved. There thus appears to be an advantage in treating the mechanism of solubility in apolar and polar systems separately.

APOLAR SYSTEMS AND HILDEBRAND'S SOLUBILITY PARAMETER

It has been known for several decades that there is a close connection between solubility and surface tension. The earliest treatment of this connection that we have found was given by Rhumbler (1898), who credited Des Coudres with a discussion in terms of the energies of cohesion of separate components together with the excess energy of adhesion of the two components. Hildebrand and Scott (1950) reported an empirical correlation between the solubility parameter, δ, and surface tension, γ:

$$\delta \alpha V^{-2/3} \gamma^{0.44} \approx V^{-2/3} \sqrt{\gamma} \qquad \text{[XXII-1]}^*$$

Girifalco and Good (1957) and Good and Girifalco (1960) showed that the equation for the interfacial tension, γ_{12}, between *apolar* phases, 1 and 2, can be put in the form:

$$\gamma_{12}^{LW} = (\sqrt{\gamma_1^{LW}} - \sqrt{\gamma_2^{LW}})^2 \qquad \text{[II-35]}$$

---

* A list of symbols can be found on pages 399–406.

Here, γ_1^{LW} and γ_2^{LW} are the surface tensions of pure phases 1 and 2. Good and Girifalco (1960) also pointed out that the right side of eq. [II-35] is closely analogous to the Hildebrand-Scatchard expression for the partial molar internal energy of mixing in regular solutions:

$$\overline{\Delta U_i^{mixing}} = \phi_j^2 V_i (\delta_i - \delta_j)^2 \qquad [XXII-2]$$

where ϕ_j is the volume fraction of j, and V_i is the molar volume of component i.

For apolar systems the correlation between solubility and the smallest possible value of $\overline{\Delta U_i^{mixing}}$ is excellent, but for polar systems that correlation deteriorates, to become significantly less than perfect to nonexistent.

POLAR SYSTEMS

In view of the close connection between solubility and the interfacial tension between solute and solvent, already apparent in the preceding section, it is not surprising, in light of the difference between eqs. [II-35] and [III-7], that the conditions for solubility in apolar and polar systems are quite different. It must be emphasized again that an important thermodynamic property in a binary system is ΔG_{121}. This is the free energy change per unit area for the process in which molecules of 1 (the solute) are initially present in phase 2, with an effectively infinite layer of phase 2 separating two surfaces of phase 1. In the process, two surfaces of 1 are brought together seamlessly. It should be recalled (cf. eqs. [II-33], [IV-17], [XVIII-1], [XIX-2]) that:

$$\Delta G_{121} = -2\gamma_{12} \qquad [XXII-3]$$

If ΔG_{121} is positive, then molecules of 1 will repel each other in liquid 2, and substance 1 will spontaneously disperse to dissolve in 2. If ΔG_{121} is negative, then, at equilibrium, molecules of solute 1 in solution will attract each other, to a greater or lesser degree, depending on the amount of negative kT's with which they attract one another.

In water, when ΔG_{iwi} is *positive*, solute, i, will be soluble in virtually all proportions; see the upper part of Table XXII-1, with ΔG_{iwi} values listed from 0 to +5.5 kT, corresponding to solubilities from a respectable 1 mol fraction to an exorbitant 244.7 mol fractions. Thus, even a small positive ΔG_{iwi} value of not more than a tiny fraction of a kT already causes a mutual repulsion between similar molecules immersed in water, giving rise to practically complete aqueous solubility.

With *negative* ΔG_{iwi} values the degree of aqueous solubility of compound, i, is much more strongly linked to the precise negative value of ΔG_{iwi}; see the lower half of Table XXII-1. For instance, both glucose and sucrose have finite but still considerable aqueous solubilities, with ΔG_{iwi} values of the order of −3 kT. Octanol and octane on the other hand are a great deal less soluble in water, with ΔG_{iwi} values of −9.4 kT and −13.0 kT respectively. Nonetheless, the difference in aqueous solubility between octanol and octane still is quite significant (octanol is 37 times more soluble

TABLE XXII-1

Aqueous Solubilities versus Values of the Interfacial (IF) Free Energy of Interaction Between Two Molecules, i, Immersed in Water, w: ΔG_{iwi}^{IF}.

Solubility		
Mol/L	Mol fraction	ΔG_{iwi}^{IF} (kT)
	244.7	+5.5	Human serum albumin with 2 hydration layers; see Table XVII-6.
	148.4	+5.0	
	68.7	+4.23	DEX T150[a]
	15.6	+2.75	PEO 6,000[b]
	7.4	+2	
	4	+1.4	
	3	+1.1	
	2	+0.7	
	1	0	
27.78*	0.5	−0.693	
5.56	0.1	−2.303	
			Glucose[c]; Sucrose[d]
5.56×10^{-1}	0.01	−4.605	
5.56×10^{-2}	0.001	−6.908	
5.56×10^{-3}	0.0001	−9.210	Octanol[e]
5.56×10^{-4}	0.00001	−11.513	
5.56×10^{-5}	0.000001	−13.816	Octane[f]
5.56×10^{-6}	0.0000001	−16.118	

[a] s = 68.7 mol fr.; ΔG_{iwi}^{IF} = +4.23 kT (van Oss and Giese, 2004).
[b] s = 15.6 mol fr.; ΔG_{iwi}^{IF} = +2.75 kT (van Oss and Giese, 2004).
[c] s = 0.048 mol fr.; ΔG_{iwi}^{IF} = −3.04 kT (van Oss and Giese, 2004).
[d] s = 0.035 mol fr.; ΔG_{iwi}^{IF} = −3.35 kT (van Oss and Giese, 2004).
[e] s = 0.000,081 mol fr.; ΔG_{iwi}^{IF} = −9.42 kT (van Oss and Good, 1996).
[f] s = 0.000,002,2 mol fr.; ΔG_{iwi}^{IF} = −13.02 kT (van Oss and Good, 1996).
* This column is not extended further upward; the next entry above this one would already amount to 55.6 Mol/L and except for water itself there are very few other compounds of which one would be able to confine 55.6 Mol, at ambient pressure, in 1 L. The aqueous solubilities given for hydrated human serum albumin and for dextran and poly (ethylene oxide) (PEO) simply indicate that these polymers are theoretically infinitely soluble in water, in practice mainly limited by excessive viscosity.

in water than octane) and for instance in the Merck Index [11th Edition (1989)] the aqueous solubility of octanol is given but octane is listed as "insoluble in water." As a rule of thumb one may assume that a compound, i, is for most practical purposes "insoluble" in water when ΔG_{iwi} is more negative than about −10 kT.

The empirically based Flory-Huggins approach (Flory, 1953) to the estimation of the thermodynamic properties of polymer solutions has had some success in correlating the solubility of polymers in various solvents. However, the crucial empirical interaction parameter, χ, employed in that approach, has traditionally been determined from the solute activity in dilute solution in the solvent (see, however, Chapter XXI and eq. [VI-8]). χ, or rather χ_{12}, is defined as:

$$\chi_{12} = 2 \, S_c \, \gamma_{12} \qquad\qquad \text{[VI-8A]}$$

so that also (cf. eq. XXII-3):

$$S_c \, \Delta G_{121} = -\chi_{12} \qquad\qquad \text{[VI-6A]}$$

and therefore (cf. eqs. XIII-2A and XIII-3):

$$S_c \, \Delta G_{121} = -\chi_{12} = kT \ln s \qquad\qquad \text{[XXII-4]}$$

where s is the solubility in mol fractions.

The parameter, χ, is usually not known in advance before any solution of a polymer in a given solvent has been prepared. There thus is a need for a combining rule by means of which the interaction of a polymeric solute, 1 and a solvent, 2, can be derived from experimental data other than the properties of the solution. As eq.VI-8A (above) gives the link between χ_{12} and γ_{12} and eq. XXII-4 provides the link between χ_{12} and ΔG_{121}, it is only necessary to obtain values for γ_{12} or for ΔG_{121} to access χ_{12}. Then, as shown in Chapters IV and XII, a rigorous characterization of the relation between the interfacial tension (γ_{12}) and the apolar (γ^{LW})surface tension component and the electron-acceptor ($\gamma^{\oplus}$) and the electron-donor ($\gamma^{\ominus}$) surface tension parameters is possible, e.g., by contact angle determinations with apolar and polar liquids.

SOLUBILITY OF APOLAR POLYMERS

Polyisobutylene (PIB), a typical apolar polymer, is soluble, *inter alia*, in decane, hexadecane, carbon tetrachloride, chloroform, tetrahydrofuran, toluene, and benzene (cf. Tables XVII-1 and 2 and XXI-1). Table XXII-2 shows the ΔG_{121} values for PIB immersed in these solvents. As the interaction with these solvents is apolar, the ΔG_{121} values can be obtained from:

$$\Delta G_{121} = -2(\sqrt{\gamma_1^{LW}} - \sqrt{\gamma_2^{LW}})^2 - 4(\sqrt{\gamma_1^{\oplus}\gamma_1^{\ominus}}$$
$$+ \sqrt{\gamma_2^{\oplus}\gamma_2^{\ominus}} - \sqrt{\gamma_1^{\oplus}\gamma_2^{\ominus}} - \sqrt{\gamma_1^{\ominus}\gamma_2^{\oplus}}) \qquad\qquad \text{[XXII-5]}$$

(cf. eq. [IV-17]), in which the $\gamma^{\oplus}$ and $\gamma^{\ominus}$ terms are set equal to zero, yielding:

$$\Delta G_{121}^{\text{apolar}} = -2(\sqrt{\gamma_1^{LW}} - \sqrt{\gamma_2^{LW}})^2 \qquad\qquad \text{[XXII-6]}$$

(cf. eqs. [II-35] and [XXII-3]).

The polar solvents listed above are monopolar, so their polar moieties neither contribute to their own cohesion, as pure liquids, nor interact with the apolar PIB (see

---

TABLE XXII-2

Values of ΔG_{121} (in mJ/m²) of Polyisobutylene (PIB) for a Number of Solvents (from eq. [XXII-6])

Solvent[a]	γ_L mJ/m²	PIB ($\gamma^{LW} = 25$)	Soluble
Decane	23.8	−0.03	+
Methyl ethyl ketone	24.6	−0.003	+
Carbon tetrachloride	26.8	−0.06	+
Chloroform	27.3	−0.10	+
Tetrahydrofuran	27.4	−0.11	+
Hexadecane	27.5	−0.12	+
Toluene	28.3	−0.20	+
Benzene	28.9	−0.28	+
α-Bromonaphthalene	44.4	−5.53	−
Water[b]	72.9	−102.22	−

[a] See Tables XVII-1 and 2.
[b] Here, eq. [XXII-4] was used.

Table XVII-5), which allows us to use eq. [XXII-6]. For PIB with α-bromonaphthalene (see Table XVII-10), $\Delta G_{121} = -5.53$ mJ/m², and PIB is insoluble in this liquid.

Now, an estimate can be made of the attraction of two solute polymer chains for each other that is relevant to the question of whether the polymer will be soluble in a given solvent. This is based on the free energy, ΔG_{121}, times the area of contact between two chains, which is equal to χ_{12}, see eq. XXII-4, above.

For two chains at right angles to each other, the minimum effective area (designated as the "contactable area," S_c) should be of the order of 1 nm². Then (see eqs. [VI-7], [VI-8] and [XXII-4]), $S_c \Delta G_{121} \approx -5.5 \times 10^{-21}$ J, which corresponds to about −1.4 kT at 20°C. If, instead of two chains being crossed at right angles, they run parallel to each other for only about 25 carbon atoms in the chains, then S_c will be about 10 nm², and $S_c \Delta G_{121} \approx -55 \times 10^{-21}$ J, which corresponds to about −14 kT. This energy is sufficiently large compared to 1 kT that insolubility may be predicted. Thus, we explain the insolubility of PIB in α-bromonaphthalene. In water, PIB would have an $S_c \Delta G_{121}$ value of approximately −125 kT, corresponding to complete insolubility. In all the other solvents shown, the ΔG_{121} values vary from −0.003 to −0.30 mJ/m² and favor solubility.

MISCIBILITY OF POLAR LIQUIDS

MISCIBILITY AND SOLUBILITY AS MICROSCOPIC-SCALE PHENOMENA

Solubility is related to miscibility but the two are by no means completely identical. Contrary to miscibility, solubility is exclusively a microscopic-scale (i.e., molecular) phenomenon which is quantitatively defined by the free energy of interaction at the closest distance between two molecules, 1, immersed in a liquid, 2, i.e., ΔG_{121} (see eq. XXII-4). As in other liquids, in water, w, the solubility of solute, i, is finite when ΔG_{iwi} is between zero and approximately -10 kT and its solubility is unlimited when ΔG_{iwi} is positive (see Table XXII-1). Every compound, i, is at least somewhat soluble in, e.g., water, but the greater the negative value of ΔG_{iwi}, the more exiguous its aqueous solubility becomes (Table XXII-1). The quantitative expressibility of the degree of solubility of exceedingly sparsely soluble compounds is only limited by the sensitivity of the available analytical methodology.

Miscibility is linked to solubility insofar that for two different liquids to be miscible, they not only need to be soluble in each other, but their mutual solubility has to be considerable. In practice, for two different liquids, 1 and 2, to be miscible, ΔG_{121} has to be positive, or only slightly on the negative side. For example, for a liquid, i, to be miscible with water, ΔG_{iwi}, when negative, has to be fairly close to zero, or have a positive value. Or (see eq. XXII-3), for miscibility with water, *its interfacial tension with water,* γ_{iw}, *has to be negative*; see also Chapter XIX. Negative interfacial tensions occur, *inter alia*: between water and glycerol ($\gamma_{iw} = -14.2$ mJ/m²), water and ethylene glycol ($\gamma_{iw} = -2.4$ mJ/m²), water and formamide ($\gamma_{iw} = -6.3$ mJ/m²) and water and ethanol ($\gamma_{iw} = -30.9$ mJ/m²) (see Chapter XVII for the surface properties of these liquids). Of these four total miscibilities with water, ethanol has a particularly elevated ΔG_{iwi} value, of $+ 61.8$ mJ/m². On the other hand, Table XXII-3, which lists ΔG_{121} values for liquids which are miscible or immiscible with n-octanol, shows that among the nine solvents interacting with octanol, ethanol has the strongest negative ΔG_{121} value, of -5.1 mJ/m², corresponding to $\Delta G_{121} = -0.77$ kT, but it still is miscible

TABLE XXII-3
ΔG_{121} (in mJ/m² and in kT) for *n*-octanol in a Number of Solvents Compared with its Miscibility in These Solvents [a]

Solvent	ΔG_{121} (in mJ/m²)	ΔG_{121} (in kT)	Miscibility
Chloroform	+9.8	+1.5	m [b]
Tetrahydrofuran	0	0	m
Methyl ethyl ketone	−0.16	−0.02	m
Decane	−0.27	−0.04	m
Ethanol	−5.1	−0.77	m
Ethylene glycol	−24.2	−3.65	i [c]
Formamide	−24.0	−3.6	i
Water	−62.3	−9.4	i
Glycerol	−44.9	−6.8	i

[a] From surface tension data given in Chapter XVII; see also Table XVII-2 for n-octanol and chloroform and Table XXI-1 for tetrahydrofuran and methyl ethyl ketone.
[b] m = miscible.
[c] i = immiscible.

with octanol. This is less contradictory than it might at first sight appear, as this negative ΔG_{121} value would indicate a solubility for octane in ethanol of 0.46 mol fractions, which corresponds to 25.7 mol/L (cf. Table XXII-1), or 3.34 kg/L, which points to a very high solubility and is entirely compatible with total miscibility. As a first approximation, one may therefore place the conditions for miscibility among liquid pairs as ranging from $\Delta G_{121} \approx -1$ kT to all positive ΔG_{121} values. *Miscibility* among two different liquids therefore prevails: a) when a slight attraction between the different molecules is counterbalanced by the diffusional free energy of Brownian motion of +1 kT (Einstein, 1907) and: b) when the molecules of the different liquids mutually repel each other when immersed in one another, i.e., when $\Delta G_{121} \geq 0$.

MISCIBILITY AND IMMISCIBILITY AS MACROSCOPIC-SCALE PHENOMENA

Immiscibility is really a macroscopic-scale phenomenon. It occurs when the molecules within each liquid attract one another cohesively to an extent which one may tentatively place at $\Delta G_{121} < -1$ kT and in practice closer to $\Delta G_{121} \leq -3$ kT. Immiscibility of two liquids is characterized by the obviously macroscopic-scale separation of the liquid pair into two distinct phases which visually closely resemble the two separate liquid phases occurring upon pouring together two volumes of the same liquid, in each of which a different polymer is dissolved; see the preceding Chapter (XXI). In the latter phenomenon phase separation occurs because the two different polymers dissolved in the same liquid *repel* one another, as this actually is a system comprising *three* different condensed-phase compounds: two different polymers and one liquid. However, in the phase separation which is characteristic of *immiscibility* among two different pure liquids, there are only *two* different condensed-phase compounds, i.e., just the two liquids, which *attract* one another. Thus, despite the visual resemblance between the immiscibility of two different liquids and the phase separation of two solutions of different polymers dissolved in the same liquid, the mechanisms are different.

It is preferable, instead, to compare *immiscibility* of two different liquids with the *insolubilization* of a solute immersed in a solvent in which it has a very low solubility. In that case the solute molecules attract each other and form insoluble precipitates. The attraction between such (e.g., hydrophobic) solute molecules may for instance have been caused by being hydrophobically squeezed together by the strong free energy of cohesion of the surrounding (polar) liquid molecules, e.g., in a solvent such as water. *Miscibility*, on the other hand, may be compared to the *solubility* of (e.g., hydrophilic) molecules (for instance proteins or polysaccharides) in water, because such molecules *repel* one another in water and thus diffuse singly into all the available liquid, keeping on average the same distance between each other.

The main difference between miscibility and solubility is that *solubility* (a microscopic-scale phenomenon) occurs at all values of ΔG_{121}: high solubility when $\Delta G_{121} > 0$ and from limited to very low solubility when $\Delta G_{121} < 0$, although, given the proper techniques some degree of solubility can practically always be demonstrated. *Miscibility* (a macroscopic-scale phenomenon) on the other hand is an all-or-nothing occurrence: Two different liquids are either miscible with one another, or they aren't. When liquids 1 and 2 are miscible, γ_{12} is close to zero or, more often, negative (see Table XXII-3, recalling also that $\Delta G_{121} = -2\gamma_{12}$). When liquids 1 and 2 are miscible, γ_{12} is positive (i.e., ΔG_{121} is negative, i.e., attractive).

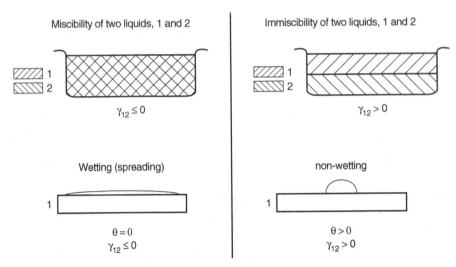

FIGURE XXII-1 Comparison between two related macroscopic-scale phenomena. *Left:* miscibility of two liquids, 1 and 2, and wetting of a flat solid surface, 1, by a single liquid, 2. *Right:* immiscibility of two liquids and non-wetting of flat surface, 1, by a liquid, 2; here the liquid forms a finite contact angle on top of the solid.

A clearer analogy of *miscibility vs. immiscibility* in two-liquid systems as a macroscopic-scale phenomenon, is *wetting vs. non-wetting* in another macroscopic-scale condensed-phase system, comprising not two liquids, but a *liquid* (1) and a *solid* (2). When $\gamma_{12} \leq 0$, the liquid wets (i.e., complete spreads over) the solid surface, indicating that the adhesion between the liquid and the solid surface is equal to or greater than the cohesion of the liquid. This strong adhesion between liquid and solid is analogous to miscibility among two liquids, due to their macroscopic-scale mutual attraction. On the other hand, when $\gamma_{12} > 0$, a liquid drop deposited on a solid surface does not wet and spread over the solid surface but remains a drop, forming a finite contact angle with the surface. Here the cohesion of the drop is greater than the adhesion of its liquid to the solid surface. This macroscopic-scale decrease in adhesion between liquid and solid is analogous to immiscibility between two liquids; see Figure XXII-1.

CONCLUSION

The comparison between miscibility and immiscibility on the one hand and wetting and non-wetting on the other, both considered as macroscopic-scale interactions between condensed-phased materials, seems to be the most appropriate one.

SOLUBILITY OF POLAR POLYMERS

In Table XXII-4 it can be seen that positive values of ΔG_{121} occur quite commonly among biopolymers and highly polar synthetic polymers immersed in water; these positive values of ΔG_{121} correlate with pronounced solubility.

TABLE XXII-4
ΔG_{121} (in mJ/m²) of a Number of Biopolymers and Strongly Polar Synthetic Polymers with Respect to Water[a]

Polymer or biopolymer	ΔG_{121}	Solubility
Polyoxyethylene (6000)	+43.6 to +52.5	+
Dextran	+23.0 to +41.2	+
Polyvinylpyrrolidone	+7.4	+
Human serum albumin (with 2 layers of hydration)	+20.4	+
Human serum albumin (dry)	−18.3	−[b]
Gelatin	−19.4	−[c]
Zein	−21.1	−

[a] From surface tension data given in Chapter XVII.
[b] It is assumed that dry HSA is insoluble in water until it becomes hydrated; see text.
[c] Soluble upon boiling but insoluble at 20°C. Boiled solutions gel upon cooling.

Gelatin, on the other hand, is insoluble in water at room temperature, and zein (a corn protein) is insoluble in water at any temperature. Curiously, human serum albumin (HSA), like other serum proteins (van Oss et al., J. Protein Chem., 1986; van Oss and Good, J. Protein Chem., 1988), is initially insoluble in water when dry but becomes very soluble upon hydration. It is, indeed, typical of many such proteins, when they are added to water as dry powders, that the dry particles first have to swell quite visibly before they undergo dissolution as a second step. Once hydrated, their solubility is very high. It is probable that freeze-dried HSA, which dissolves much more quickly in water than the air-dried material, has different surface properties from air-dried HSA. Since it is, however, not possible as yet to make a smooth, nonporous flat surface from freeze-dried proteins, no usable contact angles can be measured on freeze-dried materials. It is likely that air-drying causes a certain degree of denaturation in HSA but, in view of its unaltered properties on redissolution, that denaturation may be taken to be reversible. Upon hydration, albumin appears to re-assume the conformation of what may be likened to a monomolecular micelle. It should be noted that even in aqueous solution, serum albumin denatures reversibly at the liquid-air surface (as well as at liquid-solid interfaces) as judged, for example, by its drastic lowering of the surface tension of water (Absolom et al., 1981). The inherent water-insolubility of dry proteins is also supported by the propensity of many proteins to undergo salting out (van Oss, Moore et al., 1985) and by the possibility of precipitating them with the help of polyoxyethylene; see below.

Not all proteins, however, undergo significant (reversible) denaturation upon air-drying: fibrinogen is a case in point: contact angles measured on dried or hydrated fibrinogen are not drastically different (van Oss, J. Prot. Chem., 1990); see also Table XVII-6. Nor do most polysaccharides appear to denature upon air-drying; they are immediately soluble in water, even though the highest molecular weight materials (e.g., dextran T-2000, with $\overline{M}_w$ 2 000 000) dissolve more slowly than lower molecular weight polysaccharides.

Zein is a curious protein (found in corn) which is insoluble in water but soluble in formamide, ethylene glycol, and in a 70/30 (v/v) acetone/water mixture. The large negative ΔG_{121} of zein with water (see Table XXII-5) explains the low solubility in water. From the ΔG_{121} values, the solubility of zein in ethylene glycol and formamide would be somewhat unexpected. It may be, however, that like HSA in water, zein becomes solvated in ethylene glycol and in formamide. The solvation of zein in these solvents, unfortunately, is not easily directly measurable. Nevertheless, there is an indication for the existence of such an effect from the following observations: 10% solutions of zein in formamide and in ethylene glycol precipitate upon the addition of water (approximately 45 vol% water added to a zein solution in formamide, and less than 3 vol% water added to a zein solution in ethylene glycol). Thus, the addition of a rather minute amount of water to solutions of zein in ethylene glycol causes insolubilization, which may hint at a displacement of the solvation layer of ethylene glycol by water, which then causes precipitation through disappearance of the solvation layers. It should be noted that the affinity of zein for water is higher than for formamide and considerably higher than for ethylene glycol; see the ΔG_{12} values in Table XXII-5.

TABLE XXII-5
Values of ΔG_{121} and ΔG_{12} (in mJ/m²) of Zein [a]

Solvent	ΔG_{121}	Solubility	ΔG_{12}
Water	−21.1	−	−105.1
Ethylene glycol	−9.4	+	−86.1
Formamide	−10.5	+	−95.6

[a] From surface tension data given in Chapter XVII.

A somewhat comparable behavior may be observed with, e.g., dextran and its monomer glucose (1), both of which are *insoluble* in chloroform (2), even though the values of ΔG_{121} are strongly positive in all cases (ΔG_{121} from +42 to +53 mJ/m²). Polyoxyethylene, on the other hand, is quite soluble in chloroform, in accordance with its positive ΔG_{121} value, which is of the order of +59 mJ/m² ($\approx$ +3.1 kT). In the cases of dextran and glucose, it seems likely that the rather strongly bound residual water of hydration at the surface of the carbohydrate molecules renders them insoluble in chloroform, because of the fact that water (w) is itself largely insoluble in chloroform (2) [$\Delta G_{w2w} = -63.2$ mJ/m² ≈ -3.1 kT; see also Stephen and Stephen (1963)]. Other carbohydrates, such as cellulose, as well as proteinaceous materials, such as histological tissues (intended for imbedding and sectioning) also are unusually resistant to swelling or wetting by various organic solvents, *unless such materials are first meticulously dehydrated*; see, e.g., van Oss (1955a) and Sheehan and Hrapchak (1973).

For a comparison between the experimental solubility of polyethylene glycols and dextrans, and their solubilities calculated via two different approaches; see Table XIV-2.

SOLUBILITY IN NON-AQUEOUS POLAR LIQUIDS

Various non-aqueous polar liquids (e.g., formamide, ethylene glycol) can also dissolve biopolymers. This is due to:

a) their relatively high γ_3^{LW} component, which is close to that of most biopolymers, thus yielding a rather low (negative) value for ΔG_{131}^{LW}.

b) their $\gamma_3^{\oplus}$ parameter which, although smaller than that of water, nevertheless reacts significantly with the strong $\gamma_1^{\ominus}$ parameter of most biopolymers;

c) their relatively low (albeit non-negligible) γ_3^{AB} component, which tends to counteract solubility (see the $\sqrt{\gamma_3^{\oplus}\gamma_3^{\ominus}}$ term in eq. [IV-17], which is equal to $\frac{1}{2}\gamma_3^{AB}$; see eq. [IV-5].

In general, the best solvents for (electron-donor) monopolar solutes are those with a pronounced electron-acceptor ($\gamma_3^{\oplus}$) parameter. Apart from the dipolar solvents noted above, chloroform (which has a sizeable electron-acceptor parameter and is largely monopolar) also needs to be mentioned. For instance, chloroform dissolves PEO. (It *should* also dissolve biopolymers such as dextran, but it does not. This is most likely due to the residual hydration of dried dextran powder which is exceedingly difficult to eliminate; see the preceding section.)

Insulin is a fairly small protein (Mw ≈ 24,000), soluble in ethanol and acetone. Being a euglobulin (see sub-Section 4, on Precipitation of Euglobulins, below) it is also soluble in water at pH values that are one or two pH units removed from its isoelectric point (pH 5.3).

Due to the smaller polar energy of cohesion of non-aqueous polar liquids than of water, the net interfacial ("solvophobic") attraction of low energy solutes in such non-aqueous polar liquids is much smaller than their hydrophobic attraction in water. Thus, liquids such as formamide or ethylene glycol, with a polar energy of cohesion which is only 37% of that of water (see Table XVII-10), are solvents which are not quite as versatile as water, but they can dissolve some biopolymers which are not soluble in water, such as zein, or agarose, or gelatin* (see also the preceding section). This is due to the fact that the adhesion term $\sqrt{\gamma_3^{\oplus}\gamma_3^{\ominus}}$ still compares relatively favorably with the cohesion term $\sqrt{\gamma_3^{\oplus}\gamma_3^{\ominus}}$ (cf. eq. [IV-17]). On the other hand, the still sizeable polar energy of cohesion $-4(\sqrt{\gamma_3^{\oplus}\gamma_3^{\ominus}})$ explains the *insolubility* in these solvents of more apolar compounds such as polystyrene, which are, however, soluble in apolar solvents of similar γ_3^{LW}, such as benzene or α-bromonaphthalene; see Table XXII-6.

Nevertheless both for hydrophobic (or "solvophobic") insolubilization of mainly apolar solutes, and for hydrophilic (or "solvophilic") solubilization of strongly polar solutes, water is by far the most effective liquid. This is due to the fact that its $\gamma^{\oplus}$ parameter is higher than that of any other compound measured to date. It is largely through its high electron-acceptor capacity that water interacts so strongly with most polar organic molecules and especially with those of biological origin, which have

---

* Agarose and gelatin are soluble in water at 100°C, but not at room temperature.

TABLE XXII-6

Influence of Polar Forces, "Hydrophobic,"[a] "Solvophobic,"[b] or Lifshitz–van der Waals[e] Interactions on Solubility

SOLUTES	SOLVENTS			
	Strong self-hydrogen-bonding liquid (water)	Moderately self-hydrogen-bonding liquids (e.g., glycerol, formamide, ethylene glycol, dimethylsulfoxide)	Monopolar electron-acceptor liquids (e.g., chloroform)	Nonpolar liquids and monopolar electron-donor liquids (e.g., (a) alkanes (b) benzene, methylethyl ketone, tetrahydrofuran)
Hydrophilic solutes[c]	Very soluble[d]	Soluble[d]	Only soluble if totally non-hydrated[d]	Insoluble[a,d,e]
Somewhat polar "hydrophobic" solutes[f]	Often insoluble[a,d]	Soluble[a]	Soluble[d]	Often soluble[e]
Apolar solutes[g]	Very insoluble[a,d]	Insoluble[b,d]	Often soluble[e]	Often soluble[e]

[a] Due to hydrophobic interactions.
[b] Due to "solvophobic" interactions.
[c] Proteins, polysaccharides.
[d] Due to polar forces.
[e] Partly or mainly due to LW interactions.
[f] For example, lipoproteins, lipopolysaccharides, denatured proteins.
[g] Alkanes, lipids.

an elevated $\gamma_1^\ominus$ (electron-donor) parameter. Thus, a much greater variety of polar compounds are soluble in water than in any other liquid. At the same time, due to the very high polar energy of cohesion of water ($\Delta G_{ww}^{AB} = -4\sqrt{\gamma_w^\oplus \gamma_w^\ominus}$), apolar compounds are exceptionally *insoluble* in water; see Table XXII-6.

INSOLUBILIZATION OF BIOPOLYMERS

A number of approaches may be used to enhance the insolubilization of biopolymers.

1. *Precipitation through "salting-out," addition of ethanol, and coacervation.* "Salting-out" (van Oss, Moore *et al.*, 1985) is usually done through the admixture of fairly high concentrations of $(NH_4)_2SO_4$ (typically of the order of 1 to 2.5 M); see Chapter XXI. The cold ethanol precipitation method (Cohn *et al.*, 1946; 1965; Schultze and Heremans, 1966) operates mainly through the lack of solubility in ethanol of most proteins. It still is the principal procedure for the fractionation of plasma proteins for parenteral use. Insolubilization through phase separation (i.e., coacervation *sensu stricto*) in aqueous media is usually done by the admixture of 2 to 10% (w/v) polyethylene oxide (Schultze and Heremans, 1966; Polson (1977); see also Chapter XXI. Finally, proteins usually have a more amphoteric character than other biopolymers so that their precipitation reaches its maximum of effectivity when done at the isoelectric point of the protein in question, because the maximum of surface hydrophobicity occurs at that pH (van Oss and Good, J. Protein Chem., 1988; see also Holmes-Farley *et al.* (1985) and Wu *et al.* (1994).

2. *Precipitation through complex coacervation* is another, entirely different, approach to biopolymer insolubilization, see the right-hand side of Table XXI-9. Here use is made of the fact that most biopolymers which are not electrically neutral (e.g., many polysaccharides) usually are negatively charged (most proteins, and all nucleic acids). In such cases insolubilization generally can be achieved via the complex coacervation approach, i.e., through charge neutralization (which usually also entails an increase in hydrophobicity, see Wu *et al.*, 1994). This can be achieved by the admixture of, e.g., polycations (van Oss, 1982; van Oss, J. Dispersion Sci. Tech., 1988), or heavy metals or positively charged organic solutes, such as Rivanol (Schultze and Heremans, 1966). A more specific manner of insolubilizing selected biopolymers is by precipitation with an antibody directed to the biopolymer in question; see Chapter XXII. The drawback of the complexation approach to biopolymer insolubilization lies in the fact that the biopolymer co-precipitates with the additive, so that, much more than is the case in other approaches, after precipitation, a further purification step involving the removal of the co-precipitant is essential.

3. *Cationic proteins and peptides which avoid complex coacervation.* Positively charged small proteins and even smaller peptides can freely

circulate in mammalian (and other) body fluids, which are virtually all negatively charged, without becoming insolubilized by binding with the anionic solutes or particles comprised in these different fluids by means of complex coacervation. One of the best known small cationic proteins in this category is lysozyme, a mucolytic enzyme (Mw ≈ 14,600), which is omnipresent in body fluids (in tears; in the egg white of chicken eggs, etc. Lysozyme (discovered by Alexander Fleming) is mildly antibiotic. There also are many different smaller cationic peptides (molecular masses around 2 to 4 kD) that are intra- as well as extracellular and have antimicrobial properties (e.g., defensins, cathelidins; see Devine and Hancock, 2004). These cationic entities include lysozyme as well as the above-mentioned antimicrobial peptides (also called mammalian host defense peptides by Devine and Hancock, 2004), all of which are capable of existing in the dissolved state in most body and cytoplasmic fluids where they are exposed to the many different biopolymers, cells, tissues and subcellular particles which are negatively charged, apparently without combining with them by complex coacervation when in a resting state. This is partly due to their small size (or more precisely, to their small contactable surface area, S_c) and partly to their hydrophilicity which provides a non-electrostatic interfacial repulsion with the many different negatively charged but equally hydrophilic entities, thus easily surmounting their electrostatic attraction. However, these small cationic peptides can of course combine with their (negatively charged) receptors when activated. For lysozyme, $\Delta G_{iwi} \approx + 30$ kT and for defensins and other cationic peptides, ΔG_{iwi} may be estimated at about + 10 kT.

4. *Precipitation of euglobulins.* The name "euglobulin" (i.e., good, or real, globulin) has been given to those (plasma) proteins which become insoluble at low ionic strengths, in contrast with the other (plasma) globulins or "pseudoglobulins," which remain soluble, even in distilled water (Schultze and Heremans, 1966). The maximum of insolubility of euglobulins at low ionic strength occurs at or near their isoelectric pH. Euglobulins are somewhat more "hydrophobic" than pseudoglobulins, as they precipitate at lower $(NH_4)_2SO_4$ concentrations during "salting-out"; see above (see also Schultze and Heremans, 1966). Precipitation of euglobulins at low ionic strength is obviously not mainly due to dehydration (lowering the NaCl concentration from 10^{-1} M to, e.g., 10^{-3} M, has no direct influence on hydration), even though operating at pH values close to the isoelectric point of a protein drastically decreases its degree of hydration. Nevertheless, at NaCl concentrations of the order of 10^{-1} M, isoelectric euglobulins still are soluble in water. The crucial mechanism of insolubilization of euglobulins at low ionic strengths is complexation, or perhaps more precisely, auto-complexation. Auto-complexation operates through the attraction between the (roughly equal) positive and negative

charge sites on opposing euglobulin molecules, in the absence of the small ions that were shielding these charge sites. Euglobulins, precipitated at low ionic strengths, redissolve upon re-addition of salt ("salting-in"). A more quantitative way of interpreting the precipitation of proteins at their isoelectric point (i.e., with equal numbers of positive and negative charges) and at low ionic strength (i.e., with a large Debye length, $1/\kappa$) (see chapters V, VII and XXIII) involves the Debye-Hückel theory. Because the electrostatic interaction is at a maximum at low ionic strength (see chapters V, VII and XXIII), and considering that this particular electrostatic interaction is *attractive*, low ionic strength here gives rise to agglomeration of the euglobulin molecules, and hence to precipitation.[*] Upon the addition of salt ("salting-in") the Debye length, $1/\kappa$ (see eq. [V-2]) becomes smaller, which causes $|\Delta G^{EL}|$ to decrease (see eq. [V-5]), so that ΔG^{TOT} becomes less negative than ≈ -1 kT, and re-solubilization ensues. This method is of denaturing.

THE FLORY-HUGGINS χ-PARAMETER AND THE SOLUBILITY OF POLAR COMPOUNDS

As shown in Chapter V, the Flory-Huggins χ-parameter can be expressed as (cf. eqs. [VI-6] and [VI-8A]):

$$\chi_{12} = -S_c\Delta_{121}/kT \qquad [XXII-7]$$

Thus in the preceding sections, on the miscibility of polar liquids and the solubility of polar polymers, in all places where ΔG_{121} is used (when expressed in units of kT), that term is equivalent to $-\chi_{12}$, where χ_{12} is the Flory-Huggins parameter of interaction between solute molecules (1), in liquid (2). Thus whilst in apolar systems miscibility/solubility prevails as long as $0 < \chi_{12} < 0.5$, in polar systems miscibility occurs not only when $\chi_{12} < 0.5$, but even more so when $\chi < 0$. This implies that the intrinsic value of χ_{12} should be determined at the highest possible solute concentration (see Chapter XIX) or, even better, on samples of dry solute. The lack of information obtainable from the determination of χ_{12} at the greatest possible dilution has been discussed earlier, in Chapter XIX.

Other solubility parameters (and parachors) and their applicability to various systems have been discussed by Barton (1983) and by Yalkowsky and Banerjee (1992). For Hildebrand's solubility parameter, see the beginning of this chapter.

---

[*] See also the flocculation mechanism of kaolinite by positive and negative charge interaction (Schofield and Samson, 1954).

SOLUBILITY AND THE CMC OF NON-IONIC
SURFACTANTS

Up to the present, a satisfactory quantitative theory of the critical micelle concentration (cmc) of non-ionic surfactant solutions has not been available. A typical non-ionic surfactant consists of a hydrocarbon (HC) group (e.g., an alkane chain) and, grafted at one end, a $(OC_2H_4)_nOH$ or polyethylene oxide (PEO) chain. From data assembled by Becher (1967a), it can be seen that an increase in length of the *alkane chains* of poly(ethylene oxide) (PEO) derivatives is accompanied by a pronounced decrease in the critical micelle concentration (cmc). On the other hand, an increase in length in the *PEO moieties* of these non-ionic surfactants is accompanied by only a slight increase in the cmc (Becher, 1967a); see also Becher's Langmuir Lecture (1984). The differences in the influence of the "hydrophobic" (alkane) chains and of the hydrophilic (PEO) chains can, of course, be correlated (to a certain degree) with the hydrophile-lipophile balance (HLB) values of these systems (Becher, 1967b; 1984).

It is now possible to arrive at a quantitative expression for the total interaction energy between surfactant chains, on the basis of the surface thermodynamic treatment of attractive interactions between largely apolar moieties in polar media (Chapter XVIII), and of repulsive interactions between strongly polar moieties in polar media (Chapter XIX).

The interaction energy between surfactant chains, immersed in water, can be correlated with the critical micelle concentration (cmc), by making use of the link between the equilibrium binding constant, K_a, at the cmc, and the binding energy (ΔG_{iwi}) corresponding to that K_a value. The binding energies derived *via* these two entirely different approaches should be the same, or at least of the same order of magnitude, for each individual surfactant species. It can be shown that this is indeed the case.

The intuitive approach to the energies of interaction of non-ionic surfactants with each other and with water is to begin by stating that the HC part is "hydrophobic" and the PEO part is "hydrophilic." For the meaning of this statement, see Chapters XVIII and XIX. For use in the quantitative study of surfactant energetics, the pertinent interaction energies are:

a. the (attractive) interaction energy between hydrocarbon chains (1), immersed in water (w), i.e.; ΔG_{1w1}, and:
b. the (repulsive) interaction energy between PEO chains (2), immersed in water (w), i.e.; ΔG_{2w2}.
c. the (attractive) interaction energy between hydrocarbon and PEO chains, which may be expressed at $\Delta G_{1w2} = \gamma_{12} + 0.5\ \Delta G_{1w1} + 0.5\ \Delta G_{2w2}$. This quantity may here be neglected, as it is in all cases only about 20% of ΔG_{1w1}. (The mutual attraction between HC chains thus will be favored over the attraction between HC and PEO chains.)

This identification of these energy terms allows us to address the crux of the mechanism of micelle formation: the HC chains attract each other, in water, and thus tend to be insoluble, and the PEO chains mutually repel each other, in water, and are, therefore, very soluble. Once the limit of solubility of the entire combined HC and PEO chains is reached, micelle formation sets in. By sheltering the water-insoluble HC chains in the interior of the micelle, and by orienting the soluble PEO moieties outward, micelle formation allows the further solubilization of the multimolecules' complex, or micelle.

Let ΔG_{1w1} be the change in free energy, per unit area, that is, a measure of the attraction of two parallel surfaces of substance 1, in water (w). Thermodynamically, it is the free energy change when two unit areas of the 1-w interface are eliminated forming a continuum of substance 1. It has been shown (Chapters II, IV) that for an alkane (1) in water (w) (see eq. [IV-17]):

$$\Delta G_{1w1} = -2\gamma_{1w} \qquad\qquad [XXII\text{-}8A]$$

where γ_{1w} is the water-alkane interfacial tension. Since γ_{1w} is normally positive, ΔG_{1w1} is normally negative. For polyethylene oxide (PEO) (substance 2) we may write the corresponding relation:

$$\Delta G_{2w2} = -2\gamma_{2w} \qquad\qquad [XXII\text{-}8B]$$

PEO is very soluble in water; it has been found (see Chapter XV), that γ_{2w}, for water with PEO, is -26.2 mJ/m^2; so that here, ΔG_{2w2} is strongly positive.*

The total free energy of interaction of non-ionic surfactant molecules (which consist of a PEO chain grafted onto a hydrocarbon chain) with water can, then, be written:

$$\Delta G^{TOT} = \Delta G_{1w1} + \Delta G_{2w2} \qquad\qquad [XXII\text{-}9]$$

When applied to molecules comprising two moieties (e.g., one being hydrophobic and the other one hydrophilic) eq. [XXII-9] is only valid when all ΔG terms are expressed in units kT. Both ΔG_{1w1} and ΔG_{2w2} can be described by the equation:

$$\Delta G_{iwi} = -2\gamma_{iw} = -2(\sqrt{\gamma_i^{LW}} - \sqrt{\gamma_w^{LW}})^2 + 4(\sqrt{\gamma_i^{\oplus}\gamma_w^{\ominus}}$$
$$+ \sqrt{\gamma_i^{\ominus}\gamma_w^{\oplus}} - \sqrt{\gamma_i^{\oplus}\gamma_i^{\ominus}} - \sqrt{\gamma_w^{\oplus}\gamma_w^{\ominus}}) \qquad [IV\text{-}17A]$$

The surface tension components and parameters of various slightly polar, water-immiscible organic compounds may be derived, by using eq. [IV-9], from their interfacial tensions with water, γ_{1w}; see Tables XIII-2 and XVII-1.

PEO is a monopolar Lewis base (Chapter XIX). The free energy of interaction between PEO molecules, in water, $\Delta G_{2w2} = +52.5$ mJ/m^2. This result is obtained

---

* The electrostatic energy of repulsion may, as a first approximation, be neglected in the case of interaction between non-ionic surfactant molecules.

via eq. [IV-17A], using the following values for PEO: $\gamma^{LW} = 43$ mJ/m^2, $\gamma^{\oplus} = 0$ and $\gamma^{\ominus} = 64$ mJ/m^2 (Chapters XVII and XIX). Thus, the PEO groups, dissolved in water, mutually repel each other (when in contact) with an energy of 52.5 mJ/m^2. The interfacial free energies of attraction between alkyl groups (from hexyl to hexadecyl), immersed in water can be obtained from Table XVII-1; it may be seen that ΔG_{1w1} varies between -102.0 and -103 mJ/m^2.

The dimensions of the contactable portions (i.e., the surface area S_c along which two HC chains can achieve "contact" with each other) of the hydrocarbon chains can be found in Table XIII-1.

As shown above, whilst the hydrocarbon chains of non-ionic surfactants attract each other, the PEO chains are, in water, mutually repulsive. Thus, although two alkyl chains can approach each other closely, two PEO chains, in water (if held together at one end), will tend to diverge at first, until the point where the repulsion per chain segment drops below some minimal level. Beyond that point, they will tend to remain, on an average, at some appreciable distance from each other from there on. Their precise average configuration is fairly hard to predict. It has been suggested that they assume helical or random coil conformations. However, we may model a pair of $R(OC_2H_4)_nOH$ molecules as shown in Fig. XXII-2. This allows the calculation of the values of the repulsive interaction energies of PEO chains of various lengths. Determining the value of the effective area, S_c, for mutually repulsive chains, such as PEO moieties, is more difficult than for attracting hydrocarbon chains. The PEO chains can be treated as being in one plane, and tending to diverge from each other at an angle, starting after the first EO moiety attached to the adhering HC moieties (see Fig. XXII-2). The S_c value between two single EO moieties, in the middle of two parallel chains, is roughly estimated at $\approx 2.6 \times 3.75 = 9.73$ Å^2.

The somewhat increased uncertainty in determining the Sc value between two EO moieties that are some distance from the point where the EO chain is grafted onto the HC chain is counterbalanced by the fact that the influence of the PEO repulsion, beyond the first several EO moieties [which are assumed to diverge from each other at a 90° angle, hinging after the first EO moiety (see Fig. XXII-2), is rather small, due to the exponential rate of decay with distance of polar forces (see Chapter VII).

It is assumed here that after the first EO, the next three EO moieties diverge at an angle of 90°, and from there on run (on an average) parallel to each other, or at any rate, do not approach each other. [Other angles of divergence, which would give rise to different distances at which PEO chains start running (on average) parallel to each other, are also entirely possible; for the present model an initial angle of 90° was adopted more or less arbitrarily.] The interaction energy at a distance, ℓ, may be expressed as:

$$\Delta G_\ell^{AB} = \Delta G_{\ell_o}^{AB} \exp(\ell_o - \ell / \lambda) \qquad \text{[VII-3]}$$

where ℓ_o indicates the minimum equilibrium distance at closest contact and λ is the decay length of water. For water, $\lambda \approx 10$ Å is here assumed. At closest approach, the repulsion energy between two EO segments, for area $2.6 \times 3.75 = 9.75$ Å^2 and $\Delta G_{2w2} = +52.5$ mJ/m^2, amounts to $+1.27$ kT. Thus for 6 PEO moieties on each side,

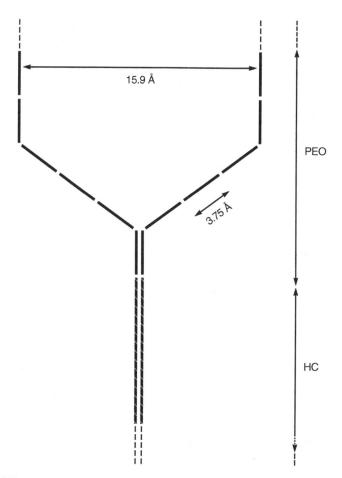

FIGURE XXII-2 Schematic presentation of the steric configuration of non-ionic alkyl-PEO surfactant molecules, dissolved in water. There is a strong *attraction* between alkyl groups ($\Delta G_{1w1} \approx -102$ mJ/m^2), which thus approach each other at the minimum equilibrium distance. And there is a strong *repulsion* between EO moieties ($\Delta G_{232} \approx +52.5$ mJ/m^2), so that these moieties must diverge. It is assumed that they diverge at an angle of 90°, after the first EO moiety. As the repulsion energy decays exponentially with distance, the repulsion is assumed to have become sufficiently feeble to allow the chains to resume a random configuration from the fourth EO moiety on, i.e., at a distance greater than about 16 Å. In the figure, a parallel configuration is shown; this is intended only to indicate that the chains (beyond the fourth EO), on an average, do not approach each other any closer than about 16 Å. (From van Oss and Good, 1991a.)

which diverge after the first pair (see Fig. XXII-2), the total repulsive interaction amounts to +2.79 kT. Using eq. [XXII-9], the total energy of interaction of various non-ionic surfactant molecule-pairs can be determined. For PEO segments longer than 4 EO moieties, the repulsion between the chains becomes so slight that the chains may be taken to resume an effectively parallel course (see Fig. XXII-2).

RELATION BETWEEN ΔG^{TOT} AND THE CMC

A comparison can now be made between the total free energies of interaction, derived from the interaction energies described above, together with the dimensions and steric configurations of the apolar and polar moieties of the surfactant molecules, and their cmc values. To that effect, a linkage can be established between the cmc (critical micelle concentration) and the equilibrium binding coefficient, K_a (where the subscript, a, denotes association). If one assumes that at the cmc, the monomeric surfactant molecules have just engaged in an incipient complex formation with another surfactant molecule,* one may express the equilibrium association constant as:

$$K_a = \frac{1}{cmc} \qquad\qquad [XXII\text{-}10]$$

where cmc is expressed in mol fractions. The free energy of interaction between two surfactant molecules, at the cmc, may be expressed as:

$$\Delta G^{TOT} = kT \ln cmc \qquad\qquad [XXII\text{-}11]$$

As the transition between monomeric surfactant molecules and complete micelle formation occurs within a narrow concentration range (Adamson, 1982, p. 448), eqs. [XXII-10] and [XXII-11] represent a plausible and legitimate approximation.

When $\Delta G^{TOT}/kT$ (eq. [XXII-11]) is plotted vs. ln cmc, the correlation between eq. [XXII-9] and eq. [XXII-11] becomes apparent. Early calculations on the PEO systems, following these rules, have been published by van Oss and Good (1991a). (The cmc is closely akin to the aqueous solubility).

SOLUBILITY AND THE CMC OF ANIONIC SURFACTANTS

In the same manner as used in calculating the cmc of non-ionic surfactants, the cmc of anionic surfactants can be approached. Here, however, in addition to the ΔG_{1w1}^{LW+AB} of the apolar tails and the ΔG_{2w2}^{LW+AB} of the polar heads, ΔG_{2w2}^{EL} of the polar must be determined. This proved feasible with the anionic surfactant, sodium dodecylsulfate (SDS). A molecular model of SDS could be made, somewhat analogous to the PEO model depicted in Fig. XXII-2, which takes the EL forces exerted by the SO_4^- heads into account, in addition to the LW and AB forces. In so doing, early results were obtained by van Oss and Costanzo (1992).

---

* Or, in other words, equating the cmc with the solubility of a single, monomeric surfactant molecule, the concentration of which remains constant at the cmc and beyond; see, e.g., Rosen (1982) and Adamson (1982, p. 448).
† SDS tail part: $\gamma_1^{LW} = 23.8$ mJ/m²; $\gamma_1^\oplus = \gamma_1^\ominus = 0$; SDS head-part: $\gamma_2^{LW} = 34.6$ mJ/m²; $\gamma_2^\oplus = 0$; $\gamma_2^\ominus = 46$ mJ/m² (van Oss and Costanzo, 1992).

From the surface tension parameters found for SDS,[†] and from the known surface tension parameters for water, cellulose, nylon, and hexadecane (Chapter XVII), the free energies of interaction between the head and tail-parts of SDS on the one hand, and polymeric surfaces of cellulose and nylon molecules, as well as "dirt" molecules (hexadecane was taken as a model for "greasy dirt"), the effectiveness of SDS in removing such dirt from cellulose and nylon could be quantified. The following conclusions could be drawn (for 20°C) (van Oss and Costanzo, 1992):

a. SDS-tails, in water, adhere to apolar molecules ("dirt") as well as to cellulose and nylon to a significant extent (work of adhesion of the order of –10 kT).
b. SDS-heads, in water adhere to none of these materials to a significant degree [work of adhesion not more than about –2 kT (to hexadecane) to close to zero kT (cellulose and nylon)].
c. Cellulose and nylon both adhere fairly strongly to "dirt" (hexadecane) molecules (work of adhesion –7 to –8 kT, respectively) but they adhere significantly more strongly to SDS-tails (see a., above).

SOLUBILITY OF ORGANIC LIQUIDS IN WATER

The solubility and/or miscibility of low molecular weight organic solutes in polar or apolar liquids obeys the general rules indicated above, even though their molecular size, and thus their contactable surface area, S_c, is significantly smaller than that of polymers. It is especially fruitful to consider the solubility or miscibility of organic liquids in water, because in many instances that solubility has been experimentally determined before, and these data are generally available. In addition, the interfacial tension data between various organic liquids and water also are available for a large number of cases (see, e.g., Girifalco and Good, 1957).

However, for even slightly polar organic liquids (e.g., benzene) the results of the experimentally obtained interfacial tensions with water (as also published by Girifalco and Good, 1957) *are not valid*; see Chapter XIII. On the other hand, Table XIII-2 furnishes a number of correct, new values of interfacial tensions between water and polar organic liquids, derived from their aqueous solubilities, using eq. XIII-3; see also eq. XXII-4.

It is possible, for apolar or polar organic solutes (1) which are not infinitely soluble in water, to predict their solubility directly from their interfacial tension with water (cf. eq. [XXII-4]); see also, eq. [XXII-11], where ΔG_{iwi}^* is expressed in mJ/m2 and s in mol fractions.

It should be noted that the best (and possibly the only) way to determine the interfacial tension with water of amphiphilic systems on a molecular scale is to dissociate the hydrophobic from the hydrophilic moiety, and to determine the

---

[*] It is crucial that, for polar compounds, ΔG_{iwi} be based on the Lewis acid-base approach. Tables XIV-2 and XIV-3 illustrate the errors that arise when instead of the AB approach, direct geometric or harmonic mean combining rules are applied.

interfacial properties of each separately, as was done for the determination of the cmc of non-ionic and anionic surfactants (see the preceding Sections), and only to add them together again after they have been expressed in kT units.

SOLUBILITY OF ELECTROLYTES

From the solubilities (Weast, 1970) of a number of the more common electrolytes [e.g., KCl, NaCl, $(NH_4)SO_4$] it can be seen that in many of these cases, ΔG_{1W2}^{TOT} has a value of the order of -1.0 kT to -3.0 kT, as they are slightly more soluble than the common sugars, glucose and sucrose (cf. Table XXII-1).

Aqueous solubility of electrolytes is less favored in the case of the heavier ions, due to a stronger LW attraction, and especially in the case of plurivalent ions, in particular when both anions and cations are plurivalent. This is due to an increase in EL attraction, concomitantly leading to an AB attraction. However, little work has been done to date on the incorporation of AB forces into anion-cation interactions in aqueous media. This is due partly to the difficulty of operating in the gray zone between macroscopic and micro-scopic interactions, and partly to the difficulty of measuring the precise electron-donor and/or electron-acceptor surface properties of anions and cations.

The effects of temperature on solubility have been treated in Chapter VIII.

CODA

The solubility, s (in mol fractions), of macro- as well as of microsolutes (i), in liquid, L, of the interfacial tension, γ_{iL}, is described by:

$$-2\,S_c\,\gamma_{iL} = -kT\,\ln s \qquad\qquad [XXII-12]$$

where S_c is the contactable surface area between two molecules i. This holds true for apolar as well as for polar and aqueous systems (including electrolytes). For asymmetrical (e.g., amphipathis) molecules, S_c and γ_{iL} must be separately determined for the apolar and the polar moieties and then added after having been expressed in kT units. In this manner the critical micelle concentration (CMC) of non-ionic as well as of anionic surfactant can be calculated (s = CMC).

The difference between the solubility (of a solute in a solvent) and miscibility of two different liquids is discussed. Solubility is clearly a microscopic-scale phenomenon, obeying eqs. [XXII-4] and [XXII-12], whilst miscibility is best understood when classified as a macroscopic-scale phenomenon. Miscibility vs. immiscibility bears a close resemblance with a liquid completely wetting a solid surface (when $\gamma_L < \gamma_S$), vs. a liquid drop forming an obvious contact angle on a solid surface (when $\gamma_L > \gamma_S$).

The influence of temperature (T) on aqueous solubility is almost completely due to the change in the $\gamma^{\oplus}/\gamma^{\ominus}$ ratio for water as a function of T and is therefore treated in Chapter VIII.

XXIII | Cell and Particle Stability

ENERGY VS. DISTANCE PLOTS IN AQUEOUS MEDIA

Particle stability, like solute solubility, is favored when, at contact, $\Delta G_{iwi}^{TOT} > -1$ kT, and especially when $\Delta G_{iwi}^{TOT} > 0$ (see Chapter XXII). However, in the case of particles with radii R $\gtrsim$ 10 to 100 nm, a certain measure of stability may occur, even if, at contact (i.e., at $\ell = \ell_o$), $\Delta G_{iwi}^{TOT} < -1$ kT. This is due to the fact that at certain interparticle distances, $\ell > \ell_o$, it may well occur that $\Delta G_{iwi}^{TOT} (\ell) > 0$. To ascertain the values which ΔG_{iwi}^{TOT} can assume, as a function of the interparticle distance, ℓ, in a given particle suspension, ΔG_{iwi} must be separately plotted vs. ℓ for all three different functions (LW, AB, EL), and then combined into a ΔG_{iwi}^{TOT} vs. ℓ plot.

This approach (comprising, however, only the LW and EL forces) was first proposed by Hamaker (1936, 1937) and subsequently further independently developed by Derjaguin and Landau (1941) and Verwey and Overbeek (1948), and became known, after the last four authors, as the DLVO theory, and the corresponding energy vs. distance plots as DLVO plots; see Chapter VII. In the absence of polar (AB) interactions (i.e., in mainly apolar media), the DLVO theory correlates admirably with the stability of particle suspensions. However, in the cases of particle suspensions in polar and especially in aqueous media, disregarding the influence of polar interactions by using simple DLVO plots leads to severely unrealistic models (van Oss, Giese and Costanzo, 1990); see also Chapter VII and compare Fig. VII-1 (A and B) with (C and D); see also Tables VII-2 and XXIII-1. In other words, in all polar systems one should take into account ΔG_{131}, as a function of ℓ, as:

$$\Delta G_{131}^{TOT}(\ell) = \Delta G_{131}^{LW}(\ell) + \Delta G_{131}^{EL}(\ell) + \Delta G_{131}^{AB}(\ell) \qquad \text{[XXIII-1]}^*$$

In water, polar (AB) interactions, whether attractive or repulsive, are *never* negligible. Either way, AB interactions in aqueous media usually are, at close range, one or several decimal orders of magnitude greater than LW or EL interactions. Some fairly typical examples of this are shown in Figs. XXIII-1 and XXIII-2. Here,

---

* A list of symbols can be found on pages 399–406.

319

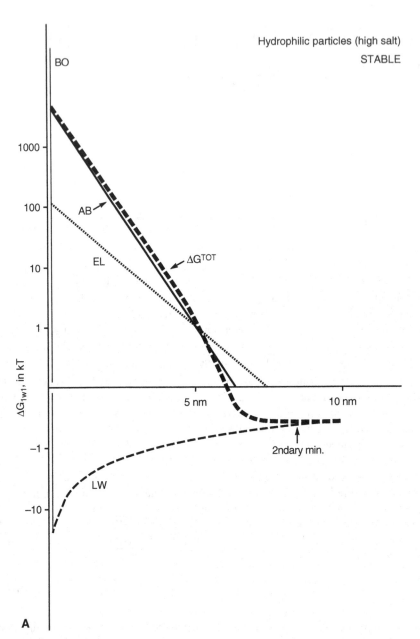

FIGURE XXIII-1 Extended DLVO energy balance of typical spherical hydrophilic particles at neutral pH: $R = 0.4\,\mu m$, $\psi_o = -20\,mV$, $A = 47 \times 10^{-23}\,J$ eq. $[\Delta G_{1w1}^{LW''}\,(\ell_o) = -0.5\,mJ/m^2]$, $\Delta G_{1w1}^{AB''}$ $(\ell_o) = +20\,mJ/m^2$, $\lambda = 0.6\,nm$. These particles are analogous to latex particles coated with, e.g., hydrated human serum albumin. (A) In the presence of 0.1 M NaCl. (B) In the presence of 0.001 M NaCl. [Free energy (ΔG_{1w1}) in kT, *vs.* distance in nm. BO = Born repulsion; LW, EL and AB indicate the interaction plots of that designation; the thick interrupted line represents the total interaction.]

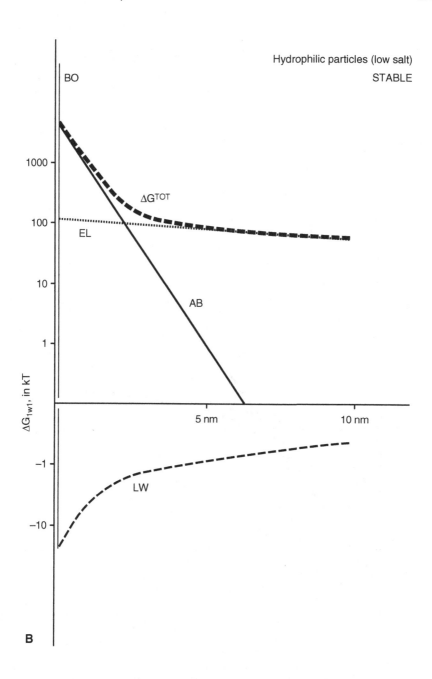

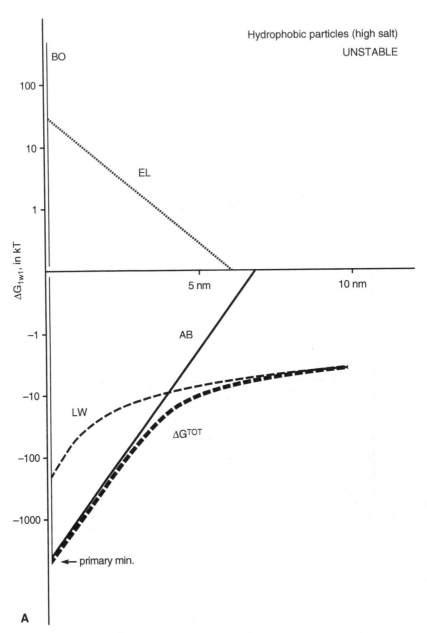

FIGURE XXIII-2 Extended DLVO energy balance of typical spherical hydrophobic particles: $R = 0.4 \, \mu m$, $\psi_o = -10 \, mV$, $A = 47 \times 10^{-22} \, J$ [$\Delta G_{1w1}^{LW''}(\ell_o) = -5.0 \, mJ/m^2$], $\Delta G_{1w1}^{AB''}(\ell_o) = -30 \, mJ/m^2$, $\lambda = 0.6 \, nm$. These particles are analogous to latex particles coated with, e.g., fibrin. (A) In the presence of 0.1 M NaCl. (B) In the presence of 0.001 M NaCl. [Free energy (ΔG_{1w1}) in kT, vs. distance in nm. BO = Born repulsion; LW, EL and AB indicate the interaction plots of that designation; the thick interrupted line represents the total interaction.]

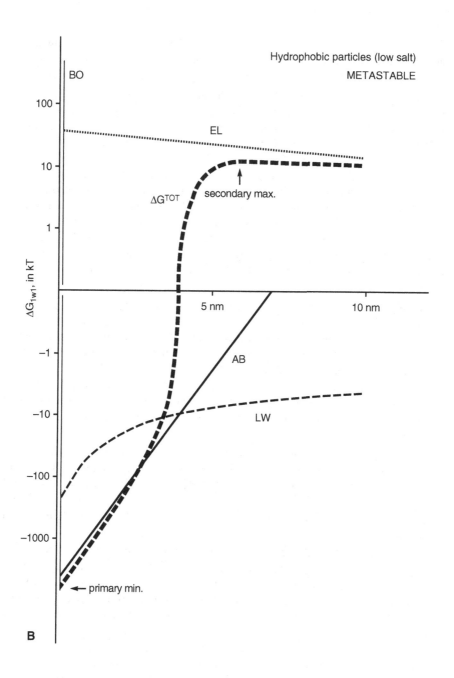

for the sake of uniformity, rather commonly used latex particles with a radius, R, of approximately 0.4 μm are chosen as an example.

In the hydrophilic case (Figs. XXIII-1A and B) they are taken to be coated with a hydrophilic material such as hydrated human serum albumin (see Table XVII-6). Human serum albumin is among the most highly (negatively) charged plasma proteins; its ψ_o-potential of –20 mV is therefore taken as a typical high value for a proteinaceous material. The Hamaker constant of hydrated albumin with respect to water (A_{iwi}) is quite low; it is only of the order of 5×10^{-22} J, due to the close resemblance between the LW properties of the outer layer of *hydrated* protein and those of the aqueous medium. The polar repulsion between hydrated albumin molecules is rather strong (see Chapter XIX); ΔG_{iwi}^{AB} is of the order of +20 mJ/m² at close range. The decay length of water is here taken to be $\lambda \approx 0.6$ nm (see below). The energy balances are shown at high salt concentrations (0.1 M NaCl), which is close to the salt concentration of physiological saline solutions (Fig. XXIII-1A), as well as at low ionic strength (0.001 M NaCl) (Fig. XXIII-1B). In both cases the suspensions are stable.

At high ionic strength there is a slight secondary minimum of attraction at about 8.5 nm, but that attraction amounts to less than 1 kT. However, for larger particles or cells of comparable properties, the attraction at the secondary minimum could easily amount to, e.g., 10 kT, as the free energy of attraction (or repulsion) is proportional to the radius, R (see, for instance, van Oss, Cell Biophys., 1989). Thus for large cells, such as blood cells, attraction at the secondary minimum may be of some importance (see below). For single protein molecules however, with a radius R that is, typically, 100 × smaller than 0.4 μm, attraction at the secondary minimum is entirely negligible.

The "high" ionic strength used here (0.1 M NaCl) should not be confused with the *very high* ionic strengths used, e.g., for "salting-out" (see Chapter XXII). At *very high* ionic strengths [e.g., 2 to 4 M $(NH_4)_2SO_4$], protein insolubilization sets in, and a ΔG_{iwi}^{AB} of the order of +20 mJ/m² (at close range) can then change to a sufficiently negative value to give rise to agglomeration, or precipitation, even of albumin (van Oss, Moore *et al.*, 1985).

For "hydrophobic" particles latex particles of the same size are taken as examples, this time coated with a moderately "hydrophobic" protein such as fibrin (see Table XVII-6), which is insoluble in water (van Oss, J. Protein Chem., 1990). Here a typically lower ψ_o-potential of ≈ -10 mV is assumed, and a Hamaker constant (with respect to water) which is ten times higher than that of a hydrated protein, as is normal for these insoluble, unhydrated, fairly hydrophobic proteinaceous materials. Thus, $A_{iwi} \approx 47 \times 10^{-22}$ J. Typically, ΔG_{iwi}^{AB} here is at least –30 mJ/m², at close range. These particles are also studied at high and at low ionic strengths; see Figs. XXIII-2A and B. At high ionic strength, where ΔG_{iwi}^{EL} decays quickly with distance, an attraction prevails at all distances, culminating in a deep primary minimum just before the Born repulsion is reached. This gives rise to total instability, i.e., to complete and unabated "hydrophobic" particle agglomeration (Fig. XXIII-2A). At low ionic strength, however, a certain degree of stability can occur with the same hydrophobic particles. This is due to the much more gradual decay of the EL repulsion with distance, at low ionic strengths (Fig. XXIII-2B). For regularly shaped spherical particles, a secondary maximum of repulsion of ≈ 10 kT should suffice to maintain stability for a fair period of time. Ultimately, however, gradual destabilization must prevail, due to the strong primary minimum of attraction, which is of the order of almost -6×10^3 kT. The DG *vs.* distance diagrams, pertaining to interactions in water, shown in Figs. XXIII-1 and 2, are designated as "extended" DLVO (XDLVO) diagrams, as they include AB interactions, in addition to LW and EL interactions.

Energy balance plots of erythrocytes are given in Chapter VI.

PARTICLE STABILITY VS. FLOCCULATION— THE SCHULZE-HARDY MECHANISM OF FLOCCULATION WITH PLURIVALENT COUNTERIONS

It was established well over a century ago that stable aqueous suspensions of electrically charged particles become destabilized (i.e., flocculated) by the addition of neutral salts with divalent, trivalent or tetravalent counterions. The concentration of added counterions needed to achieve flocculation, decreases by approximately one decimal order of magnitude with each increase in counterion valency (Schulze, 1882; 1883; Hardy, 1900); see also Overbeek (1952). It is well-established that the addition of such salts to electrically charged particles, suspended in water, causes a decrease in the ζ–potential of these particles, which gives rise to a decrease in the positive value of ΔG_{iwi}^{EL}, thus decreasing the EL-driven repulsion between particles. It is widely believed that this decrease in EL-driven repulsion then causes the always-present Lifshitz–van der Waals attraction between the particles to prevail, thus giving rise to flocculation.

This explanation is, however, quantitatively incorrect and has only persisted past the late 1980s due to a tenacious trend to disregard the polar properties of liquid water with respect to the interfacial interactions of solutes, particles or surfaces that had been immersed in it. However, with the growing experimental use of Israelachvili's force balance (see Chapter XVI) in the 1980s, in which device water tends to be the principal liquid medium, the existence of non-classical DLVO forces originating in the aqueous medium became increasingly evident (see, e.g., Claesson, 1986; Israelachvili and McGuiggan, 1988).

In experimental studies on the flocculation of the smectite clay, *hectorite*, it was found that the flocculation behavior of this clay particle in water, as a function of ionic strength, also could not be explained by the simple classical DLVO theory (using only LW *vs.* EL forces). On the other hand, the flocculation behavior of hectorite in water precisely correlated with an extended DLVO (XDLVO) theory, in which LW, EL, *as well as AB forces* were taken into account (van Oss, Giese and Costanzo, 1990; see also Wu *et al.*, 1999 and Chapter VII).

In 1994 contact angle measurements were done to obtain the surface thermodynamic properties of negatively charged ground glass, clay particles (SWy-1) and calcite particles, yielding their ΔG_{iwi}^{LW} and ΔG_{iwi}^{AB} properties, while micro-electrophoretic measurements yielded ΔG_{iwi}^{EL}, all as a function of added $CaCl_2$ or $LaCl_3$, or of the absence of these salts. At the same time flocculation of these particles or the lack thereof, was observed, all at close-to-neutral pH. Table XXIII-1 illustrates how the flocculating influence of Ca^{2+} and La^{3+} ions on negatively charged glass particles is a consequence of the large decrease in ΔG^{AB}_{iwi} (from +14,000 kT to –6,700 kT under the influence of La^{3+} ions and from +14,000 kT to –4,700 kT under the influence of Ca^{2+} ions). In other words, the glass particles changed from hydrophilic (ΔG^{TOT}_{iwi} = +17,000 kT) to hydrophobic (ΔG^{TOT}_{iwi} = –6,100 kT with La^{3+} ions and ΔG^{TOT}_{iwi} = –4,300 kT with Ca^{2+} ions) under the influence of these plurivalent counterions. In both cases however, classical DLVO analyses ($\Delta G^{LW}_{iwi} + \Delta G^{EL}_{iwi}$) still showed a net repulsion of +600 kT in the case of La^{3+} addition, an of +395 kT in the case of Ca^{2+} addition. Thus, the amplifying linkage between ΔG^{EL} and ΔG^{AB} (in favor of ΔG^{AB}) which causes a switch from (AB) hydrophilicity to (AB) hydrophobicity, triggered by a decrease in the (in this case negative) ζ–potential, is the main mechanism of the Schulze-Hardy type flocculation with plurivalent counterions (Wu *et al.*, 1994a). The increase in hydrophobicity accompanying a decrease in the particles' ζ-potential is mainly caused by a concomitant decrease in the the value of the particles' $\gamma^{\ominus}$, see Table XXIII-1; see also Schematic Chart XIX-1. The flocculation of, e.g., smectite (SWy-1) particles through the admixture of plurivalent counterions could be reversed by the addition of

---

TABLE XXIII-1
Surface Properties of Ground Glass Particles,[a] Untreated and Treated with La^{3+} or Ca^{2+} Ions, and Their Stability or Flocculation in Water, at Close to Neutral pH.

	Glass	Glass[b]	Glass[c]
$\Gamma/2 \times 10^{-3}$	15	17	21
pH	7.5	7.6	7.4
Mobility ($\mu m\ s^{-1}\ V^{-1}\ cm$)	–3.73	–1.16	–0.99
ζ (mV)	–52.7	–16.4	–14.0
Stability	Stable	Flocculated	Flocculated
Pore Size $R_{eff} \times 10^{-5}$ cm	2.83	7.44	4.11
γ^{LW}_i (mJm^{-2})	31.5	30.3	29.6
$\gamma^{\oplus}_i$ (mJm^{-2})	0.4	0.2	0.3
$\gamma^{\ominus}_i$ (mJm^{-2})	37.1	20.9	22.2
ΔG^{LW}_{iwi} (ℓ_o)(kT)	–220	–170	–145
ΔG^{AB}_{iwi} (ℓ_o)(kT)	+14000	–6700	–4700
ΔG^{EL}_{iwi} (ℓ_o)(kT)	+3000	+770	+540
ΔG^{TOT}_{iwi} (ℓ_o)(kT)	+17000	–6100	–4300

$\ell_o \approx 0.157$ nm.

[a] Suspensions (0.5% (w/v)) in 20 ml of PBS/10($\Gamma/2 = 0.015$).

[b] In the presence of 0.47 mM $LaCl_3$.

[c] In the presence of 24 mM $CaCl_2$.

Source. From Wu *et al.* (1994a), with permission.

complexing agents (Na$_3$EDTA, or Na-hexametaphosphate). Upon repeptization the particles also became once more hydrophilic. (Wu *et al.*, 1994b).

POLYMER MOLECULES, VERY SMALL PARTICLES AND PROTUBERANCES WITH SMALL RADII OF CURVATURE

As long as the energy balance plot of hydrophilic particles or macromolecules has no net primary minimum, i.e. when there is a continuously increasing repulsion with decreasing distance, such particles or macromolecules will be stable (or soluble), however large or small they may be; cf. Figs. XXIII-1 A and B.

When an energy balance plot of hydrophobic particles or macromolecules shows an attraction at all distances, with a sizeable primary minimum ($\Delta G_{iwi}^{TOT} \ll -1$ kT), instability (or insolubility) will prevail for all sizes of particles or macromolecules; cf. Fig. XXIII-2A. However, for such particles or macromolecules which are so small that at the primary minimum (i.e., at $\ell = \ell_o$), $\Delta G_{iwi}^{TOT} \geq -1$ kT, some degree of stability still can exist.

Finally, whilst relatively large hydrophobic particles (e.g., R = 0.4 µm, see Fig. XXIII-2B) can be in metastable suspension due to the existence of a sizeable secondary maximum ($\Delta G_{iwi}^{TOT} \geq + 10$ kT), smaller particles or macromolecules of the same material tend to be less stable, due to the fact that with an x times smaller particle size, the value of ΔG_{iwi}^{TOT} at the secondary maximum also becomes x times smaller (see Chapter VII). In such cases the (positive) value of ΔG_{iwi}^{TOT} may decrease to well below 1 kT, causing the attraction at the (much deeper) primary minimum to prevail, which then gives rise to instability. It is for instance known that suspensions of very small hydrophobic polystyrene microspheres (e.g., particles with a diameter smaller than 100 nm) require a higher ζ-potential than larger particles, to achieve stability (Interfacial Dynamics Co., 1991).

Although this phenomenon has not, to date, been frequently reported, a much more common occurrence is the existence of tiny surface protuberances, processes, or blips on the surface of larger particles or cells. Owing to the small radius of curvature (r) of such protuberances, the entire particle (with total radius, R) will then be able to approach other such particles or cells (or other surfaces) much more closely, as the energy barrier at the secondary maximum is then R/r times smaller; see, e.g., van Oss *et al.* (1972, 1975, 1978), van Oss and Mohn (1970), van Oss (Cell Biophys., 1989; Biophysics of the Cell Surface, 1990; J. Molec. Recognition, 2003); see also Fig. XXIII-3.

However, when the energy balance of particles of a moderately hydrophobic material is such that the value of $|\Delta G_{iwi}^{TOT}|$ at the secondary maximum of repulsion is *greater* than the value of $|\Delta G_{iwi}^{TOT}|$ at the primary minimum of attraction (contrary to the situation depicted in Fig. XXIII-2B), then smaller particles will form a more stable suspension than larger particles; This phenomenon has been observed in practice, see, e.g., Choi *et al.* (1985), who described the stability of very small latex particles obtained through emulsion polymerization.

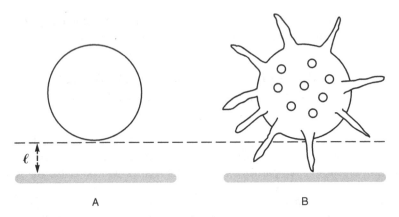

A   B

FIGURE XXIII-3 Schematic presentation of differences in accessibility of round spherical bodies and a flat plate, for a smooth sphere with a relatively large radius (left) and a similar sphere with essentially the same main radius but which is, in addition, endowed with protruding long, thin processes, each with a small radius of curvature (right). (A) The smooth hydrophilic spherical cell or particle cannot make contact with a smooth flat hydrophilic surface because their mutual aspecific, macroscopic-scale repulsion prevents a closer approach than approximately ℓ = 4 or 5 nm. (B) A similar spherical cell or particle endowed with long thin spikey processes whose ends have a small radius of curvature which can readily penetrate the macroscopic-scale repulsion field and thus achieve microscopic-scale specific contact. In both (A) and (B) the dotted line indicates the limit of closest approach (ℓ) for a smooth hydrophilic cell or particle with a relatively large radius of curvature. From van Oss, J. Molec. Recognition (2003), *16*:117, with permission from Wiley-Interscience, New York.

THE GLYCOCALYX OF MAMMALIAN CELLS

Before discussing the interactions of cells suspended in aqueous media, it should be realized that these interactions are virtually entirely governed by the properties of the glycoprotein strands which make up the glycocalyx surrounding the cells, for the very simple reason that this glycocalyx constitutes the outer layer of the cells. Many cells, and especially most peripheral blood cells, have a substantial glycocalyx. On erythrocytes the glycocalyx consists of fairly evenly spaced glycoprotein strands, in which the carbohydrate moieties are mainly distal. The most important human blood group antigenic determinants, or epitopes (A, B, H, Le, etc.), are well-characterized oligosaccharides (Cunningham, 1994); these are interspersed with sialic acid strands, which are the main bearers of the negative charge of blood cells, and especially of red blood cells.

Relatively little information exists about the length of the glycocalyx strands of blood cells, in aqueous suspension. Winzler (1972) gives data on the isolated glycocalyx glycoprotein: its sedimentation constant, s, is 1.5S and its diffusion constant D = 10.65×10^{-7} cm^2/sec. From this a molecular weight of about 9,500 can be calculated, using Svedberg's equation (eq. [VI-15]). From these values for s and D one can also obtain the friction factor ratio f/f_o = 1.507, which yields an oblong asymmetry ratio ($1/y$ = 9.3) (see Table VI-1). From the specific volume $\overline{v}$ = 0.632

(Winzler, 1972) of the material, and the molecular weight, one arrives at a cylinder with a diameter of 1.1 nm and a length, h, of 10.3 nm. In confirmation of that value for h, recent work on the influence of IgG-class antibodies on the electrophoretic mobility of various mammalian blood cells (Cohly et al., 1985a, 1985b; Cohly, 1986) shows that h is somewhat less than 13 nm. This follows from the fact that IgG-class antibodies, which are about 13 nm long (Valentine and Green, 1967), have only a slight influence on the cells' electrophoretic mobility when attached to the cell *membrane* of blood cells, i.e., when largely hidden inside the glycocalyx (Cohly, 1986). However, when a second IgG-class antibody is attached to the Fc moiety of the original antibody a reduction in cell electrophoretic mobility of about 18% results (Cohly et al., 1985a, 1985b; Cohly, 1986). From the slight (but not entirely zero) influence on the cells' electrophoretic mobility of IgG-class antibodies attached to cell membrane epitopes, it may be concluded that the length of the glycocalix strands is close to the length of unopened IgG-class antibodies, i.e., h is slightly larger than 10 nm. We may therefore adopt, as a first approximation, the value for h obtained from hydrodynamic measurements, i.e., h = 10.3 nm; see also Fig. XV-2.

BLOOD CELL STABILITY

The necessity of including polar (AB) forces in energy balance diagrams of red cell suspensions has already been pointed out in Chapter VI. In the case of erythrocytes, a certain modicum of stability still exists through their EL repulsion alone, even in the absence of polar (AB) interaction. However, especially if the various types of leukocytes of peripheral blood are also taken into consideration, it becomes clear that the inclusion of polar (AB) interactions becomes crucial (van Oss, Cell Biophys., 1989); see Table XXIII-2.

Smoothness of the cell surface also plays a role. Erythrocytes with echinocytic protuberances with a diameter of the order of a few 100 nm are able to approach each other more closely than smooth erythrocytes. Leukocytes (both lymphocytes

TABLE XXIII-2
The Components of Blood Cell Stability of Erythrocytes (ER), Lymphocytes (LY) and Polymorphonuclear Cells (PMN); ΔG Values in mJ/m²

Cell type	ψ_o-potential[a] (mV)	ΔG_{iwi}^{LW}	ΔG_{iwi}^{EL} [b]	ΔG_{iwi}^{AB}	ΔG_{iwi}^{TOT}	ΔG_{iwi}^{DLVO}
ER	−33.6	−0.6	+2.0	+25[c]	+26.4	+1.4
LY	−26.7	−0.7	+1.2	+15[d]	+15.5	+0.5
PMN	−22.4	−0.7	+0.9	+10[d]	+10.2	+0.2

[a] Cf. van Oss and Absolom, 1983.

[b] These values were somewhat underestimated in van Oss (Cell Biophys., 1989); the values given here are the more accurate ones.

[c] From van Oss and Cunningham, unpublished results.

[d] Estimated, by comparison with ER, from earlier contact angle data on LY and PMN (van Oss et al., 1975).

and granulocytes) with pseudopodia somewhat thicker than that (e.g., between 200 and 1000 nm) are also able to approach each other, or other cells, rather closely *via* the apices of their pseudopodia. However, due to the absence of a primary minimum of attraction and the presence of a very steep repulsion at close range, engendered by the AB repulsion (van Oss, Cell Biophys., 1989), it should be impossible for any of the blood cells to adhere to each other through the exertion of outside physical forces (except *via* discreet low energy patches, or by means of specific receptor and ligand sites). By ultracentrifugation at $296,000 \times g$, we have shown earlier that it is indeed impossible to cause erythrocytes to adhere to each other by the application of purely physical forces (van Oss, 1985).

ROULEAU FORMATION OF RED CELLS, OR PSEUDO-ATTACHMENT AT THE SECONDARY MINIMUM

When the energy balance shows a secondary minimum of attraction of at least 1 kT, regardless of whether a primary minimum of attraction exists (as in the hypothetical case depicted in van Oss, Cell Biophys., 1989) or not (Figs. VII-1B and VII-1D) then, on an average, the cells will not approach each other more closely than the distance at the secondary minimum. Hydrodynamic instabilities normally preclude real "adhesion" at the secondary minimum. However, with the help of crossbinding with asymmetrical polymer molecules, some modicum of "attachment" of (red) cells at the secondary minimum can be obtained. This type of "attachment" manifests itself through the formation of cylindrical stacks of red cells, or "rouleaux"* helped by the presence of crossbinding large, asymmetrically shaped, biopolymer molecules such as fibrinogen (Brooks *et al.*, 1980), polymerized albumin, macroglobulins, or dextran (see, e.g., Brooks, 1976; van Oss and Absolom, 1985). On the other hand, in saline water rouleau formation of red cells does not normally occur (Bessis, 1973). In whole fresh plasma human erythrocytes readily form rouleaux, while in fresh serum rouleau formation occurs to a much lesser extent. The prime large asymmetrical plasma protein to be implicated as the polymer bridging "glue" in rouleau formation thus would seem to be fibrinogen (Brooks *et al.*, 1980), while asymmetrical serum proteins (e.g., macroglobulins) also play a certain, although more minor role in polymer bridging (van Oss and Absolom, 1985).

The reason why the discoid-shaped red cells tend to form stacks with the flat sides facing each other, rather than forming clumps with bridging occurring between their more rounded edges, lies in the fact that, at the secondary minimum, the energy of attraction between two red cells opposing each other in the flat-flat conformation is several hundred times higher than the attractive energy between two round cell

---

* The French word "rouleau," in "rouleau formation" is generally used in English. However, the term was borrowed from the French in ignorance of the fact that a "stack" of coins, or cells, is not called "un rouleau" in French, but "une pile." In French "rouleau formation" is called "formation de piles de monnaie." Nowadays, however, most French hematologists and bloodbankers are cognizant with the English terminology, and may use both terms interchangeably.

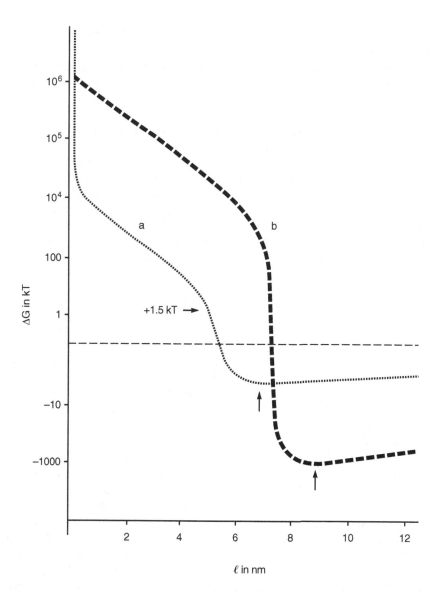

FIGURE XXIII-4 Energy balance of human erythrocytes, taking ΔG^{EL}, ΔG^{LW} *and* ΔG^{AB} into account. The decay length of water is taken to be $\lambda = 0.6$ nm (see text in the following section). The left-hand curve (a) is computed by assuming a radius of curvature at the approach of two cells of $R \approx 1.5$ μm. The right-hand curve (b) is based on the interaction between two cells in the flat parallel slab mode, with a surface area of approach $S_c = 25.9$ μm². The vertical arrows indicate the secondary minima of attraction of the two modes a and b. A horizontal arrow indicates the place on curve (a) of $\Delta G = +1.5$ kT, corresponding to $\ell = 5.1$ nm, which is the most likely minimum distance between glycocalix surfaces of two red cells approaching each other via their convex edges, under the influence of their Brownian motion. The (probably relatively slight) elastic repulsion engendered by the cell encounters at $\Delta G = +1.5$ kT is not taken into account here. (From van Oss, *Biophysics of the Cell Surface*, 1990.)

edges [van Oss and Absolom, 1983; van Oss, Biophysics of the Cell Surface, 1990)];
see Fig. XXIII-4.

The rouleau conformation is not a very stable one. Red cell rouleaux stabilized
in the flat-flat conformation through the crossbridging by, e.g., long fibrinogen
molecules, are relatively unstable: vigorous shaking of a test tube containing a
suspension of red cell rouleaux causes the rouleaux to disappear. It has been known
for some time that when mixing red blood cells of animals of different species, cell
sorting occurs, i.e., the various rouleaux that form consist of either the red cells of one
species, or of the red cells of another species, originally present in the cell mixture,
but no rouleau will form which consists of a mixture of cells of two or more species
(Sewchand and Canham, 1976). It has been shown that the cell-sorting observed in
the rouleaux formed from mixtures of, e.g., human and rabbit red cells is attributable
to the fact that the energies of attraction in the flat-flat conformation at the secondary
minimum between human red cells are quantitatively significantly different from the
energies of attraction between rabbit red cells (van Oss and Absolom, 1985).

HEMAGGLUTINATION

Hemagglutination is the clumping or flocculation of previously stable suspensions of
(usually human) red cells, which are crosslinked by antibodies that are specifically
directed to antigenic (usually blood group–specific) sites situated on the outer periphery
of the cells. Hemagglutination, when it occurs, generally is clearly visible with the naked
eye, by an increase in granulosity of the previously smooth-looking suspension, or by a
marked increase in cell settling. In cases where hemagglutination does occur, but is not
immediately visible with the naked eye, the presence (or absence) of crosslinked cells
is readily verifiable by microscopic inspection, at a magnification of $100 \times$.

Since the early days of blood transfusion, and up to the present, hemagglutination
has been and remains the principal analytical tool in immunohematology and
bloodbanking. With IgM-class antibodies, due to their size as well as to the availability
of 10 antibody sites (Edberg *et al.*, 1972) disposed at diametrical distances of about 30
nm, hemagglutination is much more readily achieved than with antibodies of the IgG
class, which have just two antibody sites, which are at most only about 12–14 nm apart
(van Oss, 1984; *Structure of Antigens*, 1992, p. 179). However, as IgG-class antibodies,
much effort has been devoted to modifications of the environment and properties of
erythrocytes, to facilitate hemagglutination with IgG. With some IgG-class blood
group antibodies (e.g., anti-A and anti-B), however, hemagglutination nonetheless is
readily achieved; the reasons for this are discussed below.

Red cells (as well as leukocytes) are not smooth little beady entities, surrounded
by just a bileaflet phospholipid membrane, but they are endowed with a short (≈ 10
nm) hairy coat of glycoprotein strands, together forming the cells' glycocalyx. It can
easily be shown (van Oss, Cell Biophys., 1989; *Biophysics of the Cell Surface*, 1990;
see also Figs. VII-1D and XXIII-4), by composing an energy balance (taking into
account the usual physiological conditions of pH and ionic strength) that the outer
edges of the glycocalyx of erythrocytes cannot approach each other more closely
than an intercellular distance, ℓ, of about 5 nm. However, this makes the minimum

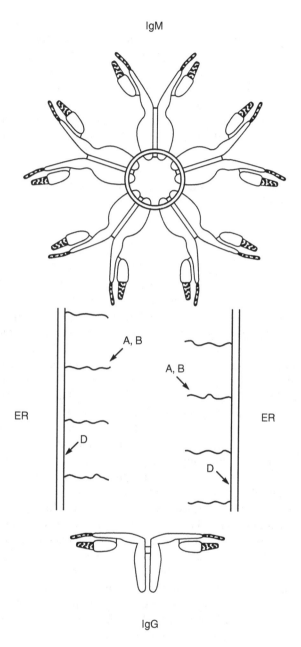

FIGURE XXIII-5 Schematic presentation of the opposing surfaces of two erythrocytes (ER) at their minimum equilibrium distance (about 5 nm between the tips of the glycocalyx strands). Drawn approximately to scale are models of IgM and IgG molecules. IgM-class antibodies clearly can cross-link all erythrocytes. IgG, in its maximally open form can bridge the distance between blood group A or B sites, or epitopes, which are incorporated on the extremities of the glycocalyx strands. However, IgG is too small to bridge the distance between $D(Rh_o)$ epitopes, which are situated on the cell membrane itself (here drawn as two parallel lines).

distance (d) between the actual cell *membranes* of two opposing erythrocytes about 15 to 16 nm (see Fig. XXIII-5), which is slightly more than the "reach" of IgG-class Ab's (d $\lesssim$ 15 nm), although quite sufficient for cross-linking by IgM-class Ab's (d $\approx$ 27 to 28 nm) (van Oss, Biophysics of the Cell Surface, 1990). Thus, IgM-class Ab's can always cross-link erythrocytes, whether the epitopes are situated on the glycocalyx strands (the tips of which are only about 5 nm apart on opposing cells) or on the cell membranes (which are at least 15 nm apart on opposing cells). On the other hand, IgG-class Ab's can only cross-link erythrocytes if the epitopes are situated near the outer edges of the glycocalices (which is the case with ABO bloodgroup Ag's), but not if the epitopes are located on the cell membranes (which is the case with Rh bloodgroup Ag's). Therefore, to facilitate hemagglutination with anti-Rh Ab's (which tend to be of the IgG-class), two types of measures can be taken:

1. The cells may be brought closer together, or
2. The "reach" of IgG-class Ab's may be extended.

Measure #1: There are numerous ways by which red cells can be brought closer together (van Oss, Structure of Antigens, 1992, p.179): centrifugation at about 15,000 × g (van Oss, 1985); decrease of $|\Delta G^{EL}|$ *and* $|\Delta G^{AB}|$ and simultaneously, by partially destroying the glycocalyx glycoprotein strands enzymatically (using neuraminidase, bromelin or papain), decrease of $|\Delta G^{AB}|$ by the addition of plurivalent cations.* Another approach to decreasing just the intercellular distance, ℓ, is by pushing the cells closer together through exterior polar (AB) forces, caused by a high concentration of extracellular polar polymers. Serum albumin (in concentrations of 12 to 20%, w/v) is the most frequently used polymer for this purpose, but dextran, polyvinyl pyrrolidone and other water-soluble polymers are also effective (van Oss *et al.*, 1978). Yet another method for bringing the cells, or at least parts of the cells closer together, is by creating bumpy irregularities, or "spicules" on the cells' surfaces. Such irregularities or spicules are processes of a much smaller radius of curvature than the cells themselves, and thus undergo a much weaker repulsion than the smooth parts of the cell surface. Now, antibodies to blood groups A and B (of the IgM, as well as of the IgG-class) upon binding to red cells tend to spiculate them. Thus hemagglutination with anti-blood group A or B antibodies of both the IgM and the IgG classes is greatly facilitated by the closer approach of the sites near the tips of the spicules (van Oss and Mohn, 1970; see also Rebuck, 1953; Salsbury and Clarke, 1968). Relatively low molecular weight dextran (MW 40,000) also spiculates red cells (van Oss *et al.*, 1978). However, attachment of anti-Rh$_0$ (D) antibodies to red cells leaves them completely smooth and unspiculated (van Oss and Mohn, 1970; Rebuck, 1953; Salsbury and Clarke, 1968). This observation, plus the fact that the

---

* Lerche *et al.* (1979) studied the effects of LaCl$_3$ on hemagglutination. Van Oss (1994a) showed that hemagglutination with LaCl$_3$ is virtually solely due to the effect of La^{+++} ions on ΔG^{AB}_{1w1} which changes from strongly positive values (in the absence of La^{+++}) to strongly negative values (0.5 mm LaCl$_3$), thus changing the strongly stabilizing AB repulsion into a flocculating AB attraction. Similar effects were observed in the flocculation of montmorillonite particles (Wu *et al.*, 1994a).

Rh$_o$ (D) antigenic determinant is not situated at the outer edges of the glycocalyx strands (like the blood group A and B determinants), but at the cell membrane itself, makes it impossible for IgG-class anti-Rh$_o$ (D) antibodies to cross-link Rh$_o$ (D)-positive red cells (see Fig. XXIII-5).

Finally, another approach to bringing the cells more closely together is by cross-binding them with polymer molecules (see also Chapter XX). This can be done in three ways.

a. Cross-linking by high molecular weight, very asymmetrical polymers: Such polymers cross-bind cells at rather low concentrations, and the higher their molecular weight, the lower the concentration needed for cross-binding. For example, 1% dextran (Mw ≈ 100,000) gives rise to clumping, while only 0.4% dextran (Mw ≈ 270,000) is needed to obtain the same effect (Mollison, 1972). The clumping of red cells which is brought about by cross-binding with asymmetrical polymers usually occurs in the guise of rouleau formation; see the preceding section. The cross-binding induced by dextran can be inhibited by glucose (van Oss et al., 1978), which is indicative of the existence of a lectin-like peptide in the cells' glycocalyx, with glucose specificity (van Oss, 1992).

b. Cross-linking by adsorbed euglobulins: Cross-binding of red cells can also be effected by making use of the fact that, in vivo, red cells quite commonly aspecifically adsorb a certain amount of plasma euglobulin (see Chapter XXII), most of which consists of immunoglobulins of the IgM class. Thus, by lowering the ionic strength of the suspending medium (while maintaining isotonicity through the admixture of, e.g., glucose), the adsorbed euglobulins become insoluble (i.e., they precipitate), thus cross-binding the cells. Adding salt causes them to go back into suspension (van Oss and Buenting, 1967; van Oss et al., 1978).

c. Cross-binding by basic polymers: Finally, cross-binding of (negatively charged) red cells can be done by the admixture of positively charged polymers. Basic polymers of this type are polybrene [poly(hexadimethrine bromide)] (Lalezari, 1968), protamine and polylysine (Greenwalt and Steane, 1973; van Oss and Coakley, 1988). Cross-binding with basic polymers such as polybrene is most effective at low ionic strength, and can usually be reversed by the addition of salt. The clumping of red cells by means of positively charged polymers is not only due to the cross-binding activity of these polymers. It should be kept in mind that neutralization of negatively charged sites on the glycocalyx also makes the outer cell surface more "hydrophobic" (see Chapter XIX), which further enhances the mutual attraction between such cells.

Measure #2: An entirely different approach to facilitating hemagglutination by means of IgG-class antibodies is by extending the "reach" of the IgG molecules. One way of doing this is by opening up a few of the interchain S-S bonds by mild reduction and alkylation (Romans et al., 1977). Another and much more widely used approach is by cross-linking the (human) IgG-class antibodies with (rabbit) anti-human IgG antibodies (Coombs *et al.*, 1945), which approximately triples their "reach."

In practice, the presence of human anti-Rh_o (D) antibodies (and other IgG-class antibodies) is determined by using each of three different approaches in concert: (1) enzymatic treatment of the red cells, which shortens the glycocalyx strands, decreases their ζ-potential and makes them more hydrophobic; (2) adding (usually bovine) serum albumin in high concentrations, which pushes the cells closer together, mainly by polar (AB) forces; and (3) using an antiglobulin, or "Coombs" antibody, which extends the "reach" of the IgG-class antibodies.

THE IgG MOLECULE AS A PROBE FOR DETERMINING THE INTERCELLULAR DISTANCE, ℓ, AT THE SECONDARY MINIMUM; ESTIMATION OF THE DECAY LENGTH OF WATER

The rates of decay with distance, ℓ, of ΔG^{EL} and ΔG^{LW} are well known (see Chapter VII). The equation for the rate of decay with distance of ΔG^{AB} is also known (eq. [VII-4]), but the parameter λ, the decay length of water at 20°C, which is of the order of 1.0 nm, has not yet been determined more precisely. Now, the distance to which two erythrocytes can approach one another is, given the importance of ΔG^{AB} in that interaction, strongly linked to the value of λ. Thus, if the distance, ℓ, can be established with some precision under a given set of conditions, it may be possible also to obtain a better estimate of λ. The distance, ℓ, at the secondary minimum of attraction between two human red blood cells can be estimated, by means of the yardstick provided by IgG molecules. Antibodies of the IgG-class, directed to the Rh_o (D) epitope (antigenic determinant) of Rh_o (D)-positive human erythrocytes (ER), have a distance between the extremities of their two paratopes (antibody-active sites) that is just too small to achieve cross-linking two Rh_o (D)-positive ER. Given that the known maximum distance between the two paratopes of IgG is about 15 nm (Valentine and Green, 1967), and given that IgG molecules are only a little bit short of being able to cross-link two ER (see the preceding section), the distance of closest approach between the cell *membranes* of two ER must be slightly more than 15 nm; see Fig. XXIII-6. It should be kept in mind that the energy balance between two ER is determined primarily between the EL, LW and AB interactions *between the distal edges of their glycocalices*. It thus is indispensable to take the thickness of the glycocalyx into account, in determining the closest distance of approach between red cells. The distance, ℓ, between glycocalices is governed by the three components of ΔG_{131}^{TOT} (see eq. [XXIII-1]), and their rates of decay with distance (Chapter VII). The closest distance of approach may be taken to be the distance at which the repulsive energy between two cells is equal to their energy of Brownian motion, i.e., $\Delta G_{131}^{TOT} = +1.5$ kT. At that distance, the repulsive energy between opposing glycoprotein strands, in terms of energy per glycoprotein pair, is still relatively slight, but of sufficient importance to cause a reduction in the maximum length, h, of the glycocalyx strands, by folding. Given that h is approximately 10.3 nm (van Oss, *Biophysics of the Cell Surface*, 1990), and that the minimum length of ℓ, even at the

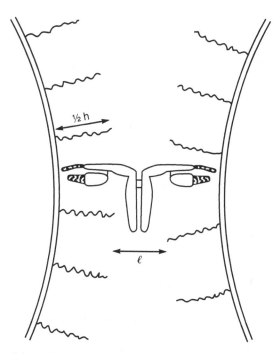

FIGURE XXIII-6   Schematic presentation of parts of two human red cells (A, Rh_o-positive), approaching each other to a minimum distance (between cell *membranes*) of slightly over 15 nm. Under these conditions the glycoprotein strands of the glycocalyx (originally about $h \approx 10.3$ nm long) are compressed to about half that length (= ½ h), leaving a minimum distance between the tips of the glycoprotein strands of about $\ell \approx 5$ nm. The dimensions of a completely opened IgG-class antibody molecule, with a maximum "reach" of 15 nm are just insufficient to allow cross-linking.

lowest value for λ (see Fig. XXIII-7), is 3.7 nm, and that only a slight decrease in ℓ suffices to allow IgG (with a "reach" of 15 nm, see Fig. XXIII-6) to cross-link two red cells, the distance $\ell + 2h$ must be at least reduced by 8 or 9 nm (at the smallest value for λ), at closest normal encounter between two erythrocytes. At a minimum, this is about a 40% reduction of h. At a maximum, the strands are unlikely to become more shortened than to their radius of gyration (Israelachvili, 1985), or be more reduced than to $1/\sqrt{12}$ times their original length (Hiemenz, 1984), i.e., a reduction of 71%. A reduction of 71% (which is only attainable under conditions of very high outside pressure; see Israelachvili, 1985), leading to h = 3 nm, would give rise to a total length of only slightly more than $\ell \approx 9$ nm, which indicates that for water, $\lambda = 1.0$ nm would be the highest possible value; see Fig. XXIII-7.

IgM, with a "reach" of about 27.4 nm (Dorrington and Mihaesco, 1970), is able to cross-bind two erythrocytes under any circumstances (see Fig. XXIII-5). Thus only IgG, and not IgM, can be used as a feasible yardstick for erythrocyte interactions; it allows to estimate the upper limit of λ for water (i.e., 1.0 nm). The lower limit of λ for water is harder to measure. Nevertheless, a reasonable estimate

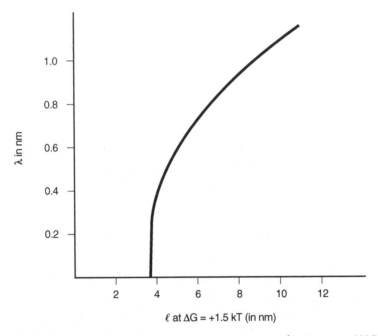

FIGURE XXIII-7   Plot of the possible values of the decay length (λ) of water at 20°C *vs.* the corresponding distance ℓ between the outer edges of the glycocalices of human erythrocytes, where ℓ is computed taking ΔG^{EL}, ΔG^{LW} *and* ΔG^{AB} into account at $\Delta G = +1.5$ kT; see Figure XXIII-4. From hemagglutination experiments with IgG-class antibodies, ℓ may be estimated as being just over 5 nm, which yields a most likely lower value for λ of ≈ 0.6 nm (see text).

can be made of its most likely value. As the minimum reduction of the length, h, of the glycocalyx strands (under the conditions of mean closest encounter of two erythrocytes, at $\Delta G_{121} \approx +1.5$ kT) is 40%, and the maximum reduction of h is 71%, an actual reduction of h of $\approx 50\%$ appears plausible; see Fig. XXIII-4. This could give rise to $\ell \geq 15 -\frac{1}{2} \times 2$ h $\geq 15 - 10.3 \geq 4.7$ nm. Thus a value for λ (water) = 0.6 nm, which corresponds to $\ell \approx 5$ nm (see Figs. XXIII-4, 6 and 7) appears the most plausible. A value of λ (water) ≈ 0.6 nm is also plausible in view of the structure of liquid water. From the molecular dimensions of solitary water molecules,[*] a value of λ (water) ≈ 0.2 nm would seem appropriate (Chan *et al.*, 1979; Parsegian *et al.*, 1979). However, considering that at least 10% of the water molecules are linked by hydrogen bonds to one or several other water molecules (which causes the energy of cohesion of water at 20°C to be $\gamma_w^{AB}/\gamma_w = 51/72.8 = 70\%$ due to hydrogen-bonding), a considerable

---

[*] As $\gamma_w^{AB} = 51$ mJ/m², $\gamma_{ww}^{AB} = -102$ mJ/m² (cf. eq. [II-33]), thus the hydrogen-bonding energy of cohesion for water at 20°C, which corresponds to -9.85×10^{-21} J/molecule, or -2.4 kT, or ≈ -1.4 kcal/mole. We also know that the free energy of one hydrogen bond, for water, has been estimated to be of the order of ≈ -6.0 kcal/mole (Eisenberg and Kauzmann, 1969, p. 145) to -12.6 kcal/mole (*ibid.*, p. 269). Thus at any one moment, more than 10% and possibly even more than 20% of the water molecules in liquid water are engaged in cohesive hydrogen-bonding.

proportion of the water molecules in liquid water must at any one time form an array of several water molecules. Thus λ (water) ≈ 0.6 nm is a minimum value which also appears reasonable from the viewpoint of cohering water molecules. Israelachvili and Pashley (1984) arrived at a value $\lambda \approx 1.0$ nm, for "hydrophobic" attractions, using Israelachvili's force balance (see also Chapters VII and XVI). In conclusion $\lambda \approx 1.0$ nm appears to be the most acceptable mean value for water at 20°C.

Considerably higher values quoted for λ, obtained with the force balance (see, e.g., Christenson, 1988), were influenced by the presence of micro or macromolecules which had not achieved a complete and irreversible attachment to the mica surfaces used in these devices; however, *see Chapter VII*. For uncoated mica surfaces, unusually high values for λ (water) have not been reported.

CELL AND PARTICLE STABILITY IN THE ABSENCE AND IN THE PRESENCE OF POLYMERS

STABILITY IN THE ABSENCE OF POLYMERS

Smooth particles of micron or submicron dimensions, whether spherical or asymmetrical, when suspended in water, can owe their stability to two mechanisms:

1. Electrostatic repulsion (positive ΔG^{EL}), due to their ζ-potential; and
2. Net hydrogen-bonding repulsion (positive ΔG^{AB}), generally due to their strong electron-donicity.

The LW interaction energy between similar particles immersed in water practically always has a negative value. Thus, to attain stability, ΔG^{EL} and/or ΔG^{AB} have to be sufficiently positive to overcome the LW attraction. In liquids, this attraction is rarely enormous, but it is always there, except in cases of very low-energy particles, such as droplets of octane or nonane suspended in water.

Mechanism #1, EL repulsion, in aqueous media, is almost always accompanied by an AB repulsion (see Chapter XIX), and in many cases a reduction in ζ-potential (e.g., by adjusting the pH of media containing amphoteric materials to their isoelectric point, or by the addition of plurivalent counterions) will simultaneously obliterate the EL and the AB repulsion. This holds for amphoteric polymers such as proteins, as well as for other charged materials such as phospholipids, or clay particles.

Mechanism #2, AB repulsion in aqueous media, can, however, act on its own: it is not necessarily accompanied by an EL repulsion. Various materials, such as neutral polysaccharides (e.g., dextran, ficoll, carbohydrate glycocalyx moieties), polyethylene oxide, polyvinyl alcohol, etc., manifest a strong mutual AB repulsion in water, in the absence of a significant ζ-potential.

One fundamental problem which is inherent in the EL and/or AB-driven stabilization of particles (or cells) immersed in a liquid, lies in the fact that the very molecules situated at the particle (or cell) interface with the liquid (furnishing the net mutual repulsion which causes the stability) also *must be soluble in that liquid* (see Chapter XX). Thus, unless some special linkage of the stabilizing molecules is provided, they could be prone to become eroded away by dissolving into the liquid. However, various cases are known of stability of suspensions of perfectly *smooth* particles: charged (EL + AB forces) as well as uncharged (AB forces alone) phospholipid vesicles (Ohki *et al.*, 1982), polystyrene latex particles stabilized by imbedded persulfate moieties (Interfacial Dynamics Co., 1991) (EL + AB forces), and montmorillonite particles, whose stability is due for 85% to AB and only for 15% to EL forces (van Oss, Giese and Wu, 1993).

However, when more "hydrophobic" (and thus insoluble) particles must be stabilized in aqueous suspension, peripheral attachment of water-soluble appendages to such particles is the preferred solution. This attachment can be achieved by non-covalent adsorption, or by covalent binding; see the following section. It must, however, be reiterated that the sole function of attaching water-soluble polymers or surfactants to "hydrophobic" particles is not to do something "steric" which somehow lends them some degree of stability (see Chapter XX). The polymers function by continuing to exert their mutual repulsion (which was the reason for their aqueous solubility in the first place) which, transferred to the particles, keeps these in stable aqueous suspension; see below.

STABILITY INDUCED BY ATTACHED LYOPHILIC POLYMERS; DEPLETION PHENOMENA

Polymers Attached by Adsorption

The simplest and easiest way of stabilizing lyophobic (e.g., hydrophobic) particles is by dissolving a fair amount of lyophilic (e.g., hydrophilic) polymers into the liquid. Usually, a proportion of the dissolved polymer molecules will then adsorb onto the particles, and stability ensues (Napper, 1983). This has widespread utilization, e.g., for the stabilization of latex-based paints. It was the discovery that latices could be stabilized through the admixture of *non-ionic* surfactants, which gave rise to the complicated purely qualitative theories of "steric" stabilization (Napper, 1983),[*] that were felt to be needed to explain the stabilizing power of uncharged polymers, in the days before it was realized that even uncharged polymer molecules, dissolved in water, repel each other (see Chapter XX).

There is one drawback to stabilization of particles by means of adsorption of dissolved polymers or surfactants: this is the risk of flocculating the particles at bulk polymer concentrations that are high enough to cause a phase separation (see Chapter XXI), i.e., high enough to exert a stronger pressure on the polymer-coated particles than these particles can exert on the bulk polymer molecules, or on each other. At even higher polymer concentrations the situation may be reversed, [i.e., when the concentration of adsorbed polymer locally surpasses the bulk polymer concentration; see van Oss, Arnold and Coakley (1990)], and particle stability can once more

---

[*] Napper (1983) alludes to "The misnomer 'steric stabilization'" on pp. 26–27 of his work. Napper also indicates (*ibid.*, p. 13) that it is a requirement for stabilizing polymers "that they can generate repulsion," but he does not specify the surface properties which are essential for (e.g., non-ionic) polymers to generate such a repulsion.

ensue. These two phenomena have been designated as "depletion flocculation" and "depletion stabilization" respectively (see, e.g., Napper, 1983). There is indeed a layer between the particle surface (with adsorbed polymer molecules) and the bulk polymer solution, which is largely devoid of polymer. The thickness of the depleted layer is of the order of the radius of gyration of the polymer molecules (see, e.g., Bäumler and Donath, 1987; also, Chapter XX). The depletion layer however is an *effect* of the repulsion which causes the phase separation; it is not its *cause* (van Oss, Arnold and Coakley, 1990; see also Chapter XX).

Amphipathic polymers (e.g., surfactants) are the most efficient stabilizing molecules that can be attached to lyophobic particles by simple adsorption. For instance, in aqueous media the hydrophobic tails of surfactant molecules will strongly adsorb onto the surface of hydrophobic particles, immersed in water, leaving the hydrophilic heads more free to protrude and thus to generate the required interparticle repulsion (see, e.g., Ottewill and Walker, 1986; Napper, 1983, p. 28). Whilst hydrophilic compounds (such as polyethylene oxide) still can adsorb in water onto low-energy surfaces of particles such as polystyrene, their free energy of adsorption is not nearly as high as that of, e.g., polyethylene oxide chains linked to an alkyl group. Even stronger attachment can however be ensured by other means; see below.

Proteins can be adsorbed onto latex particles (van Oss and Singer, 1966; Norde and Lyklema, 1978); such proteins will then either mask or displace the surfactants or polymers initially used for stabilization (see also Chapter XXIV). The ultimate stabilization is then incumbent on the adsorbed protein molecules. Important applications of protein-coated latex particles is in the field of agglutination, used for testing the presence of antibodies (when the antigen is coated onto the particles) or of antigens (when the particles are coated with antibody). Important examples are the "latex fixation" agglutination test for the presence of rheumatoid factor in patients' sera (Singer, 1961), and various pregnancy tests (where the level of chorionic gonadotrophin in putative pregnancy urine is assayed by inhibition of latex agglutination). For various other applications of coated latex particles, see Rembaum and Tökés (1988) and for information on latices in general, see Bangs (1984).

Polymers Attached by Anchorage or Covalent Bonds

For the stabilization of (usually polystyrene) latex particles, the most efficient approach is to add the stabilizing polymer already during the process of latex particle formation by means of emulsion polymerization (Napper, 1983, p. 30). For the stabilization of silica particles the stabilizing polymer can be covalently grafted onto the particles *via* siloxane bridges (Bridger *et al.*, 1979; Clarke and Vincent, 1981).

Enzymes can be covalently bound to polymeric particles *via* a number of approaches, among which are, e.g., diazotization between aromatic amine groups on the particles and amino groups of the proteins, coupling between carboxyl groups on the particles and proteins with carbodiimide or Woodward's Reagent-K, etc. (see Bangs, 1984). The use of enzymes immobilized on latex particles is extensive (Lindsey, 1970). Latex particles comprising both polybutadiene and polystyrene (in a proportion of about 5 to 95) can be radioiodinated (Singer *et al.*, 1969).

Peripheral blood cells are mainly stabilized by the polymeric sialoglycoprotein strands which are anchored in the cell membrane by means of proximal hydrophobic moieties, and which form the glycocalyx (see above). "Naked" cells, with a purely phospholipid bilayer (with one layer of hydrophilic heads oriented toward the water interface) can be perfectly stable (see above), but they do not normally occur *in vivo*. The reasons for this are two-fold:

a. "Naked" cells, with just a phospholipid bilayer at the water interface, when exposed to the vagaries of the exceedingly dynamic system of blood circulation, are too fragile, too easy to lyse and especially too prone to interact with the phagocytic cells of the organism's reticuloendothelial system, which markedly reduces their half-life. An example of this can be found in the work by Lasic *et al.* (1991), who compared the clearance *in vivo* of various types of liposomes with that of liposomes provided with protruding polyoxyethylene strands: the latter had a strikingly longer half-life.

b. In the absence of glycocalices comprising a great many different glycoprotein strands, immunological recognition of either "self" or "nonself" would be severely curtailed if not virtually impossible. It should be kept in mind that almost all the immunological markers on circulating (and sessile) leukocytes and erythrocytes are part of these cells' glycocalices. These include, e.g., the histocompatibility markers, B-cell paratopes, T-cell receptors, numerous other receptors, blood group epitopes, etc.; see, e.g., Chapters 2–4, 23, 26–29 in Atassi *et al.* (1984), and Zucker-Franklin *et al.* (1988).

The communication among blood cells, as well as between blood cells (especially the various types of leukocytes) and other cells (foreign as well as "self") of all vertebrate organisms takes place *via* receptors which are mostly part of the cells' glycocalices (see also Chapter XXIV).

CODA

Using the energy *vs.* distance equations given in Chapter VII, energy *vs.* distance plots are given for typical spherical particles ("hydrophobic" and hydrophilic, at low and high ionic strengths). In aqueous media, the predominant importance of AB forces is demonstrated.

It could be shown that in the destabilization (i.e., flocculation) of, e.g., negatively charged particles by plurivalent counterions (here cations), called the Schulze-Hardy phenomenon, it is not the decrease in ζ-potential that is the direct cause of the flocculation, but rather the *hydrophobization of the particles* concomitant to the decrease in ζ-potential, which is the real cause of the particles' destabilization.

Practical examples are given for blood cell (especially red cell) stability; rouleau formation, and particularly hemagglutination phenomena are treated. Using the immunoglobulin molecule, IgG, as a yardstick, in energy *vs.* distance plots of red cells, an estimation could be made of the decay length of water, λ, as about 1.0 nm.

The influence of dissolved polymers on particle stability is discussed. When the repulsive energy between polymers and particles is greater than that between the particles, the particles are destabilized. This phenomenon is called "depletion" flocculation on account of a depletion layer surrounding the particles, which is devoid of polymer molecules; however, this depletion layer is an effect of the mutual repulsion between polymer molecules and the particles, and not its cause. When the repulsive energy between particles is greater than that between particles and polymer molecules, the particles are stable ("depletion" stabilization).

Whilst surfactants, polymers, or glycoprotein strands (e.g., in the glycocalices of cells) are helpful in stabilizing particles, there are various classes of particles with a negligible or zero ζ-potential which are "naked" and stable, mainly or wholly as a consequence of AB-repulsions. There are of course some obvious advantages of "steric" stabilization by means of attached polymer strands, for synthetic particles as well as for living cells. However, one must not lose sight of the fact that the stability of such particles then is governed by the physicochemical properties of the attached polymers, which then shifts the treatment of the properties of the particles' surface to that of the properties of the polymers' surface. In addition, it should not be forgotten that stabilization by polymers is only possible with polymers that are *soluble* in the liquid medium in which the particles are suspended. In water such polymers, to be soluble, must repel each other (Chapter XXII). In polymer-mediated stabilization of aqueous particle suspensions, these soluble polymers therefore simply confer their mutual repulsive capacity to the particles to which they are attached.

XXIV Adsorption and Adhesion in Aqueous Media, Including Ligand-Receptor Interactions

INTERACTION BETWEEN TWO DIFFERENT MATERIALS IMMERSED IN WATER

Both adhesion between two different solid materials immersed in water, and adsorption of solutes dissolved in water onto solid bodies are aspects of the same phenomenon and are governed by the same equations:

$$\Delta G_{1W2}^{TOT} = \Delta G_{1W2}^{LW} + \Delta G_{1W2}^{AB} + \Delta G_{1W2}^{EL} \qquad [\text{XXIV-1}]^*$$

where $\Delta G_{1W2}^{LW} + \Delta G_{1W2}^{AB} = \Delta G_{1W2}^{IF}$, representing the free energy of interfacial interaction (cf. eq. [III-16], based upon the Dupré equation):

$$\Delta G_{1W2}^{IF} = 2[\sqrt{\gamma_1^{LW}\gamma_W^{LW}} + \sqrt{\gamma_2^{LW}\gamma_W^{LW}} - \sqrt{\gamma_1^{LW}\gamma_2^{LW}} - \gamma_W^{LW}$$

$$+ \sqrt{\gamma_W^{\oplus}}(\sqrt{\gamma_1^{\ominus}} + \sqrt{\gamma_2^{\ominus}} - \sqrt{\gamma_W^{\ominus}}) + \sqrt{\gamma_W^{\ominus}}(\sqrt{\gamma_1^{\oplus}} \qquad [\text{XXIV-2}]$$

$$+ \sqrt{\gamma_2^{\oplus}} - \sqrt{\gamma_W^{\oplus}}) - \sqrt{\gamma_1^{\oplus}\gamma_2^{\ominus}} - \sqrt{\gamma_1^{\ominus}\gamma_2^{\oplus}}]$$

and the expressions for ΔG_{1W2}^{EL} are given in eqs. [V-3–6]; see also Chapter VII.

For adhesion or adsorption to take place, ΔG_{1W2}^{TOT} (eq. [XXIV-1]) must have a negative value. For greater accuracy in predicting the degree of adhesion or adsorption

---

* A list of symbols can be found on pages 399–406.

it is of course desirable to elaborate energy balance plots of ΔG_{1W2}^{TOT} vs. distance (ℓ) (cf. Chapters VII and XXIII). However, it usually suffices to know that $\Delta G_{1W2}^{TOT} < 0$, at $\ell = \ell_o$. It should be kept in mind that whilst the negative value of ΔG_{1W2}^{TOT} is a fair measure of the degree of adhesion or adsorption one may expect, the quantitative prediction of the amount adhering or adsorbed is linked to the equilibrium constant, K_{ass} (cf. eqs. [XIX-9,10 and 10A]):

$$S_c \cdot \Delta G_{1W2}^{TOT}/kT = -\ln K_{ass} \qquad \text{[XXIV-3]}$$

where $\Delta G_{1W2}^{TOT}/kT$ is expressed in energy units per contactable surface area (S_c) per molecule (see eqs. [VI-6, 7 and 8]), and:

$$K_{ass} = \frac{[S_b]}{[S_{f1}][S_{f2}]} \qquad \text{[XXIV-4]}$$

(cf. eq. [XIV-8]) and where the symbols in brackets indicate concentrations (in mol fractions): S_b indicates bound sites and S_{f1} and S_{f2} free sites [for material (1) and solute or material (2)]. K_{ass} can be derived from the initial slope of the Langmuir plot of the adhesion or adsorption system under study (Hiemenz, 1986, p. 401); thus ΔG_{1W2}^{TOT} is proportional to the logarithm of that slope.

MACROSCOPIC AND MICROSCOPIC-SCALE INTERACTIONS

In the simplest case of monosized hydrophobic particles (e.g., polystyrene particles), interacting with a hydrophobic surface (e.g., teflon), when immersed in water, the extended DLVO (XDLVO) energy balance diagram of ΔG_{1W2}^{TOT} (= $\Delta G_{1W2}^{LW} + \Delta G_{1W2}^{AB} + \Delta G_{1W2}^{EL}$), vs. ℓ, will approximately resemble one of the two XDLVO diagrams shown in Figs. XXIII-2 A or 2 B, depending upon the salt content of the aqueous medium. Here the entire ΔG_{1w2} interaction as a function of distance, ℓ, can be described as a single, macroscopic-scale phenomenon. However, in an only slightly more complicated and still quite common phenomenon, i.e., that of adsorption (in water) of a dissolved protein onto a silica surface, one is faced with: 1) The occurrence of a macroscopic-scale, net repulsive interaction at a distance between each whole protein molecule and the silica surface (whose ΔG_{1W2}^{TOT} vs. distance diagram would approximately resemble either Fig. XXIII-1A or XXIII-1B, depending on the salt content) and: 2) The simultaneous occurrence of a microscopic-scale, net attractive interaction at a distance between a distally located peptide chain moiety, with a small radius of curvature, protruding from the protein molecule, which is attracted to a discrete metal ionic site (e.g., Si^+), imbedded in the silica surface. This microscopic-scale, localized ΔG_{1W2}^{TOT} attraction vs. distance plot would roughly have the same shape as the XDLVO diagram depicted in Fig. XXIII-2A, except that here the ΔG_{1W2}^{EL} part would also be negative (i.e., attractive).

The macroscopic-scale (repulsive) interaction between a whole protein molecule and the silica surface is usually markedly stronger than the microscopic-scale (attractive) interaction between a protruding pointy peptide moiety on the protein molecule and a discrete cationic site on the silica surface. Nonetheless, despite the dominance of this macroscopic-scale repulsion field, a number of favorably oriented protein molecules ususally succeed in piercing it with one of their pointy processes and thus in adsorbing onto one of the discrete cationic sites on the silica surface (Docoslis *et al.*, 1999). It is only at pH 10.5 and above that the macroscopic-scale repulsion becomes so great that proteins such as human serum albumin are no longer able to adsorb onto silica at all (Docoslis *et al.*, 2001b). The connection between the macroscopic-scale XDLVO analysis of the ΔG_{1W2}^{TOT} *vs.* ℓ force balance and its microscopic-scale counterpart, plays an important role in the study of the *kinetics* of protein adsorption onto metal oxide surfaces such as silica; see Chapter XXV, below.

ADSORPTION

INFLUENCE OF THE SIZE OF THE SOLUTE

Even when ΔG_{1W2} (eq. [XXIV-2]) has a negative value in terms of mJ/m^2, that does not necessarily signify that a measurable degree of adsorption will take place. The size of the solute molecule plays a crucial role here. It should be realized that, to judge the degree of adsorption, ΔG_{1W2} must first be expressed in units of kT, according to eqs. [XXIV-3] and [XXII-4], as: ΔG_{1W2} (in mJ/m^2) $\times S_c/kT$. Once ΔG_{1W2} is expressed in kT, the equilibrium adsorption constant, K_{ass}, can be found (eq. [XXIV-3]). K_{ass} is a fairly reliable measure of the degree of adsorption which will occur: normally K_{ass}-values which are not at least of the order of 10^2 L/mole or 5.6×10^3 (mol fractions)$^{-1}$ can be considered to be negligible, or at least, hardly measurable. Table XXIV-1 gives a list of K_{ass}-values as a function of the free energy of adsorption, ΔG_{1W2}, expressed in kT.

As an example one may take the adsorption from an aqueous solution of vinyl-pyrrolidone and its polymers (PVP). Assuming that the surface tension components and parameters of PVP do not change with size, they may be taken as: $\gamma_1^{LW} = 43.4$ mJ/m^2, $\gamma_1^{\oplus} = 0$ and $\gamma_1^{\ominus} = 29.7$ mJ/m^2 (Table XVII-5). As adsorbent, polymethyl methacrylate (PMMA) may be taken, with the surface properties: $\gamma_2^{LW} = 42.7$ mJ/m^2, $\gamma_2^{\oplus} = 0$ and $\gamma_2^{\ominus} = 20.4$ mJ/m^2 (van Oss, Giese, Wentzek *et al.*, 1992). The influence of the size of PVP, when adsorbing onto PMMA, in water, is shown in Table XXIV-2: the crucial importance of the value of S_c will be apparent. It can be seen that due to this factor, measurable adsorption of PVP is unlikely to be encountered with PVP with polymerization numbers much less than 20.

In the same manner it is easy to explain why various proteins adsorb more strongly (i.e., with a higher K_{ass}) onto solid surfaces from dilute than from more concentrated solutions; see, e.g., Corsel *et al.* (1986); Cuypers *et al.* (1986). It is only from more dilute solutions that proteins are able to interact with a flat surface with their maximum surface area. Thus at the greatest dilution, the effective S_c of a protein is at a maximum, which gives rise to a maximum value of K_{ass}, according to eq. [XXIV-3].

BLOTTING

As already suggested (van Oss *et al.*, J. Chromatog., 1987), the "Southern blotting" method (Southern, 1975) of transferring previously electrophoretically separated

TABLE XXIV-1

Equilibrium Association Constant, K_{ass}, as a Function of the Free Energy of Adsorption, ΔG_{1W2}

ΔG_{1W2} (kT)	K_{ass} (eqs. [XXIV-3 and 4]) (L/mole)	$K_{ass}{}^a$
0	1.0	0.018
−2	7.4	0.13
−4	54.6	0.98
−6	403.4	7.3
−8	3.0×10^3	53.7
−10	2.2×10^4	396.5
−12	1.6×10^5	2.9×10^3
−14	1.2×10^6	2.2×10^4
−16	8.9×10^5	1.6×10^5
−18	6.6×10^7	1.2×10^6
−20	4.6×10^8	8.7×10^6
−22	3.6×10^9	6.5×10^7
−24	2.6×10^{10}	4.8×10^8
−26	2.0×10^{11}	3.5×10^9
−28	1.4×10^{12}	2.6×10^{10}
−30	1.1×10^{13}	1.9×10^{11}

[a] K_{ass} in this column is given in (mol fractions)$^{-1}$, in the middle column, K_{ass} in L/mole, although not to be used in eqs. [XXIV-3 and 4] with these units, is also given here, for comparison with the K_{ass} dimensions with which the majority of users is most familiar.

TABLE XXIV-2

Adsorption of Monomer and Polymers of Polyvinyl-pyrrolidone (PVP) onto Polymethyl-methacrylate (PMMA), in Water, as a Function of the Degree of Polymerization of PVP

Degree of polymerization	S_c (nm²)	$\Delta G_{1W2}{}^a$ (kT)	K_{ass} (mol fractions)$^{-1}$
1	0.35	−0.74	2.1
5	1.75	−3.71	40.85
10	3.50	−7.41	1.65×10^3
20	≈5.5[b]	−11.64	1.15×10^5
100	≈10.0[b]	−21.85	3.09×10^9

[a] Using eq. [XXIV-2] ΔG_{1W2}, (PVP-Water-PMMA) = −8.47 mJ/m².
[b] Estimated: at higher molecular weights, the entire theoretical contactable surface area is not available for molecular contact, for statistical reasons.

DNA fractions onto cellulose nitrate membranes, for the detection of specific DNA sequences by hybridization with RNA, is based upon the property of DNA to adhere to cellulose nitrate, while RNA does not adhere to this material. Thus *when* RNA binds to such DNA-treated membranes, this is solely due to the specific adherence of RNA to its complementary DNA counterpart. The reason why DNA binds to cellulose nitrate, while RNA does not, lies in the considerably higher $\gamma^{\ominus}$ value of RNA, as compared to that of DNA (see Table XVII-9). Thus DNA is attracted to cellulose nitrate and RNA is repelled; see Table XXIV-3.

Similarly, "Western Blotting" (Towbin *et al.*, 1979) for the attachment of proteins to cellulose nitrate, and other types of membranes, is also based on the net attraction between most proteins and cellulosic polymers (van Oss *et al.*, J. Chromatog., 1987); see also Table XXIV-3.

PROTEIN ADSORPTION

Adsorption of dissolved proteins to lower-energy surfaces immersed in aqueous media is a common occurrence. The adsorption of (radioiodinated) plasma proteins onto polystyrene (PST) surfaces (latex particles as well as flat plate surfaces) was first studied by van Oss and Singer (1966). In particular, the degree of adsorption (at different protein concentrations) onto PST surfaces was investigated as a function of pH, and of the presence of various surfactants. As the overall aim of the NIH grants which funded this project was the study of the ultimate localization in the reticuloendothelial system of variously treated PST latex particles after intravenous injection in experimental animals, the final report of this work was published in 1966 in the *Journal of the Reticuloendothelial Society* (in 1984 renamed *Journal of Leukocyte Biology*). Due to the predominantly noncolloid or surface science nature of the usual subject matter treated in that journal, our early paper on protein

TABLE XXIV-3
Free Energies of Interaction Between Cellulose Nitrate[a] and Nucleic Acids or Proteins, Immersed in Aqueous Media

	ΔG_{1W2} (in mJ/m^2)[b]
Nucleic Acid	
DNA[c]	−24.0
RNA[c]	+17.5
Protein	
HSA[d]	−25.1
IgG[d]	−40.2

[a] See Table XVII-8.
[b] Eq. [XXIV-2].
[c] See Table XVII-9.
[d] See Table XVII-6.

adsorption remained largely unnoticed by the colloid and surface science community. MacRitchie (1972) published a lucid and important paper on the adsorption of (bovine) serum albumin at the solid/liquid (hydrophobic as well as hydrophilic solid surfaces) and at the liquid/air interface, as a function of pH and taking adsorption as well as desorption isotherms into account. Norde and Lyklema (1978) subsequently reported on their extensive investigations on the adsorption of (normally anionic) human serum albumin, and (cationic) bovine pancreas ribonuclease, onto PST particle surfaces. These are usually quoted as the first published quantitative studies of protein adsorption onto low-energy surfaces. This work gave rise to many further studies by the Dutch school on the underlying mechanisms of the adsorption of proteins (and other polymers), based on the Flory-Huggins-type interactions (Cohen Stuart *et al.*, 1984), and on statistical theories; see Scheutjens and Fleer (1979, 1980, 1982) and Fleer and Scheutjens (1982), and see also Vincent's overview (1984) on polymer adsorption to particles and their effect on particle stability; see also the review by Norde (1986), and the more recent analysis by Norde (1992) on the energy and entropy of protein adsorption. Earlier papers on protein adsorption have been reviewed by Baier (1975).

Adsorption of polymers from solution is most conveniently treated *via* an interfacial approach which includes statistical effects (eqs. [XXIV-3 and 4]) as well as interactions of the liquid molecules (eqs. [XXIV-1 and 2]). The Flory-Huggins χ-parameter is closely related to ΔG^{IF}; cf. Chapter XIX. Thus, in the absence of significant electrostatic interactions, adsorption is entirely due to interfacial, or 'hydrophobic" attraction, i.e., $\Delta G^{IF}_{1W2} = (\Delta G^{LW}_{1W2} + \Delta G^{AB}_{1W2})$ (see eq. [XXIV-2]) which then has a negative value. It should be remembered that all but the most hydrophilic materials (1) can undergo a net attraction to hydrophobic (2) surfaces when immersed in water (Chapter XVIII). Only the most hydrophilic proteins (such as hydrated IgA; see van Oss, Moore *et al.*, 1985, and see also the sub-section on Liquid Chromatography, this chapter, below) will not adsorb to low-energy surfaces from an aqueous solution. The ΔG^{IF}_{1W2}-values (see eq. [IV-16]) for a number of plasma proteins (Table XVII-6), interacting with various polymer surfaces (Table XVII-5), in water, are given in Table XXIV-4.

It can be seen from Table XXIV-4 that for all proteins dissolved in water, there is a maximum of adhesion to polymers with a γ-value of around 30 mJ/m^2 (cf. polyethylene, Table XVII-5). This is because up to $\gamma \approx 30$ to 33 mJ/m^2 one can find completely apolar polymers; the adsorption of proteins then is at a maximum for the polymer with the highest γ_2^{LW} value, which gives rise to the highest negative value for ΔG^{LW}_{1W2} (because, with $\gamma_2^\oplus = \gamma_2^\ominus = 0$, ΔG^{LW}_{1W2} is invariable as far as the apolar polymer is concerned and depends only on the surface properties of the protein and the liquid). The higher-energy polymers ($\gamma_2 > 33$ mJ/m^2) have a tendency (which becomes more pronounced with higher γ_2 values) to be increasingly polar, which gives rise to less negative values of ΔG^{AB}_{1W2}, thus resulting in a decrease in protein adsorption. To each given polymer surface, the least hydrophilic proteins (dissolved in water) adhere the most strongly (Table XXIV-4): this is mainly due to the contribution of ΔG^{AB}_{1W2} (van Oss, Biofouling, 1991). For the adsorption of IgG onto PST, see Martin *et al.* (1992).

It is interesting to note from Table XXIV-4 that the free energy of attraction of proteins to the air-water interface is smaller than to apolar polymer surfaces, even though the air interface may be considered as *the* most "hydrophobic" of all surfaces.

TABLE XXIV-4

Interfacial Free Energies of Adsorption (ΔG^{IF}_{1W2}) (see eq. [XXIV-2]), in mJ/m², of Various Human Plasma Proteins (1) onto Different Low-Energy Polymer Surfaces as Well as to the Air Interface (2), in Water (W). Also Given is the Free Energy of Hydration (ΔG^{IF}_{1W}) in mJ/m², of These Low-Energy Surfaces[a]

Protein[c] (in decreasing order of hydrophilicity)[d]	Air interface	Teflon	Polypropylene	Polyethylene	Polystyrene	Polymethyl-methacrylate
IgA (hydrated)	+0.13	-4.2	-4.9	-5.7	+1.2	+2.86
Human serum albumin (hydrated)	-0.07	-4.4	-5.3	-6.0	-1.4	+10.9
Fibronectin	-1.47	-7.9	-9.1	-10.2	-3.9	+9.5
IgG (hydrated)	-7.65	-17.5	-19.4	-21.0	-14.7	+4.0
Fibrinogen	-11.27	-25.7	-28.5	-30.8	-22.7	+2.0
Human serum albumin (dry)	-40.20	-55.3	-58.2	-60.2	-46.6	-27.7
IgG (dry)	-49.76	-65.1	-68.1	-70.6	-63.8	-41.7
Free energy of hydration (ΔG^{IF}_{1W})[a,d]	0	-39.6	-47.3	-53.6	-71.1	-94.5

ΔG^{IF}_{1W2} [b]

[a] A number of these data may also be found in van Oss, Biofouling, 1990.
[b] Surface tension component data from Table XVII-5.
[c] Surface tension component data from Table XVII-6 (albumin data with two layers of hydration).
[d] See Chapter XIX.

For instance, its free energy of hydration, ΔG_{1W}^{IF}, equals zero and its free energy of "hydrophobic" interaction, in water, $\Delta G_{1W1}^{IF} = -145.6$ mJ/m^2. The reason for the relatively low attraction for proteins to the air–water interface lies in the fact that in this case the liquid medium (water) has a Lifshitz–van der Waals surface tension component, $\gamma_W^{LW} = 21.8$ mJ/m^2, which lies in between that of the protein (γ_1^{LW}, from 27.0 mJ/m^2, for hydrated protein, to 42.0 mJ/m^2, for dry protein) and air ($\gamma_2^{LW} = 0$). Thus a net Lifshitz–van der Waals *repulsion* occurs (see Chapter IV), which is of course more than counter-balanced by an AB (hydrophobic) attraction, but which still causes the total attraction to be less than in the cases of the apolar polymers (cf. MacRitchie, 1972).

It can also be seen from Table XXIV-4 that the most hydrophilic proteins (especially hydrated IgA and serum albumin) should adhere very little, or not all all, to the highest energy polymer surfaces. In practice this is indeed the case: IgA does not adhere to most hydrophobic chromatography beads unless the protein molecules are first made more hydrophobic through the admixture of considerable amounts of salt (see the sub-section on Liquid Chromatography, this chapter, below).

On the other hand, a rather hydrophilic protein such as serum albumin, at pH values under 8[*] (MacRitchie, 1972), does adsorb onto *clean glass surfaces*, notwithstanding a positive macroscopic ΔG_{iW2} value (Michaeli *et al.*, 1980; Silberberg, 1984; Absolom *et al.*, 1981, 1984), as does, e.g., IgG (Lutanie *et al.*, 1982). New microscope slides, washed in ethanol (see Michaeli *et al.*, 1980) and then air-dried would, with non-hydrated bovine serum albumin (assuming it to have about the same properties as human serum albumin; see Table XIII-6), manifest a ΔG_{1W2}^{IF}-value of ≈ -0.7 mJ/m^2. This does not take a certainly non-negligible positive ΔG^{EL} into account, so that a significant macroscopic positive ΔG_{1W2}^{TOT} must have prevailed. Similarly, chromic acid-washed glass also adsorbed serum albumin (see Absolom *et al.*, 1981). In this case the value of ΔG_{1W2}^{IF} would amount to +36.2 mJ/m^2, so that, by also taking ΔG_{1W2}^{IF} into account, ΔG_{1W2}^{TOT} would be even more strongly positive. In spite of these (macroscopically) net average repulsion forces, albumin adsorption takes place. The reason for this must be sought in the properties of the glass surface on a microscopic level. In other words, whilst from a macroscopic viewpoint, glass only has a small measurable electron-acceptor parameter ($\gamma^\oplus$ 1.2 mJ/m^2) compared to its large electron-donor parameter ($\gamma^\ominus$ 50 mJ/m^2) in individual microscopic patches, the electron-acceptor sites (due to the presence of plurivalent cations in the glass) will bind to the electron-donor sites (e.g., the carbonyl groups) of the protein; see Morrissey and Stromberg (1974). To that effect it is of course necessary for the protein molecules first to be able to penetrate the net average repulsion field. This is most readily done by those protein moieties which have the smallest radius of curvature (see Chapter XXIII; see also the Section on Macroscopic and Microscopic-Scale Interaction, above, as well as Chapter XXV).

The X-ray diffraction studies on human serum albumin (HSA) by He and Carter (1992), which elucidated its three-dimensional configuration, showed HSA

---

[*] Serum albumin only ceases to adsorb onto pure silica surfaces at pH > 10 (Docoslis *et al.*, 2001b); see also the Section on Macroscopic and Microscopic-Scale Interactions, above.

to have a fairly contorted serpentinoid V-like structure. Thus HSA has one major "elbow" with a small radius (R) of curvature at the bottom of the V, with four minor "elbows," plus N- and C-terminals with about the same R. R is about 1.3 nm. By energy-balance analysis (van Oss, 1994b), which also takes the considerable macroscopic repulsions into account (cf. Chapter XXIII), it can be shown that of these five "elbows" plus the N- and C-terminals, all with a low radius of curvature, the "elbow" at the lower point of the V is the only one which can easily penetrate the macroscopic repulsion field between itself and the glass surface deeply enough to make contact with the microscopic attractors in the glass, and thus adhere to its surface; see Fig. XXIV-1.

I

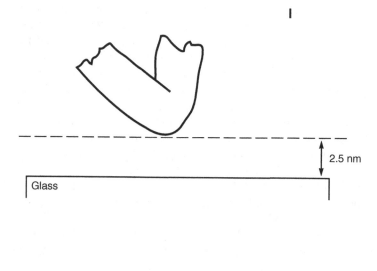

2.5 nm

Glass

II

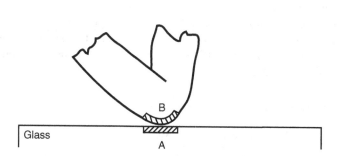

B

Glass

A

FIGURE XXIV-1   Schematic presentation of the mechanism of adsorption of human serum albumin (HSA) onto glass. In I only the averaged macroscopic repulsion between one "elbow" of the HSA molecule and the glass surface is operative: even a moiety with a small radius of curvature cannot approach the glass surface more closely than to $\ell \approx 2.5$ nm. In II it is shown that the attraction between a microscopic attractor site (A) imbedded in the glass surface and an HSA site on a point of maximum curvature (B) or "elbow" can surmount the macroscopic repulsion and lead to a local point of attachment where B meets A; see also Fig. XXIII-3.

This condition correlates well with the finding that the molecules of the first layer of HSA adsorb onto glass with the long axes of the V's almost perpendicular to the glass (Absolom *et al.*, 1981, 1984), or mica surfaces (Fitzpatrick *et al.*, 1972). It is curious to note that whilst most of the HSA remains adsorbed to the glass, the adsorbed molecules could apparently freely move parallel to the glass surface while remaining attached. This was determined by diffusion measurements (Michaeli *et al.*, 1980), as well as by electrophoresis (Absolom *et al.*, 1981). The electrophoresis experiments showed that the migration of the HSA molecules obeyed the laws of charged cylinders migrating perpendicularly to the applied electric field.

For the much more ready desorbability of HSA from hydrophilic silica than from hydrophobic surfaces, see MacRitchie (1972). The contrast between macroscopic and microscopic interfacial interactions is also of crucial import in specific interactions (treated below in this chapter; see also van Oss, J. Molec. Recogn., 1990).

In the 1980s further reports on protein adsorption were collected by Goddard and Vincent (1984) and by Brash and Horbett (1987). One of the more important developments, of great impact on our understanding of protein adsorption *in vivo*, the *Vroman Effect*, was treated in the proceedings edited by Baier (1975) and Brash and Horbett (1987). The Vroman Effect (see, e.g., Vroman *et al.*, 1975, for the earlier observations) pertains to the sequential adsorption of proteins from protein mixtures (such as blood plasma). As stated by Vroman and Adams (1987): "plasma deposits a sequence of proteins, the more abundant being adsorbed first and then being replaced by less abundant ones." Protein adsorption from complex mixtures was also studied by Corsel *et al.* (1986), Horbett (1987) and Eberhart *et al.* (1987). Particular stress on the adsorption of serum complement from complex systems is given by Ward (1987) and by Elwing *et al.* (1987).

A strong decrease in the adsorptivity of proteins can be achieved by covalent attachment of polyethylene oxide (PEO) to their amino groups (Eppstein, 1988), which is logical, given the unusually strong $\gamma^{\ominus}$ monopolar properties of PEO molecules (Table XVII-5) and their concomitantly pronounced polar (AB) repulsive capacity (Chapter XIX).

In recent years it has become possible to study the properties of protein layers adsorbed onto mica surfaces, and the interaction between two such protein-coated surfaces as a function of distance (Blomberg *et al.*, 1991) by using Israelachvili's force balance (Chapter XVI).

CHANGE IN CONFIGURATION AND DENATURATION

Changes in Protein Configuration upon Adsorption

Much has been written on the presence, or absence, of changes in protein configuration upon their adsorption onto high- or low-energy surfaces. It is useful to distinguish between protein adsorption onto: a) high-energy surfaces, b) low-energy surfaces and c) the super-low-energy air/water interface.

a. *High-energy surfaces.* Under most conditions the adsorption of proteins onto high-energy surfaces is less than to low-energy surfaces, or to the air/water interface (MacRitchie, 1972). Also, the adsorption of proteins onto high-energy surfaces is much more pH-dependent than adsorption onto low-energy surfaces (MacRitchie, 1972; see also Norde and Lyklema, 1978; Norde, 1986). Two things should be kept in mind about the adsorption of proteins onto high-energy surfaces: (1) The adsorption usually is strongest at pH values closest to the isoelectric point (see the references quoted above, and also van Oss and Singer, 1966); in this context it should be noted that, at their isoelectric point, proteins also orient their most hydrophobic moieties to the interface with water (van Oss and Good, J. Protein Sci., 1988; see also Chapter XIX). Under such conditions the electrostatic interaction (attractive or repulsive) is at a minimum, while the hydrophobic attraction is at a maximum. (2) The adsorption of hydrophilic proteins onto hydrophilic surfaces (should be considered on a *microscopic* level (see this chapter, above) so that the surface of attachment is small. This agrees well with the observation that only minimal change (if any) in protein configuration is usually observed in these cases; see, e.g., Morrissey and Stromberg (1974).

b. *Low-energy surfaces.* Contrary to adsorption of proteins onto high-energy surfaces, adsorption on low-energy surfaces tends to give rise to a change in configuration; see, e.g., Kochwa *et al.* (1967), Lee *et al.* (1973) and Kim and Lee (1975). The mechanism of this change of configuration becomes clearer from the observation by Lee *et al.* (1973) that a protein adsorbed onto a hydrophobic polymer surface tends to expose its more hydrophilic moiety to the protein/water interface, when a hydrophobic attachment occurs between the hydrophobic moiety of the protein and the low-energy surface. This is quantitatively best understood when one expresses ΔG_{1W2}^{IF} of the interfacial interaction between a protein (1) and a low-energy surface (2), immersed in water, *in separate terms* for the hydrophilic, and for the hydrophobic sites of a typical protein, vis-à-vis a low-energy surface or the air interface. It is a simple matter to compare, e.g., the value for ΔG_{1W2} for hydrated serum albumin vis-à-vis polyethylene, with ΔG_{1W2} for dried serum albumin. These are: $\Delta G_{1W2} = -5.3$ and $\Delta G_{1W2} = -58.2$ mJ/m² (see Table XXIV-4). Clearly the more hydrophobic moieties of the dried protein adhere a decimal order of magnitude more strongly to polyethylene than the hydrophilic surface of the same protein, in the hydrated state.

c. *The air/water interface.* The above is a fairly good model of what happens upon actual adsorption onto low-energy surfaces, as "dried" protein may be approximated to protein adsorbed at the water/air interface (cf. Table XXIV-4), and then fixed in its orientation with its more "hydrophobic" moieties pointing outward. This is substantiated by the fact that the γ values found for dried proteins (Table XIII-6) are only slightly lower than the surface tensions (γ_L) measured for dilute protein *solutions* (Absolom, van Oss *et al.*, 1981), where the lowest γ_L values were found at the higher protein concentrations (going, e.g., from 0.005 to 0.01 to 1% protein). It should be kept in mind that when measuring the surface tension of a protein solution, one mainly measures the surface tension of the more hydrophobic chains of the protein, which orient toward the water/air interface.

Denaturation/Hysteresis

A case can be made for considering a protein which upon adsorption onto a low-energy surface orients its more hydrophobic moieties toward that surface, and its more hydrophilic aspects away from it, as *denatured*. Thus in many cases, dried protein may be considered to be to some extent denatured. Nevertheless, that denaturation usually is largely reversible upon complete redissolution in the aqueous solvent.

Hysteresis is the strengthening of the free energy of adsorption as a function of time elapsed since the beginning of the adsorption process. In the adsorption of globular proteins such as serum albumin (HSA) onto a hydrophobic surface (including a water-air interface, see above), the adsorption energy is enhanced by a change in the tertiary structure of the protein which allows some of its normally internalized hydrophobic moieties to externalize and bind to the hydrophobic surface to which the protein then becomes increasingly strongly adsorbed (as a function of time), further aided by a concomitant increase in the surface area of actual adsorptive contact (van Oss, Docoslis and Giese, 2001).

In the adsorption of, e.g., hydrophilic HSA onto a hydrophilic surface such as silica (which has discrete cationic adsorption sites on its surface), one of the earliest hysteretic mechanisms (with respect to time elapsed since the first adsorptive encounter) is the expulsion of water between protein site (1) and the specific metal ion site (2) on the silica, when immersed in water (w). This locally gives rise to: $\Delta G_{1w2} \rightarrow \Delta G_{12}$, where: $|\Delta G_{12}| > |\Delta G_{1w2}|$, denoting an increase in free energy of attraction between 1 and 2, upon the (usually only partial) loss of interstitial water. With time, secondarily to this mechanism of stronger attachment of the protein aided by the partial loss of interstitial water, the surface area of attachment may become enlarged by encroachment of more of the protein molecule's surface toward neighboring adsorptive cationic sites (van Oss, Docoslis and Giese, 2001).

The mechanism of adsorption of hydrophilic HSA onto talc (which is hydrophobic) is somewhat different from the adsorption of HSA onto silica (which is hydrophilic). Despite the strong hydrophobicity of talc but due to the even stronger hydrophilicity of HSA, there still is a (more modest) macroscopic-scale repulsion between HSA and talc, immersed in water (van Oss, Docoslis and Giese, 2001). Similar to silica, talc also has cationic attractive sites by which HSA are first attracted to the talc surface, on a microscopic level. Once attached in this manner, the HSA molecules undergo, in addition, a hydrophobic attraction close to the first point of attraction, which then locally induces a change in the tertiary structure of the HSA molecules. This in turn orients the now more hydrophobic HSA moieties that were closest to the original site of attraction, more tightly toward the hydrophobic talc surface, thus hydrophobically enlarging and reinforcing the hysteretic bond between HSA and talc (van Oss, Docoslis and Giese, 2001).

With the exception of purely electrostatic (EL) low- and medium-affinity bonds between DNA and auto-immune anti-DNA antibodies (Smeenk et al, 1983; van Oss,

Smeenk and Aarden, 1985), virtually all other types of specific and aspecific non-covalent bonds in aqueous media are unavoidably exceedingly prone to hysteresis, i.e., to bond strengthening as a function of time. This bond strengthening is noticeable within less than one second (Docoslis *et al.* 1999). Thus the measurement results of kinetic association rate constants (k_a) virtually always appear slower than they are in reality, unless care is taken to measure k_a within only a fraction of a second after the start of the adsorption (Docoslis *et al.*, 1999); see also Chapter XXV. Hysteresis unavoidably influences the measurement of kinetic dissociation constants (k_d), as well as that of equilibrium association constants (K_{ass}) and equilibrium dissociation constants (K_{diss}) (Docoslis *et al.*, 2001a). Indeed, whilst for the measurement of k_a one can obviate the influence of hysteresis by exceedingly fast and early measurements, for the measurement of other constants, such as k_d, K_{ass} and K_{diss}, where prior adsorption during many minutes and even up to one hour is unavoidable, a large but often unsuspected influence of hysteresis always exists. In these cases special experimental stratagems are required to circumvent, or at least to minimize the influence of early binding hysteresis (Docoslis *et al.*, 2001a); see also Chapter XXV.

LIQUID CHROMATOGRAPHY

With the exception of pore exclusion chromatography, all other modes of liquid chromatography—reversed-phase (RPLC), hydrophobic interaction (HILC), ion exchange (IELC) and even affinity (AC) liquid chromatography—depend on negative values of ΔG_{132} for binding and on positive values of ΔG_{132} for the elution step. [In pore exclusion liquid chromatography (PELC) one wishes to avoid binding of the molecules to be separated to the stationary phase surface, so that one ideally aims at a zero or slightly positive ΔG_{132} value.]

Reversed-phase liquid chromatography (RPLC) makes use of the attraction of hydrophobic surfaces (in aqueous media) for macromolecules (and other solutes) which are anywhere from fairly hydrophilic to distinctly hydrophobic, i.e., the system has a negative ΔG_{132} (or in this mode, ΔG_{1W2}) value, for the attachment step. For the elution step, the properties of the liquid medium are modified, to cause the value of ΔG_{132} to become positive, which leads to detachment. To effect this, the water in the initial aqueous medium is (usually gradually) replaced by water/organic solvent mixtures of increasing organic solvent content, until complete elution is achieved. The organic solvents used must be (1) miscible with water, and (2) less self-hydrogen bonding than water (i.e., γ_3^{AB} of the liquid mixture must be significantly smaller than 51 mJ/m², which is the case with practically all organic solvents). A typical case of RPLC was described for the adsorption (in an aqueous medium) of human immunoglobulin G (IgG) onto phenyl sepharose beads, and its subsequent elution by means of the admixture of increasing amounts of ethylene glycol to the aqueous medium (van Oss, Israel J. Chem., 1990). Here the adsorption step cumulates in a fairly strongly negative value of ΔG_{132} ($\Delta G_{132} = -8.2$ mJ/m², which for IgG molecules with an estimated contactable surface area, $S_c \approx 10$ nm², corresponds to $\Delta G_{132} \approx -20$ kT). Then, at about 10% (v/v) ethylene glycol (EG) $\Delta G_{132} \approx 0$, at which point IgG *begins* to elute. Most of the IgG is eluted by the point where the EG concentration in the mobile phase has reached 45%, corresponding to $\Delta G_{132} = +6.5$ mJ/m^2. When organic (water-miscible) solvents other than EG are used the elution pattern may be somewhat different from that obtained with EG. For instance, with acetonitrile the energy of adsorption of a protein onto the hydrophobic carrier may increase at first (due to the dehydrating action of acetonitrile), showing, with human serum albumin, an adsorption maximum with 25% (v/v) acetonitrile (Place *et al.*, 1992). However, at higher acetonitrile concentrations (from 25 to 40%), elution can be achieved.

"Hydrophobic" interaction liquid chromatography (HILC) differs from RPLC only in that the solute (usually a hydrophilic protein) is so hydrophilic that it will not spontaneously adsorb onto the hydrophobic carrier in aqueous media, i.e., $\Delta G_{1W2} \geq 0$. To induce the hydrophilic solute to adsorb onto the carrier it is necessary to decrease its hydrophilicity. This is done by increasing the hydrophobizing capacity of the solvent by the admixture of fairly large amounts of salt [typically of the order of 1 M (NH$_4$)$_2$SO$_4$] (van Oss, Moore *et al.*, 1985; van Oss, Good and Chaudhury, Separ. Sci. Tech., 1987; see also Chapter IX). Once adsorption is achieved, elution is effected by gradually decreasing the ionic strength of the mobile phase (i.e., of the liquid medium).

Ion exchange liquid chromatography (IELC) is principally based upon the attraction between charged sites of one sign of charge on the carrier, and ionized sites of the opposite sign of charge on the solute molecules. Elution is effected by exchanging the attached solute ions against other ions which can take their place on the carrier, through the admixture of salts, or an acid (or a base). In IELC, EL forces are rarely the only ones that play a role in the process. In aqueous media, attractive LW forces between solute and carrier, while not enormous, virtually always exist and must be taken into account. And, especially, polar (AB) forces play an important role in aqueous media. When attractive (due to hydrophobic interactions), AB forces will tend to resist elution through the admixture of salts (see the preceding sub-section on HILC) and even of acids or bases when used within moderate pH ranges. On the other hand, when repulsive, AB forces will tend to counteract attachment by EL forces. And situations may well occur where at a certain distance, ℓ, a repulsion exists between carrier and solute due to a secondary maximum (see Chapter XXIII), while at $\ell = \ell_o$, an attraction at the primary minimum prevails. It thus may be desirable in certain cases to establish energy *vs.* distance diagrams to ascertain the conditions under which adsorption, or elution, will be favored (Chapter XXIII). This was first realized by Ruckenstein and Prieve (1976), in the context of IELC, although they then still only considered EL and LW forces, i.e., the classical DLVO theory which does not account for the AB forces that are predominant in aqueous media (see Chapters VII and XXIII). In 1982, Ruckenstein and Srinivasan recognized the existence of AB repulsion (calling this "hydration repulsion"), but did not take it into account quantitatively; see also Lesius and Ruckenstein (1984), Jun and Ruckenstein (1984) and Ruckenstein and Chillakuru (1990). Ruckenstein and collaborators also included the Born repulsion in their treatment. However, all the Born repulsion does is to prevent the interpenetration of the electrons of the atoms which constitute the interacting molecules, in non-covalent interactions. But if one wishes to take the Born repulsion into account, it becomes necessary to utilize somewhat different proportionality constants, such as the ratio between ΔG^{LW} and the Hamaker constant, A. This then impinges upon the value of ℓ_o to be used in the applicable equations (van Oss and Good, 1984; van Oss *et al.*, Chem. Rev., 1988; see also Chapter III). On the whole there is little advantage in considering the Born repulsion when elaborating energy balances; in determining LW interaction energies, it suffices to use the appropriate equation (e.g., eq. [II-32], instead of eq. [III-3]).

Affinity chromatography (AC) makes use of the exceedingly specific attraction between small specific sites, such as the epitope and paratope on antigen and antibody molecules, the specific sites on lectins and their corresponding carbohydrates, ligands and their specific receptors, and enzymes and their substrates; see the section on Aspecific and Specific Interactions, below. The operative forces are also just the LW, AB and EL interactions: in specific interactions they are always attractive under physiological conditions of pH and ionic strength. In AC given molecules are selectively and specifically bound, e.g., by an appropriate antibody-active site (paratope), bound to the chromatographic carrier. Elution then is subsequently achieved by effecting changes in the properties of the liquid medium (pH, ionic strength, admixture of hydrogen-bond breaking water-miscible organic solvents, etc.) (Mayers and van Oss, 1992); see also the section on Aspecific and Specific Interactions, below.

Pore exclusion liquid chromatography (PELC) is not directly based on LW, AB or EL attraction or repulsion between the (macro-) molecules that are to be separated and the carrier beads, but rather on the relative ease with which macromolecular or smaller solutes can become temporarily entrapped in pores inside the beads. Thus in their course through a column of such porous beads, smaller solutes undergo greater delay than larger ones, which allows fractionation according to molecular size. It remains, however, important for the separation to proceed purely as a function of size, so that undue LW, AB or EL attractions or repulsions be avoided. With porous beads made of agarose (see Table XVII-8) and most hydrated proteins (see Tables XVII-6 and 7), there is little likelihood of an attraction; repulsions will prevail. The repulsions between hydrated proteins and agarose surfaces are the greatest in the case of the largest proteins (which will be excluded by the pores anyway). With smaller proteins and peptides the repulsion by agarose surfaces becomes less important. This is because whilst in terms of mJ/m^2, ΔG_{1W2} may remain strongly positive, in terms of kT, ΔG_{1W2} becomes more negligible with the smaller proteins and peptides, on account of their smaller contactable surface areas, S_c (see also Chapter XXIII). Thus the occurrence of repulsions will not unduly impede PELC. Attractions must, however, be avoided at all cost because, if they occur they also will be strongest will the largest proteins (which in addition have a larger S_c in the attractive than in the repulsive mode), thus giving rise to the strongest chromatographic retardation with exactly those macro-molecules which are intended to be the least retarded in PELC.

CELL ADHESION

The adhesion of cells to solid surfaces, particles or other cells is an exceedingly complex phenomenon* which is strongly influenced by:

1. The manifold biological properties and capabilities of cells, which include: surface heterogeneity, motility, endocytosis, exocytosis, pinocytosis, spiculation, extension of pseudopodia, synthesis and excretion of biopolymers and other solutes.
2. The fact that most solid surfaces usually adsorb proteins and other biopolymers (which were present in the liquid medium, or even were exuded by the cells themselves) prior to engaging in cell adhesion.
3. The significantly different mechanism of adhesion (and adsorption) to high-energy compared to low-energy surfaces.
4. The very different processes of aspecific (macroscopic) and specific (microscopic) adhesion which, however, often occur simultaneously.

Even if one neglects point no. 1 (which is also tacitly disregarded by practically all other authors), the variegated influences of the factors mentioned under points nos. 2–4 amply explain why we are still as far removed as ever from any unifying understanding of cell adhesion, in spite of the impressive amount of literature that has been written on the subject. However, points nos. 2–4 are discussed in this section, in the hope that some pattern may emerge from their separate treatment.

INFLUENCE OF PROTEIN ADSORPTION

In the absence of proteins or other biopolymers, bacteria as well as mammalian cells suspended in aqueous media should, exactly like proteins (cf. Table XXIV-4), *adhere most strongly* to polymer surfaces of medium energy ($\gamma_s \approx 30$ mJ/m^2); see van Oss (Biofouling, 1990). In actual fact, however, it is well known that, *in vivo*, cells *adhere least* to medium energy polymer surfaces ($\gamma_s \approx 30$ mJ/m^2) (Baier *et al.*, 1984). The explanation for these seemingly contradictory observations is simple: under *in vivo* conditions cells are always accompanied by extracellular proteins, polysaccharides or biosurfactants (already present, or exuded by the cells themselves) (Schakenraad and Busscher, 1989; Pratt-Terpstra *et al.*, 1987; Busscher *et al.*, 1990; Pratt-Terpstra and Busscher, 1991). Owing to their small size (relative to cells), such proteins, polysaccharides and other biosurfactants adsorb onto polymer surfaces before cells adhere. Thus, *in vivo*, cells may practically always be considered to adhere to polymer surfaces that are precoated with protein. And because medium-energy polymer surfaces become *maximally* coated with protein, the degree of adherence of cells to such surfaces tends to be at a *minimum* (van Oss, Biofouling, 1991), as

---

* Many reviews have been written on the subject, e.g., Weiss (1960, 1961); Curtis (1962, 1973); Pethica (1961); Hubbe (1981); Bell *et al.* (1984); Bongrand *et al.* (1982); Bongrand (1988); Glaser and Gingell (1990); all comprising a physical or physicochemical approach. On microbial cell adhesion the following volumes should be mentioned: Berkeley *et al.* (1980) and Bitton and Marshall (1980).

a consequence of the hydrophilic nature of the hydrated surface of the adsorbed protein molecules exposed to the interface with the aqueous medium. The minimum adhesion of cells to medium-energy polymer surfaces was first reported by Baier *et al.* (1984). As all cells (especially mammalian cells) exude some proteins and/or other biopolymers, even *in vitro*, one must take this phenomenon into account in all experimental studies on cell adhesion, *in vitro* as well as *in vivo*. Nevertheless, *total* coverage of polymer surfaces by protein molecules need not always exist and adsorbed proteins may well be displaced by adhering cellular appendages, in the same manner by which different protein molecules can displace each other (see the section on the Vroman Effect, above).

ADHESION TO HIGH- AND LOW-ENERGY SURFACES

In the same manner as hydrophilic proteins, which can adsorb to equally hydrophilic glass surfaces (see above), hydrophilic cells will adhere to glass (see, e.g., Busscher *et al.*, 1986). Here too, it must be concluded that electron-donor sites on the hydrophilic cell glycocalices can bind to discrete electron-acceptor sites (and/or to plurivalent cations) present in the glass surface, by a microscopic mechanism which cannot be readily studied *via* macroscopic approaches. Thus, cell adhesion to high-energy surfaces in *in vivo* implants differs drastically in most histological and morphological aspects from cell adhesion to low-energy surfaces (Baier *et al.*, 1984). For cell adhesion used in tissue culture *in vitro*, one usually employs polystyrene surfaces, which have been treated (e.g., by sulfonization) to make them more hydrophilic, i.e., more "high-energy," for optimal cell adhesion.

Weiss (1960) pointed out the importance of the influence of calcium ions (in sub-millimole concentrations) on the dramatic enhancement of cell adhesion to glass. The role of Ca^{++} in the binding between two hydrophilic, negatively charged entities, immersed in water, can be of great importance. Ca^{++} (as well as other plurivalent counterions) can provide not only electrostatic crosslinking, but also serves as an electron-acceptor which after binding to negatively charged entities, identifiable with or in the vicinity of hydrophilic electron-donors, neutralizes these electron-donors and thus renders the surface more hydrophobic, and provides an additional attraction energy, of the hydrophobic variety (cf. Chapter XIX, and the section on Cell Fusion, this chapter, below).

PREVENTION OF ADHESION

The crucial problem of designing artificial surfaces to which cells will not adhere when implanted *in vivo* has not yet been solved. The solution of this problem would be extremely important for the use of, e.g., artificial organs, and even of temporary devices such as catheters, as well as in targeted drug delivery systems. In all of these cases a total absence of cell adhesion is desirable. However, as shown above, even high-energy surfaces can adsorb proteins and other biopolymers, as well as adhere to cells. A promising chance for a solution may lie in the direction of exploiting the mechanism which peripheral blood cells use to avoid adhering to each other. This is achieved by a combination of hydrogen-bonding (AB) and electrostatic (EL) repulsions between opposing cell surfaces. For all human blood cells (erythrocytes, leukocytes, etc.) the ratio between the AB and EL energies of repulsion, under physiological conditions, is about 12; see Table XXIII-2. That repulsion is brought about by the ($\gamma^{\ominus}$) monopolar glycoprotein strands of these cells' glycocalices. In some cases (e.g., in the case of erythrocytes) these strands contain a sizeable proportion of sialic acid and are thus moderately negatively charged, in other cases (e.g., leukocytes) the electrostatic contribution is smaller. (The AB/EL ratio remains about the same, because the leukocytes' AB repulsion is also about 40% smaller than that of erythrocytes; see Table XXIII-2). The best synthetic (i.e., enzymatically non-degradable) monopolar ($\gamma^{\ominus}$) material at our disposal is undoubtedly polyethylene oxide (PEO) (cf. Table XVII-5 and Chapter XIX). After coating inert, hydrophobic, particles with, e.g., PEO, their half-life *in vivo* could be increased from less than one minute (Davis and Illum, 1988), to several hours, or even days. This is approaching the half-life of most circulating leukocytes, but still much shorter than that of erythrocytes ($\approx$60 days).

Chapman *et al.* (1999) treated the antibody-active moieties of IgG class antibody molecules (Fab') with PEO (called PEG by these authors), which significantly enhanced their bioavailablity, largely by increasing their time-span of *in vivo* circulation. Gref *et al.* (1994) describes the use of "PEGylated" biodegradable nanospheres.

Antigenic molecules ("*antigens*"), normally used to elicit the formation of specific antibodies against their antigenic sites, can be transformed into "*toleragens*" (thus eliciting, instead, the creation of specific *tolerance* vis-à-vis these antigens) by tagging them with PEO (PEG) (Lee and Sehon, 1977; Sehon *et al.*, 1987; Sehon, 1989). Using this approach with PEO-tagged (or "PEGylated") antigens, Atassi *et al.* (1992) and Oshima and Atassi (2000) succeeded in significantly diminishing the auto-immune response in mice against experimental auto-immune *myasthenia gravis* (induced by the injection of acetylcholine reseptor molecules). The effect appeared to be a consequence of a diminished T-cell involvement. A possible mechanism for this is that "PEGylated" antigens, once attached to, e.g., antigen-presenting cells (APC) helped *repelling*, or at least significantly *decreased attracting* T-cells than occurs normally with untreated antigen molecules, thus resulting in a reduced immune response to the auto-antigen in question.

O'Mullane *et al.* (1988) showed that PEO-treated (via poloxamer 338) polystyrene latex particles adsorbed significantly less fibrinogen than untreated particles and concluded that the prolonged survival *in vivo* of such particles is due to their diminished adhesion to monocytes and macrophages. Holmberg *et al.* (1992) described techniques for PEO (PEG) grafting to surfaces, as well as a number of applications of "PEGylation" of biologically active molecules. Croyle *et al.* (2001) described how to improve the *in vivo* circulation time of "PEGylated" adenoviruses by protecting them from elimination by anti-adenovirus antibodies, with an aim to using such "PEGylated" viruses as carriers for gene delivery *in vivo*.

THE INFLUENCE OF CELL-SHAPE ON ADHESIVENESS

The *adhesiveness, or "stickiness" of cells* to solid surfaces is inversely proportional to their smoothness, i.e., to the local radius (R) of curvature. Smooth spherical cells (large R) do not adhere nearly as much to various surfaces than spiculated cells (cells with many protruding processes with a small R; see Fig. XXIII-3). Extracellular glucose concentrations strongly influence the shape of phagocytic leukocytes (van Oss, Gillman and Good, 1972; van Oss *et al.*, 1975, p. 115). Higher than physiologically normal glucose levels (normally between 0.6 and 1.2 mg/ml) cause phagocytes to lose their propensity for pseudopod formation (small R) for osmotic reasons, and become largely spherical (large R), with concomitantly much decreased powers of adhesion to other surfaces. Glucose has no influence on either the surface tension components, or the ζ-potential of phagocytic cells; this effect is mainly osmotic.

Blood *platelets become "sticky"* under the influence of adenosine diphosphate (ADP). The only noticeable change in platelets, made "sticky" with ADP, is that they become spiculated (White, 1968; van Oss *et al.*, 1975, p. 123); thus here also cell adhesiveness is favored by the existence of processes with a small R. For the effect of spiculation on cell-cell interactions with erythrocytes, see Chapter XXIII, and for the influence of R on the closeness of cell and particle approach, see also Chapter VII; see Fig. XXIII-3. Finally it should be recalled that viruses (which of course are smaller than cells) tend to adhere to cells *via* their spicules, which have a very small R. Examples of this are numerous; one may refer to many bacteriophages, as well as to, e.g., adenoviruses (van Oss, Gillman and Good, 1972), and retroviruses. The closer-range mechanism of adhesion in the case of viruses is usually by way of special sites on the extremities of the viral spicules, which specifically interact with receptors on the cell surface (see below for Specific Interactions); see also Fig. XXIII-3.

CELL ADHESION TO LOW-ENERGY SURFACES

In *in vivo* adhesion of cells to low-energy (hydrophobic) surfaces one almost invariably encounters the Vroman Effect (see above under Influence of Protein Adhesion), i.e., proteins (of different varieties) adsorb first, so that often cell adhesion onto low-energy surfaces can be fairly slight, but the cells that do adhere are sitting on, and are partly surrounded by, a substantial biopolymer film (Baier *et al.*, 1984). For the role of microbial surface hydrophobicity in microbial adhesion, see also Doyle and Rosenberg (1990).

ASPECIFIC AND SPECIFIC INTERACTIONS IN MICROBIAL ADHESION

Much of cell adhesion, and especially bacterial cell adhesion with a view to pathogenicity, is effected *via* specific biopolymers (lectins, adhesins, integrins, etc.), which can adhere (or are attached) to various solid substrates and which can bind to specific receptors on the bacterial (or other) cell surface. For reviews, see, e.g., Isberg (1991) for bacterial adhesion to host cells, Shimizu and Shaw (1991) for lymphocyte interactions with extracellular matrix, Edelman (1983) for cell adhesion molecules, and Bell *et al.* (1984) for the contrast between aspecific repulsion and immunologically mediated specific binding. Interactions between cells, e.g., between potentially pathogenic bacteria and phagocytic leukocytes are also frequently mediated *via* specific antibodies (immunoglobulins) to the bacterial surface, which then adhere (equally specifically) with their Fc-extremity to the Fc-specific receptors on the phagocytic leukocytes' surface (see, e.g., van Oss, 1986). A few examples of specifically adhering bacteria are: various *Streptococcus* species, some of which may preferentially adhere to tooth enamel surfaces, others to epithelial surfaces, others again to underlying bacteria; *Escherichia coli*: intestinal lining cells, and other sub-species to the lining of the urinary bladder; *Shigella* to the intestinal epithelium; *Vibrio cholerae* to the mucous coating of the intestines; *Mycoplasma pneumoniae* to the tracheal epithelium; *Neisseria gonorrhoeae* to the urethral epithelium; etc. (Davis *et al.*, 1980, p. 556). Specific bacterial adhesion most often occurs *via* bacterial pili, which can most readily penetrate the aspecific polar (AB) repulsion field due to their small R; see Fig. XXIII-3. The role of specific cell adhesion in microbial pathogenicity is further treated in the following two sub-sections; the mechanism of specific attachment in general is discussed in a later section of this chapter; see also Wilson (2002) and Wilson and Devine (2003) for recent reviews on microbial adhesion.

MICROBIAL PATHOGENICITY AND NEGATIVE AND POSITIVE CELL ADHESION

Microbes, and especially bacteria, can be successful pathogens by being able to adhere (specifically or aspecifically) to various other cells or tissues (or other biological surfaces, such as teeth), but they also may be successful pathogens by the very fact that they can (aspecifically) *avoid* adhesion to other (especially phagocytic) cells. Thus an important category of pathogenic bacteria consists of bacteria which can surround themselves with a thick, slimy, exceedingly hydrophilic, hydrated protein or polysaccharide layer, called a *capsule*. Encapsulated bacteria, such as *Streptococcus pneumoniae, Streptococcus pyogenes, Staphylococcus aureus* (S. Smith), *Escherichia coli* (type 0111), *Klebsiella pneumoniae, Salmonella typhimurium* ("smooth" strains), and *Haemophilus influenzae* (group B) (van Oss and Gillman, 1972a; Cunningham *et al.*, 1975; van Oss *et al.*, 1975, p. 29) all are able to avoid becoming phagocytized, owing to the capacity of their capsules to repel all blood cells, including phagocytic leukocytes, principally (i.e., for at least 90%) through a net hydrogen-bonding (AB) repulsion. It should be recalled (cf. Table XXIII-1) that leukocytes have a rather low (negative) surface potential, so that their electrostatic repulsive interaction with other cells (whether these have a significant negative surface potential or not) remains quite minor. In this manner all the above-mentioned pathogenic microorganisms can evade phagocytosis and multiply unhindered in the host. If, however, the host has formed "opsonizing" antibodies (see below) against the capsular substance of such bacteria, through earlier encounters with these organisms, or through prior immunization (vaccination) with just the capsular substance, phagocytosis can ensue. Before the advent of antibiotics the morbidity and mortality from the above-mentioned pathogenic bacteria were very high, because it takes about a week before sufficient numbers of effective antibodies are formed against bacterial antigens which have not been encountered before by the host. However, most other bacteria which one may encounter, and which have no capsules, are usually much more hydrophobic; they thus become quickly phagocytized and are, therefore, in general non-pathogenic (van Oss and Gillman, 1972a; van Oss *et al.*, 1975.)*

On the other hand, if and when a sufficient number of microorganisms have evaded phagocytic uptake (allowing them to continue their pathogenic aggression), they depend in many if not in most cases, upon adhesion to their preferred cells or tissues. As the best way to avoid phagocytic uptake (especially for bacteria) is *via* aspecific repulsion through mainly polar (AB) forces, microorganisms following this pathway usually cannot at the same time aspecifically adhere to other surfaces. They thus depend on *specific* adherence to their tissues of choice, *via* the lectins, adhesins, etc., already mentioned in the preceding sub-section.

OPSONIZATION AND PHAGOCYTOSIS

The vertebrate organism, however, is not powerless in arming itself against invading microorganisms, even if the latter can escape the first line of phagocytic defense, because the vertebrate, and especially the mammalian and avian systems, can mobilize a sophisticated array of specific and aspecific immunological recognition and defense measures to deal with foreign particles and macromolecules. To that effect "opsonizing" (from the Greek, το οψον, a spice, a seasoning) antibodies, usually of the IgG1 and IgG3 classes are employed. Once specific antibodies (Ab) have been elicited to the capsular biopolymers of encapsulated bacteria, these Abs will bind to the outer periphery of the capsules with their Ab-specific site attached to the capsule and their Fc-tail protruding away from it. The Fc-tail then can bind to the Fc-receptor on the surface of phagocytic cells, and phagocytosis will ensue. After specific binding of Abs of the classes IgG1, IgG3 and IgM, another opsonin, in the guise of serum complement factor C3b (an aspecific serum factor, which, however, will specifically bind to IgG and IgM, once these immunoglobulins are bound to their specific Ags) can further enhance binding to the phagocytic cell surface, *via* specific C3b receptors. Thus, especially after opsonizing Abs have been formed to the capsular material of pathogenic bacteria, a very efficient defense is mounted and

---

* Several decades ago it was still generally believed that interfacial tensions always had to have a positive value and that surface tensions were single indivisible entities which could, in all cases, be determined by means of contact angle measurement with a single liquid, e.g., water (cf. Chapter XIV). Thus the interaction, in water, between all particles or cells always was thought to be attractive, even if rather feeble in the cases of exceedingly hydrophilic cells (cf. van Oss *et al.*, 1975, p. 95). In actual fact, given the surface properties of phagocytic cells (Table XXIII-2) as well as of the only slightly less hydrophilic bacteria such as noncapsulated *Staphylococcus epidermidis* [$\gamma^{LW} = 35.1$, $\gamma^{\oplus} = 0$ and $\gamma^{\ominus} = 50$ mJ/m^2, cf. van Oss (Bioadhesion, 1991)], phagocytic cells still repel these relatively hydrophobic micro-organisms. The main mechanism for the establishment of contact between phagocyte and bacterium is to be found in the adsorption of IgG$_1$ and IgG$_3$ molecules (given the surface properties of IgG listed in Table XVII-6). Thus, IgG should adsorb onto S. *epidermidis* (noncapsulated) ($\Delta G_{1W2} = -7.6$ mJ/m^2), but not onto S. *epidermidis* (capsulated) ($\gamma^{LW} = 25.6$, $\gamma^{\oplus} = 6.1$ and $\gamma^{\ominus} = 47.2$ mJ/m^2, so that $\Delta G_{1W2} = +4.1$ mJ/m^2). It was indeed observed that non-immune polyclonal IgG, as well as monoclonal IgG$_3$ adsorb significantly onto S. *epidermidis* (non-capsulated) (Absolom *et al.*, 1982); see also the following section.

the bacteria are eliminated (Stinson and van Oss, 1971; van Oss and Gillman, 1972b; van Oss et al., 1975; Bell et al., 1984).

Less pathogenic, or non-pathogenic bacteria (which represent the vast majority) mainly adhere to phagocytic cells, because their surfaces are more hydrophobic, which causes them to adsorb serum IgG1 and IgG3 aspecifically, upon which these immunoglobulins specifically bind to phagocytic cells via the latter's Fc-receptors (Absolom et al., 1982); see also footnote on page 365.

Thus in most cases opsonization occurs through either the specific binding, or the aspecific adsorption of immunoglobulins (usually IgG1 and IgG3), which then bind to the Fc-receptors on phagocytic cells, which results in phagocytosis of the microorganisms and their subsequent digestion and disposal. However, here also the first contact is usually made via those Fc-receptors on phagocytic leukocytes which are situated on the ends of pseudopodic processes with a small radius of curvature, R, which serve to overcome the aspecific polar (AB) repulsion between cells (see above). However, in the presence of somewhat higher than normal physiological (0.06 to 0.12%) glucose concentrations, i.e., at 0.2 to 0.8% glucose, phagocytic uptake of bacteria is severely diminished (van Oss, 1971). The reason for this is the same as for the decrease in cell adhesiveness at elevated glucose concentrations discussed earlier: at these glucose concentrations phagocytic leukocytes (granulocytes, monocytes and macrophages) become spherical and are no longer able to extend pseudopodia with a small R (van Oss, Gillman and Good, 1974), and thus have increasing difficulties of overcoming the aspecific polar (AB) intercellular repulsion field. These phenomena may explain the increased propensity for bacterial infections among less than optimally controlled diabetics (van Oss, 1971). The mechanism by which such increased glucose levels induce cell sphericity and impede pseudopod formation is osmotic.

However, in some cases the vertebrate organism is better served when foreign invading microorganisms (especially viruses) can be prevented from adhering to (and being internalized by) cells, because, e.g., viruses can only multiply and be infectious when they can get into host cells. The defense here is effected by another type of antibody, i.e., IgA and especially secretory IgA (sIgA), which is even more hydrophilic than the other Abs and for which phagocytic cells have few or no receptors. Thus, e.g., viruses, specifically coated with IgA-class Abs, are repelled (again mainly through polar, AB forces) by phagocytic and other cells and can be harmlessly excreted. This line of defense is especially important in the upper respiratory (removal by sneezing, coughing, etc.) and the lower digestive tracts.

CELL FUSION

For two cells to be able to fuse they must be able to approach each other closely and their surfaces must, at least locally, attract each other more strongly than the attraction each one of them has for water (see also Ohki, 1988). Thus:

$$W_{iwi} > W_{iw} \qquad\qquad [XXIV\text{-}5]$$

so that (cf. eqs. [IV-17] and [IV-18]):

$$2\gamma_{iw} > (\gamma_i + \gamma_w - \gamma_{1w}) \qquad\qquad \text{[XXIV-6]}$$

or:

$$3\gamma_{iw} > (\gamma_i + \gamma_w) \qquad\qquad \text{[XXIV-6A]}$$

which is the same prerequisite as the one found for the occurrence of cavitation in "hydrophobic" interactions (Chapter XVIII). This is reasonable, because the condition favoring cavitation is also that the attraction between two ("hydrophobic") surfaces, immersed in water, be greater than the attraction between each such surface, and water. However, there is an important difference between cavitation and the conditions which prevail in the preliminary stages of cell fusion: when two cells approach each other closely (driven by the W_{iwi} attraction), interstitial water molecules can be expelled even when $W_{iw} > W_{iwi}$ (expressed in energy per unit surface area). This is because for the expulsion of single water molecules of hydration, the work of hydration and the work of adhesion (W_{iw} and W_{iwi}) is more properly expressed in energy units per molecule or particle, e.g., in kT units. Thus, when $W_{iwi} \approx 50$ mJ/m^2 and $W_{iwi} \approx 120$ mJ/m^2 for the interaction between two cells with an incipient contactable surface area, $S_c \approx 10$ nm^2, while for individual water molecules $S_c \approx 0.1$ nm^2, it can be seen that whilst the intercellular energy of adhesion, $W_{iwi} \approx 123$ kT, the free energy of hydration, $W_{iw} \approx 3$ kT, which allows the water molecules of hydration to be expelled individually. (It should be noted that $W = -\Delta G$).

One factor which is crucial for cell fusion is the presence of Ca^{++}. Ca^{++} ions here play the role of plurivalent counterions, vis-à-vis the negatively charged cell surfaces: Ca^{++} decreases the (negative) ζ-potential of the cells, while at the same time they function as electron-acceptors, thus causing a decrease in the cells' (monopolar) $\gamma_i^{\ominus}$ (see Chapter XIX for the linkage between EL and AB forces). This also causes the cells' surfaces to become more hydrophobic, so that instead of repelling, the cells will attract each other (van Oss *et al.*, in *Molecular Mechanisms of Cell Fusion*, 1988). The presence of extracellular polyethylene oxide further helps Ca^{++}-induced cell fusion (Arnold *et al.*, 1988), by exerting outward pressure on the cells (in the manner of aqueous phase separation; see Chapter XXI), which forces them more closely together.

There is one further condition for cell fusion, i.e.,

$$W_{ii} > W_{iwi} \qquad\qquad \text{[XXIV-7]}$$

or:

$$\gamma_i > \gamma_{iw} \qquad\qquad \text{[XXIV-8]}$$

This is because ultimately, for fusion to take place, there must be a direct linking between the phospholipids of two cell membranes (governed by W_{ii}), rather than just an attraction between the phospholipids of two cell membranes, while still having a layer of water between them (W_{iwi}), however exiguous that layer of water may be (van

Oss and Good, 1984). In practice the condition expressed in eq. [XXIV-8] is satisfied for all phospholipids containing alkyl groups with $\gamma_1 \geq 21.8$ mJ/m^2 ($\approx \gamma_w^{LW}$), in other words, comprising nonyl groups, or larger. This is always the case, because with phospholipids comprising octyl groups or smaller, it would, for the same reasons, be impossible to form coherent cell membranes, in water.

CELL FREEZING AND NEGATIVE AND POSITIVE ADHESION TO ADVANCING ICE FRONTS

As shown by van Oss, Giese, Wentzek *et al.* (1992), relatively slowly advancing freezing fronts *repel* hydrophilic micro and macro-solutes (including electrolytes), as well as suspended hydrophilic particles. This also applies to cells (e.g., peripheral blood cells) suspended in plasma, or in various isoosmotic buffers. Thus, as has been shown theoretically (and confirmed experimentally), cells (e.g., red blood cells) are *excluded* by advancing freezing fronts, together with the surrounding micro- and macro-solutes (van Oss, Giese and Norris, 1992; see also Kuivenhoven, 1966). This puts the cells under considerable osmotic stress and represents the principal source of cell destruction upon freezing. It has also been shown theoretically and experimentally that in aqueous suspensions, admixed with appropriate concentrations of a cryoprotectant (e.g., glycerol), cells are *engulfed* by advancing freezing fronts. When cells become engulfed by the freezing front, they do not undergo any osmotic stress and remain undamaged when frozen. In ordinary isotonic aqueous media, when attempts are made to freeze cells, these cells, as well as the surrounding solutes, are compressed in a continuously shrinking compartment through *exclusion* by the advancing ice front, so that the extracellular osmotic pressure increases considerably. The normal extracellular osmotic pressure of about seven bars can increase tenfold during the freezing process, before the counter-pressure of about 75 bars, exerted on the cells (at close range) by the freezing front is overcome and cell inclusion commences. Cell *exclusion* puts a considerable strain on the cells, as the osmotic stress becomes compounded by a purely mechanical compression. On the other hand, when the polar properties of the liquid medium are changed through the admixture of polar water-soluble solvents or solutes with a smaller γ^{AB} than that of water, the liquid medium can be sufficiently modified so as to favor *inclusion* of the cells by the advancing ice front during freezing. Water-soluble polar solvents or solutes most used for this purpose are: glycerol, dimethyl sulfoxide (DMSO), formamide and polyvinyl pyrrolidone (PVP); see Tables XXIV-5 and 6.

Positive values for ΔG_{1W2} (see Table XXIV-6) indicate that in water, erythrocytes are *excluded* by the advancing ice front, whilst negative values for ΔG_{1W2} signify that (e.g., in glycerol) erythrocytes should be *engulfed* by the advancing freezing front and that even at glycerol concentrations down to about 40% (assuming a linear dependence of ΔG_{1W2} on glycerol concentration), the cells would still be engulfed. Freezing experiments with (glutaraldehyde-fixed) erythrocytes, in water and in an aqueous solution of 50% (v/v) glycerol, confirmed these predictions; see van Oss, Giese and Norris (1992). It can also be seen in Table XXIV-6, that formamide should be slightly less effective as a cryoprotectant for erythrocytes than glycerol, while DMSO should be more effective, as is indeed the case (Meryman, 1966). Polyvinyl pyrrolidone (PVP) is not normally a liquid, but may be regarded as contributing to the liquid properties of a mixture, when present in solution. It would then appear to have to be highly effective according to Table XXIV-6, but one must not lose

TABLE XXIV-5

Surface Tension Components and Parameters of Ice, Erythrocytes (glutaraldehyde-fixed), Water, Glycerol, DMSO, and Formamide, at 0°C, in mJ/m^2

Compound	γ	γ^{LW}	γ^{AB}	$\gamma^{\oplus}$	$\gamma^{\ominus}$
Ice[a]	69.2	29.6	39.60	14.00	28.0
Erythrocytes[b]	36.6	35.2	1.36	0.01	46.2
Lymphocytes[c]	27.7	27.7	0.00	0.00	32.2
Water[a,d]	75.8	22.8	53.00	26.50	26.5
Glycerol[d,e]	65.8	35.0	30.80	4.02	58.9
DMSO[d]	45.1	36.9	8.21	0.51	33.0
Formamide[d,e]	59.5	40.0	19.50	2.38	40.0
Polyvinyl pyrrolidone (PVP)	43.4	43.4[f]	0.00	0.00	29.7[g]

[a] From van Oss, Giese, Wentzek et al., 1992.

[b] From $\theta_{DIM} = 49°$, $\theta_{\alpha\ Br\text{-}naphthalene} = 38°$, yielding γ^{LW}, and from $\theta_{glycerol} = 51°$, $\theta_{formamide} = 51°$ and $\theta_{ethylene\ glycol} = 47°$, yielding γ approx 0 and $\gamma = 46.2$, using the Young equation for polar systems.

[c] See also Table XXIII-2.

[d] By extrapolation to 0° and upon assuming that the ratios γ^{LW}/γ^{AB} and γ/γ, valid at 20°C, remain unchanged; DMSO = dimethylsulfoxide.

[e] See Table XVII-10; extrapolated to 0°C.

[f] From $\theta_{DIM} = 30°$ and $\theta_{\alpha\ Br\text{-}naphthalene} = 17°$.

[g] From $\theta_{H2O} = 56.6°$, $\theta_{glycerol} = 51\text{-}8°$ and $\theta_{formamide} = 49.5°$; see Table XVII-5.

From van Oss, Giese and Norris, 1992.

TABLE XXIV-6

ΔG_{1W2} Values (in mJ/m^2) for Interactions Between Erythrocytes and Ice and Between Lymphocytes and Ice in Various Aqueous Liquids at 0°C

	ΔG_{1W2} (in mJ/m^2)	
	Erythrocytes and ice, in water	Lymphocytes and ice, in water
H$_2$O	+4.5	+2.3
Glycerol	−6.1	−3.3
Formamide	−5.8	−2.2
DMSO[a]	−7.1	−1.3
PVP[b]	−11.6	−4.8

[a] DMSO (dimethylsulfoxide).

[b] PVP (polyvinyl pyrrolidone) is not normally a liquid, but it can be used in aqueous solutions.

From van Oss, Giese and Norris, 1992.

sight of the fact that this polymer (as well as other polymers) tends to adsorb to the surface of cells, thus changing the cell surface properties, which would decrease the negative value of ΔG_{1W2}. Lymphocytes (Table XXIV-6) are less repelled by ice than erythrocytes and thus require lower concentrations of additives for cryoprotection, as is indeed well known (Farrant and Knight, 1978).

It could be demonstrated experimentally (van Oss, Giese and Norris, 1992) that at the slow freezing rate of less than 2.75 μm/second initially, and only 50% of that value soon thereafter, i.e., under conditions where surface thermodynamic interactions may be held to be the principal driving force (Neumann *et al.*, 1979, 1982), human erythrocytes are completely *excluded* by the advancing freezing front. On the other hand, in a 50% (v/v) glycerol/water mixture, the erythrocytes are totally *included* by the freezing front. It should, in addition, be realized that the electrostatic interaction between the cells and the advancing ice front may be considered to be negligible. Even the electrostatic free energy of interaction *between cells* (see Table XXIII-2) is less than 10% of the AB interaction, under physiological conditions, and the ζ-potential at the ice-water interface—while not known with precision—may be estimated to be quite low, so that the total cell-ice EL interaction, in water, may be taken to be small.

The influence of the freezing velocity is of some interest in cell freezing. It has been noted earlier (Omenyi, 1978; Neumann *et al.*, 1979, 1982) that at high velocities (u), an advancing freezing front can *engulf* particles or cells even though a static repulsion exists between the advancing solidification front and the particles or cells. At high velocities (u > 5 μm/s), the advancing freezing front can engulf cells too fast for the repulsive forces to prevail. For every system, there is a "critical solidification velocity" beyond which the latter situation is favored (Omenyi, 1978). However, an advancing freezing front velocity of the order of approximately 2 μm/s or less is held to be sufficiently slow to be safely below that critical velocity, so that then, for all practical purposes, thermodynamic equilibrium conditions dominate. The experimental conditions described by van Oss, Giese and Norris (1992) therefore fall into the category of "slow freezing." It thus becomes understandable why when "fast freezing" is practiced, lower concentrations of cryoprotectant additives suffice to safeguard against cell damage than with slow freezing (Seidl, 1978). During "slow freezing," maximum repulsive AB forces prevail between the advancing ice front and the cells, so that the full theoretical amount of cryoprotectant is required to reverse the repulsion. But when "fast freezing" is practiced, the very speed of the advancing freezing front favors cell or particle engulfment, so that in such cases somewhat lower concentrations of cryoprotectants suffice to achieve engulfment. Indeed, when rapid freezing is adopted, up to 70–85% cell recovery is possible even without cryoprotectants (Meryman, 1966).

In view of the important role played by the concept of the "critical freezing velocity" (Omenyi, 1978; Neumann *et al.* 1979, 1982), it would seem more useful to express cell freezing rates in units of μm/s, rather than in degrees C/min. (van Oss, Giese and Norris, 1992).

The order of cryoprotectant effectiveness appears to be DMSO > glycerol > formamide with the cryoprotectants that are most frequently mentioned (Meryman,

1966). This also follows from the ΔG_{1W2} values shown in Table XXIV-6. When used with slow freezing, the cryoprotectant action of PVP is not very great; the reason for this probably lies in the tendency of PVP to adsorb onto the cell surface, which reduces the negative value of ΔG_{1W2}. However, when applied in the fast freezing mode, PVP (at concentrations as low at 7.5%) still is an effective cryoprotectant for erythrocytes (Meryman, 1966). This is most likely due to the fact that even at such low concentrations, PVP still strongly adsorbs onto the cells' surfaces, which reduces the repulsive value of ΔG_{1W2} for erythrocytes in water from +4.5 to +1.4 mJ/m². Thus, fast freezing then may completely overcome the slight residual repulsion, allowing the cells to be completely engulfed by a rapidly advancing freezing front.

The major function of cryoprotectants thus turns out to be their ability to modify the surface thermodynamic properties of the *liquid*, so as to allow the cells to be *engulfed* by the advancing freezing front. It is therefore not surprising that nonpenetrating additives can be quite as effective as cryoprotectants that can readily penetrate the cell membrane (Farrant and Woolgar, 1969) and enter the cell. In principle, nonpenetrating cryoprotectants should be even more efficient in the case of concentrated cell suspensions because their total activity can be applied outside the cells, where it is needed, without any loss through penetration inside the cells. It should be stressed that sugars and polysaccharides, and even peptides and proteins, whether penetrating or otherwise, have much the same properties as glycerol, as far as their capacity for reducing the polar (AB) free energy of cohesion of aqueous media is concerned. For polypeptide cryoprotectants, see Yang *et al.* (1988), and for the cryoprotectant action of a disaccharide (trehalose), see Tsvetkov *et al.* (1989).

ASPECIFIC AND SPECIFIC INTERACTIONS

Specific interactions between: *antigens* (Ag) and *antibodies* (Ab), *lectins* and *carbohydrates*, *enzymes* and *substrates*,[*] and in general, *ligands* and *receptors* only involve the non-covalent forces (the LW, AB, EL and BR forces, discussed in Chapters II to VI, and XX). Thus, the same type of forces which *in vivo* are sufficiently *repulsive* to keep cells and biopolymers in suspension or solution are involved in the *attraction* between the same cells and/or biopolymers when they become engaged in specific interactions between receptors and their corresponding ligands. In general, there is a need for, e.g., blood cells, to repel each other to remain in stable suspension, and for, e.g., plasma proteins to repel each other to remain in solution. But in relatively exceptional and well-defined cases, the need can arise for individual cells and/or biopolymers to adhere to each other, e.g., to trigger a specific chain of events, such as those leading to the disposal of pathogenic microorganisms. As in the general case the interactions are repulsive, while in the special circumstance they must be attractive, safeguards are needed to insure that macromolecules and/or cells that *can* attract each other only do so in the exceptional condition where the attraction must be allowed to override the normally prevailing repulsion. This is achieved *in vivo* thanks to the fact that:

---

[*] In some enzyme-substrate interactions, disulfide bonds may also occur.

1. aspecific repulsive forces operate between much larger surface areas than specific attractive forces, and:

2. specifically reacting molecular chains (e.g., immunoglobulin paratopes) only specifically react with their corresponding counterparts (e.g., antigenic sites, or epitopes), when the latter are actually present, which normally happens infrequently.

The crucial advantage of the smallness of the specifically reacting moieties or macromolecules is that they can overcome the normally prevailing repulsion fields any time that aspecific repulsion is locally smaller than the specific attraction; see Fig. XXIV-1 and also Fig. XXIII-3. Preferential attraction between receptors (e.g., Fc receptors on cells) and ligands (e.g., Fc tails of antibody molecules) occurs when after complex formation between several Ab molecules and an Ag, *several* Fc tails can attach to Fc receptors on one cell (van Oss, Absolom and Michaeli, 1985). Thus if the binding constant of one Fc tail to one Fc receptor, $K_{ass} = 10^4$ L/M, then, for the interaction of an Ag/Ab complex which can attach to a cell with three Fc tails simultaneously, the binding constant K_{ass} for the complex jumps to 10^{12} L/M (cf. eq. [XXIV-3]). This amplification mechanism explains why Ag/Ab complexes with one or two Ab molecules remain largely unnoticed by the phagocytic system, and only Ag/Ab complexes with three or more Ab molecules readily become phagocytized (van Oss, Absolom and Neumann, 1984). For the same reason biological recognition molecules for Fc have an advantage in being multivalent: the first serum complement factor, Clq, is hexavalent, i.e., it has six separate sites, each one of which can bind one Fc moiety of an immunoglobulin molecule. This is essential, because Clq molecules circulate freely in the bloodstream, surrounded by an excess of free immunoglobulin molecules, which it must not, in the normal course of events, bind irreversibly. Only a much more irreversible binding of complexes comprising immunoglobulins, to two or more of the binding sites of Clq activates the complement cascade (Walport, 1989; Cooper, 1985; Morgan, 1990).

SMALLNESS OF SPECIFIC SITES

Contrary to aspecific interactions, the contributions of the separate constituent chains cannot be measured directly by contact angle or electrokinetic determinations. This is due to the smallness of the actual specifically interacting determinants, which usually have an interacting surface area of the order of only 0.5–7.5 nm², representing from 0.3 to 5.0% of the total surface area of a specific biopolymer (e.g., the two paratopes of an antibody of the IgG class), or of the biopolymer which it can specifically recognize (e.g., one epitope of a serum albumin molecule). Also, even if in some cases one can readily isolate the specific epitope* (or more rarely, paratope*) molecules, due to the pronounced variability of the detailed molecular composition

---

* An epitope is one antigenic determinant (often one among several different ones) of an antigen molecule or particle, and a paratope is one of the (usually two) identical antibody-active sites of an antibody molecule. In a given Ag-Ab system, an epitope of the Ag binds specifically to one of the paratopes of the Ab.

of the epitope of paratope, contact angle and/or electrokinetic measurements on such isolated moieties will usually still yield little information on the type and strength of the binding modes of each of the individual constituent monomers. In addition, decomposing the epitopes or paratopes into their constituent monomers will yield neither the relative contribution of each type of bond, nor the quantitative bond strength of each monomer (e.g., of each amino acid in a peptide) vis-à-vis the opposing monomer. Even if one could synthesize monotonous polypeptides for each amino acid of a given epitope, and measure their γ^{LW}, $\gamma^\oplus$, $\gamma^\ominus$ values and their ζ-potential, to obtain their energetic contributions to the total specific bond, the problem still would not be solved. This is because one cannot add up all these contributions to obtain the total binding energy of the epitope, without situating each of them precisely into the three-dimensional configuration of the epitope within the complete antigenic macro-molecule. Such three-dimensional configurations are known for only a handful of antigens. In addition, all these properties also have to be ascertained for the corresponding paratope and the two (epitope and paratope) then must be precisely fitted together in a three-dimensional array. Unfortunately, the three-dimensional (and even the linear) configurations of paratopes are known in only a few cases. Work is proceeding slowly to unravel the three-dimensional configuration of a few intermolecular epitope/paratope interactions (Amit *et al.*, 1986), but little, if anything, has been done up to now to measure the value of each of its different contributing physical interaction forces. For the time being, indirect approaches remain the sole means by which one can estimate the contribution to specific interactions of the different types of intermolecular forces (van Oss, J. Molec. Recognition, 1990).

INTERFACIAL (LW + AB) SPECIFIC INTERACTIONS

By one such indirect approach it is possible in certain cases to estimate the total interfacial (IF) contribution to a specific interaction (IF = LW + AB). The IF interaction energy can be determined, e.g., in cases where either ligand or receptor is known to have no, or only a negligible electrostatic potential, and where the surface properties of the active sites can be estimated. An example of such a case is the dextran/anti-dextran system. Dextran is one of the least charged biopolymers: its ζ-potential is –0.05 mV at ionic strength 0.15, and –0.55 mV at ionic strength 0.015 (van Oss *et al.* 1974). In practice, ζ-potentials smaller than $\approx$ 10 mV have no measurable influence on intermolecular or interparticle interactions. Thus, electrostatic interactions play no significant role in this system. The association constant found in the dextran (rabbit IgG) anti-dextran system is (Edberg *et al.*, 1972; Kabat, 1976, p. 126): $K_{ass} \approx 10^5$ L/mole. Thus, according to eq. [XXIV-3], $\Delta G\gamma^{IF}_{1W2} = -15.5$ kT. If we make the assumption that the paratopic cleft in the IgG (anti-dextran) molecule has the approximate hydrophobicity of dry IgG, at pH 7 (cf. Table XVII-6), with $\gamma^{LW}_{IgG} = 42$ mJ/m², $\gamma^\oplus_{IgG} = 0.3$ mJ/m² and $\gamma^\ominus_{IgG} = 8.7$ mJ/m², and for dextran (Table XVII-8), $\gamma^{LW}_{DEX} = 41.8$ mJ/m², $\gamma^\oplus_{DEX} = 1$ mJ/m² and $\gamma^\ominus_{DEX} = 47.2$ mJ/m², then (cf. eq. [XXIV-2]), $\Delta G^{IF}_{1W2} = -7.13$ mJ/m², so that in units of kT [assuming the contactable surface area, S_c, of the paratopic cleft to be about 8 nm² (see, e.g., Kabat, 1976, p. 124)], $\Delta G^{IF}_{1W2} = -14.1$ kT per paratope/dextran epitope pair. This correlates reasonably well with the value of -15.5 kT found *via* the association constant of 10^5 L/mole. There thus does not appear to be significant scope for the formation of a further, secondary, *direct* hydrogen bond to occur between one of the glucose moieties of the isomaltohexaose epitope and a paratopic amino acid. It is important to note that there is a significant interfacial primary *attraction* between epitope and paratope. In this case the attraction (ΔG^{IF}_{1W2}) is for 91% due to LW forces, dextran being so hydrophilic, that the AB contribution to its attraction to a moderately "hydrophobic" surface is relatively small. Thus, while there is a sizeable close-range attraction, the attraction at a distance is fairly slight: at $\ell = 1.0$ nm it already is down to about one kT unit. This is why dextran, notwithstanding its high molecular weight, does not easily give rise to the formation of antibodies: it behaves more like a hapten than an antigen, i.e., it can only elicit a strong immune response when several dextran molecules are linked to a carrier macromolecule (Edberg *et al.*, 1972).

ROLE OF DIRECT HYDROGEN BONDING IN SPECIFIC INTERACTIONS

There is no indication from the Ag-Ab system described above, or from other Ag-Ab systems which might be putative candidates for direct H-bonding mechanisms, that *direct* H-bonding plays a significant role in Ag-Ab reactions. One further reason for this lies in the fact that *direct* H-bonds only occur when both the precise bond angle and the exact bond distance can be achieved. The direct H-bond thus cannot exert its attraction at distances significantly exceeding 0.2 nm, and as such cannot play a role as the *primary* attractive force in Ag-Ab interactions.

Direct H-bonds may, on occasion, occur as *secondary* Ag-Ab bonds, either between epitope and paratope, or between opposing chains in the vicinity of epitope and paratope, which are brought closely enough together through the epitope/paratope interaction to form aspecific (secondary) bonds (van Oss, *Structure of Antigens*, 1992, Ch. 6), one possible example of secondary direct H-bond formation occurring within the epitope/paratope interface is the reaction between (negatively charged) double-stranded DNA and high-avidity (positively charged) anti-DNA Abs (van Oss, Smeenk and Aarden, 1985). The primary DNA–anti-DNA bond, which is purely electrostatic, could be prevented from forming in 0.2 M to 1.0 M NaCl, but once formed, could not be dissociated even in 5.0 M NaCl. However, under conditions of pH 11 to 12, these DNA–anti-DNA complexes could equally readily be prevented from forming, as they could be dissociated, once formed; see also Smeenk *et al.* (1983). The impossibility to dissociate these Ag/Ab bonds, once formed, at salt concentrations which easily could prevent them from forming in the first place has prompted the speculation that the *electrostatic* Ag/Ab bonds, after formation, might evolve into *direct H-bonds*, as the former can, but the latter cannot be dissociated at high ionic strengths (van Oss, Smeenk and Aarden, 1985). However, in the light of more recent findings on the linkage between electrostatic and Lewis acid-base interactions (cf. Chapter XIX), these observations can be equally plausibly explained by a hydrophobic interaction resulting from the neutralization of the negatively charged epitope by the positively charged paratope in this system; see the following sub-section.

H-Bonds Used in Ag-Ab Modeling

Apart from the fact that direct H-bonding between Ag and Ab is an appealing idea, and that many authors can be quoted who have expressed that idea, there are no hard experimental data to substantiate it. The widespread practice of optimally fitting direct H-bonds between epitope and paratope, among practitioners of computer modeling of Ag-Ab and similar ligand-receptor interactions, may therefore have little relevance to the Ag-Ab bonds which occur in real life. On the other hand, much can still be learned from the many careful experimental measurements done on a vast array of actual hapten-Ab and Ag-Ab interactions, by Pressman and Grossberg (1968), and many others (see, e.g., Nisonoff *et al.*, 1975, Ch. 2; Glynn and Steward, 1976; Kabat, 1976).

Whilst there are no strong theoretical or experimental arguments *for* an important role of H-bonds in Ag-Ab bonding, serious doubts can be raised *against* their widespread occurrence. The two main ones are: (1) direct H-bonds are too short-range to be able to attract Ag to Ab (and to help to overcome the normally prevailing macroscopic repulsion), and (2) direct H-bonds require a very narrow range of precise bond angles, which are unlikely to occur frequently in the relatively haphazard apposition of the various amino acids (or other moieties) of epitope and paratope. Argument 1 largely excludes direct H-bonding from playing an important role in the primary Ag-Ab interaction, while argument 2 would indicate that whilst direct H-bonding in the secondary, close-range interactions between Ag and Ab could occasionally be possible, it is unlikely to be predominant.

Role of H-Bonding in Specific Interactions in General

Thus, whilst *direct H-bonding* is unlikely to occur in *primary* specific interactions and only occasionally plays a role in *secondary* interactions, this should not detract from the fact that hydrophobic bonds, which are also principally due to *hydrogen-bonding*, i.e., to the hydrogen-bonding energy of cohesion of the surrounding water molecules (cf. Chapter XVIII), often are important in primary Ag-Ab formation, such as in the cases of epitopes which contain dinitrobenzene or dinitrophenyl, or trinitrobenzene or trinitrophenyl moieties, and also with epitopes comprising azopyridine, halopyridines, azotoluidine, azobenzene, etc. (Pressman and Grossberg, 1968). It should again be stressed that "hydrophobic" interactions can readily occur between one "hydrophobic" and one hydrophilic moiety, immersed in water (cf. Chapter XVIII). In most cases in this category, the hydrophobic surface of the specifically interaction pair is located on the paratope or, more precisely, in the *paratopic cleft*. (In other cases, it can be the *epitopic cleft* which is hydrophobic.) The reason for this is easy to understand: Due to the overall macroscopic repulsion between biopolymers and/or cells *in vivo*, the specific recognition moieties of at least one of the Ag-Ab or other ligand-receptor partners must be located on a protruding site with a small radius of curvature. But the placement of such protruding hydrophobic sites on, e.g., circulating immunoglobulin molecules would cause them to adhere aspecifically to many other hydrophilic biopolymers and/or cells, and thus to form precipitates, agglutinates or other life-threatening thrombi. However, the location of hydrophobic paratopic recognition sites inside the *clefts* of the antibody-active moieties of immunoglobulin molecules (or of the epitopes) effectively obviates this problem.

Secondary bonds, forming subsequent to the primary specific bond formation (be it interfacial or electrostatic, or a combination of both), practically always are of the hydrophobic (AB) variety (van Oss, *Structure of Antigens*, 1992, Ch. 6).

SPECIFIC ELECTROSTATIC INTERACTIONS

Electrostatic interactions causing the specific attraction between epitope and paratope are quite common. EL interactions play a role with the majority of protein Ag's and with a great many haptens, such as benzoates, arsonates, amines, phosphonates, anilates, hippurates, phthalates, propionates, etc. (Pressman and Grossberg, 1968).

The larger antigen, bovine serum albumin (BSA), which has six immunodominant surface epitopes (Atassi, 1984), also is specifically attracted to the anti-BSA Ab through EL forces. Epitopes 1 to 3 (and especially 1 and 2) show rather strong similarity; among other points of similarity, each has three neighboring basic amino acids, which would tend to be attracted to acidic amino acids in the paratope. This agrees well with the fact that whilst BSA binds strongly to anti-BSA Abs at neutral pH, at pH 9.5 virtually no reaction between BSA and anti-BSA Abs can be observed (van Oss, Absolom and Bronson, 1982). The primary specific attraction is therefore clearly electrostatic: at pH 9.5 the basic epitope moieties which consist of Arg-Arg-His (1 and 2) and of Lys-His-Lys (3) (Atassi, 1984), would lose most (if not all) of their dissociated basicity. However, once BSA–anti-BSA complexes have been formed at neutral pH, increasing the pH to 9.5 is to no avail for dissociating them. Thus other, non-electrostatic, secondary bonds form, once contact is made through EL interactions, which are impervious to such an increase in pH. These are due in part to the hydrophobic attraction between the paratope and the more apolar peptide moieties of the epitopes: Leu-Tyr-Glu-Ile-Ala (epitope 1), Leu-Tyr-Glu-Tyr-Ser (epitope 2), and Leu-Val-Glu-Leu-Leu (epitope 3) (Atassi, 1984). Another contribution to a secondary hydrophobic (AB) attraction is furnished by the interaction between chains close to the epitope and paratope chains which are brought more closely together through the primary reaction. Finally, it might be hypothesized that the mutual neutralization of the ionized moieties upon the encounter between epitope and paratope could not only lead to the local abolition of their respective ζ-potentials but also, concomitantly, to a strong increase in their hydrophobicity (see Chapter XIX for the linkage between EL and AB interactions).

BRIDGING WITH PLURIVALENT COUNTERIONS—CALCIUM BRIDGING

When both epitope and paratope are negatively charged, they can bond together in the presence of plurivalent cations, e.g., Ca^{++} (see, e.g., Liberti, 1975). In these cases it is not only bridging through the Epitope– Ca^{++} – Paratope⁻ which occurs, but upon interaction with Ca^{++} both epitope and paratope become more hydrophobic, through the neutralization of the Lewis basicity of these entities by the Ca^{++} ions which are Lewis acids. Thus admixture of Ca^{++} secondarily brings about an interfacial hydrophobic

(AB) attraction; see also the section on Cell Fusion, above, and Chapter XIX.

Plurivalent cations also play a role in other specific interactions; see, e.g., the role of Ca^{++} and of Mg^{++} in different parts of the complement cascade (Cooper, 1985; Morgan, 1990), and the requirement for Ca^{++} in enzymatic reactions (e.g., phosphorylase kinase), and the blood clotting cascade (prothrombin) (see, e.g., Stryer, 1981).

ANTIGEN-ANTIBODY BINDING HYSTERESIS

The hysteresis in Ag-Ab binding, which is equivalent to $\Delta G_{\text{secondary}}$, is the difference between $\Delta G_{\text{dissociation}}$ and $\Delta G_{\text{primary}}$, where $\Delta G_{\text{primary}} = \Delta G_{\text{prevention of association}}$ (van Oss, *Structure of Antigens*, 1992, Ch. 6). Thus:

$$\Delta G_{\text{hysteresis}} = \Delta G_{\text{secondary}} = \Delta G_{\text{dissociation}}$$
$$- \Delta G_{\text{prevention of association}} = \Delta G_{\text{total}} - \Delta G_{\text{primary}} \qquad \text{[XXIV-9]}$$

EL bonds do not often occur as secondary bonds. For instance, in the purely electrostatic system, DNA-anti-DNA, the fact that the same increases in pH that can prevent association also suffice to cause dissociation is indicative of an absence of secondary electrostatic bonds in this system (Smeenk *et al.*, 1983). Secondary EL bonds are likely to be rare in any event, because the statistical likelihood is very small for the occurrence of basic amino acids on Ag or Ab, outside of the primary binding sites, to be situated precisely opposite acidic amino acids, and in the rare instances where this might happen, such a secondary binding phenomenon would be indistinguishable from primary binding.

The secondary bonds most often encountered are:

a. Hydrophobic (AB) bonds between some of the lower-energy moieties within, as well as immediately neighboring the epitope and paratope, and:
b. Interfacial bonds strengthened through dehydration.

It is not easy to distinguish experimentally between these two mechanisms of secondary bond formation, but there are strong indications that the second mechanism (b) is the more important one: Whilst some incipient dehydration (and thus some slight randomization of water molecules) also occurs as a consequence of mechanism (a), the marked increase in entropy typically occurring in Ag-Ab interactions (Absolom and van Oss, 1986; van Oss, *Structure of Antigens*, 1992, Ch. 6) is mainly due to the randomization of the most *strongly oriented* water molecules from the very first layer of hydration of the biopolymers (van Oss and Good, J. Protein Chem., 1988; Chapter XIX); see the following two sections; see also the earlier sub-section on Change in Configuration and Denaturation, above.

ROLE OF HYDRATION

All surfaces in contact with water (with the exception of air) become hydrated with *energies of hydration*, ΔG_{iw}, varying from -40 to -112 mJ/m^2 for hydrophobic, i.e., water-insoluble compounds, to −114 to about −142 mJ/m² for hydrophilic, water-soluble materials (cf. Table XIX-1). In terms of free energy per unit surface the usual $|\Delta G_{iw}|$ values for epitopes and paratopes are higher than the epitope/paratope binding energy, $|\Delta G_{1w2}|$, so that one might at first sight believe that both epitope and paratope have to keep at least one molecular layer of water of hydration even after binding. This is not so, and if it were the case, the epitope-paratope binding energy (ΔG_{1w2}), at $\ell \geq 0.7$ nm, would be very small, e.g., of the order of only −1 kT, which would be exceedingly ineffective. However, hydration energies must be expressed in units of kT to understand the fate of the individual water molecules of hydration (see also the section on Cell Fusion, above).

First, it should be noted that for even the most hydrophilic materials known, the free energy of hydration (ΔG_{iw}) is always smaller than the free energy of cohesion between the water molecules themselves: the strongest energy of hydration known, i.e., that of PEO, with $\Delta G_{iw} = -142$ mJ/m² (Table XIX-1), still is smaller than the energy of cohesion of water, (ΔG_{ww}), which is equalt to −145.6 mJ/m². But given a contactable surface area, S_c, for single water molecules of about 0.10 nm², $\Delta G_{ww} = -145.6$ mJ/m² still equals only −3.5 kT. All energies of hydration thus are smaller than 3.5 kT. At the same time, the free energies of binding between epitopes and paratopes generally is from at least −7.5 to −10, to as much as −12 to −25 kT (corresponding to K_{ass} value of 1.8×10^4 to 7.2×10^{10} L/M; see eq. [XXIV-3]). Thus upon epitope-paratope binding, water of hydration can be largely extruded, if their "fit" is good enough; at a less than perfect "fit" however, water will not be expelled from gaps where otherwise a vacuum would form. Because the water of hydration, especially on hydrophilic molecules, tends to be strongly oriented (see Chapter XIX), loss of such oriented water of hydration into the randomly oriented bulk liquid gives rise to a marked increase in the entropy of the system. This is the main cause of the typical increase in entropy observed in many Ag-Ab reactions (Absolom and van Oss, 1986; van Oss, *Structure of Antigens*, 1992, Ch. 6). However, all water molecules of hydration are not necessarily always extruded. Some water may remain in cases of less than perfect "fit", and in other cases tightly bound water molecules have been (erroneously) thought to participate in the achievement of an optimal "fit" (Pressman and Goldberg, 1968, pp. 132–137). In general, however, the presence of residual interstitial water will locally reduce the binding energy between epitope and paratope rather severely, because they then interact at a greater than minimum distance.

With increasing temperature there is little change in the total Ag/Ab binding energy, but there is an increase in TΔS, which compensates almost completely for a concomitant decrease in |ΔH| (eq. [XVIII-3]) (van Oss, Absolom and Bronson, 1982; Mukkur, 1984). The mechanism of this type of enthalpy/entropy compensation in antigen-antibody systems would be as follows: With an increase in temperature (T) the cluster size of water decreases, leaving an increase of monomeric water molecules. This increases the entropy of the system while at the same time diminishing the distance between epitope and paratope, which then strengthens the bond between the two. However, with an increase in T the hydration of epitope and paratope increases (via term V of eq. [IV-7], as displayed in Table IV-2), which counteracts the bond strength increase between epitope and paratope, leaving $\Delta G_{\text{epitope-w-paratope}}$ largely constant, while at the same time $T\Delta S$ increases and ΔH decreases..

THE $\Delta G_{1w2} \rightarrow \Delta G_{12}$ TRANSITION

Dehydration of the epitope/paratope interface, as exemplified by the $\Delta G_{1W2} \rightarrow \Delta G_{12}$ transition, and as manifested by the increase in entropy typically observed in Ag/Ab and other specific interactions, is one of the most important aspects of the secondary (or "ripening") part of the specific bond. As a consequence of dehydration, and of an optimum "fit" between epitope and paratope, the necessary binding energy of the order of 10 kT or more can be reached even with a small surface area of the specific epitope/paratope interface.

Thus in many, if not in most cases, the secondary binding due to dehydration (i.e., the $\Delta G_{1W2} \rightarrow \Delta G_{12}$ transition) is crucial for the ultimate strength of the specific bond. As an example we may again revert to the dextran–anti-dextran system, already discussed above in terms of ΔG_{1W2}, where at least four of the glucose units in the α-(1→6) linkage are of primary importance in the Ag/Ab bond, and two more are needed to achieve the total binding strength (Kabat, 1976, p. 126). However, it should also be realized that with dextran as well as with other polysaccharides (branched or otherwise) the terminal non-reducing sugar residue often is an even more important *immunodominant moiety* (Kabat, 1976, p. 126). The non-reducing half of glucose is probably the "primary point of attachment" to the epitope (Kabat, 1976, p. 125). It also is the less hydrophilic (or the more hydrophobic) moiety of the terminal glucose molecule, as well as significantly less hydrophilic than its polymer, dextran (cf. Table XVII-8; see also Docoslis *et al.*, 2000).

In any event, the optimal "fit" between epitope and paratope still is the most important prerequisite for the formation of a high-energy–specific Ag-Ab bond. Thus not only were Emil Fischer (1894) with his lock-and-key proposal, and Paul Ehrlich (1990) (for more historical details, see Silverstein, 1989) with his side-chain theory exceedingly right in their insight in the importance of the "goodness of fit" principle, but all experimental observations reported since point to the unique importance of the best steric fit between epitope and paratope as a *conditio sine qua non* for obtaining the highest Ag/Ab binding energy: see, e.g., Landsteiner (1936, 1962), Pressman and Grossberg (1968).

SPECIFIC RECOMBINATION DNA-DNA INTERACTIONS

The linkage between two complementary strands of poly(deoxyribonucleic acid) (DNA) are a curious example of the interaction between two ribbon-shaped polymeric molecules of which one side (the deoxyribose side) *repels* the corresponding (deoxyribose) side of the other ribbon, whilst the opposing bases *attract* each other, pair-wise. From contact angle measurements on hydrated DNA (Table XVII-9) it can be seen that the hydrophilic interfacial repulsion between DNA ribbons, immersed in water is considerable: $\Delta G_{iwi}^{IF} + 16$ mJ/m^2. For a mutual contactable surface area between two such opposing strands, $S_c \approx 50$ Å^2, this repulsive energy $\Delta G_{iwi}^{IF} \approx 2$ kT, per close encounter between two such strands, at an angle of about 90°. From an estimated electrophoretic mobility of 1.45×10^{-4} cm^2 sec^{-1} V^{-1} for nucleic acids (Chrambach and Rodbard, 1971), at ionic strength, $\mu = 0.02$, a value can be estimated of $\psi_o \approx -42$ mV. At the physiological value of $\mu = 0.15$, $\Delta G_{iwi}^{EL} \approx +0.1$ kT, at close range. Then, for the total repulsive energy of the hydrophilic part of the DNA ribbon, ΔG_{iwi}^{TOT} +2.1 kT.

The specifically interacting (nucleoside) sites of DNA ribbons are much more hydrophobic than the (deoxyribose phosphate) moieties. This is a consequence of the fact that on the nucleosides the electron-donor and the electron-acceptor sites *alternate*, with one of each on adenine and thymine, and one electron-donor and two electron-acceptors on guanine and cytosine (see, e.g., Stryer, 1981). As discussed in Chapter XIX, on a macroscopic level, the presence of roughly equal numbers of electron-donor and electron-acceptor sites favors hydrophobicity. Thus DNA is amphipathic in the unusual sense that one edge of the polymeric ribbon is hydrophilic and the other hydrophobic (much like polyethylene oxide at temperatures higher than the θ-point). It thus is reasonable, as a first approximation, to take the surface properties measured on hydrated DNA (Table XVII-9) as those of the deoxyribose phosphate edges (see above), and to assimilate the surface properties measured on dried DNA (Table XVII-9) as those of the hydrophobic nucleoside edges. In so doing, one finds for the interaction energy between two opposing nucleoside moieties: $\Delta G_{iwi}^{IF} = -16.65$ mJ/m^2, or (also for an Sc estimated at 50 Å^2) $\Delta G_{iwi}^{IF} = -2.1$ kT. (For the hydrophobic edge, ΔG_{iwi}^{EL} may be neglected as too low to play any role.) Whilst the macroscopic attraction energy between the proximal hydrophobic edges is about the same as the repulsion between the distal hydrophilic edges, a balance of force *vs.* distance plot can show that a net attraction still must prevail between the DNA ribbons, resulting in a close approach between the hydrophobic nucleoside edges. The hydrophilic edges of two interacting DNA ribbons then orient outward and with regard to the greater distance between these exterior hydrophilic edges, their net repulsive energy at the close approach level between the two ribbons decays to less than $\approx + 1$ kT, which leaves a small, total net attraction between two such DNA ribbons of about -1 kT (per Sc $\approx$ 50 Å^2, or per ribbon length of about 6 à 7 Å). It should be noted that other configurations are not favored; e.g., the interaction between a hydrophilic and a hydrophobic moiety would be mildly repulsive, being of the order of + 0.6 kT.

Taking all these quantitative estimations into account, the mutual repulsion of the outer (hydrophilic) edges of two DNA ribbons causes them to orient at an angle of roughly 90° to each other (van Oss, Arnold *et al.*, 1990, see also Chapter XIX), *thus forcing them to assume a double helical configuration*, such as the one which is now familiar to everyone (see, e.g., Stryker, 1981).

The mechanism proposed above only serves to bring two (any two) opposing DNA ribbons close enough together, in an aspecific manner, to allow them to come into the range where *specific* H-bonding interactions can take over, but only of course in those cases where the sequences are exactly such that they will give rise to the required specificity of precise nucleoside complementarity. Then, to take a fairly high average value for the energy of a direct H-bond (e.g., -4 kT per bond; see Eisenberg and Kauzmann, 1969), one obtains $\Delta G^{specific} \approx -8$ kT for the adenine-thymine bond, and $\Delta G^{specific} \approx -12$ kT for the guanine-cytosine bond. These bonds are fairly strong and do not readily lend themselves to dissociation, in contrast with the $\Delta G^{aspecific} \approx -1$ kT found earlier; the latter only permits the first tentative approach between two DNA strands but does not suffice for permanent attachment when no specific bonding ensues.

The specific bonds of 8 to 12 kT per site (corresponding to K_{ass} values of 3×10^3 to 1.6×10^5 L/M; cf. eq. [XXIV-3]) are multiplied many-fold per DNA-pair forming a double helix, giving rise to bonds of hundreds or thousands of kT units, and to very large K_{ass} values. In contrast, nonspecific loosely connected DNA ribbon-pairs, while apparently attached with energies of multiples of –1 kT, nevertheless can detach again with fair ease, because each monomeric unit can still undergo some Brownian movement, of the order of $\Delta G^{BR} \approx +1$ kT, which roughly counterbalances the small, net, hydrophobic attractions.

Summary of the Mechanism of Specific Ligand-Receptor Interactions

The following steps are involved in specific ligand-receptor interactions, taking antigen/antibody (Ag/Ab) interactions as an example (and keeping in mind that the Ag-active site is the *epitope* and the Ab-active site the *paratope*):

1. *Primary Interaction*: The normal (mainly AB) macroscopic repulsion between Ag and Ab must be overcome. This is achieved through the placement of the epitopes (and especially the immunodominant epitope) on prominent, protruding parts of the Ag molecule or particle, with a small radius of curvature and an overall small total surface area of 0.4 to 8.0 nm². The primary bond energy may be expressed as ΔG_{1W2} and is significantly attractive at a distance of 2 to 3 nm. Primary bond energies can be of the LW + AB, or of the LW + EL + AB classes (usually the LW interaction is by far the feeblest). The attractive energies decay with distance according to the rules outlined in Tables VII-1 to 3, for LW, AB and EL attractions, respectively.

Direct H-bonds play no significant role in the primary interaction.

2. *Secondary Interaction*: Once the first (primary) contact is made between epitope and paratope two things happen: (a) After the first contact between the (still hydrated) epitope and paratope, their water molecules of hydration become expelled into the bulk liquid, thus increasing the entropy of the system, because previously oriented water molecules of hydration become randomized upon expulsion. If the "fit" between epitope and paratope is optimal, the final dehydration step gives rise to a switch in free energy designation from ΔG_{1W2} to ΔG_{12} which among biopolymers generally entails a significant increase in bond energy, allowing sizeable attractive energies to occur with small epitope/paratope interfacial areas. Whilst ΔG_{1W2} usually is principally represented by ΔG_{1W2}^{AB}, ΔG_{12} comprises mainly LW interactions, due to the monopolarity of 1 and 2, which entails a virtual zero value for γ_1^{AB} and γ_2^{AB}. Upon the mutual neutralization of oppositely charged sites on epitope and paratope, a further strengthening of the bond also occurs after expulsion of water, but only in the highest energy EL bonds. (b) Between the Ag and Ab chains immediately neighboring the epitope and paratope, which are brought closely together, through

the primary attraction, secondary largely non-specific bonds of the LW and AB variety tend to form, which further increase the overall bond energy (Absolom and van Oss, 1986).

Direct H-bonds and direct EL interactions rarely play a role in the secondary interactions.

It is only when the epitope fits the paratope precisely that all interstitial water of hydration is expelled, so that everywhere at the interface $\ell = \ell_o$, which favors the strongest possible bond. The principal prerequisite for a strong Ag/Ab (or other ligand/receptor) bond is therefore an optimal fit between the Born-Kihara shells (see Chapter III) of epitope and paratope, regardless of their mode(s) of interfacial interaction.

CODA

It is stressed again that a hydrophobic attraction between a hydrophobic surface and a hydrophilic particle or macromolecule, immersed (or dissolved) in water, is a frequent occurrence: it is the principal mechanism of protein adsorption onto low-energy surfaces (which include the water/air interface).

In all such cases, $\Delta G_{1W2} < 0$. However, hydrophilic proteins, in aqueous solution, also adsorb to some extent onto hydrophilic surfaces, such as clean glass, notwithstanding the fact that here, $\Delta G_{1W2} > 0$ (and even $\Delta G_{1W2}^{EL} > 0$). This can be explained by a *microscopic attraction* between, e.g., electron-donor sites on peptide-bends with a small radius of curvature, and discreet electron-acceptor sites disseminated in the glass surface, in such a manner that, locally, the microscopic attraction can prevail over the *macroscopic repulsion*.

Various modes of liquid chromatography also depend on hydrophobic attractions, which can be switched to hydrophilic repulsions. These include the reversed phase, hydrophobic interaction, ion-exchange and affinity modes of liquid chromatography.

Cell adhesion is based on the same principles as, e.g., protein adsorption, but the scale is different and there is the added complication that it is virtually always necessary when studying cell adhesion, to take prior protein adsorption into account. This is further complicated by the fact that in accordance with the Vroman effect, different proteins adsorb consecutively, and to different degrees. Thus, those low-energy surfaces which adsorb proteins most strongly cause the least amount of cell adhesion (on account of the fact that such hydrophobic surfaces quickly become coated with hydrophilic proteins) even though in the (ideal) absence of proteins, cells should adhere most to these surfaces. This principle also applies to cell-cell interactions, which often occur through the (specific or aspecific) prior attachment of proteins to at least one of the cell types. Only in the case of the most hydrophilic cells (e.g., blood cells) does an overall AB-repulsion prevail. Hydrophilically encapsulated bacteria thus can thwart contact and subsequent ingestion by phagocytic leukocytes; this is an important mechanism for one type of bacterial pathogenicity. Specific antibodies directed against beacterial capsular material however can help surmount this bacterial defense mechanism.

Cell fusion requires that the cells, or at least a part of the cells be able to attract and approach each other. This is usually accomplished by the admixture of Ca^{++} ions, which render the cell surface more hydrophobic and which can also cross-bind two negatively charged cells.

Ice *repels* cells as well as proteins and other micro- and macro-solutes dissolved in the aqueous medium. Thus, upon freezing, an advancing ice-front pushes cells *and* solutes into an ever-decreasing, as yet non-frozen volume, which causes osmotic damage. To obviate this effect, water-miscible polar solvents (e.g., glycerol) may be dissolved in the liquid medium. The properties of the liquid are then changed to such an extent that now the cells and the solutes are *attracted* by the advancing ice-front, and engulfed upon freezing. This freezing method causes no damage to cells.

Specific, microscopic *attractions* such as those between antibodies (Ab) and antigens (Ag), lectins and carbohydrates, enzymes and their substrates, and receptors and ligands in general, operate with the same types of forces (LW, AB and EL) which also drive the aspecific, macroscopic *repulsions* that keep cells in suspension and biopolymers in solution. The difference between the two is mainly one of size, and radius of curvature. Specific attractions always operate between small sites (surface areas between 0.3 and 3 nm^2) and the sites frequently are found on processes with a small radius of curvature (typically, $R \leq 1$ nm). Aspecific repulsions operate between much larger surface areas (e.g., 10 to 10,000 nm^2), with a large R. Whilst AB interactions in aqueous media are of course mainly the consequence of net hydrogen-bonding interactions (which can be attractive or repulsive), direct H-bonding only rarely plays a role in specific (e.g., Ag-Ab) binding.

XXV Kinetics and Energetics of Protein Adsorption onto Metal Oxide Surfaces

MEASUREMENT OF THE KINETIC ON-RATE CONSTANT OF PROTEIN ADSORPTION ONTO METAL OXIDE SURFACES—EXPERIMENTAL CONSTRAINTS TO BE OBSERVED

There are varous physical and physico-chemical pitfalls that can drastically diminish the value of the observed kinetic rate constant of attachment (k_a) in protein adsorption to solid surfaces, which are frequently ignored or disregarded. These are of two different categories:

1. Pitfals pertaining to mass transport and to steric hindrance among dissolved protein molecules causing competition for adsorption sites, both of which can be avoided by taking simple experimental precautions, and:

2. Usually unsuspected pitfalls that are inherent in (often unavoidable) experimental conditions, but which can be taken into account, to yield data which allow the derivation of the real molecular k_a value. Category 1 pitfalls are treated in this Section; category 2 pitfalls are treated in the two following Sections.

Mass Transport

For the kinetics equations (treated in the following two Sections) to be applicable, it is essential that one provides adequate mass transport of the dissolved protein toward the adsorbent surface (Docoslis *et al.*, 1999; van Oss *et al.*, 2004), so that the diffusion dependence of the observed kinetic on-rate constant, k_a,will remain unimpeded. To that effect one needs to establish a liquid flow regimen of incipient turbulence, attainable at Reynolds numbers (Re) of 2000 to 3500:

$$Re = (dv\rho)/\eta \qquad [XXV\text{-}1]$$

385

where d is the diameter of the lumen of the tube (when streaming takes place in a cylindrical tube), v the velocity of the liquid solution, ρ the density of the solution and η the viscosity of the liquid. For instance with water flowing in a tube of 2.5 cm (one inch) inside diameter, at 0.34 m/sec, at 20°C, one attains 2100 Re (van Oss, 1955). However, to reach the same Re number (at which incipient turbulence occurs), using an interior cross section of, e.g., 0.1 mm, one would require 85 m/sec, or about 306 km/h (i.e. 191 mph), which is not practically attainable within such a small lumen. Thus, by streaming aqueous solutions inside a flat cell of 0.05 mm inside height, there is even less hope of reaching turbulence, so that it is impossible to attain anything close to diffusion dependence or adequate mass transport in such a structure (van Oss *et al.*, 2004); see, e.g., the flow cell depicted by Fägerstam and Karlsson (1994).

In a small magnetically stirred vertical cylindrical device with an inside diameter of 1.2 cm and a height of 0.5 cm one can attain 2100 Re (Docoslis *et al.* 1999):

$$Re = (Nd^2\eta)/\rho \qquad [XXV\text{-}2]$$

where N indicates stirring rotations per minute (1300 rpm in this case). Under these conditions it should be noted that the effective diffusion coefficient of the protein [e.g., human serum albumin (HSA)] has to be mutiplied by a factor 62.7 (Docoslis *et al.*, 1999; van Oss *et al.*, 2003; 2004).

Steric Hindrance by Dissolved Protein Molecules

A relatively large protein such as HSA, with a molecular weight of about 69,000, can only be soluble in water if the protein molecules repel one another when immersed in water; see Chapter XXII. This mutual repulsion is mainly driven by the extremely strong hydrophilicity of HSA, caused almost solely by $\Delta G_{iwi}^{AB} = +86.9$ mJ/m2, or +203 kT, the latter being based on an averaged radius for hydrated HSA (with one layer of tightly bound water of hydration, see Table XVII-6), of 3.0 nm. For HSA, $\Delta G_{iwi}^{LW} = -0.48$ mJ/m², or −0.176 kT, which at a distance $\ell = 6.0$ nm (see below) becomes negligibly small. As for ΔG_{iwi}^{EL}, HSA in 10 × diluted phosphate-buffered saline (PBS/10), with a double-layer thickness, $1/\kappa = 2.53$ nm, has a ψ_o value of −35.9 mV (van Oss *et al.*, 2001), yielding a ΔG_{iwi}^{EL} value of +4.16 kT (i.e., the free energy of EL interaction between two spherical molecules (see Table VII-3). XDLVO analysis, using both ΔG_{iwi}^{AB} and ΔG_{iwi}^{EL}, as a function of distance, ℓ, between two HSA spheres in aqueous solution, shows that at $\ell = 6$ nm, $\Delta G_{iwi}^{AB} = +0.48$ kT and $\Delta G_{iwi}^{EL} = 0.03$ kT. Thus at a distance of 6 nm between two hydrated HSA molecules, their mutual repulsion, ΔG_{iwi}^{TOT}, is reduced to +0.5 kT. (It should be noted that at ionic strength 0.015, beyond that distance, at repulsive energies smaller than 0.5 kT, the interaction energies between two such protein molecules cease to be significant). However, in de-ionized water, $1/\kappa = 1000$ nm, so that at $\ell = 6$ nm, ΔG_{iwi}^{EL} still is +11.8 kT and even at $\ell = 100$ nm, ΔG_{iwi}^{EL}, which decays exceedingly slowly in pure water, still equals +11.0 kT.

This means that hydrated HSA molecules (at ionic strenght 0.015) each can occupy a flat surface of about 12.2 × 12.2 nm² (the diameter of each hydrated HSA molecule being 6.0 + 6.2 = 12.2 nm, which includes the hydrated HSA molecule itself, surrounded by a repulsion ring of 3.1 nm thickness of buffered water). However, when HSA molecules are present in solution at a very low concentration, each single one of them can occupy a significantly larger surface area of silica than 12.2 × 12.2 nm², mainly by flattening, i.e., by stretching out laterally in two directions, driven by hysteretic attraction to several more cationic attractive sites imbedded in the silica, in addition to the initial cationic attractive site. The result of this spreading, at low concentrations of dissolved HSA (Docoslis *et al.*, 1999; van Oss *et al.*, 2001) is that the attraction of further HSA molecules from solution quickly decreases, because the repulsion of dissolved HSA by the already attached (and flattened) HSA molecules continues, while the specific attraction of dissolved HSA molecules by cationic sites diminishes through their obliteration by already adsorbed HSA; see also the last Section of this Chapter, treating hysteresis. In de-ionized water the repulsion field between HSA molecules reaches considerably farther. Thus, either in an aqueous buffer of ionic strength 0.015, or in de-ionized water of zero ionic strength, if one wishes to measure the real adsorption rate constant (k_a) of HSA onto, e.g., a silica surface, one must do this under conditions:

1. Of exceedingly close to zero HSA concentration and:
2. Of starting observations within the shortest possible time lapse after the start of the adsorption process.

Condition 1 can be fulfilled by using a number of extremely low but still measurable dissolved protein concentrations and extrapolating to zero protein concentration. Similarly, condition 2 can be complied with by doing the measurements after a number of extremely short but feasible time lapses after time zero and then extrapolating back to time zero. For both procedures, see Docoslis *et al.* (1999; 2001a). The errors one makes in measuring k_a values of protein binding when not complying with rules 1 and 2, are quite large. For instance at HSA concentrations between 7.3 and 71.5 nM HSA, the k_a values found for the adsorption of HSA onto monosized silica particles were 3.7×10^6 and 0.5×10^6 L/M.sec respectively, with the final value (using the extrapolated to zero HSA concentration) amounting to $k_a = 4.53 \times 10^6$ L/M.sec (Docoslis *et al.*, 1999). Thus a 10-fold greater protein dilution already yielded a 7.4-fold greater k_a value, whilst the the final k_a value obtained via extrapolation to zero concentration amounted to a value which was 9 times greater than when based on a free protein concentration of 71.5 nM. Similarly, at a constant HSA concentration of 7.3 nM, measured 20 sec after the start of the adsorption, k_a was observed at 0.6×10^6 L/M.sec whilst at t = 0, k_a was found to be 3.6×10^6 L/M.sec, i.e., a 6-fold improvement; see Docoslis *et al.* (1999), with whose methodology t and k_a could be measured every 0.1 sec, starting with t = 0.1 sec. Determining k_a values at about one minute post-inception of protein adsorption yields k_a values that are of the order of 30 to 40-fold too low. It is useful to compare this important time window, of the order of 0.1 sec, with the time lapses used in some other methodologies, using several minutes (Fägerstam and Karlsson, 1994) and even 10 to 20 minites (Malmquist, 1993). In the latter case decreases in measured k_a values to several orders of magnitude lower than the real k_a values of the system are quite common; see also van Oss (1999).

DECREASE IN THE ADSORPTION OF PROTEIN ONTO METAL OXIDE
SURFACES WHEN DISSOLVED IN SOME COMMON
BUFFER SOLUTIONS

The presence of phosphate ions in, e.g., phosphate buffered saline (PBS), or even in 10 times diluted PBS (PBS/10), corresponding to an ionic stregth of 0.015, significantly reduces the amount of protein that becomes adsorbed (Docoslis et al., 2001b). The competition between dissolved proteins and cation-complexing agents (such as EDTA or phosphates) for metal ion adsorptive sites is well-documented (van Oss, Wu and Giese, 1995; van Oss, Wu et al., 1995) This competition, occurring in PBS as well as in PBS/10, also diminishes the value of the observed adsorption rate constants (k_a) of, e.g., HSA onto metal oxide surfaces suh as silica. Therefore, if one wishes to determine the k_a value that is characteristic for a given protein when adsorbing to, e.g., silica, it is best to avoid the competition for adsorptive sites altogether, while simply dissolving the protein in pure deionized water where, if needed, the pH can be adjusted with small amounts of NaOH or HCl (Docoslis et al., 2001b). It is not, of course always possible to dissolve a protein in de-ionized water, but there the addition of NaCl, plus small amounts of either NaOH or HCl to reach the desired pH, will still allow one to work with almost all proteins, while still avoiding buffers with competing anions.

MACROSCOPIC-SCALE REPULSION BETWEEN
ALBUMIN AND GLASS OR SILICA

Given the occurrence of macroscopic-scale repulsion between dissolved hydrophilic proteins and the surfaces of equally hydrophilic metal oxide surfaces, such as glass or silica, these repulsions have to be measured and taken into account. When calculating the likelihood of the adsorption of human serum albumin (HSA) molecules from an aqueous solution onto clean glass, one would predict from the macroscopic-scale surface thermodynamic properties of hydrated HSA and of clean glass (van Oss, Wu and Giese, 1995; Docoslis et al., 1999) that hydrophilic HSA molecules, even when oriented in the most favorable manner with respect to an equally hydrophilic glass surface, would be repelled by the glass surface, when immersed in water. The repulsive free energy would amount to +40 to 65 kT per molecule, when approaching the glass surface via a spherically shaped moiety with a radius of 1.3 nm, facing the glass, and to +100 to 170 kT in the case of an R = 3 nm cylindrical peptide chain facing the glass in a parallel configuration, all at neutral pH; see Fig. XXIV-1. This surely would at first sight allow one to assert that HSA should be unable to approach the glass surface sufficiently closely to achieve adsorption. However, one would be quite wrong in such an assertion: HSA adsorbs very well to glass from aqueous solutions at neutral pH. In general albumin adsorption to clean glass and other hydrophilic metal oxides can only be avoided at pH values greater than 10. The reason for this seemingly paradoxical adsorption of albumin onto glass (or silica) at neutral pH, is that macroscopic-scale repulsion is not the only force involving dissolved albumin molecules in the vicinity of glass or silica. At a range of about

5 nm or less, favorably oriented protein moieties *of a small radius of curvature* can locally pierce the macroscopic-scale repulsion field and undergo a microscopic-scale attraction to cationic sites imbedded in the glass (or silica, etc.); see van Oss, Wu and Giese (1995); see also Fig. XXIV-1.

MICROSCOPIC-SCALE ATTRACTION BETWEEN PROTEINS AND DISCRETE CATIONIC SITES IMBEDDED IN HYDROPHILIC METAL OXIDE SURFACES

The attractive sites for protein adsorption in the surfaces of hydrophilic metal oxides such as glass, silica and montmorillonite particles (van Oss, Wu and Giese, 1995), as well as hydrophobic metal oxides, comprising talc, tin oxide and zirconia (van Oss, Wu *et al.*, 1995) have been found to be imbedded cationic loci. In glass these are most commonly Ca ions plus some charged Si sites, and in silica virtually only charged Si sites, whilst in talc they wouild be mainly charged Si sites, plus some Mg ions, and in tin oxide and zirconia they would be incompletely neutralized Sn and Zr ions respectively. In all these cases the protein could be eluted with the metal ion complexing agent, EDTA (usually in the form of Na_2EDTA); see van Oss, Wu and Giese (1995) and van Oss, Wu *et al.* (1995). In all cases tried, Na hexametaphosphate (also a metal ion complexing agent) gave the same results as EDTA. The attraction between these cationic sites imbedded in the surfaces of the metal oxide particles, and the (negatively charged) peptide moieties on the small-radius, roughly spherical protrusions of protein molecules, is not only due to electrical double layer interactions, but it is further amplified by a Lewis acid-base attraction between the (cationic) electron-accepting site and the electron-donating peptide moiety (van Oss, Wu and Giese, 1995).

EXTENDED DLVO (XDLVO) ANALYSIS OF PROTEIN ADSORPTION AS A FUNCTION OF DISTANCE AND GEOMETRIC SHAPE

In Chapter VII it is shown that in the encounter between a spherical object (1) of radius, R, and a flat plate (2), immersed in water (w), as a function of distance, ℓ, the free energy of interaction, ΔG_{1W2}^{TOT} (which is composed of $\Delta G_{1W2}^{LW} + \Delta G_{1W2}^{AB} + \Delta G_{1W2}^{EL}$), is directly proportional to R, for LW, AB as well as for EL interactions (see Tables VII – 1, 2 and 3). This means that if one compares the interaction between a large sphere of radius, R and a flat surface, with the interaction between a large body with long, narrow, distal protrusions with a small radius, r, only the long narrow protrusions will be able to pierce the repulsion field; see Fig. XXIII-3. This is why, e.g., mammalian phagocytic cells, which can stick out long thin processes, can readily make contact with, e.g., invading bacteria and subsequently endocytize and destroy them; see van Oss *et al.* (1975, pp. 112–113). It is also why protein molecules are most likely to make adsorptive contact with an attractive site on a flat surface (which otherwise macroscopically repels the protein molecules) via distal processes (or "elbows") protruding from the protein molecules; see also Figs. XXIII-3 and XXIV-1. In van Oss, Wu and Giese (1995) the different energy *vs.* distance diagrams (= XDLVO plots; see also Chapters VII and XXIV) are shown of: 1) The macroscopic-scale interaction between a parallel cylindrical domain and a glass surface; 2) The macroscopic-scale interaction between and HSA "elbow" and a glass surface; 3) Microscopic-scale attraction *vs.* macroscopic-scale repulsion between a parallel cylindrical HSA domain and a glass surface, where the composite curve still indicates a net repulsion; 4) Microscopic-scale attraction *vs.* macroscopic repulsion between an HSA "elbow" and a glass surface. In the "elbow" position the composite curve shows that an attraction between microscopic attractor sites on the "elbow" and the glass surface now prevails. This type of XDLVO analysis can serve to identify which part of a protein such as HSA will remain unattracted by an (e.g., hydrophilic) surface such as glass, in a parallel position, whilst a different orientation (e.g., with a small radius of curvature "elbow" of such a protein pointing toward the glass) permits an attraction to occur toward a cationic site imbedded in the glass, which then gives rise to HSA adsorption onto the glass, at that site.

This type of XDLVO analysis also plays a role in the determination of the kinetic association rate constant (k_a) in the adsorption of protein molecules onto, e.g., hydrophilic metal oxide surfaces, when using von Smoluchowski's (1917) mathematical treatment (originally designed for a study of the kinetics of flocculation); see the following Section.

INCORPORATION OF BOTH MACROSCOPIC-SCALE REPULSION AND MICROSCOPIC-SCALE ATTRACTION ENERGIES IN THE ANALYSIS OF MEASURED KINETIC ASSOCIATION RATE CONSTANTS, USING VON SMOLUCHOWSKI'S FORMALISM

VON SMOLUCHOWSKI'S EQUATION

In the year of his death (1917), M. von Smoluchowski's flocculation kinetics paper was published (insofar that on account of the War, it actually appeared in print in 1918). Von Smoluchowski's formalism applies equally well to the adsorption of hydrophilic biopolymers from aqueous solution onto hydrophilic metal oxide surfaces such as glass or silica. It is even more felicitous that von Smoluchowski's equation applies remarkably aptly to the interplay between the aspecific, macroscopic-scale repulsion (as between, e.g., whole dissolved protein molecules and silica, immersed in water) on the one hand, and the specific, microscopic-scale attraction between given small peptide moieties of protein molecules and discrete cationic sites imbedded in the silica surface on the other hand. In von Smoluchowski's terminology, the aspecific macroscopic-scale repulsion would be equivalent to his "improbability"

(Unwahrscheinlichkeit), whilst the specific microscopic-scale attraction in a small ligand-receptor system can be likened to his "probability" (Wahrscheinlichkeit), see below, under von Smoluchowski's **f** factor.

Hammes (1978) gave the following version of von Smoluchowski's equation for the kinetic association rate constant, k_a, pertaining to the interaction between molecules, A and B:

$$k_a = 4\pi r \, (D_A + D_B) \, \mathbf{f} \, (N/1000) \qquad [\text{XXV-3}]$$

where r is the shortest distance between the centers of molecules A and B, D_A and D_B are the diffusion constants of A and B, **f** is the factor introduced by von Smoluchowski (1917), expressing the summed average of the contributions of all interactions and N is Avogadro's number, indicating the number of molecules in 1 Mol ($N = 6.02 \times 10^{23}$). However, at all macroscopic and most microscopic scales (larger than, e.g., those depicting interactions between simple atoms in a dilute gas), the expression of the distance, r, as that between the centers of two atoms is no longer tenable, especially in the case of an adsorptive interaction between a macromolecule and a flat surface. Also, in the latter case, there is only one diffusion constant D, i.e., that of the macromolecule. Von Smoluchowski's equation then becomes (van Oss, 1997):

$$k_a = 4 \, \pi \, \ell_o \, D \, \mathbf{f} \, (6.02 \times 10^{20}) \qquad [\text{XXV-4}]$$

where ℓ_o (which is a constant) is the minimum equilibrium distance from the edge of the macromolecule to the surface of a flat plate, at 0.157 nm (see Chapter III); D is expressed in $cm^2.sec^{-1}$ and k_a in $cm^3.mMol^{-1}.sec^{-1}$ ($= L. \, Mol^{-1}.sec^{-1}$); for human serum albumin (HSA), $D = 6.1 \times 10^{-7} \, cm^2/sec$. The factor, **f**, comprises an averaging over all distances, ℓ, from $\ell = \ell_o$ to $\ell = \infty$, so that k_a is the association velocity parameter incorporating all events at all distances, ℓ, as well as at all orientations, φ, of the macromolecule, if the latter's surface is non-homogeneous and/or non-spherical; see eq. XXV- 5, below.

VON SMOLUCHOWSKI'S F FACTOR

Von Smoluchowski's **f** factor is expressed as:

$$\mathbf{f} = \int_{\varphi} \exp\left[\frac{1}{\ell} \int_{\ell=\ell_o}^{\ell=\infty} \left(\frac{-\Delta G_{1W2}}{kT}\right) d\ell \right] d\varphi \qquad [\text{XXV-5}]$$

where φ indicates all orientations of the dissolved potentially interacting macromolecules, as either unfavorable (i.e., repulsive), or favorable (i.e., attractive), such that the ratio of favorable (a) to unfavorable orientations) $(1 - a)$, φ, is (Docoslis, 2000; Docoslis et al., 2001a; van Oss et al., 2004):

$$\varphi = a/(1 - a) \qquad [\text{XXV-6}]$$

Given the very short timespan after which k_a is (or ought to be) measured, and the very low solute concentrations used, the different possible orientations can be legitimately subdivided into either just the unfavorable of just the favorable ones, as is done in eq. XXV-6. We can now subdivide the total interaction energy, X^{TOT} as follows:

$$X^{TOT} = a.X^{mac} + (1 - a).X^{mic} \qquad [XXV-7]$$

where:

$$X^{mac} = 1/\ell \int_{\ell=\ell_o}^{\ell=\infty} \left(\frac{\Delta G_{1W2}^{mac}}{kT} \right) d\ell \qquad [XXV-8A]$$

and:

$$X^{mic} = 1/\ell \int_{\ell=\ell_o}^{\ell=\infty} \left(\frac{\Delta G_{1W2}^{mic}}{kT} \right) d\ell \qquad [XXV-8B]$$

The factor, **f**, can then be rewritten as:

$$\mathbf{f} = -\exp X^{TOT} = -\exp [a.X^{mac} + (1 - a).X^{mic}] \qquad [XXV-9]$$

and:

$$\ln \mathbf{f} = \ln \mathbf{f}^{mac} + \ln \mathbf{f}^{mic} \qquad [XXV-9A]$$

or:

$$\mathbf{f} = \mathbf{f}^{mac} \times \mathbf{f}^{mic} \qquad [XXV-9B]$$

Once X^{mac} and X^{mic} are known, as well as X^{TOT} (eq. XXV-7), one can obtain the value for a [and thus also for $(1 - a)$], using eq. XXV-7. Then, also when X^{mac} and X^{mic} are known, and using eq. XXV-9, above:

$$\mathbf{f}^{mac} = -\exp [a.X^{mac}] \qquad [XXV-10A]$$

and:

$$\mathbf{f}^{mic} = -\exp [(1 - a).X^{mic}] \qquad [XXV-10B]$$

the values for $\mathbf{f}^{mac}$ and $\mathbf{f}^{mic}$ can be obtained.

DETERMINATION OF X^{MAC} AND X^{MIC}

X^{mac} and X^{mic} are obtained from from the ΔG^{mac} and ΔG^{mic} values, respectively (at $\ell = \ell_o$; following eqs. [XXV-8A and 8B]; see also Docoslis *et al.,* 2001a), followed by XDLVO treatment for all distances, ℓ (see Chapter VII, eqs. [VII-1, 2, and 3]; see also Docoslis et al., 2001a). The distance, ℓ, in eqs. [XXV-5], [XXV-8A] and [XXV-8B] is the greatest distance utilized; usually $\ell = 10$ nm suffices, but for interactions in de-ionized water, ℓ is much longer.

Docoslis et al. (2001a) shows an example of the XDLVO plots pertaining to one case of X^{mac} and X^{mic}, see eqs. [XXV-8A and B], where for the greatest accuracy ΔG^{mac} and ΔG^{mic}_{1W2} should be decomposed as, e.g., ΔG^{mac-LW}_{1W2}, ΔG^{mac-AB}_{1W2} and ΔG^{mac-EL}_{1W2}, and idem for ΔG^{mic}_{1W2}. ΔG^{mac}_{1W2} is derived from the macroscopic-scale LW, AB and EL properties of the macromolecules and of the adsorbing surface material employed, whilst ΔG^{mic}_{1W2} is best derived from K_a, using (Docoslis et al., 2001b):

$$55.56 \cdot K_a = \exp [- (\Delta G^{mic}_{1W2} \cdot S_c)/kT)] \qquad [XXV-11]$$

where S_c is the contactable surface area between macromolecule (1) and an attractive surface site on the adsorbing surface (2), immersed in water (w). Here the K_a value used is K_a ($t \to 0$), as described in a subsequent Section, below. The connection between K_a and k_a and k_d is expressed as:

$$K_a = k_a^*/k_d \qquad [XXV-12]$$

where k_a^* is the measured k_a value, corrected for non-stirred conditions and k_d is the kinetic dissociation rate constant.

DETERMINATION AND SIGNIFICANCE OF k_a^{mic}

A principal aim of measuring the k_a value for non-covalent interactions is the determination of the microscopic-scale k_a^{mic} of the interaction, that is unencumbered by extraneous but usually unavoidable macroscopic-scale interactions, which are repulsive.

Here one uses:

$$k_a^{mic} = <k_a> \cdot f^{mic} \qquad [XXV-13]$$

where $<k_a>$ is the value for k_a from von Smoluchowsiki's equation in the form of eq. [XXV-4], when $f = 1$ (Docoslis et al., 2001a; van Oss et al., 2004). and the derivation of f^{mic} is given in eq. [XXV-10B].

In Table XXV-1 numerical examples are given of the k_a, K_a, k_a^{mic} and k_d^{mic} values, as well as of the various intermediate parameters used, from experimental results obtained for the adsorption of HSA onto monosized amorphous silica spheres of 2.1 µm diameter, obtained from Bangs Laboratories (Fishers, IN) (Docoslis et al., 2001a; van Oss et al., 2004).

The microscopic scale, k_a^{mic}, is one of the two most crucial parameters pertaining to the molecular scale binding kinetics of, in the present example, the adsorption of HSA onto pure, smooth, amorphous silica. The other crucial parameter is k_d^{mic}, which is the kinetic dissociation rate constant, also at the molecular (microscopic) level. These two rate constants are related via the equilibrium association constant, K_a, see eq. [XXV-12]. As the equilibrium constant, K_a, is linked to the microscopic-scale interactions, and thus to k_a^{mic} and k_d^{mic} (eq. [XXV-12]), it could also have been named K_a^{mic}, but as there is only one K_a, or more precisely, K_a ($t \to 0$), there seems to be no need for that. It is true that the value for K_a in the case of regular monosized silica spheres is considerably higher than with the carboxylated monosized silica spheres (see Table XXV-2). This can, however, be ascribed to the fact that the number of attractive metal oxide sites that are available on the carboxylated spheres' surfaces, has become considerably less than on the regular silica spheres, due to a significant portion of the surface area that has been replaced by carboxyl groups.

In using k_a^{mic} and K_a ($t \to 0$), where k_a^{mic} was based on k_a, (which was measured at very low concentrations as well as at $t \to 0$, to obtain the value for k_d^{mic}, the latter may thus be taken as equivalent to k_d^{mic}($t \to 0$) (Docoslis et $al.$ 2001a; 2002). The two kinetic rate constants thus obtained: k_a^{mic} and k_d^{mic}, describe the two different and independent aspects of the protein adsorption rate constants. k_a is mainly linked to just the diffusion constant (D) of, in the case of the present example, the protein HSA, when dissolved in water; see von Smoluchowski's equation (eq. [XXV-4]). However, k_a^{mic} would only be $solely$ dependent on D in the case where $\mathbf{f} = 1$, in which case $X^{TOT} = 0$ (eq. [XXV-9]), so that then: a.X^{mac} = (1 – a).X^{mic}, and: $\mathbf{f}^{mac} = \mathbf{f}^{mic}$ (cf. eqs. [XXV-10A and B]). In actual practice, however, when one actually determines k^{mac},

---

TABLE XXV-1

Numerical Examples Pertaining to the Derivation of k_a, k_a^*, K_a, k_d^{mic} and k_d^{mic}, as Well as of the f Factors, φ, X^{TOT}, X^{mac} and X^{mic}, from Experimentally Obtained Data on the Adsorption of Human Serum Albumin onto Monosized Amorphous Silica Spheres of 2.1 µm Diameter.

Symbol	Value	Remarks
k_a	3.4×10^6 L/M.s	K_a is the measured value at $t \to 0$ and $C \to 0$ [a]
k_a^*	0.55×10^5 L/M.s	k_a^* is the non-stirred k_a value [b]
k_a^{mic}	2.66×10^8 L/M.s	see eq. [XXV-13]
K_a($t \to 0$)	2.78×10^8 L/M	see eq. [XXV-11]
k_d^{mic}	0.96 s^{-1}	see eq. [XXV-12]
<ka>	7.2×10^7 L/M.s	k_a from eq. [XXV-4], with $\mathbf{f} = 1$
f	757×10^{-6}	see eq. [XXV-5 and 9]
$\mathbf{f}^{mac}$	192×10^{-6}	see eq. [XXV-10A]
$\mathbf{f}^{mic}$	3.94	see eq. [XXV-10B]
X^{TOT}	7.187 kT	see eq. [XXV-7]
X^{mac}	14.83 kT	see eqs. [XXV-8A and 10A]
X^{mic}	–3.24 kT	see eqs. [XXV-8B and 10B]
φ	1.364	see eq. [XXV-6]

[a] C is the protein concentration used in measuring k_a.
[b] k_a^* = the k_a value divided by a factor 62.7 to reduce k_a to a stationary, non-stirred condition; see eq. [XXV-12].
From van Oss et $al.$, J. Biol. Phys. Chem. 4 (2004) 145, with permission.

TABLE XXV-2

Comparison Between the Kinetic and Energetic Data Pertaining to the Adsorption of HSA onto Monosized Amorphous Silica Spheres of 2.1 µm Diameter, and the Adsorption of HSA on the Same Spheres Which Have Been Carboxylated. (For more detailed explanations, see Table XXV-1.)

	Values	
Symbol	Monosized SiO$_2$ Spheres	Monosized carboxylated SiO$_2$ Spheres
k_a	3.4×10^6 L/M.s	0.075×10^5 L/M.s
k_a^*	0.55×10^5 L/M.s	0.0012×10^5 L/M.s
k_a^{mic}	2.66×10^8 L/M.s	1.71×10^8 L/M.s
$K_a(t \rightarrow 0)$	2.78×10^8 L/M	4.2×10^6 L/M
k_d^{mic}	0.96 s^{-1}	40.7 s^{-1}
f	757×10^{-6}	1.29×10^{-6}
f^{mac}	192×10^{-6}	0.545×10^{-6}
f^{mic}	3.94	2.37
X^{TOT}	7.187 kT	13.34 kT
X^{mac}	14.83 kT	31.24 kT
X^{mic}	-3.24 kT	-2.01 kT
φ	1.364	0.848

From van Oss *et al.*, J. Biol. Phys. Chem. 4 (2004) 145, with permission; see also Docoslis *et al.* (2001a, 2002).

it correlates with $f^{mac} < 1$; see Table XXV-1. Also, the f values for the carboxylated silica spheres are considerably smaller than for the regular, non-carboxylated silica spheres, whilst concomitantly, X^{TOT} and X^{mac} are considerably larger than for the non-carboxylated spheres, see Table XXV-2.

Whilst k_a^{mic} [in conjunction with K_a (t → 0)] is an indispensable datum, needed for the determination of k_d^{mic}, it (k_a^{mic}) does not divulge a great deal about the binding kinetics of HSA to silica, as it is mainly (although not quite solely) a function of the diffusion constant (eq. [XXV-4]). It should also be emphasized here how much the actual molecular k_a^{mic} value is greater than the total measured k_a^* value (i.e., the k_a value, corrected for the increase in the diffusion constant, D, caused by stirring); see Table XXV-2. Thus even carefully measured k_a values are on a molecular level virtually meaningless and much lower than the k_a^{mic} values. Finally, whilst it is necessary to measure the k_a values, as raw material for obtaining k_a^{mic}, even the resulting k_a^{mic} values are not by themselves terribly informative about the interacting system, because as mentioned before, k_a^{mic} values are, for the greater part mainly linked to the diffusion constant, D, of the adsorbing macromolecule. Instead, it is the k_d^{mic} value [in conjunction with K_a (t → 0), which is needed, together with k_a^{mic}, to obtain k_d^{mic}], which is the most accurate parameter for the quantitative expression of the kinetics of adsorption or adhesion. This is because k_a^{mic} only indicates the *rate of arrival*, which is mainly a function of the diffusion constant, D, of the molecule to be adsorbed. What decides the strength of adsorption is the *rate at which that molecule is allowed to re-depart from the adsorbing surface*. The more energetic the K_a of the arriving molecule, the slower its departure, so that it is the speed of that departure, expressed as the *kinetic dissociation rate constant, k_d*, which, paradoxically, best characterizes the strength of that molecule's *attachment*, from a kinetics point of view. Purely energetically speaking, it is of course the equilibrium association constant, K_a, which characterizes the free energy of association; cf. eq. [XXV-11].

Thus k^{mic} is mainly important as a necessary factor in obtaining k^{mic} [with the additional help of K_a ($t \to 0$)]. On the other hand, k_a or k_a^*, when obtained alone without further serving to extract k^{mic} and k^{mac}, are meaningless data, heavily impacted by many experimental factors which usually remain ingnored by the vast majority of workers. In the example given in Table XXV-1, listing the various experimental data obtained for the adsorption of HSA onto monosized amorphous silica spheres of 2.1 μm diameter, one notes the large difference between k_a^* [the measured k_a value, corrected for non-stirring, using the method given by Docoslis *et al.* (1999)] and k^{mic}, where k^{mic} is 4836 times larger than k_a^*, which is for the greater part due to the fact that f^{mic} is 5202 times larger than f. These differences are a consequence of the various above-mentioned aspecific macroscopic-scale repulsions; see also van Oss (1999) and van Oss *et al.* (2003), but they can become even more enormous when using monosized amorphous silica spheres which were purpously made even more repulsive than the untreated silica spheres, by carboxylation. This is exemplified in Table XXV-2, where one notes that for the COO^- treated spheres k_a, k_a^* and K_a ($t \to 0$) are much smaller than for the untreated silica spheres, and where the k^{mic} value is only somewhat lower, whilst k^{mic} is much higher. Similarly, for the COO^- spheres f is 573 times smaller and f^{mac} is 352 times smaller, whilst the f^{mic} values do not differ greatly. X^{TOT} is almost doubled for the COO^- spheres, whilst X^{mac} is slightly more than doubled and X^{mic} is about 38% smaller.

The differences in K_a ($t \to 0$) and k^{mic} are not so much due to the greater repulsion by the COO^- treated spheres, but are rather a consequence of the greatly reduced number of attractive sites on the treated spheres, due to the fact that the COO^- groups replaced an important proportion of the attractive metal oxide sites.

The importance of k_d^{mic}, as compared to k_a^{mic}, for the characterization of the kinetic aspects of bond energies has been known for some time. Absolom and van Oss (1986) showed that among 14 different hapten/anti-hapten antibody systems the k_a values varied only 77-fold, between 6.2×10^8 and 8.0×10^6 L/M.sec, whilst the k_d values varied tremendously (by a factor 1.76×10^7: between 3.4×10^{-4} and 6000 sec^{-1}). Concomitantly, K_a also varied strongly, i.e., by a factor 8×10^6; from 8×10^{10} to 1×10^4 L/M.

The determinations of both k_d and K_a are hampered by the fact that they are necessarily measured after the onset of hysteresis. However, in the following Section an approach is shown which helps to obviate these effects.

INFLUENCE OF HYSTERESIS ON THE DETERMINATION OF K_a AND k_d AND METHOD FOR OBVIATING ITS EFFECTS

The Greek word hysteresis (see Chapter XII) describes, *inter alia*, the strengthening, with time, of a bond. Even among non-covalent bonds occurring in physisorption and also in ligand-receptor (e.g., antigen-antibody) bonds one almost always encounters a non-negligible degree of bond energy hysteresis as a function of time (see also Chapters XII and XXIV). If done properly, the measurement of k_a, within a fraction of a second and at very low solute concentrations, and further corrected by extrapolation to zero time as well as to zero concentration (Docoslis *et al.*, 1999; 2001b), is not affected by binding hysteresis. However, when determining K_a or k_d which both require pre-adsorption during many minutes or even hours, binding hysteresis is impossible to avoid and few authors have even attempted to do so. Thus virtually all published K_a and k_d values are fatally flawed because due to the occurrence of hysteresis they both correlate with a much stronger bond than existed when k_a was measured (if done properly); see Docoslis *et al.* (2001b). One exception to, e.g., antigen-antibody bond hysteresis occurs in DNA/anti-DNA systems in low- or intermediate-affinity systems, which are mainly of an electrostatic nature; see Chapter XXIV, and Smeenk *et al.* (1983); van Oss, Smeenk and Aarden (1985).

It is nonetheless possible in virtually all systems to obtain a K_a value which comes very close to being unaffected by hysteresis. This is based on the determination of K_a (t → 0), by deriving Ka via Langmuir adsorption curves taken at a number of different times, t, followed by extrapolation of the K_a values thus obtained, to t = 0 (Docoslis *et al.*, 2001a), which then allows one to obtain a correct, pre-hysteretic value for K_a, as well as for k_d, for which eq. [XXV-12] should be used. The values for K_a (t → 0) and k_d^{mic} shown in Tables XXV-1 and 2 were obtained in this manner.

CODA

In treating some of the conditions that play a role in the measurement of kinetic association rate constants (k_a) of the adsorption of proteins onto silica surfaces, a number of precautionary measures are described which should be observed in order to avoid the occurrence of large experimental errors. These measures include the requirement for incipient turbulence, to avoid inadequate mass transport, as well as the necessity to operate under conditions of extrapolated to zero time, together with the need to use extrapolated to zero protein concentrations. Less obvious to most practicians but equally important if one wishes to measure just the k_a value which is uniquely relevant to the specific molecular bond formation between adsorbate and adsorbent (or, e.g., between ligand and receptor), is the methodology for the isolation of the virtually always present macroscopic-scale, aspecific and strongly repulsive interaction forces, from the microscopic-scale specific attractions, where only the latter represents the real molecular k_a. To that effect, use of the formalism originally proposed in 1917 by M. von Smoluchowski for the analysis of the kinetics of flocculation, was also found to be remarkably applicable for treating the repulsion *versus* attraction dichotomy which almost invariably accompanies protein adsorption onto hydrophilic surfaces.

Also treated are the errors caused by the well-nigh unavoidable hysteresis accompanying the measurements of the equilibrium association constant (K_a), as well as of the kinetic dissociation rate constant (k_d), of adsorption (or e.g., ligand-receptor) interactions, caused by the need to pre-adsorb protein onto the substratum during a sugnificant amount of time, prior to the measurement of either parameter. One solution for resolving this quandary is to determine K_a by means of multiple Langmuir isotherms, measured after different time-spans of pre-adsorption, followed by extrapolation to time zero, thus yielding K_a ($t \rightarrow 0$). Then, also using k_a values, obtained after extrapolation to time zero, and applying: k_a/K_a ($t \rightarrow 0$) = k_d, kinetic *dissociation* rate constant (k_d) values can be obtained which also are unaffected by adsorption hysteresis.

Finally, some practical data pertaining to the adsorption of protein [here, human serum albumin (HSA)] onto monosized silica spheres, as well as on monosized and carboxylated silica spheres are given. A comparison between the adsorption of HSA onto moderately repulsive hydrophilic silica surfaces and the adsorption of HSA onto strongly repulsive (hydroxylated) hydrophilic silica surfaces, serves to illustrate the strong influence of extraneous macroscopic-scale aspecific hydrophilic repulsion on the various kinetic and energetic constants.

List of Symbols Used

Latin Characters

A	Hamaker constant (eq. [II-5]; Chapter II, III); specific surface area (eq. [XII-11]).
A_{11} or A_{ii}	Hamaker constant pertaining to compound 1 or i (eq. [II-5]).
A_{131}	Hamaker constant of interaction between molecules or particles, 1, immersed in liquid, 3 (eq. [II-10]).
A_{132}	Hamaker constant of interaction between molecules or particles 1 and 2, immersed in liquid, 3 (eq. [II-11]).
a	Radius of particle (eq. [V-1]); proportion of favorable orientations (eqs. [XXV-6, 10A]).
(a-1)	Proportion of unfavorable orientations (eqs. [XXV-6, 10B]).
AB	Lewis acid-base (see also superscripts).
Ab	Antibody (Chapter XXIV).
B_2	Constant in second virial coefficient (eq. [VI-3]).
B_3	*Idem* in third virial coefficient (eq. [VI-4]).
BR	Brownian movement (see also superscripts).
C	Concentration (eq. [VI-2]).
c	Velocity of light in vacuo (eq. [II-18]); concentration as volume fraction (eq. [VI-1]).
cmc	Critical micelle concentration (eqs. [XXII-10, 11]).

Most of the usual mathematical and physical symbols (e.g., i, π, Σ, etc.) and chemical symbols are not listed here. As far as possible the equations are identified in which the listed symbol is first used.

D	Diffusion coefficient (eq. [VI-15]).
d	Relative density of polymer with respect to solvent (eq. [VI-2]); diameter of tube (eq. [XXV-1, 2]).
d_e, d_s	Diameters of hanging drop (eq. [XII-13]).
DIM	Diiodomethane (Tables XII-2, XVII-10).
DLVO	Initials of Derjaguin, Landau, Verwey and Overbeek; pertaining to the stability of particles, taking into consideration just the LW and EL interaction energies, as a function of distance (Chapters VII, XXIII, and XXIV).
E	Term used for electrophoretic mobility (eq. [V-17]; Table V-4; Fig. V-8).
e	Charge of electron (eq. [V-2]).
EL	Electrostatic (see also superscripts).
ER	Erythrocytes (Chapters VII, XXIII).
F	Force (Tables VII-1, 2, 3)
f	Friction factor (eq. [VI-17]; Tables VI-1, 2).
f_o	Friction factor of spherical particle (eq. [VI-18]; Tables VI-1, 2).
f	von Smoluchowski's **f**-factor (eqs. [XXV-4, 5, 9, 9A, 9B, 10A]).
fmac	Macroscopic-scale (repulsive; aspecific) part of **f** (eq. [XXV-10A]).
fmic	Microscopic-scale (attractive, specific) part of **f** (eq. [XXV-10B]).
f_1, f_2	Surface fraction of materials, 1 or 2 (eq. [XII-2]).
Fc	Carboxy terminal chains of immunoglobulins (Chapter XXIV).
FO	Formamide (Table XII-2).
g	Acceleration of gravity (eq. [XII-13]).
GLY	Glycerol (Table XII-2).
ΔG	Free energy difference (eqs. [II-13, 14, 15] etc.).
ΔG_{11}; ΔG_{ii}	Free energy of cohesion (eq. [II-33]).
ΔG_{12}	Free energy of adhesion between molecules or particles, 1 and 2, *in vacuo* (eqs. [II-36], [IV-6]).
ΔG_{121}; ΔG_{131}	Free energy of interaction between molecules or particles 1, immersed in liquid 2 (or 3) (eqs. [II-33A], [IV-17]).
ΔG_{SL}	As ΔG_{12} between a solid S and a liquid L (eq. [IV-6B]).

H	Shape-dependent parameter (eq. [XII-13]).
HSA	Human serum albumin (Table XIX-4).
h	Planck's constant (eqs. [II-4, 21]); height of capillary rise (eq. [XII-9]).
ΔH	Enthalpy component of ΔG (eqs. [VI-5], [XVIII-3, 4]).
HLB	Hydrophile-lipophile balance (Chapter XXII).
IF	Interfacial, i.e., AB + LW (see also superscripts).
IgA, IgG, IgM	Immunoglobulins A, G, M (Chapter XXIV).
j	Number of different types of forces (eq. [II-43]).
K, K_a, K_{ass}	Equilibrium constant (eqs. [XIV-8], [XXIV-3, 4], [XXV-11, 12]; Table XXIV-1).
K_a	$K_a = k_a/k_d$ (eq. [XXV-12]).
$K_a (t \to 0)$	K_a as measured via Langmuir isotherms, extrapolated to zero time (Tables XXV-1, 2).
k_a	Kinetic rate constant of adsorption (eqs. [XXV-3, 4]).
k_a^{mac}	Macroscopic-scale (repulsive, aspecific) part of k_a (eq. [XXV-.....]).
k_a^{mic}	Microscopic-scale (attractive, specific) part of k_a (eq. [XXV-13]).
$<k_a>$	Value for k_a as given in eq. [XXV-13], when $f = 1$: Table XXV-1.
k_d	Kinetic rate constant of desorption (eq. [XXV-12]), Tables XXV-1, 2.
k	Boltzmann's constant = 1.38×10^{-23} J deg^{-1} (eq. [II-2]).
ℓ	Distance between the outer shells of atoms, molecules or particles (eqs. [II-6], [II-32], [III-3]).
ℓ_o, ℓ_{eq}	Minimum equilibrium distance = 0.157 ± 0.009 nm (eqs. [II-32], [III-3]).
LW	Lifshitz–van der Waals (see also superscripts).
M, Mw	Molecular weight (weight average) (eqs. [VI-15, 16, 18]).
M, Mn	Number average molecular weight (eqs. [VI-2, 3, 4]).
m_i	Molar concentration of species, i (Table V-2).
N	Avogadro's number = 6×10^{23} (eq. [VI-18], [XXV-2]).
n, n_i	Quantum number (eq. [II-21]); number of ions of each ionic species, i, per cm^3 (eq. [V-2]); number of surface charges per particle or molecule (eq. [V-8]); number of monomers per polymer molecule (eq. [VI-3]).

n_o	Refractive index in visible range (eq. [II-28]).
OS	Osmotic (see also superscripts).
P	Polar (see also superscripts); pressure (eq. [II-1]); integration parameter (eq. [II-18]).
[P]	Protein concentration (eq. [XIV-8]).
p	Periphery of Wilhelmy plate (eq. [XII-12]).
ΔP	Capillary pressure (eq. [XII-10]).
PEO	Polyethylene oxide (Chapter XIX; Table XIX-5).
PIB	Polyisobutylene (Table XXII-2).
q	Number of atoms per unit volume (eq. [II-5]).
R	Radius of sphere (eq. [V-5]; Tables VII-1, 2, 3); radius of curvature (Chapters XXIII, XXIV); average pore radius (eqs. [XII-9, 10, 11]); gas constant (eq. [VI-1]).
Re	Reynolds number (eqs. [XXV-1, 2]).
Rg	Radius of gyration of polymer molecules (eqs. [VI-11, 12], [VII-8, 11]).
r	Mean-square end-to-end length of a random coil in solution (eq. [VI-12]); radius of curvature of small cellular processes (Chapter XXIII).
r_o	Maximum radius of rotating drop (eq. [XIII-1]).
S	Spreading coefficient; usually equal to π_e (Chapter XII, footnote p. 137); Svedbergs; $1S = 10^{-13}$ sec (Table VI-2).
[S]	Concentration of adsorption sites (eqs. [XIV-8], [XXIV-4]).
Sc; S_c	Contactable surface area between chains or molecules (eqs. [VI-6A, 8A], [XXII-4, 7, 12], [XXV-11]).
S.D.	Standard deviation (eqs. [III-2A, B]).
SDS	Sodium dodecyl sulfate (Chapters XV and XXII).
s	Sedimentation coefficient (eq. [VI-15]); solubility (eq. [XIV-7]; Chapter XXII).
ΔS	Entropy (eq. [XVIII-3]; Table XVIII-1).
T	Absolute temperature, in degrees Kelvin (eqs. [II-2], [XVIII-2, 3, 4]; Table XVIII-1).
t	Thickness (eq. [II-15]); time, in seconds (eq. [XII-9]).
U	Electrophoretic velocity (eqs. [V-12, 14, 16, 17]).

u	Electrophoretic mobility (eqs. [V-11, 12, 14, 16]; Table V-3).
ΔU	Energy (eq. [XXII-2]).
$v, \overline{V}$	Atomic or molar volume (eq. [II-1]).
v	Valency of counterions (eq. [V-3]), velocity of liquid (eq. [XXV-1]).
v_o	Volume of 1 mole of solvent (eqs. [VI-3, 4])
v_i	Valency of ionic species, i (eq. [V-2]).
$\overline{v}$	Specific volume (eq. [VI-16]).
W	Work of adhesion ($W = -\Delta G$) (eq. [XXIV-7]).
W_{1w1}; W_{iwi}	Work of adhesion (Chapter XVIII).
ΔW	Additional weight (eq. [XII-12]).
X_o	Defined in eq. [II-23].
XDLVO	Extended DLVO approach (see DLVO), taking the AB interaction energy into account in addition to the LW and EL energies (Chapters VII, XXIII, XXIV and XXV).
Y_o	Surface potential, as defined in eq. [V-18] (Table V-4; Fig. V-8).
y	Asymmetry factor (Table VI-1).
z	Distance from particle surface to slipping plane (eq. [V-1]).

Superscripts

AB	Lewis acid-base (eq. [IV-1, 2, 3, 4]).
BR	Brownian movement (eq. [VII-13 to 17]; Chapter VI).
coh	Cohesion (eq. [II-33B]).
D	Debye: induction forces (eq. [II-40]).
EL	Electrostatic; pertaining to electrical double layer interactions (eq. [V-3]).
IF	Interfacial: LW + AB (Chapter XXIV).
j	Component of γ arising from the j'th type of interaction (eq. [II-34]).
K	Keesom: orientation forces (eq. [II-40]).
L	London: dispersion forces (eq. [II-40]).
LW	Lifshitz–van der Waals (eq. [II-35]).
M	Mixing (eq. [VI-5]).
OS	Osmotic (eq. [VI-9]).

P	Polar: superscript P is generally used erroneously (eq. [XIV-5]; Table XIV-2).
TOT	Total: including LW + AB + EL (+ BR where appropriate) (eq. [VII-16]).

Subscripts

A	θ_A: aggregate contact angle (eq. [XII-2]); γ_A: aggregate surface tension components or parameter (Chapter XII—Cassie's equation: eqs. [XII-2, 4]).
a	θ_a: advancing contact angle (Chapter XII—hysteresis).
c	γ_c: critical surface tension (Chapter IV).
G	Glycerol (eq. [IV-21]).
IR	Infrared (eq. [II-28]).
i, j	Generalization for subscripts 1 or 2.
j	Integral numbers, 1, 2, 3 . . . (eqs. [II-22, 34]).
L	Liquid (eqs. [IV-6B, 11]).
ℓ, ℓ_o	Pertaining to ℓ, ℓ_o (Tables VII-1, 2, 3).
MW	Microwave (eq. [II-28]).
n	Quantum number (eq. [II-26]).
o	Oil (eq. [XII-5, 6, 7, 7A, 8]).
r	θ_r: retreating contact angle (Chapter XII—hysteresis).
S	Solid (eq. [IV-10]).
SL	Solid/liquid (eq. [IV-10]).
UV	Ultraviolet (eq. [II-28]).
V	γ_{LV}: Vapor given off by the liquid (eq. [XII-1]).
W, w	Water (eqs. [IV-19A, B]).

Greek Characters

α	Polarizability (eqs. [II-3, 4]); Cole-Cole parameter (eq. [II-29]).
αBrN	α-Bromonaphthalene (Table XII-2).
β	See eqs. [II-5, 6, 7].
γ	Surface tension.
$\gamma_1, \gamma_2, \gamma_i$	Surface tension of materials 1, 2 or i (eq. [II-33]).

$\gamma^\oplus, \gamma^\ominus$	Electron-acceptor and electron-donor parameters of the surface tension (eq. [IV-3]).
γ_o	Defined in eq. [V-4] (eq. [V-3]; Table V-3).
Δ	Defined in eqs. [II-19, 20]).
δ	Hildebrand's solubility parameter (proportional to $\sqrt{\gamma^{LW}}$) (eqs. [XXII-1, 2]).
$\delta_{iw}^\oplus$	$\sqrt{\gamma_i^\oplus/\gamma_w^\oplus}$ (eq. [IV-19A]).
$\delta_{iw}^\ominus$	$\sqrt{\gamma_i^\ominus/\gamma_w^\ominus}$ (eq. [IV-19B]).
ε	Dielectric constant (eqs. [II-18, 19, 20], [V-2, 5, 6, 10, 11, 13, 14]; Table VII-3).
ζ	Electrokinetic potential, in mV (eqs. [V-1, 10, 11, 14]).
η	Viscosity of liquids (eqs. [VI-18], [VII-9], [XXV-1, 2]; Table XVII-10).
θ	Contact angle, in degrees (eqs. [IV-10, 13], Chapter XII); θ-temperature (Chapter VI).
κ	Inverse thickness of diffuse ionic double layer, or inverse Debye length (eq. [V-2]).
λ	Characteristic length, or decay length, of liquids (here generally of water) (eq. [VII-3]; Table VII-2; Chapter XXIII).
μ	Dipole moment (eqs. [II-2, 3]); ionic strength (Chapters XXIII, XXIV).
ν	Main dispersion frequency (eq. [II-4]).
Π	Osmotic pressure (eq. [VI-1]).
π_e	Equilibrium spreading pressure, sometimes appended to Young's equation (eq. [VII-1]).
ρ	Specific density (eqs. [VI-15], [XXV-1, 2]).
$\Delta\rho$	Density difference (eqs. [XII-13, 14]).
σ	Surface charge density (eq. [V-13]).
Φ	Good's interaction parameter; in completely apolar systems, $\Phi = 1$ (eq. [II-41]); volume fraction) eq. [XII-11]).
ϕ	Degrees of orientation (eqs. [XXV-5, 6]).
ϕ_1, ϕ_2	Volume fractions of condensed-phase materials, 1 and 2 (eqs. [VII-7, 8]).

χ Electric field strength (eq. [V-8]); Flory-Huggins parameter (eqs. [VI-5, 6, 8, 9, 10]; Chapters XIX, XXII).

ψ_o Electric surface potential, in mV (eq. [V-1]); Table VII-3.

ω Frequency (eq. [II-21]); speed of revolution in radians/sec. (eq. [XII-14]).

χ^{TOT} Sum total of interaction energies, in kT units (eq. [XXV-7]).

χ^{mac} Macroscopic-scale (repulsive, aspecific) part of the interaction energies, in kT units (eqs. [XXV-8A, 9, 10A]).

χ^{mic} Microscopic-scale (attractive, specific) part of the interaction energies, in kT units (eqs. [XXV-8B, 9, 10B]).

References

Abramson, H.A., Moyer, L.S., and Gorin, M.H. (1942). *Electrophoresis of Proteins*, Reinhold, New York; Reprinted: Hafner, New York (1964).

Abramson, H.A., Moyer, L.S., and Voet, A. (1936). J. Amer. Chem. Soc. *58*, 2362.

Absolom, D.R., Michaeli, I., and van Oss, C.J. (1981). Electrophoresis 2, 273.

Absolom, D.R., Neumann, A.W., and van Oss, C.J. (1984). ACS Symp. Series *240*, 169.

Absolom, D.R., and van Oss, C.J. (1986). Crit. Rev. Immunol. *6*, 1.

Absolom, D.R., van Oss, C.J., Zingg, W., and Neumann, A.W. (1981). Biochim. Biophys. Acta *670*, 74.

Absolom, D.R., van Oss, C.J., Zingg, W., and Neumann, A.W. (1982). J. Reticuloendothelial Soc. *31*, 59.

Adamson, A.W. (1982; 1990). *Physical Chemistry of Surfaces*, Wiley-Interscience, New York.

Albertsson, P.Å. (1971, 1986). *Partition of Cell Particles and Macromolecules*, Wiley-Interscience, New York.

Albini, B., Fagundus, A.M., and Vladutiu, A.O. (1984). In: *Molecular Immunology*, M.Z. Atassi, C.J. van Oss and D.R. Absolom, Eds., Marcel Dekker, New York, p. 381.

Allen, R.E., Rhodes, P.H., Snyder, R.S., Barlow, G.H., Bier, M., Bigazzi, P.E., van Oss, C.J., Knox, R.J., Seaman, G.V.F., Micale, F.J., and Vanderhoff, J.W. (1977). Separ. Purif. Meth. *6*, 1.

Alty, T. (1924). Proc. Roy. Soc. Lond., Ser. A. *106*, 315.

Alty, T. (1926). Proc. Roy. Soc. Lond., Ser. A. *112*, 235.

Amit, A.G., Mariussa, R.A., Phillips, S.E.V., and Poljak, R.J. (1986). Science *233*, 747.

Ananathapadmanabhan, K.P., and Goddard, E.D. (1987). Colloids Surfaces *25*, 393.

Anderson, N.G., and Anderson, L. (1982). Clin. Chem. *28*, 739.

Andrade, J.D., Ma, S.M., King, R.N., and Gregonis, D.E. (1979). J. Colloid Interface Sci. *72*, 488.

Arlinger, L. (1976). J. Chromatog. *119*, 9.

I must apologize for, or at least explain, the fact that quotations of my own work and the work of my close collaborators may seem referenced with unusual frequency. The explanation is two-fold: (1) even though I spend considerable time trying to keep up with the literature I still am, like most authors, more familiar with my own than with everybody else's work, and (2) the substance of this book deals with the nature and the effects of interfacial interactions in polar, and especially in aqueous media, mainly resulting from my own and my close collaborators' experimental and theoretical endeavors, which began with some of us as early as the late 1950s, but which has intensified considerably since 1985.

Arnold, K., Hermann, A., Gawrisch, K., and Pratsch, L. (1988). In: *Molecular Mechanisms of Membrane Fusion*, S. Ohki, D. Doyle, T.D. Flanagan, S.W. Hui and E. Mayhew, Eds., Plenum Press, New York, p. 255.

Arnold, K., Zschoernig, O., Barthel, D., and Harold, W. (1990). Biochim. Biophys. Acta. *1022*, 303.

Ataman, M., and Boucher, E.A. (1982). J. Polymer Sci. *20*, 1585.

Atassi, M.Z. (1984). In: Atassi *et al.*, 1984, p. 15.

Atassi, M.Z., Ruan, K.H., Jinnai, K., Oshima, M. and Ashizawa, T. (1992). Proc. Natl Acad. Sci. USA, *89*, 5852.

Atassi, M.Z., van Oss, C.J., and Absolom, D.R., Eds. (1984). *Molecular Immunology*, Marcel Dekker, New York.

Ausubel, F.M., Brent, R., Kingston, R.E., Moore, D.D., Smith, J.A., Seidman, T.G., and Stzuhl, K. (1987). *Current Protocols in Molecular Biology*, Greene Publ. Assoc., Brooklyn, NY, Section 7.1.

Baier, R.E., Ed. (1975). *Applied Chemistry at Protein Interfaces*, Advan. Chem. Series *145*.

Baier, R.E., Meyer, A.E., Natiella, J.R., Natiella, R.R., and Carter, J.M. (1984). J. Biomed. Materials Res. *18*, 337.

Bangham, A.D., Pethica, B.A., and Seaman, G.V.F. (1958). Biochem. *69*, 12.

Bangham, D.H., and Razouk, R.I. (1937). Trans. Faraday Soc. *33*, 1459.

Bangs, L.B. (1984). *Uniform Latex Particles*, Seragen Diagnostics, Indianapolis, IN.

Bank, M., Leffingwell, J., and Thies, C. (1971). Macromolecules *4*, 43.

Barclay, L., Harrington, A., and Ottewill, R.H. (1972). Kolloid-Z. Z. Polymere *250*, 655.

Barnett, W., and Spragg, S.P. (1971). Nature New Biol. *234*, 191.

Barry, P.H., and Hope, A.B. (1969). Biophys. J. *9*, 700; *9*, 729.

Bartell, F.E., and Whitney, C.E. (1932). J. Phys. Chem. *326*, 3115.

Barton, A.F.M. (1983). *Handbook of Solubility and other Cohesion Parameters*, CRC Press, Boca Raton, FL.

Bäumler, H., and Donath, E. (1987). Studia Biophys. *120*, 113.

Baxter, S., and Cassie, A.B.D. (1945). J. Text. Inst. *36*, T67.

Becher, P. (1967a) In: *Nonionic Surfactants*, M.J. Schick, Ed., Marcel Dekker, New York, pp. 478–515.

Becher, P. (1967b). *Idem*, pp. 604–626.

Becher, P. (1984). J. Dispersion Sci. Technol. *5*, 81.

Beijerinck, M.W. (1896). Zentralbl. Bakt. *2*, 627.

Beijerinck, M.W. (1910). Kolloid-Z. *7*, 16.

Bell, G.I., Dembo, M., and Bongrand, P. (1984). Biophys. J. *45*, 1051.

Bellqvist, B., Ek, K., Righetti, P.G., Gianazza, E., Görg, A., Westermeier, R., and Postel, W. (1982). J. Biochem. Biophys. Meth. *6*, 317.

Berkeley, R.C.W., Lynch, J.M., Melling, J., Rutter, P.R., and Vincent, B., Eds. (1980). *Microbial Adhesion to Surfaces*, Ellis Horwood, Chichester/Wiley, London.

Besseling, N.A.M. and Lyklema, J. (1997). J. Phys. Chem. *101*, 7604.

Bessis, M. (1973). *Living Blood Cells and their Ultrastructure*, Springer-Verlag, New York, pp. 141–143.

Bidwell, E., Dike, G.W.R., and Denson, K.W.E. (1966). Br. J. Haematol. *12*, 583.

Bier, M., and Egen, N.B. (1979). In: *Electrofocus/78*, H. Haglund, J.B. Westerfeld and J.T. Ball, Eds., North-Holland, New York, p. 35.

Bier, M., Egen, N.B., Allgyer, T.T., Twitty, G.E., and Mosher, R.A. (1979). In: *Peptides: Structure and Biological Function*, E. Gron and J. Meienhofer, Eds., Pierce Chemical Co., Rockford, IL, p. 79.

Binnig, G., Quate, C.F. and Gerber, C. (1986). Phys. Rev. Lett. *56*, 930.

Bitton, G., and Marshall, K.C., Eds. (1980). *Adsorption of Microorganisms to Surfaces*, Wiley, New York.

Bixler, H.J., and Michaels, A.S. (1969). Encycl. Polymer Sci. *10*, 765.

Black, W., de Jongh, J.G.V., Overbeek, J.Th.G., and Sparnaay, M.J. (1960). Trans. Faraday Soc. *56*, 1597.

Blomberg, E., Claesson, P.M., and Christenson, H.K. (1990). J. Colloid Interface Sci. *138*, 291.

Blomberg, E., Claesson, P.M., and Gölander, C.G. (1991). Biocolloids Biosurfaces *1*, 179; in J. Dispersion Sci. Tech. *12*, 179.

Boltz, R.C., Todd, P., Gaines, R.A., Bilito, R.P., Kocherty, J.J., Thompson, C.J., Notter, M.F.D., Richardson, L.S., and Mortel, R. (1976). J. Histochem. Cytochem. *24*, 16.

Boltz, R.C., Todd, P., Hammerstedt, R.H., Hymer, W.C., Thompson, C.J., and Docherty, J. (1977). In: *Cell Separation Methods*, H. Bloemendal, Ed., Elsevier North-Holland Publ. Co., Amsterdam, p. 145.

Boltz, R.C., Todd, P., Streibel, M.J., and Louie, M.K. (1973). Prep. Biochem. *3*, 383.

Bongrand, P. (1988). *Physical Basis of Cell-Cell Adhesion*, CRC Press, Boca Raton, FL.

Bongrand, P., Capo, C., and Depieds, R. (1982). Progr. Surface Sci. *12*, 217.

Booth, F. (1948). Nature *161*, 83.

Booth, F. (1950). Proc. Roy. Soc. Lond. A*203*, 514.

Boucher, E.A., and Hines, P.M. (1976). J. Polymer Sci. *14*, 2241.

Boucher, E.A., and Hines, P.M. (1978). J. Polymer Sci. *16*, 501.

Boumans, A.A. (1957). *Streaming Currents in Turbulent Flows and Metal Capillaries*, Ph.D. thesis, Utrecht; Physica, 1007, 1027, 1038, 1047.

Braithwaite, G.J.C., Howe, A. and Luckham, P.F. (1996). Langmuir *12*, 4224.

Brash, J.L., and Horbett, T.A., Eds. (1987). *Proteins at Interfaces. Physicochemical and Biochemical Studies*, ACS Symp. Series *343*.

Bridger, K., Fairhurst, D., and Vincent, B. (1979). J. Colloid Interface Sci. *68*, 190.

Broers, G. (1986). *Microemulsion Studies-Phase Behavior and Anomalous Sedimentation*, M.Sc. thesis, SUNY, Buffalo.

Brooks, D.E. (1973). J. Colloid Interface Sci. *43*, 687, 700, 714.

Brooks, D.E. (1976). In: *Microcirculation*, Vol. 1, Grayson, J., and Zingg, W., Eds., Plenum, New York, pp. 33–52.

Brooks, D.E., Greig, R.G., and Janzen, J. (1980) In: *Erythrocyte Mechanics and Blood Flow*, G.R. Cocelet, H.J. Meiselman and D.E. Brooks, Eds., A.R. Liss, New York, pp. 119–140.

Brooks, D.E., and Seaman, G.V.F. (1973). J. Colloid Interface Sci. *43*, 670.

Bungenberg de Jong, H.G. (1949). In: *Colloid Science*, Vol. II, H.R. Kruyt, Ed., Elsevier, Amsterdam, pp. 232, 335, 433.

Bungenberg de Jong, H.G., and Kruyt, H.R. (1930). Kolloid-Z. *50*, 39.

Bunn, C.W. (1939). Trans. Faraday Soc. *35*, 482.

Busscher, H.J., Bellon-Fontaine, M.N., Mozes, N., van der Mei, H.C., Sjollema, J., Cerf, O., and Rouxhet, P.G. (1990). Biofouling *2*, 55.

Busscher, H.J., Kip, G.A.M., van Silfhout, A., and Arends, J. (1986). J. Colloid Interface Sci. *114*, 307.

Busscher, H.J., Stokroos, I., van der Mei, H.C., Rouxhet, P.G. and Schakenraad, J.M. (1992), J. Adhesion Sci. Technol. *6*, 347.

Busscher, H.J., Uyen, H.M.J.C., Weerkamp, A.H., Postma, W.J., and Arends, J. (1986). FEMS Microbiol. Lett. *35*, 303.

Cain, F.W., Ottewill, R.H., and Smitham, J.B. (1978). Disc. Faraday Soc. *65*, 33.

Cairns, R.J.R., Ottewill, R.H., Osmond, D.W.J., and Wagstaff, I. (1976). J. Colloid Interface Sci. *54*, 45.

Caldwell, K.D., Kesner, L.F., Myers, M.N., and Giddings, J.C. (1972). *176*, 296.

Cann, J.R. (1979). In: Righetti *et al.* (1979), p. 369.

Cantor, C.R., and Schimmel, P.R. (1980). *Biophysical Chemistry*, 3 Vols., W.H. Greeman, San Francisco.

Cantor, C.R., Warburton, P., Smith, C.L., and Gaal, A. (1984). *Electrophoresis '86*, M.J. Dunn, Ed., Verlag Chemie, Weinheim, p. 161.

Carle, G.F., and Olson, M.V. (1984). Nucl. Acids Res. *12*, 5647.

Casimir, H.B.G., and Polder, D. (1948). Phys. Rev. *73*, 360.

Cassie, A.B.D. (1948). Discuss. Faraday Soc. *3*, 11.

Cassie, A.B.D., and Baxter, S. (1944). Trans. Faraday Soc. *40*, 546.

Catsimpoolas, N. (1973). Analyt. Biochem. *54*, 66; *54*, 79; *54*, 88.

Catsimpoolas, N., Ed. (1976). In: *Isoelectric Focusing*, Academic Press, New York, p. 229.

Chan, D.Y.C., Mitchell, D.J., Ninham, B.W., and Pailthorpe, B.A. (1979). In: *Water*, Vol. 6, Franks, F., Ed., Plenum, New York, p. 239.

Chapman, A.P., Antoniw, P., Spitali, M., West, S. and King, D.J. (1999). Nature Biotechnol. *17*, 780.

Chaudhury, M.K. (1984). *Short-range and Long-range Forces in Colloidal and Macroscopic Systems*, Ph.D. dissertation, SUNY Buffalo.

Chaudhury, M.K. (1987). J. Colloid Interface Sci. *119*, 174.

Chaudhury, M.K., and Good, R.J. (1983). J. Colloid Interface Sci. *94*, 292.

Chien, S., Simchon, S., Abbott, R.E., and Jan, K.M. (1977). J. Colloid Interface Sci. *62*, 461.

Chirife, J., and Fontan, C.F. (1980). J. Food Sci. *12*, 65.

Chrambach, A., and Rodbard, D. (1971). Science *172*, 440.

Christenson, H.K. (1988). J. Disp. Sci. Tech. *9*, 171.

Christenson, H.K. (1992). In: *Modern Approaches to Wettability*, M.E. Schrader, and G.I. Loeb, Eds., Plenum, New York, p. 29.

Christenson, H.K., and Claesson, P.M. (1988). Science *239*, 390.

Chwastiak, S. (1973). J. Colloid Interface Sci. *42*, 299.

Claesson, P.M. (1986). *Forces Between Surfaces Immersed in Aqueous Solutions*, Ph.D. dissertation, Royal Institute of Technology, Stockholm.

Claesson, P.M., Blom, C.E., Herder, P.C., and Ninham, B.W. (1986). J. Colloid Interface Sci. *114*, 234.

Claesson, S., and Claesson, I.M. (1961). In: *A Laboratory Manual of Analytical Methods of Protein Chemistry*, Vol. 3, P. Alexander and R.J. Block, Eds., Pergamon Press, New York, p. 119.

Clarke, J., and Vincent B. (1981). J. Colloid Interface Sci. *82*, 208.

Clausen, J. (1979). In: Righetti *et al.*, 1979, p. 55.

Cohen Stuart, M.A., Scheutjens, J.M.H.M., and Fleer, G.J. (1984). ACS Symp. Series *240*, 53.

Cohly, H. (1986). *Cell Separation by Preparative Electrophoresis and Double Antibody Tagging*, Ph.D. dissertation, SUNY, Buffalo.

Cohly, H., van Oss, C.J., Weiser, M., and Albini, B. (1985a). In: *Cell Electrophoresis*, W. Schütt and H. Klinkmann, H., Eds., de Gruyter, Berlin, p. 603.

Cohly, H., Albini, B., Weiser, M., Green, K., and van Oss, C.J. (1985b). In: *Cell Electrophoresis*, W. Schütt and H. Klinkmann, Eds., de Gruyter, Berlin, p. 611.

Coombs, R.R.A., Mourant, A.E. and Race, R.R. (1945). Br. J. Exp. Pathol. *26*, 255.

Cooper, N.R. (1985). Advan. Immunol. *37*, 151.

Corsel, J.W., Willems, G.M., Kop, J.M.M., Cuypers, P.A., and Hermens, W.T. (1986). J. Colloid Interface Sci. *111*, 544.

Costanzo, P.M., Giese, R.F., and van Oss, C.J. (1990). J. Adhesion Sci. Technol. *4*, 267.

Costanzo, P.M., Giese, R.F., and van Oss, C.J. (1991). In: *Advances in Measurement and Control of Colloidal Processes*, R.A. Williams and N.C. de Jaeger, Eds., Butterworth, London, p. 223.

Cowell, C., Li-In-On, R., and Vincent, B. (1978). J. Chem. Soc. Faraday Trans.-I, *74*, 337.

Cowley, A.C., Fuller, N.L., Rand, R.P., and Parsegian, V.A. (1978). Biochem. *17*, 3163.

Craig, V.S.J., Ninham, B.W. and Pashley, R.M. (1993). J. Phys. Chem. *97*, 10192.

Croyle, M.A., Chirmule, N., Zhang, Y. and Wilson, J.M. (2001). J. Virol. *75*, 4792.

Cunningham, R.K. (1994). In: *Immunochemistry*, C.J. van Oss and M.H.V. Van Regenmortel, Eds., Marcel Dekker, New York, Ch. 14.

Curtis, A.S.G. (1962). Biol. Rev. *37*, 82.

Curtis, A.S.G. (1973). Progr. Biophys. Molec. Biol. *27*, 317.

Cuypers, P.A., Corsel, J.W., Willems, G.M., and Hermens, W.T. (1986). Abstr. Colloid Div. 192nd Am. Chem. Soc. Natl. Mg., Anaheim, CA., Sept.'86, No. 7.

Dalal, E.D. (1987). Langmuir *3*, 1009.

Davidson, D.W. (1993). In: *Water*, Vol. 2, F. Franks, Ed., Plenum Press, New York, p. 115.

Davidson, H.N. (1977). *The Biochemistry of the Nucleic Acids*, Academic Press, New York.

Davis, B.D., Dulbecco, R., Eisen, H.N., and Ginsberg, H.S. (1980). *Microbiology* (3rd. ed.), Harper & Row, Hagerstown, MD.

Davis, B.J. (1964). Ann. N.Y. Acad. Sci. *121*, 404.

Davis, S.S., and Illum, L. (1988). In: *Targeting of Drugs*, G. Gregoriadis and G. Poste, Eds., Plenum Press, New York, p. 177.

de Boer, J.H. (1936). Trans. Faraday Soc. *32*, 21.

Debye, P. (1920). Physik. Z. *21*, 178.

Debye, P. (1921). Physik. Z. *22*, 302.

de Gennes, P.G. (1982). Macromolecules *15*, 492.

de Gennes, P.G. (1985). Rev. Mod. Phys. *57*, 827.

de Jongh, J.G.V. (1958). *Measurements of Retarded van der Waals Forces*, Ph.D. dissertation, Utrecht.

Delmotte, P. (1977). Science Tools *24*, 34.

De Palma, V.A. (1976). *Correlation of Surface Electric Properties with Initial Events of Biological Adhesion*, Ph.D. dissertation, SUNY Buffalo.

De Palma, V.A., Baier, R.E., and Gott, V.L. (1977). Report for Biomaterials Program, Natl. Heart, Lung, Blood Inst., U.S. Publ. Health Serv., Washington, DC.

Derjaguin, B.V. (1934). Kolloid Zh. *69*, 155.

Derjaguin, B.V. (1939). Acta Physicochimica URSS *10*, 333.

Derjaguin, B.V. (1940). Trans. Faraday Soc. *36*, 203.

Derjaguin, B.V. (1940a). Acta Physicochimica URSS *12*, 181.

Derjaguin, B.V. (1940b). *Ibid. 12*, 314.

Derjaguin, B.V. (1954). Disc. Faraday Soc. *18*, 85.

Derjaguin, B.V. (1966). Disc. Faraday Soc. *42*, 109, 134.

Derjaguin, B.V. (1989). *Theory of Stability of Colloids and Thin Films*, Consultants Bureau/ Plenum Press, New York.

Derjaguin, B.V., and Abrikossova, I.I. (1951). Zh. Eksperim. Teor. Fiz. *21*, 945; (1956). *Ibid. 30*, 993; C.R. Acad. Sci. URSS (1953). *90*, 1055.

Derjaguin, B.V., and Churaev, N.V. (1989). Colloids Surfaces *41*, 223.

Derjaguin, B.V., Churaev, N.V., and Muller, W.M. (1987). *Surface Forces*, Plenum Press, New York.

Derjaguin, B.V., Churaev, N.V., Fedyakin, N.N., Talayev, M.V., and Yershova, I.G. (1967). Bull. Acad. Sci. USSR, Chem. Ser. *10*, 2178.

Derjaguin, B.V., and Dukkin, S.S. (1974). Surface Colloid Sci. *7*, 273.

Derjaguin, B.V., and Landau, L.D. (1941). Acta Physicochimica URSS *14*, 633.

Derjaguin, B.V., Rabinovich, Y.I., and Churaev, N.V. (1978). Nature *272*, 313.

Derjaguin, B.V., Titijevskaia, A.S., Abricossova, I.I., and Malkina, A.D. (1954). Disc. Faraday Soc. *18*, 24.

Derjaguin, B.V., Voropayeva, T.N., Kabanov, B.N., and Titiyevskaya, A.S. (1964). J. Colloid Sci. *19*, 113.

Devine, D.A. and Hancock, R.E.W., Eds. (2004). *Mammalian Host Defense Peptides*. Cambridge University Press, Cambrtidge, UK.

Diao, J., and Fuerstenau, D.W. (1991). J. Colloid Interface Sci. *60*, 145.

Dobry, A. (1938). J. Chim. Phys. *35*, 387.

Dobry, A. (1939). J. Chim. Phys. *36*, 102.

Dobry, A. (1948). Bull. Soc. Chim. Belg. *57*, 280.

Dobry, A., and Boyer-Kawenoki, F. (1947). J. Polymer Sci. *2*, 90.

Docoslis, A. (2000). *Adsorption Kinetics and Energetics of Human Serum Albumin onto Metal Oxide Microparticles*. Doctoral Thesis, Department of Chemical Engineering, University at Buffalo, State University oif New York.

Docoslis, A., Giese, R.F. and van Oss, C.J. (2000). Colloids Surfaces-B: Biointerfaces *19*, 147.

Docoslis, A., Wu, W., Giese, R.F. and van Oss, C.J. (2001-a). Colloids Surfaces-B: Biointerfaces *22*, 205.

Docoslis, A., Rusinski, L.A., Giese, R.F. and van Oss, C.J. (2001-b). Colloids Surfaces-B: Biointerfaces *22*, 267.

Docoslis, A., Wu, W., Giese, R.F. and van Oss, C.J. (1999). Colloids Surfaces-B: Biointerfaces *13*, 83.

Docoslis, A., Wu, W. Giese, R.F. and van Oss, C.J. (2002). Colloids Surfaces-B: Biointerfaces *25*, 97.

Doren, A., Lemaitre, J., and Rouxhet, P.G. (1989). J. Colloid Interface Sci. *130*, 146.

Doyle, R.J., and Rosenberg, M., Eds. (1990). *Microbial Cell Surface Hydrophobicity*, American Society for Microbiology, Washington, DC.

Drago, R.S., Vogel, G.C., and Needham, T.E. (1971). J. Am. Chem. Soc. *93*, 6014.

Drelich, J., and Miller, J.D. (1993). Langmuir *9*, 619.

Ducker, W.A., Senden, T.J. and Pashley, R.M. (1992). Langmuir *8*, 1831.

Dukkin, S.S., and Derjaguin, B.V. (1974). Surface Colloid Sci. *7*, 49.

Dunn, M.J., and Patel, K. (1986). In: *Electrophoresis '86*, M.J. Dunn, Ed., VCH, Weinheim, p. 574.

Dunstan, D., and White, L.R. (1986). J. Colloid Interface Sci. *111*, 60.

Dupré, A. (1869). *Théorie Mécanique de la Chaleur*, Gauthier-Villars, Paris, pp. 367–370.

Durrum, E.L. (1951). J. Am. Chem. Soc. *73*, 4875.

Dzyaloshinskii, I.E., Lifshitz, E.M., and Pitaevskii, L.P. (1961). Adv. Phys. *10*, 165.

Eberhart, R.C., Munro, M.S., Frantschi, J.R., and Sevastianov, V.I. (1987). ACS Symp. Series *343*, 378.

Edberg, S.C. (1971). *The Valency of IgM and IgG Anti-Dextran Antibody*, Ph.D. dissertation, SUNY Buffalo.

Edberg, S.C., Bronson, P.M., and van Oss, C.J. (1972). Immunochemistry *9*, 273.

Edelman, G.M. (1983). Science *219*, 450.

Edgell, M.H., Hutchinson, C.A., and Sinsheimer, R.L. (1969). In: *Progress in Separation and Purification*, Vol. II, T. Gerritsen, Ed., Wiley-Interscience, New York, p. 1.

Edmond, E., and Ogston, A.G. (1968). Biochem. J. *109*, 569.

Ehrlich, P. (1900). Proc. Roy. Soc. Lon. *66*, 424.

Einstein, A. (1907). Z. Elektrochemie *13*, 41.

Eisenberg, D., and Kanzmann, W. (1969). *The Structure and Properties of Water*. Clarendon Press, Oxford, pp. 145, 169.

Elwing, H., Askenda, A., Ivarsson, B., Nilsson, U., Welin, S., and Lundström, I. (1987). ACS Symp. Series. *343*, 468.

Eppstein, D.A. (1988). In: *Targeting of Drugs*, G. Gregoriadis and G. Poste, Eds., Plenum, New York, p. 189.

Erbil, H.Y., Demirel, A.L., Yonca, A. and Mert, O. (2003). Science *299*, 1377.

Evans, D.F. and Wennerström, H. (1999). *The Colloidal Domain*, Wiley, New York, p. 282.

Evans, E., and Needham, D. (1988). In: *Molecular Mechanisms of Membrane Fusion*, S. Ohki, D. Doyle, T.D. Flanagan, S.W. Hui, and E. Mayhew, Eds., Plenum Press, New York, p. 83.

Evans, E., Needham, D., and Janzen, J. (1988). In: *Proteins at Interfaces*, J.L. Brash and T.A. Horbett, Eds., ACS Symp. Series *343*, 88.

Everaerts, F.M., Mikkers, F.E.P., and Verheggen, T.P.E.M. (1977). Separ. Purif. Meth. *6*, 287.

Everaerts, F.M., and Verheggen, T.P.E.M. (1975). In: Righetti, 1975, p. 309.

Fägerstam, L.G. and Karlsson, R. (1994). In: *Immunochemistry*, C.J. van Oss, and M.H.V. van Regenmortel, Eds., Marcel Dekker, New York, p. 949.

Farrant, J., and Knight, S.C. (1978). In: *Cell-Separation and Cryobiology*, H. Rainer, H. Borberg, J.M. Mischler and N. Schafer, Eds., F.K. Schattauer Verlag, New York, p. 211.

Farrant, J., and Woolgar, A.E. (1969). In: *The Frozen Cell*, G.E.W. Wolstenholme and M. O'Connor, Eds., Churchill, London, p. 95.

Fawcett (1977). In: Radola and Graeselin, 1977, p. 59.

Fawcett, J.S., and Chrambach, A. (1986). In: *Electrophoresis '86*, M. Dunn, Ed., VCH, Weinheim, p. 569.

Fedyakin, N.N. (1962). Colloid J. USSR *24*, 425.

Fergin, R.I., and Napper, D.H. (1980). J. Colloid Interface Sci. *75*, 525.

Fensom, D.S., Ursino, D.J., and Nelson, C.D. (1967). Can. J. Bot. *45*, 1267.

Fike, R.M., and van Oss, C.J. (1973). Prep. Biochem. *3*, 183.

Fike, R.M., and van Oss, C.J. (1976). In Vitro *12*, 428.

Fischer, E. (1894). Ber. Dtsch. Chem. Ges. *27*, 2992.

Fitzpatrick, H., Luckham, P.F., Eriksen, S., and Hammond, K. (1992). Colloids Surfaces *65*, 43.

Fleer, G.J., and Scheutjens, J.M.H.M. (1982). Advan. Colloid Interface Sci. *16*, 341.

Flory, P.J. (1953). *Principles of Polymer Chemistry*, Cornell University Press, Ithaca, NY.

Flory, P.J., and Krigbaum, W.R. (1950). J. Chem. Phys. *18*, 1086.

Fokkink, L.G.J., and Ralston, J. (1989). Colloids Surfaces *36*, 69.

Fowkes, F.M. (1962). J. Phys. Chem. *66*, 382.

Fowkes, F.M. (1963). J. Phys. Chem. *67*, 2538.

Fowkes, F.M. (1964). Ind. Eng. Chem. (Dec.), 40; see also Fowkes, 1965.

Fowkes, F.M. (1965). In: *Chemistry and Physics at Interfaces*, S. Ross, Ed., Am. Chem. Soc. Publ., Washington, DC, p. 1.

Fowkes, F.M. (1967). In: *Surfaces and Interfaces*, J.J. Burke, Ed., Syracuse University Press, New York, p. 199.

Fowkes, F.M. (1972). J. Adhesion *4*, 155.

Fowkes, F.M. (1983). In: *Physico-Chemical Aspects of Polymer Surfaces*, Vol. 2, K.L. Mittal, Ed., Plenum, New York, p. 583.

Fowkes, F.M. (1987). J. Adhesion Sci. Tech. *1*, 7.

Fowkes, F.M. (1988). Abstr. 3rd Chem. Congr. N. America, Toronto, Canada; Colloid Surface Chem. #114.

Fowkes, F.M., McCarth, D.C., and Mostafa, M.A. (1980). J. Colloid Interface Sci. *78*, 200.

Fowkes, F.M., Jinnai, H., Mostafa, M.A., Anderson, F.W., and Moore, R.J. (1982). In: *Colloids and Surfaces in Reprographic Technology*, M. Hair and M.D. Croucher, Eds., Am. Chem. Soc. Symposium Series *200*, 307.

Fowkes, F.M., Chen, W.J., Fluck, D.J., and Johnson, R.E. (1988). Abstr. Colloid and Surface Chem. Div., Third Chem. Congr. N. Amer., Toronto, No. 8.

Franglen, G., and Gosselin, C. (1958). Nature *181*, 1152.

Frank, H.S., and Evans, M.W. (1945). J. Chem. Phys. *13*, 507.

Franks, F. (1975). In: *Water*, Vol. 4, F. Franks, Ed., Plenum Press, New York, pp. 1–94.

Franks, F. (1981). *Polywater*, MIT Press, Cambridge, MA.

Franks, F. (1984). *Water*, Royal Soc. Chemistry, London.

Fridborg, K., Hjertén, S. Höglund, S., Liljas, A., Lundberg, B.K.S., Oxelfelt, E., Philipson, L., and Strandberg, A. (1965). Proc. Natl. Acad. Sci. USA *54*, 513.

Fuerstenau, D.W. (1956). Trans. Am. Inst. Min. Engrs. *205*, 8345; Min. Engrs. *8*, 834.

Fuerstenau, D.W., Diao, J., and Williams, M.C. (1991). Colloid Interface Sci. *60*, 127.

Gallez, D., Prévost, M., and Sanfeld, A. (1984). Colloids Surfaces *10*, 123. Gauthier, V.J., Mannik, M., and Striker, G.E. (1982). J. Exp. Med. *156*, 766.

Gawrisch, K. (1986). *Molekulare Mechanismen und Membranveränderungen bei der durch Polyethylenglykol induzierten Zellfusion*, thesis (Dr. Sc. Nat.) Leipzig University.

Gawrisch, K., Arnold, K., Dietze, K., and Schulze, U. (1988). In: *Electromagnetic Fields and Biomembranes*, M. Markov and M. Blank, Eds., Plenum Press, New York, p. 9.

Gerson, D.F. (1982). Colloid Polymer Sci. *260*, 539.

Gianazza, E., Astrua-Testori, S., Caccia, P., Quaglia, L., and Righetti, P.G. (1986). In: *Electrophoresis '86*, M.J. Dunn, Ed., VCH, Weinheim, p. 563.

Gianazza, E., and Righetti, P.G. (1979). In: Righetti *et al.*, 1979, p. 293.

Giddings, J.C. (1966). Separ. Sci. *1*, 123.

Giddings, J.C. (1968). J. Chem. Phys. *49*, 81.

Giese, R.F., Costanzo, P.M., and van Oss, C.J. (1991). Phys. Chem. Miner. *17*, 64.

Giese, R.F. and van Oss, C.J. (2002). *Colloid and Surface Properties of Clays and Related Minerals*. Marcel Dekker, New York.

Giese, R.F., van Oss, C.J., Norris, J., and Costanzo, P.M. (1990). *Proc. 9th Intl. Clay Conf.*, Strasbourg, 1989, V.C. Farmer and Y. Tardy, Eds., Sci. Géol. Mém., Strasbourg *86*, p. 33.

Giese, R.F., Wu, W. and van Oss, C.J. (1996). J. Dispersion Sci. Technol. *17*, 527.

Gill, S.J., Nichols, N.F. and Wadsö, I. (1976). J. Chem. Thermodynamics *8*, 445.

Gillman, C.F., Bigazzi, P.E., Bronson, P.M., and van Oss, C.J. (1974). Prep. Biochem. *4*, 457.

Gillman, C.F., and van Oss, C.J. (1971). Abstr. 15th Natl. Biophys. Mg., New Orleans; Biophys. J. *11*, 962.

Girifalco, L.A., and Good, R.J. (1957). J. Phys. Chem. *61*, 904.

Glaser, R., and Gingell, D., Eds. (1990). *Biophysics of the Cell Surface*, Springer-Verlag, New York.

Glynn, L.E., and Steward, M.W., Eds. (1977). *Immunochemistry*, Wiley, New York.

Goddard, E.D., and Vincent, B., Eds. (1984). *Polymer Adsorption and Dispersion Stability*, ACS Symp-Series *240*.

Good, R.J. (1967a). SCI Monograph *25*, 328.

Good, R.J. (1967b). In: *Treatise on Adhesion and Adhesives*, R.L. Patrick, Ed., Marcel Dekker, New York, p. 46.

Good, R.J. (1977). J. Colloid Interface Sci. *59*, 398.

Good, R.J. (1979). Surface Colloid Sci. *11*, 1.

Good, R.J., Chaudhury, M.K., and van Oss, C.J. (1991). In: *Fundamentals of Adhesion*, L.H. Lee, Ed., Plenum, New York, p. 153.

Good, R.J., and Girifalco, L.A. (1960). J. Phys. Chem. *64*, 561.

Good, R.J., and Elbing, E. (1971). In: *Physics of Interfaces*, Vol. 2, S. Ross, Ed., Amer. Chem. Soc. Pub., Washington, DC, p. 72; (March 1970). Ind. Eng. Chem. *62*, 54.

Good, R.J., and Kotsidas, E.D. (1979). J. Adhesion *10,* 17.

Good, R.J., and van Oss, C.J. (1983). Abstr. 57th Colloid Surface Sci. Symp., Toronto.

Good, R.J., van Oss, .C.J., Ho., J.T., Yang, X., Broers, G., and Cheng, M. (1986). Colloids Surfaces *20*, 187.

Gouy, G. (1910). J. Phys. Radium *9*, 457.

Grabar, P., and Burtin, P., Eds. (1964). *Immuno-Eectrophoretic Analysis*, Elsevier, Amsterdam.

Grabar, P., and Williams, C.A. (1953). Biochim. Biophys. Acta *10*, 193.

Graciaa, A., Morel, G., Saulner, P., Lachaise, J. and Schlechter, R.S. (1995). J. Colloid Interface Sci. *172*, 131.

Grassman, W., and Hannig, K. (1949). W. German Patent No. 805399.

Grassman, W., and Hannig, K. (1950). Naturwissenschaften *37*, 397.

Greenwalt, T.J. and Steane, E.A. (1973). Br. J. Haematol. *25*, 227.

Gref, R., Minamitake, M.T., Peracchia, V., Trubetskoy, V., Torchilin, I. and Lauger, R. (1994) Science *263*, 1600.

Gregg, S.J., and Sing, K.S.W. (1982). *Adsorption, Surface Area and Porosity*, Academic Press, New York, pp. 79–83.

Griffith, A.L., Catsimpoolas, N., and Wortis, H.H. (1975). Life Sci. *16*, 1693.

Gross, D. (1961). J. Chromatog. *5*, 194.

Gross, D. (1955). Nature *176*, 72.

Gruen, D.W.R., Marcelja, S. and Parsegian, V.A. (1984). In: *Cell Surface Dynamics*, A.S. Perelson, C. DeLisi and F.W. Wiegel, Eds., Marcel Dekker, New York, p. 59.

Gutmann, V. (1976). Coord. Chem. Rev. *18*, 225.

Gutmann, V. (1978). *The Donor-Acceptor Approach to Molecular Interactions*, Plenum Press, New York.

Gygi, S.P., Corthals, G.U., Zhang, Y., Rochon, Y. and Hebersold, R. (2000). Proc. Natl. Acad. Sci. USA *97*, 9390.

Haglund, H. (1970), Science Tools *17*, 1.

Hamaker, H.C. (1936). Rec. Trav. Chim. Pays-Bas *55*, 1015.

Hamaker, H.C. (1937a). Physica *4*, 1058.

Hamaker, H.C. (1937b). Rec. Trav. Chim. Pays-Bas *56*, 3.

Hamaker, H.C. (1937c). *Ibid. 56*, 727.

Hamilton, W.C. (1974). J. Colloid Interface Sci. *47*, 672.

Hammes, G.G. (1978). *Principles of Chemical Kinetics*, Academic Press, New York, p. 63.

Hannig, K. (1961). Z. Anal. Chem. *181*, 244.

Hannig, K. (1964). Z. Physiol. Chem. *338*, 211.

Hannig, K. (1969). In: *Progr. Separ. Purif.*, Vol. 2, T. Gerritsen, Ed., Wiley-Interscience, New York, p. 45.

Hannig, K., and Heidrich, H.G. (1977). In: *Cell Separation Methods*, H. Bloemendal, Ed., Elsevier/North Holland, Amsterdam, p. 95.

Hannig, K., Wirth, H., Meyer, B.H., and Zeiller, K. (1975). Hoppe-Seyler Z. Physiol.Chem. *356*, 1209.

Hansen, C.M. (1967). J. Paint Technol. *39*, 104.

Hansen, C.M. (1969). Ind. Eng. Chem. Prod. Res. Dev. *8*, 2.

Hansen, C.M. (1970). J. Paint Technol. *42*, 660.

Hansen, C.M., and Beerbower, A. (1971). In: *Kirk-Othmer Encyclopedia of Chemical Technology*, Suppl. Vol., 2nd ed., A. Standen, Ed., Wiley-Interscience, New York, p. 889.

Hansen, E., and Hannig, K. (1983). In: *Electrophoresis '82*, D. Stathakos, Ed., de Gruyter, Berlin, p. 313.

Harkins, W.D. (1952). *The Physical Chemistry of Surface Films*, Reinhold, New York, p. 25.

Harrison, R.G., Todd, P., Rudge, S.R. and Petrides, D. (2003). *Bioseparations Science and Engineering*, Oxford University Press, New York, p. 247.

Hardy, W.B. (1900). Proc. Roy. Soc. London *66*, 110.

Hauxwell, F., and Ottewill, R.H. (1970). J. Colloid Interface Sci., *34*, 473.

Häyry, P., Pentiner, K., and Saxen, E. (1965). Ann. Med. Exp. Fenn. *13*, 91.

He, X.M., and Carter, D.C. (1992). Nature *358*, 209.

Helm, C.A., Israelachvili, J.N., and P.M. McGuiggan (1989). Science *246*, 919.

Henry, D.C. (1931). Proc. Roy. Soc. Lond. A*133*, 106.

Herder, P.C. (1990). J. Colloid Interface Sci. *134*, 336, 346.

Hermans, J.J. (1949). In: *Colloid Science*, Vol. II, H.R. Kruyt, Ed., Elsevier, Amsterdam, p. 49.

Herrmann, A., Arnold, K., and Pratsch, L. (1985). Bioscience Reports *5*, 689.

Hiemenz, P.C. (1984). *Polymer Chemistry*, Marcel Dekker, New York.

Hiemenz, P.C. (1986). *Principles of Colloid and Surface Chemistry*, Marcel Dekker, New York.

Hiemenz, P.C. and Rajagopalam, R. (1997). *Principles of Colloid and Surface Chemistry*, Marcel Dekker, New York, p. 35.

Hildebrand, J.H. (1979). Proc. Natl. Acad. Sci. USA *76*, 194.

Hildebrand, J.H., and Scott, R.L. (1950). *Solubility of Nonelectrolytes*, 3rd ed., Reinhold, New York (1964). Dover Publications, New York.

Hirtzel, C.S., and Rajagopalan, R. (1985). *Advanced Topics in Colloidal Phenomena*, Noyer Publ., Park Ridge, NJ.

Hjertén, S. (1967). Chromatog. Rev. *9*, 122.

Hjertén, S. (1967). *Free Zone Electrophoresis*, Ph.D. thesis, Univ. Uppsala, Almqvist and Wiksell, Uppsala.

Hjertén, S. (1985a). J. Chromatog. *347*, 157.

Hjertén, S. (1985b). J. Chromatog. *347*. 191.

Hjertén, S., Höglund, S. and Hellman, B. (1964) In: *The Structure and Metabolism of Pancreatic Islets*, S.E. Brolin, B. Hellman and K. Knutson, Eds., Pergamon Press, London, p. 223.

Ho, T.L. (1975). Chem. Revs. *75*, 1.

Hofstee, B.H.J. (1973). Anal. Biochem. *52*, 430.

Hofstee, B.H.J. (1976). J. Macromolec. Sci.-Chem. A*10*, 111.

Holmes-Farley, S.R., Reamey, R.H., McCarthy, T.J., Dent, J., and Whitesides, G.M. (1985). Langmuir *1*, 725.

Homola, A.M., and Robertsen, A.A. (1976). J. Colloid Interface Sci. *54*, 286.

Horbett, T.A. (1987). ACS Symp. Series *343*, 239.

Horn, R.G., and Israelachvili, J.N. (1981). J. Chem. Phys. *75*, 1400.

Hough, D.B., and White, L.R. (1980). Advan. Colloid Interface Sci. *14*, 3.

Howard, G.A., and Martin, A.J.P. (1950). Biochem. J. *46*, 532.

Hückel, E. (1924). Physik. Z. *25*, 204.

Huggins, M.L. (1941). J. Chem. Phys. *9*, 440.

Huggins, M.L. (1942). Ann. N.Y. Acad. Sci. *43*, 1.

Hui, S.W., Isac, T., Boni, L.T., and Sen, A. (1985). J. Membr. Biol. *84*, 137.

Hunter, R.J. (1981). *Zeta Potential in Colloid Science*, Academic Press, London.

Hunter, R.J. (1987). *Foundation of Colloid Science*, Vol. 1, Oxford University Press, Oxford.

Hunter, R.J. (1989). *Foundation of Colloid Science*, Vol. 2, Oxford University Press, Oxford.

Hvidt, A. (1983). Ann. Rev. Biophys. Bioeng. *12*, 1.

Interfacial Dynamics Co. (1991). *IDC Ultraclean Uniform Latex Microspheres*, Portland, OR, Catalogue, p. 3.

Isberg, R. (1991). Science 252, 934.

Israelachvili, J.N. (1974). Quart. Rev. Biophys. *6*, 341.

Israelachvili, J.N. (1982). In: *Principles of Colloid and Surface Chemistry*, R.J. Hunter, Ed., Roy. Austr. Chem. Soc., Sydney.

Israelachvili, J.N. (1985). *Intermolecular and Surface Forces*, Academic Press, London.

Israelachvili, J.N. (1987). Proc. Natl. Acad. Sci. USA *84*, 4722.

Israelachvili, J.N. (1991). *Intermolecular and Surface Forces*, Academic Press, London, 2nd Edition.

Israelachvili, J.N., and Adams, G.E. (1978). J. Chem. Soc. Faraday Trans. I *74*, 975.

Israelachvili, J.N., Kott, S.J., McGee, M.L., and Witten, T.A. (1989). Langmuir *5*, 1111.

Israelachvili, J.N., and McGee, M.L. (1989). Langmuir *5*, 288.

Israelachvili, J.N., and McGuiggan, P.M. (1988). Science *241*, 795.

Israelachvili, J.N., and Pashley, R. (1982). Nature *300*, 342.

Israelachvili, J.N., and Pashley, R.M. (1984). J. Colloid Interface Sci. *98*, 500.

Israelachvili, J.N., and Tabor, D. (1972). Proc. Roy. Soc. Lond. *A331*, 19.

Janczuk, B., Chibowski, E., Choma, I., Davidowicz, A.L., and Bialopiotrowicz, T. (1990). Materials Chem. Phys. *25*, 185.

Janczuk, B., and Bialopiotrowicz, T. (1989). J. Colloid Interface Sci. *127*, 189.

Janzen, J., and Brooks, D.E. (1988). In: *Interfacial Phenomena in Biological Systems*, M. Bender, Ed., Marcel Dekker, New York.

Jasper, J.J. (1972). J. Phys. Chem. Ref. Data *1*, 841.

Jensen, W.B. (1980). *The Lewis Acid-Base Concepts*, Wiley-Interscience, New York.

Johansson, G., Abusugra, I., Lövgren, K., and Morein, B. (1988). Prep. Biochem. *18*, #4.

Johnson, R.E., and Dettre, R.H. (1989). Langmuir *5*, 293.

Jun, S.H., and Ruckenstein, E. (1984). Separ. Sci. Technol. *19*, 531.

Just, W.W., Leon, J.O., and Werner, G. (1975). Analyt. Biochem. *67*, 590.

Just, W.W., and Werner, G. (1979). In: Righetti *et al.*, 1979, p. 143.

Kabat, E.A. (1968). *Structural Concepts in Immunology and Immunochemistry*, Holt, Rinehart and Winston, New York.

Kabat, E.A. (1976). *Structural Concepts in Immunology and Immunochemistry*, 2nd ed., Holt, Rinehart and Winston, New York.

Kabat, E.A. and Mayer, M.M. (1961). *Experimental Immunochemistry*. C.C. Thomas, Springfield, IL, p. 159.

Kaelble, D.H. (1970). J. Adhesion 2, 66.

Katz, A.M., Dreyer, W.J., and Anfinsen, C.B. (1959). J. Biol. Chem. 234, 2897.

Kavanau, J.L. (1964). *Water and Solute-Water Interactions*, Holden-Day, New York, pp. 22–28.

Kawaguchi, M., Suzuki, C., and Takahashi, A. (1988). J. Colloid Interface Sci. *121*, 585.

Keesom, W.H. (1915). Proc. Roy. Neth. Acad. Sci. *18*, 636.

Keesom, W.H. (1920). Proc. Roy. Neth. Acad. Sci. *23*, 939.

Keesom, W.H. (1921). Physik. Z. *22*, 129, 643.

Ketcham, W.M. and Hobbs, P.V. (1969). Phil. Mag. *19*, 1161.

Kim, S.W., and Lee, R.G. (1975). Adv. Chem. Series *145*, 218.

Kitahara, A., and Watanabe, A. (1984). *Electrical Phenomena at Interfaces*, Marcel Dekker, New York.

Klein, J. (1980). Nature *288*, 248.

Klein, J. (1990). Science *250*, 640.

Klein, J., and Luckham, P.F. (1984). Macromolecules *17*, 1041.

Kochwa, S., Brownell, M., Rosenfield, R.E., and Wasserman, L.R. (1967). J. Immunol. *99*, 981.

Kolin, A. (1966). Proc. Natl. Acad. Sci. USA *56*, 1051.

Kolin, A. (1967). J. Chromatog. *26*, 164; *26*, 180.

Kolin, A. (1976). In: Catsimpoolas, 1976, p. 1.

Kolin, A. (1977). In: Radola and Graeselin, 1977, p. 3.

Kolin, A., and Luner, S.J. (1971). In: *Progress in Separ. and Purif.*, Vol. 4, E. Perry and C.J. van Oss, Eds., Wiley/Interscience, New York, p. 93.

Kollman, P. (1977). J. Am. Chem. Soc. *99*, 4875.

Korpi, G.K., and de Bruyn, P.L. (1972). J. Colloid Interface Sci. *40*, 263.

Krstulović, A.M., and Brown, P.R. (1982). *Reversed-Phase Liquid Chromatography*, Wiley, New York.

Krupp, H. (1967). Advan. Colloid Interface Sci. *1*, 1.

Kruyt, H.R., Ed. (1949). *Colloid Science*, Vol. 2, Elsevier, Amsterdam.

Kruyt, H.R., Ed. (1952). *Colloid Science*, Vol. 1, Elsevier, Amsterdam.

Ku, C.A., Henry, J.D., Siriwardane, R., and Roberts, L. (1985). J. Colloid Interface Sci. *106*, 377.

Kuivenhoven, A.C.J. (1966). Transfusion (Paris) *9*, 118.

Kurihara, K., and Kunitake, T. (1992). J. Am. Chem. Soc. *114*, 10927.

Kuznetsov, Y. G., Malkin, A.J., Lucas, R.W. and McPherson, A. (2000). Colloids Surfaces-B: Biointerfaces *19*, 333.

Labib, M.E., and Williams, R. (1984). J. Colloid Interface Sci. *97*, 356.

Labib, M.E., and Williams, R. (1986). Colloid Polymer Sci. *264*, 533.

Labib, M.E., and Williams, R. (1987). J. Colloid Interface Sci. *115*, 330.

Lalezari, P. (1968). Transfusion *8*, 372.

Lancaster, M.V., and Sprouse, R.F. (1976). Infect. Immun. *13*, 758.

Landsteiner, K. (1936). *The Specificity of Serological Reactions*, C.C. Thomas, Springfield, IL; Repr. Dover, New York (1962), p. 156.

Lasic, D.D., Martin, F.J., Gabizon, A., Huang, S.K., and Papahadjapoulos, D. (1991). Biochim. Biophys. Acta *1070*, 187.

Lee, C.S., and Belfort, G. (1989). Proc. Natl. Acad. Sci. USA, *86*, 8392.

Lee, G.U., Chrisey, L.A. and Colton, R.J. (1994). Science *266*, 771.

Lee, R.G., Adamson, C., Kim, S.W., and Lyman, D.J. (1973). Thrombosis Res. *3*, 87.

Lee, W.Y. and Sehon, A.H. (1977). Nature *267*, 618.

Leise, E.M., and LeSane, F. (1974). Prep. Biochem. *4*, 395.

Leneveu, D.M., Rand, R.P., and Parsegian, V.A. (1977). Biophys. J. *18*, 209.

Lerche, D., Hessel, E., and Donath, E. (1979). Stud. Biophys. *78*, 95.

Lesius, V., and Ruckenstein, E. (1984). Separ. Sci. Technol. *19*, 219.

Lewis, G. N. (1923). *Valence and the Structure of Atoms and Molecules*. American Chemical Society; Chemical Catalog Co., New York, p. 138 ff.

Liberti, P.A. (1975). Immunochemistry *12*, 303.

Li-In-On, F.K.R., Vincent, B., and Waite, F.A. (1975). ACS Symposium Series 9, 165.

Lifshitz, E.M. (1955). Zh. Eksp. Teor. Fiz. 29, 94.

Lindsey, A.S. (1970). In: Reviews in Macromolecular Chemistry, Marcel Dekker, New York, p. 1.

Ling, G.N. (1972). In: *Water and Aqueous Solutions*, R.A. Horne, Ed., Wiley-Interscience, New York, pp. 663–700.

Lis, L.J., McAlister, M., Fuller, N., Rand, R.P., and Parsegian, V.A. (1982). Biophys. J. 37, 657.

Loeb, A.L., Overbeek, J.Th.G., and Wiersema, P.H. (1961). The Electrical Double Layer around a Spherical Colloid Particle, MIT Press, Cambridge, MA.

Loening, U.E. (1967). Biochem. J. 102, 251.

London, F. (1930). Z. Physik. 63, 245.

Lucassen, J. (1979). J. Colloid Interface Sci. 70, 355.

Luckham, P.F., Bailey, A.I., Afshar-Rad, T., Chapman, D., and MacNaughton, V. (1986). Abstr. Conf. Colloid Surface Chem. Group, Soc. Chem. Ind., Imp. Coll., Lon.

Luckham, P.F., and Klein, J. (1985). Macromolecules 18, 721.

Lutanie, E., Schaaf, .P., Schmitt, A., Voegel, J.C., Freund, M., and Cazenave, J.P. (1992). Biocolloids Biosurfaces 2; in J. Dispersion Sci. Tech. 13, 379.

Lutanie, E., Schaaf, P., Schmitt, A., Voegel, J.C., Freund, M., and Cazenave, J.P. (1992). Biocolloids Biosurfaces: J. Dispersion Sci. Tech. 13, 379.

Luzar, A. and Chandler, D. (1996). Nature 379, 55.

Lyklema, J. (1968). Advan. Colloid Interface Sci. 2, 65.

Mackor, E.L. (1951). J. Colloid Sci. 6, 492.

Mackor, E.L., and van der Waals, J.H. (1952). J. Colloid Sci. 7, 535.

MacRitchie, F. (1972). J. Colloid Interface Sci. 38, 484.

Mahanty, J., and Ninham, B.W. (1976). Dispersion Forces, Academic Press, New York.

Malcolm, G.N., and Rowlinson, J.S. (1957). Trans. Faraday Soc. 53, 921.

Malmquist, M. (1993). Nature 361, 186.

Maniatis, T., Fritsch, E.F., and Sambrook, J. (1982). *Molecular Cloning*, Cold Spring Harbor Laboratory, Cold Spring Harbor, NY, pp. 150–185.

Manne, S., Cleveland, J.P., Gaub, H.E., Stucky, G.D. and Hansma, P.K. (1994). Langmuir 10, 4409.

Mazur, P., and Overbeek, J.Th.G. (1951). Rec. Trav. Chim. Pays-Bas 70, 83.

Margolis, J., and Kendrick, K.G. (1968). Analyt. Biochem. 25, 347.

Marra, J. (1985). J. Colloid Interface Sci. 107, 446.

Marra, J. (1986). J. Colloid Interface Sci. 109, 11; Biophys. J. 50, 815.

Marra, J., and Israelachvili, J.N. (1985). Biochem. 24, 4608.

Marshall, T., and Williams, K.M. (1986). In: *Electrophoresis '86*, M.J. Dunn, Ed., VCH, Weinheim, p. 523.

Martin, A., Puig, J., Galisteo, F., Serra, J., and Hidalgo-Alvarez, R. (1992). Biocolloids Biosurfaces 2; in J. Dispersion Sci. Tech. 13, 399.

Mashburn, T.A., and Hoffman, P. (1966). Analyt. Biochem. 16, 267.

Mattock, P., Aitchison, G.F., and Thomson, A.R. (1980). Separ. Purif. Meth. 9, 1.

Maugh, T.H. (1983). Science 222, 259.

Mayers, G.L., and van Oss, C.J. (1992). In: *Encyclopedia of Immunology*, I.M. Roitt and P.J. Delves, Eds., Saunders, Orlando, FL.

Mazur, P., and Overbeek, J.Th.G. (1951). Rec. Trav. Chim. Pays-Bas 70, 83.

McCann, G.D., Vanderhoff, J.W., Strickler, A., and Sacks, T.I. (1973). Separ. Purif. Meth. 2, 153.

McIver, D., and Schürch, S. (1982). Biochim. Biophys. Acta 691, 52.

McTaggart, H.A. (1914). Phil. Mag. *27*, 297; *28*, 367.

McTaggart, H.A. (1922). Phil. Mag. *44*, 386.

Mel, H.C. (1959) J. Chem. Phys. *31*, 559.

Mel, H.C. (1960). Science *132*, 1255.

Mel, H.C. (1964). J. Theoret. Biol. *6*, 159, 181, 307.

Mel, H.C. (1970). In: *Myeloproliferative Disorders of Animals and Man*, W.J. Clarke, E.B. Howard and P.L. Hackett, Eds., U.S. Atomic Energy Commission, Div. of Technical Information, Washington, DC, p. 665.

Mel, H.C., Mitchell, L.T., and Thorell, B. (1965). Blood *25*, 63.

Micale, F.J., Vanderhoff, J.W., and Snyder, R.S. (1976). Separ. Purif. Meth. *5*, 361.

Meryman, H.T. (1966). In: *Cryobiology*, H.T. Meryman, Ed., Academic Press, New York, p. 63.

Michaeli, I., Absolom, D.R., and van Oss, C.J. (1980). J. Colloid Interface Sci. *77*, 586.

Michaels, A.S., and Dean, S.W. (1962). J. Phys. Chem. *61*, 904.

Michaels, A.S., and Miekka, R.G. (1981). J. Phys. Chem. *65*, 1765.

Michel, B.E., and Kaufman, M.R. (1973). Plant Physiol. *51*, 814.

Mijnlieff, P.F. (1958). *Sedimentation of Colloidal Electrolytes*, Ph.D. thesis, Utrecht.

Mirza, M., Guo, Y., Arnold, K., van Oss, C.J. and Ohki, S. (1998). J. Dispersion Sci. Technol. *19*, 951.

Mollison, P.L. (1972). *Blood Transfusion in Clinical Medicine*, Blackwell, Oxford, pp. 150; 384.

Molyneux, P. (1983). *Water Soluble Synthetic Polymers*, CRC Press, Boca Raton, FL.

Mooney, M. (1931). J. Phys. Chem. *35*, 331.

Morgan, B.P. (1990). *Complement*, Academic Press, London.

Morrison, I.D. (1989). Langmuir *5*, 543.

Morrissey, B.W., and Stromberg, R.R. (1974). J. Colloid Interface Sci. *46*, 152.

Mould, D.L., and Synge, R.L.M. (1954). Biochem. J. *58*, 571.

Mukkur, T.K.S. (1980). Trends Biochem. Sci. *5*(3), 72.

Mukkur, T.K.S. (1984). Crit. Rev. Biochem. *16*, 133.

Muller, N. (1990). Acc. Chem. Res. *23*, 23.

Napper, D.H. (1983). *Polymeric Stabilization of Colloidal Dispersions*. Academic Press, New York.

Napper, D.H. (1987). In: *Foundations of Colloid Science*, R.J. Hunter, Ed., Vol. 1, Clarendon Press, Oxford, p. 473.

Neumann, A.W., Absolom, D.R., Francis, D.W., Zingg, W., and van Oss, C.J. (1982). Cell Biophys. *4*, 285.

Neumann, A.W., Absolom, D.R., van Oss, C.J., and Zingg, W. (1979). Cell Biophys. *1*, 79.

Neumann, A.W., and Francis, D.W. (1983). Private communication.

Neumann, A.W., and Good, R.J. (1979). Surface Colloid Sci. *11*, 31.

Neumann, A.W., Good, R.J., Hope, C.J., and Sejpal, M. (1974). J. Colloid Interface Sci. *49*: 291.

Neumann, A.W., Omenyi, S.N., and van Oss, C.J. (1979). Colloid Polymer Sci. *257*, 413.

Neumann, A.W., Omenyi, S.N., and van Oss, C.J. (1982). J. Phys. Chem. *86*, 1267.

Neumann, A.W., Visser, J., Smith, R.P., Omenyi, S.N., Francis, D.W., Spelt, J.K., Vargha-Butler, E.B., Zingg, W., van Oss, C.J., and Absolom, D.R. (1984). Powder Technol. *37*, 229.

Ninham, B.W., and Parsegian, V.A. (1970). Biophys. J. *10*, 646.

Nir, S. (1976). Progr. Surface Sci. *8*, 1.

Nisonoff, A., Hopper, J.E., and Spring, S.B. (1975). *The Antibody Molecule*, Wiley, New York.

Nobel, P.S., and Mel, H.C. (1966). Arch. Biochem. Biophys. *113*, 695.

Norde, W. (1986). Advan. Colloid Interface Sci. *25*, 267.

Norde, W. (1992). Biocolloids Biosurfaces 2; in J. Dispersion Sci. Tech. *13*, 363.

Norde, W., and Lyklema, J. (1978). J. Colloid Interface Sci. *66*, 257; 266; 277; 285; 295.

Norrish, R.S. (1966). J. Food Technol. *1*, 25.

Nyilas, E., Morton, W.A., Cumming, R.D., Leberman, D.M., Chiu, T.H., and Baier, R.E. (1977). J. Biomed. Mater. Res. *8*, 51.

Ockham, W. (ca. 1320). *Philosophical Writings*, P. Boehner, Ed., Nelson, London, 1957, p. xxi. ("Plurality is not to be assumed without necessity").

O'Farrell, P.H. (1975). J. Biol. Chem. *250*, 4007.

Ohki, S. (1982). Biochim. Biophys. Acta *689*, 1.

Ohki, S. (1988). In: *Molecular Mechanisms of Membrane Fusion*, S. Ohki, D. Doyle, T.D. Flanagan, S.W. Hui and E. Mayhew, Eds., Plenum Press, New York, p. 123.

Ohki, S., Doyle, D., Flanagan, T.D., Hui, S.W. and Mayhew, E., Eds. (1988). *Molecular Mechanisms of Membrane Fusion*, Plenum, New York.

Ohki, S., Düzgüne, N., and Leonards, K. (1982). Biochem. *21*, 2127.

Omenyi, S.N. (1978). *Attraction and Repulsion of Particles by Solidifying Melts*, Ph.D. thesis, University of Toronto.

Omenyi, S.N., and Neumann, A.W. (1976). J. Appl. Phys. *47*, 3956.

Omenyi, S.N., Neumann, A.W., Martin, W.W., Lespinard, G.M., and Smith, R.P. (1981). J. Appl. Phys. *52*, 796.

Omenyi, S.N., Snyder, R.S., van Oss, C.J., Absolom, D.R., and Neumann, A.W. (1981). J. Colloid Interface Sci. *81*, 402.

Omenyi, S.N., Snyder, R.S., Absolom, D.R., van Oss, C.J., and Neumann, A.W. (1982). J. Dispersion Sci. Technol. *3*, 307.

Omenyi, S.N., Snyder, R.S., and van Oss, C.J. (1985). J. Dispersion Sci. Technol. *6*, 391.

Omenyi, S.N., Snyder, R.S., Tipps, R., Absolom, D.R., and van Oss, C.J. (1986). J. Dispersion Sci. Technol. *7*, 1.

O'Mullane, J.E., Davison, C.J., Petrak, K. and Tomlinson, E. (1988). Biomaterials *9*, 203.

Onda, T., Shibuichi, S., Satoh, N. and Tsujii, K. (1996). Langmuir *12*, 2126.

Ornstein, L. (1964). Ann. N.Y. Acad. Sci. *121*, 321.

Oshima, M. and Atassi, M.Z. (2000). Autoimmunity *32*, 45.

Ottewill, R.H. (1976). Progr. Colloid Polymer Sci. *59*, 14.

Ottewill, R.H. (1967). In: *Nonionic Surfactants*, M.J. Schick, Ed., Marcel Dekker, New York, p. 627.

Ottewill, R.H., and Shaw, J. N. (1967). Kolloid Z. *218*, 34.

Ottewill, R.H., and Walker, T. (1968). Koll. Z., Z. Polymere *227*, 108.

Overbeek, J.Th.G. (1943). Kolloid Beihefte *54*, 287.

Overbeek, J.Th.G. (1952). In: *Colloid Science*, Vol. 1, H.R. Kruyt, Ed., Elsevier, Amsterdam.

Overbeek, J.Th.G. (1988). J. Chem. Soc.-Faraday Trans.-I *84*, 3079.

Overbeek J.Th.G., and Sparnaay, M.J. (1952). J. Colloid Sci. *7*, 343.

Overbeek, J.Th.G., and Wiersema, P.H. (1967). In: *Electrophoresis*, Vol. II, M. Bier, Ed., Academic Press, New York, p. 1.

Owens, D.K., and Wendt, R.C. (1969). J. Appl. Polymer Sci. *13*, 1741.

Parker, J.L., Christenson, H.K., and Ninham, B.W. (1989). Rev. Sci. Instrum. *60*, 3135.

Parsegian, V.A. (1973). Ann. Rev. Biophys. Bioeng. 2, 221.

Parsegian, V.A. (1975). In: *Physical Chemistry: Enriching Topics from Colloid and Surface Science*, H. van Olphen and K.J. Mysels, Eds., Theorex, La Jolla, CA, p. 27.

Parsegian, V.A., Fuller, N., and Rand, R.P. (1979). Proc. Natl. Acad. Sci. USA *76*, 2750.

Parsegian, V.A., Rand, R.P., Fuller, N.L., and Rau, D.C. (1986). Meth. Enzymology *127*, 400.

Parsegian, V.A., Rand, R.P., and Rau, D.C. (1985). Chem. Scripta 25, 28.

Parsegian, V.A., Rand, R.P., and Rau, D.C. (1987). In: *Proc. Symp. Complex and Supermolecular Fluids*, S. Satran and N. Clark, Eds., Wiley, New York.

Pashley, R.M. (1981). J. Colloid Interface Sci. 83, 531.

Pashley, R.M. (1982). Advan. Colloid Interface Sci. 16, 57.

Pashley, R.M., and Israelachvili, J.N. (1984a). J. Colloid Interface Sci. 97, 446.

Pashley, R.M., and Israelachvili, J.N. (1984a). J. Colloid Interface Sci. 101, 511.

Pashley, R.M., McGuiggan, P.M., Ninham, B.W., and Evans, D.F. (1985). Science 229, 1088.

Pauling, L., and Hayward, R. (1964). *The Architecture of Molecules*, W.H. Freeman, San Francisco, p. 7.

Patterson, D. (1969). Macromolecules 2, 672.

Perez, E., and Proust, J.E. (1987). J. Colloid Interface Sci. 118, 182.

Pethica, B.A. (1961). Exptl. Cell Res., Suppl. 8, 123.

Philpot, J.St.L. (1938). Nature 141, 283.

Philpot, J.St.L. (1980). *Methodological Developments in Biochemistry*, Vol. 2, *Preparative Techniques*, E. Reid, Ed., Longmans, London, p. 81.

Pidot, A.L., and Diamond, J.M. (1964). Nature 201, 701.

Pistenma, D.A., Mel, H.C., and Snapir, N. (1971). Reprod. Fert. 24, 161.

Place, H., Sébille, B., and Vidal-Madjar, C. (1992). Biocolloids Biosurfaces 2; in J. Dispersion Sci. Tech. 13, 417.

Podesta, R.B., Boyce, J.F., Schürch, S., and McDiarmid, S.S. (1987). Cell Biophysics 10, 23.

Polson, A. (1977). Prep. Biochem. 7, 129.

Pouchert, C.S., Ed. (1975). *The Aldrich Library of Infrared Spectra*, Aldrich Chemical Co., Milwaukee.

Pratt-Terpstra, I.H., and Busscher, H.J. (1991). Biofouling 3, 199.

Pratt-Terpstra, I.H., Weekamp, A.H., and Busscher, H.J. (1987). J. Gen. Microbiol. 133, 3199.

Prausnitz, J.M., Lichtenthaler, R., and Gomez de Azvedo, E. (1986). *Molecular Thermodynamics of Fluid Phase Equilibria*, Prentice Hall, New York.

Pressman, D., and Grossberg, A.L. (1968). *The Structural Basis of Antibody Specificity*, W.A. Benjamin, Reading, MA.

Pretorius, V., Hopkins, B.J., and Schieke, J.D. (1974). J. Chromatog. 99, 23.

Prévost, M., and Gallez, D. (1984). J. Chem. Sco. Faraday Trans. 2 80, 517.

Prince, L.M. (1967). J. Colloid Interface Sci. 23, 165.

Princen, H.M., Zia, I.Y.Z., and Mason, S.G. (1967). J. Colloid Interface Sci. 23, 99.

Privalov, P.L., and Gill, S.J. (1988). Adv. Protein Chem. 40, 191.

Privalov, P.L., and Gill, S.J. (1989). Pure Appl. Chem. 61, 1097.

Quast, R. (1979). In: Righetti *et al.*, (1979), p. 221.

Rabinovich, Ya.I., and Derjaguin, B.V. (1988). Colloids Surfaces 30, 243.

Radola, B.J., and Graeselin, D. (1977). *Electrofocusing and Isotachophoresis*, de Gruyter, New York.

Raghavachari, M., Tsai, H.M., Kottke-Marchant, K. and Marchant, R.E. (2000). Colloids Surfaces-B: Biointerfaces 19, 315.

Rand, R.P., and Parsegian, V.A. (1984). Can. J. Biochem. Cell Biol. 62, 752.

Rao, C.N.R. (1972). In: *Water*, Vol. 1, F. Franks, Ed., Plenum Press, New York, p. 93.

Raymond, S., and Weintraub, L. (1959). Science 130, 711.

Rebuck, J.W. (1953). Anat. Rec. 115, 591.

Rembaum, A., and Tökés, Z.A. (1988). *Microspheres: Medical and Biological Applications*. CRC Press, Boca Raton, FL.

Resnick, M.A., Tippetts, R.D., and Mortimer, R.K. (1967). Science 158, 803.

Rhumbler, L. (1898). Arch. Entwiklungsmech. Organ *7*, 103.

Richards, E.G., Coll, J.A., and Gratzer, W.B. (1965). Analyt. Biochem. *12*, 452.

Richmond, P., Ninham, B.H., and Ottewill, R.H. (1973). J. Colloid Interface Sci. *45*, 69.

Righetti, P.G. (1975). *Progress in Isoelectric Focusing and Isotachophoresis*, Elservier/North Holland Pub. Co., New York.

Righetti, P.G. (1984). J. Chromatog. *300*, 165.

Righetti, P.G., and Drysdale, D., Eds. (1977). *Electrofocusing and Isotachophoresis*, de Gruyter, New York.

Righetti, P.G., Gianazza, E., and Gelfi, C. (1987). Separ. Purif. Meth. *16*, 105.

Righetti, P.G., van Oss, C.J., and Vanderhoff, J.W., Eds. (1979). *Electrokinetic Separation Methods*, Elsevier, Amsterdam.

Rilbe, H. (1976). In: Catsimpoolas, 1976, p. 13.

Rilbe, H. (1977). In: Radola and Graeselin, 1977, p. 34.

Rogers, J.A., and Tam, T. (1977). Can. J. Pharm. Sci. *12*, 65.

Romans, D.G., Tilley, C.H., Crookston, M.C., *et al.* (1977). Proc. Natl. Acad. Sci. USA *74*, 2531.

Rose, N.R. (1979). In: *Principles of Immunology*, N.R. Rose, F. Milgrom and C.J. van Oss, Eds., Macmillan, New York, p. 277.

Rosen, M.J. (1982). (1982). In: *Surfactants and Interfacial Phenomena*, Wiley, New York, pp. 83–122.

Rosengren, J., Påhlman, S., Glad, M., and Hjertén, S. (1975). Biochim, Biophys. Acta *412*, 51.

Rouweler, G.C.J. (1972). *Measurement of van der Waals Forces*, Ph.D. thesis, University of Utrecht.

Ruckenstein, E. (1992). Colloids Surfaces *65*, 95.

Ruckenstein, E., and Chillakuru, R. (1990). Separ. Sci. Tech. *25*, 207.

Ruckenstein, E., and Prieve, D.C. (1976). AIChE J. *22*, 276; 1145.

Ruckenstein, E., and Churaev, N. (1991). J. Colloid Interface Sci. *147*, 535.

Ruckenstein, E., and Srinivasan, R. (1982). Separ. Sci. Tech. *17*, 763.

Ruhenstroth-Bauer, G. (1965). In: *Cell Electrophoresis*, E.J. Ambrose, Ed., Little Brown, Boston, p. 66.

Russell, W.B. (1987). *The Dynamics of Colloidal Systems*, University of Wisconsin Press, Madison, WI.

Rutgers, A.J., and de Smet, M. (1947). Trans. Faraday Soc. *43*, 102.

Rytov, S.M. (1953). *Theory of Electric Fluctuations and Thermal Radiations*, Moscow Acad. Sci. Press, Moscow.

Salsbury, A.J. and Clarke, J.A. (1967). Rev. Franç. Etud. Clin. Biol. *12*, 981.

Sargent, J.R., and George, S.G. (1975). *Methods in Zone Electrophoresis*, BDH Chemicals, Poole, UK.

Sato, T., and Ruch, R. (1980). *Stabilization of Colloidal Dispersions by Polymer Adsorption*, Marcel Dekker, New York.

Schakenraad, J.M., and Busscher, H.J. (1989). Colloids Surfaces *42*, 331.

Schenkel, J.H., and Kitchener, J.A. (1960). Trans. Faraday Soc. *56*, 161.

Scheutjens, J.M.H.M., and Fleer, G.J. (1982). Advan. Colloid Interface Sci. *16*, 360.

Scheutjens, J.M.H.M., and Fleer, G.J. (1979). J. Phys. Chem. *83*, 1619; *ibid.* (1980) *84*, 178.

Schmitz, A. and Gulas, D.J. (1978). Nucleic Acid Res. *5*, 3157.

Schulze, H. (1882). Z. f. praktische Chemie *25*, 431.

Schulze, H. (1883). Z. f. praktische Chemie *27*, 320.

Schultze, H.E. and Heremans, J.F. (1966). *Molecular Biology of Human Proteins*, Elsevier, New York.

Schürch, S., Gerson, D.F., and McIver, J.L. (1981). Biochim. Biophys. Acta *640*, 577.

Schwartz, D.C., and Cantor, C.R. (1984). Cell *37*, 67.

Schwartz, D.C., Saffran, W., Welsh, J., Haas, R., Goldenberg, M., and Cantor, C.R. (1983). Cold Spring Harbor Symp. Quant. Biol. *47*, 189.

Seaman, C.V.F., and Brooks, D.E. (1979). In: Righetti *et al.*, 1979, p. 95.

Sehon, A.H. (1989). In: *Immunobiology of Proteins and Peptides,* Vol. V – *Vaccines*, M.Z. Atassi, Ed., Plenum Press, New York, p. 341.

Sehon, A.H., Jackson, C.J.C., Holford-Stevens, V., Wilkinson, I., Maiti, P.K. and Lang, G. (1987). *The Pharmacology and Toxicology of Proteins*, J.S. Holdenberg and J.L. Winkelhake, Eds., Alan Liss, p. 206.

Seidl, S. (1978). In: *Cell Separation and Cryobiology*, H. Rainer, H. Borger, J.M. Mishler and U. Schafter, Eds., F.K. Schattauer Verlag, New York, p. 211.

Seymour, R.B., and Carraker, C.E. (1988). *Polymer Chemistry*, Marcel Dekker, New York.

Sewchand, L., and Canham, P.B. (1976). Can. J. Physiol. Pharmacol. *54*, 437.

Shapiro, A.L., Viñuela, E., and Maizel, J.V. (1967). Biochem. Biophys. Res. Comm. *28*, 815.

Shaw, D.J. (1969). *Electrophoresis*, Academic Press, New York.

Shaw, J.N., and Ottewill, R.H. (1965). Nature *208*, 681.

Sheehan, D.C., and Hrapchak, B.B. (1973). *Theory and Practice of Histotechnology*, C.V. Mosby Co., St. Louis, MO, pp. 27–40.

Shimizu, Y., and Shaw, S. (1991). FASEB J. *5*, 2292.

Shu, L.K. (1991). *Contact Angles and Determination of the Surface Energy of Polymer Surfaces*, Ph.D. Dissertation, SUNY, Buffalo.

Siebenlist, U. and Gilbert, W. (1980). Proc. Natl Acad. Sci. USA *77*, 122.

Silberberg, A. (1984). ACS Symp. Series *240*, 161.

Silverstein, A.M. (1989). *A History of Immunology*, Academic Press, New York.

Singer, J.M. (1961). Am. J. Med. *31*, 766.

Singer, J.M., van Oss, C.J., and Vanderhoff, J.W. (1969). J. Reticuloendothelial Soc. *6*, 281.

Small, P.A. (1953). J. Appl. Chem. *3*, 71.

Smeenk, R.J.T, Aarden, L.A., and van Oss, C.J. (1983). Immunol. Commun. *12*, 177.

Smith, I. (1979). In: Righetti *et al.*, 1979, p. 33.

Smith, K.L., Ninslow, A.E., and Petersen, D.E. (1959). Ind. Eng. Chem. *51*, 13161.

Smithies, O. (1955). Biochem. J. *61*, 629.

Smithies, O. (1959). Advan. Protein Chem. *14*, 65.

Smithies, O. (1962). Arch. Biochem. Suppl. *1*, 125.

Snyder, R.S., Bier, M., Griffin, R.N., Johnson, A.J., Leidheiser, H., Micale, F.J., Ross, S., and van Oss, C.J. (1973). Separ. Purif. Meth. *2*, 259.

Sober, H., Ed. (1968). *Handbook of Biochemistry*, CRC Co., Cleveland, OH, pp. C10, C36.

Somasundaran, P. (1968). J. Colloid Interface Sci. *27*, 659.

Somasundaran, P. (1972). Separ. Purif. Meth. *1*, 117.

Somasundaran, P., and Kulkarni, R.D. (1973). J. Colloid Interface Sci. *45*, 591.

Southern, E.M. (1975). J. Mol. Biol. *98*, 503.

Sparnaay, M.J. (1952). *Direct Measurements of van der Waals Forces* (in Dutch), Ph.D. thesis, Utrecht.

Sparnaay, M.J. (1972). *The Electrical Double Layer*, Pergamon Press, Oxford.

Spelt, J.K. (1985). *Solid Surface Tension: The Equation of State Approach and the Theory of Surface Tension Components*, dissertation, University of Toronto.

Spelt, J.K., Absolom, D.R., and Neumann, A.W. (1986). Langmuir *2*, 620.

Spelt, J.K., Smith, R.P., and Neumann, A.W. (1987). Colloids Surfaces *28*, 85.

Spencer, E.W., Ingram, V.M., and Levinthal, C. (1966). Science *152*, 1722.

Steane, E.A., and Greenwalt, T.J. (1974). Transfusion *14*, 501.

Stephen, H., and Stephen, T. (1963). *Solubilities of Inorganic and Organic Compounds*, Vol. I, Part 1, Pergamon Press, Oxford.

Stinson, M.W., and van Oss, C.J. (1971). J. Reticuloendothelial Soc. *9*, 503.

Strickler, A., and Sacks, T.I. (1973). Prep. Biochem. *3*, 269; Ann. N.Y. Acad. Sci. *209*, 497.

Stromberg, R.L. (1967). In: *Treatise on Adhesion and Adhesives*, Vol. 1, R.L. Patrick, Ed., Marcel Dekker, New York, p. 69.

Stegemann, H. (1979). In: *Electrokinetic Separation Methods*, P.G. Righetti, C.J. van Oss and J.W. Vanderhoff, Eds., Elsevier, Amsterdam, p. 313.

Ström, G., Frederiksson, M., and Klason, T. (1988). J. Colloid Interface Sci. *123*, 324.

Stryer, L. (1981). *Biochemistry*, W.H. Freeman, San Francisco, pp. 175, 371, 566.

Svedberg, T., and Pedersen, K.O. (1940). *The Ultracentrifuge*, Clarendon Press, Oxford; Johnson, Reprint Corp., New York, 1959.

Svensson, H. (1939). Kolloid Z. *87*, 181.

Svensson, H., and Brattsten, L. (1949). Arkiv. Kemi *1*, 401.

Svenesson, H., and Thompson, T.E. (1961). In: *A Laboratory Manual of Analytical Methods of Protein Chemistry*, Vol. 3, P. Alexander and R.J. Block, Eds., Pergamon Press, New York, p. 57.

Synge, R.L.M., and Tiselius, A. (1950). Biochem. J. *46*, xii.

Szejtli, J. (1988). *Cyclodextrin Technology*, Kluwer, Boston.

Tabor, D., and Winterton, R.H.S. (1969). Proc. Roy. Soc. Lond. A*312*, 435.

Tanford, C. (1973, 1980). *The Hydrophobic Effect*, Wiley, New York.

Tanford, C. (1979). Proc. Natl. Acad. Sci. USA *76*, 4175.

Tippets, R.D., Mel, H.C., and Nichols, A.V. (1967). In: *Chemical Engineering in Medicine and Biology*, D. Hershey, Ed., Plenum Press, New York, p. 505.

Tiselius, A. (1937). Trans. Faraday Soc. *33*, 524.

Tiselius, A., and Flodin, P. (1953). Advan. Protein Chem. *8*, 461.

Tiselius, A., Hjertén, S., and Jerstedt, S. (1965). Arch. Gesamte Virusf. *17*, 512.

Towbin, H., Staehelin, T., and Gordon, J. (1979). Proc Natl. Acad. Sci. USA *76*, 4350.

Tsvetkov, T.D., Tsonev, L.I., Tsvetkova, N.M., Koynova, R.D., and Tenchov, B.G., (1989). Cryobiology *26*, 162.

Tuman V.S. (1963). J. Appl. Phys. *34*, 2014.

Uriel, J. (1966). Bull. Soc. Chim. Biol. *48*, 969.

Usui, S., and Sasaki, H. (1978). J. Colloid Interface Sci. *65*, 36.

Vaidhyanathan, V.S. (1982). J. Biol. Phys. *20*, 153.

Vaidhyanathan, V.S. (1985). Studia Biophys. *110*, 29.

Vaidhyanathan, V.S. (1986). In: *Proc. Symp. Bioelectrochem.*, Plenum Press, New York, p. 30.

Vaidhyanathan, V.S. (1988). *Interfacial Phenomena in Biotechnology and Materials Processing*, Y.A. Attia, B.M. Moudgil and S. Chander, Eds., Elsevier, Amsterdam, p. 27.

Valentine, R.C. and Green, N.M. (1967). J. Mol. Biol. *27*, 615.

van Blokland, P.H.G.M. (1977). *Direct Measurement of van der Waals Forces*, Ph.D. thesis, University of Utrecht.

van Blokland, P.H.G.M., and Overbeek, J.Th.G. (1979). J. Colloid Interface Sci. *68*, 96.

van Brakel, J., and Heertjes, P.M. (1975). Nature *254*, 585.

van den Winkel, P., Mertens, J., and Massare, D.L. (1974). Analyt. Chem. *46*, 1765.

Van der Ven, Th.G.M. (1989). *Colloidal Hydrohynamics*, Academic Press, London; New York.

Van der Ven, Th.G.M., Smith, P.G., Cox, R.G., and Mason, S.G. (1983). J. Colloid Interface Sci. *91*, 298.

Vanderhoff, J.W., Micale, F.J., and Krumrine, P.H. (1977). Separ. Purif. Meth. *5*, 61.

Vanderhoff, J.W., Micale, F.J., and Krumrine, P.H. (1979). In: Righetti *et al.*, 1979, p. 121.

Vanderhoff, J.W., Vitkuske, J.F., Bradford, E.B., and Alfrey, T. (1956). J. Polymer Sci. *20*, 225.

van der Waals, J.D. (1873). *Over de continiteit van den gas-en vloeistoftoestand*, dissertation, Leiden.

van der Waals, J.D. (1899). *Die Kontinuität des gasformigen und flüssigen Zustände*, Leipzig.

Van Lier, J.A. (1959). *The Solubility of Quartz*, Ph.D. dissertation, University of Utrecht.

van Oss, C.J. (1955a). *Recherches Physiques sur l'Utilisation de l'Ultrafiltration dans le Domaine de la Chimie Biologique*. Doctoral Thesis, University of Paris, pp. 43, 49.

van Oss, C.J. (1955b). *L'Influence de la Forme et de la Dimension des Molécules sur leur Mobilité en Electrophorèse*, second thesis, University of Paris.

van Oss, C.J. (1963). Science *139*, 1123.

van Oss, C.J. (1968). In: *Progress in Separation and Purification*, E.S. Perry, Ed., Vol. 1, Wiley/Interscience, New York, 187.

van Oss, C.J. (1971). Infect. Immun. *4*, 54.

van Oss, C.J. (1972). In: *Techniques of Surface and Colloid Chemistry and Physics*, R.J. Good, R.R. Stromberg, and R.L. Patrick, Eds., Marcel Dekker, New York, p. 213.

van Oss, C.J. (1975). Separ. Purif. Meth. *4*, 167.

van Oss, C.J. (1979). Separ. Purif. Meth. *8*, 119.

van Oss, C.J. (1982). Separ. Purif. Meth. *11*, 131.

van Oss, C.J. (1984). In: Atassi *et al.*, 1984, p. 361.

van Oss, C.J. (1985). J. Dispersion Sci. Tech. *6*, 139.

van Oss, C.J. (1986). Meth. Enzymol. *132*, 3.

van Oss, C.J. (1988). J. Dispersion Sci. Technol. *9*, 561.

van Oss, C.J. (1989a). Cell Biohhys. *14*, 1.

van Oss, C.J. (1989b). Abstr. 6th International Conf. on Partitioning in Aqueous Two-Phase Systems, Assmannshausen, Germany, p. 27.

van Oss, C.J. (1989c). J. Protein Chem. *8*, 661.

van Oss, C.J. (1990a). In: *Biophysics of the Cell Surface*, R. Glaser and D. Gingell, Eds., Springer-Verlag, New York, pp. 131–152.

van Oss, C.J. (1990b). Israel J. Chem. *30*, 251.

van Oss, C.J. (1990c). J. Molec. Recognition *3*, 128.

van Oss, C.J. (1990d). J. Protein Chem. *9*, 487.

van Oss, C.J. (1990e). J. Dispersion Sci. Tech. *11*, 491.

van Oss, C.J. (1991a). Biocolloids Biosurfaces *1* (J. Dispersion Sci. Tech. *12*, 201.

van Oss, C.J. (1991b). Biofouling *4*, 25.

van Oss, C.J. (1991c). Polymer Preprints *32*, 598.

van Oss, C.J. (1992a). In: *Structure of Antigens*, Vol. 1, M.H.V. Van Regenmortel, Ed., CRC Press, Boca Raton, FL., p. 99.

van Oss, C.J. (1992b). *Ibid.*, p. 179.

van Oss, C.J. (1992c). In: *Encyclopedia of Immunology*, I.M. Roitt and P.J. Delves, Eds., Saunders, Orlando, FL.

van Oss, C.J. (1993). In: *Water and Biological Macromolecules*, E. Westhof, Ed., Macmillan, London, p. 393.

van Oss, C.J. (1994a). In: *Cell Electrophoresis*, J. Bauer, Ed., CRC Press, Boca Raton, FL, p. 219.

van Oss, C.J. (1994b). Abstr. 207th Am. Chem. Soc. Mg., San Diego.

van Oss, C.J. (1994c). *Interfacial Forces in Aqueous Media*, Marcel Dekker, New York.

van Oss, C.J. (1997). J. Molec. Recognition *10*, 203.

van Oss, C.J. (1999). Internatl. J. Biochromatog. *4*, 139.

van Oss, C.J. (2003). J. Molec. Recognition *16*, 177.

van Oss, C.J., and Absolom, D.R. (1985). J. Dispersion Sci. Technol. *6*, 131.

van Oss, C.J., Absolom, D.R., and Bronson, P.M. (1982). Immunol. Commun. *11*, 139.

van Oss, C.J., Absolom, D.R., and Michaeli, I. (1985). Immunol. Invest. *14*, 167.

van Oss, C.J., Absolom, D.R., and Neumann, A.W. (1979a). Separ. Sci. Tech. *14*, 365.

van Oss, C.J., Absolom, D.R., and Neumann, A.W. (1983). Ann. NY Acad. Sci. *416*, 332.

van Oss, C.J., Absolom, D.R., and Neumann, A.W. (1984). In: *The Reticuloen-dothelial System*, Vol. 7a, S.M. Reichard and J.P. Filkins, Eds., Plenum, New York, p. 3.

van Oss, C.J., Absolom, D.R., Neumann, A.W., and Zingg, W. (1981). Biochimica Biophysica Acta. *670*, 64.

van Oss, C.J., Arnold, K., and Coakley, W.T. (1990a). Cell Biophys. *17*, 1.

van Oss, C.J., Arnold, K., Good, R.J., Gawrisch, K., and Ohki, S. (1990b). J. Macromolec. Sci. Chem. *A27*, 563.

van Oss, C.J., and Bartholomew, W. (1980). In: *Methods in Immuno-diagnosis*, N.R. Rose and P.E. Bigazzi, Eds., Wiley, New York, p. 65.

van Oss, C.J., and Beyrard, N.R. (1963). J. Chim. Phys. *60*, 648.

van Oss, C.J., Beyrard, N.R., de Mende, S., and Bonnemay, M. (1959). Comptes Rendus Acad. Sci., Paris, *248*, 223.

van Oss, C.J., Bigazzi, P.E., and Gillman, C.F. (1974). Proc. 12th AIAA Aerospace Sci. Mg., Washington, DC, AIAA paper No. 74–211, AIAA Library, New York.

van Oss, C.J., and Bronson, P.M. (1979). In: Righetti *et al.*, 1979, p. 251.

van Oss, C.J., and Buenting, S.B. (1967). Transfusion *7*, 77.

van Oss, C.J., Chaudhury, M.K., and Good, R.J. (1987a). Advan. Colloid Interface Sci. *28*, 35.

van Oss, C.J., Chaudhury, M.K., and Good, R.J. (1987b). Separ. Sci. Technol. *22*, 1515.

van Oss, C.J., Chaudhury, M.K., and Good, R.J. (1988a). Chem. Rev. *88*, 927.

van Oss, C.J., Chaudhury, M.K., and Good, R.J. (1988b). In: *Molecular Mechanics of Membrane Fusion*, S. Ohki, D. Doyle, T.D. Flanagan, S.W. Hui and E. Mayhew, Eds., Plenum Press, New York, p. 113.

van Oss, C.J., Chaudhury, M.K., and Good, R.J. (1989). Separ. Sci. Tech. *24*, 15.

van Oss, C.J., and Cixous, N. (1963). Bull. Soc. Chim. Biol. *45*, 1211.

van Oss, C.J., and Coakley, W.T. (1988). Cell Biophys. *13*, 141.

van Oss, C.J., and Costanzo, P.M. (1992). J. Adhesion Sci. Tech. *6*, 477.

van Oss, C.J., Docoslis, A. and Giese, R.F. (2001). Colloids Surfaces-B:Biointerfaces *22*, 285.

van Oss, C.J., Docoslis, A., Wu, W. and Giese, R.F. (1999). Colloids Surfaces-B: Biointerfaces *14*, 99.

van Oss, C.J., and Fike, R.M. (1979). In: Righetti *et al.*, 1979, p. 111.

van Oss, C.J., Fike, R.M., Good, R.J., and Reinig, J.M. (1974). Analyt. Biochem. *60*, 242.

van Oss, C.J. and Giese, R.F. (1995). Clays Clay Miner. *42*. 474.

van Oss, C.J. and Giese, R.F. (2003). J. Dispersion Sci. Technol. *24*, 363.

van Oss, C.J. and Giese, R.F. (2004). J. Dispersion Sci. Technol. *25*, 631.

van Oss, C.J. and Giese, R.F. (2005). J. Adhesion *81*, 237.

van Oss, C.J., Giese, R.F., Bronson, P.M., Docoslis, A., Edwards, P. and Ruyechan, W.T. (2003). Colloids Surfaces-B:Biointerfaces *30*, 25.

van Oss, C.J., Giese, R.F., and Costanzo, P.M. (1990). Clays Clay Miner. *38*, 151.

van Oss, C.J., Giese, R.F. and Docoslis, A. (2004). J. Biol. Phys. Chem. *4*, 145.

van Oss, C.J., Giese, R.F. and Good, R.J. (2002). J. Dispersion Sci. Technol. *23*, 455.

van Oss, C.J., Giese, R.F., Li, Z., Murphy, K., Norris, J., Chaudhury, M.K., and Good, R.J. (1992). J. Adhesion Sci. Tech. *6*, 413.

van Oss, C.J., Giese, R.F., and Norris, J. (1992). Cell Biophys. *20*, 253.

van Oss, C.J., Giese, R.F., Wentzek, R., Norris, J., and Chuvilin, E.M. (1992). J. Adhesion Sci. Tech. 6, 503.

van Oss, C.J., Giese, R.F., and Wu, W. (1993). Abstr. 67th Colloid and Surface Science Symp., Toronto, No. 217.

van Oss, C.J., Giese, R.F. and Wu, W. (1998). J. Dispersion Sci. Technol. 19, 1221.

van Oss, C.J., and Gillman, C.F. (1972a). J. Reticuloendothelial Soc. 12, 283.

van Oss, C.J., and Gillman, C.F. (1972b). J. Reticuloendothelial Soc. 12, 497.

van Oss, C.J., Gillman, C.F., and Good, R.J. (1972). Immunol. Commun. 1, 627.

van Oss, C.J., Gillman, C.F., and Neumann, A.W. (1975). Phagocytic Engulfment and Cell Adhesiveness, Marcel Dekker, New York.

van Oss, C.J., and Good, R.J. (1984). Colloids Surfaces 8, 373.

van Oss, C.J., and Good, R.J. (1988a). J. Protein Chem. 7, 179.

van Oss, C.J., and Good, R.J. (1988b). J. Dispersion Sci. Tech. 9, 355.

van Oss, C.J., and Good, R.J. (1989). J. Macromolec. Sci. Chem. A26, 1183.

van Oss, C.J., and Good, R.J. (1991a). J. Dispersion Sci. Tech. 12, 95.

van Oss, C.J., and Good, R.J. (1991b). J. Dispersion Sci. Tech. 12, 273.

van Oss, C.J., and Good, R.J. (1992). Langmuir 8, 2877.

van Oss, C.J. and Good, R.J. (1996). J. Dispersion Sci. Technol. 17, 433.

van Oss, C.J., Good, R.J., and Busscher, H.J. (1990). J. Dispersion Sci. Technol. 11, 75.

van Oss, C.J., Good, R.J., and Chaudhury, M.K. (1986a). J. Chromatog. 376, 111.

van Oss, C.J., Good, .RJ., and Chaudhury, M.K. (1986b). J. Colloid Interface Sci. 111, 378.

van Oss, C.J., Good, R.J., and Chaudhury, M.K. (1986c). J. Protein Chem. 5, 385.

van Oss, C.J., Good, R.J., and Chaudhury, M.K. (1987c). Separ. Sci. Technol. 22, 1.

van Oss, C.J., Good, R.J., and Chaudhury, M.K. (1987d). J. Chromatog. 391, 53.

van Oss, C.J., Good, R.J., and Chaudhury, M.K. (1988c). Langmuir 4, 884.

van Oss, C.J., Good, R.J., Neumann, A.W., Wieser, J.D., and Rosenberg, P. (1977). J. Colloid Interface Sci. 59, 505.

van Oss, C.J., Ju, L., Chaudhury, M.K., and Good, R.J. (1989). J. Colloid Interf. Sci. 128, 313.

van Oss, C.J., and Mohn, J.F. (1970). Vox Sang. 19, 432.

van Oss, C.J., Mohn, J.F., and Cunningham, R.K. (1978). Vox Sang. 34, 351.

van Oss, C.J., Moore, L.L., Good, R.J., and Chaudhury, M.K. (1985). J. Protein Chem. 4, 245.

van Oss, C.J., Omenyi, S.N., and Neumann, A.W. (1979b). Colloid Polymer Sci. 257, 737.

van Oss, C.J., Roberts, M.J., Good, R.J., and Chaudhury, M.K. (1987e). Colloids Surfaces 23, 369.

van Oss, C.J., and Singer, J.M. (1966). J. Reticuloendothelial Soc. 3, 29.

van Oss, C.J., and Singer, J.M. (1966). J. Colloid Interface Sci. 21, 117.

van Oss, C.J., Smeenk, R.J.T., and Aarden, L.A. (1985). Immunol. Invest. 14, 245.

van Oss, C.J., Wu, W., Docoslis, A. and Giese, R.F. (2001). Colloids Surfaces-B: Biointerfaces 20, 87.

van Oss, C.J., Wu, W. and Giese, R.F. (1995). In: Proteins at Interfaces II, T.A. Horbett and J.L. Brash, Eds., ACS Symposium Series 602, American Chemical Society, Washington, DC, p. 80.

van Oss, C.J., Wu, W., Giese, R.F. and Naim, J.O. (1995). Colloids Surfaces-B: Biointerfaces 4, 185.

van Silfhout, A. (1966). Dispersion Forces Between Macroscopic Objects, dissertation, University of Utrecht.

van Thoor, T.J.W., Ed. (1968). Materials and Technology, Vol.1, Longmans, London, p. 86 / Chemical Technology, Vol. 1, Barnes and Noble, New York, p. 86.

van Wagenen, R.A., Andrade, J.D., and Hibbs, J.B. (1976). J. Electrochem. Soc. 123, 1438.

Verwey, E.J.W., and Overbeek, J.Th.G. (1946). Trans. Faraday Soc. *42B*, 117.

Verwey, E.J.W., and Overbeek, J.Th.G. (1948). *Theory of the Stability of Lyophobic Colloids*, Elsevier, Amsterdam.

Vesterberg, O. (1968). Sv. Kem. Tidskr. *80*, 213.

Vesterberg, O. (1969). Acta Chim. Scand. *23*, 2653.

Vesterberg, O. (1976). In: Catsimpoolas, 1976, p. 53.

Vincent, B. (1984). ACS Symp. Series *240*, 3.

Visser, J. (1968). Rep. Progr. Appl. Chem. *53*, 714.

Visser, J. (1970). J. Colloid Interface Sci. *34*, 26.

Visser, J. (1972). Advan. Colloid Interface Sci. *3*, 331.

Visser, J. (1976a). J. Colloid Interface Sci. *55*, 664.

Visser, J. (1976b). Surface Colloid Sci. *8*, 3.

von Boehmer, H., Shortman, K., and Nossal, G.J.V. (1974). J. Cell Physiol. *83*, 231.

von Hevesy, G. (1913). Physik. Z. *14*, 49.

von Hevesy, G. (1917). Kolloid Z. *21*, 129.

Vonnegut, B. (1942). Rev. Sci. Instr. *13*, 6.

von Smoluchowski, M. (1917). Z. phys. Chem. *92*, 129. (Published in 1918).

von Smoluchowski, M.Z. (1921). In: *Handbuch der Electrizität*, Vol. II, L. Graetz, Ed., Barth, Leipzig, p. 374.

Voorn, M.J. (1956). Rec. Trav. Chim. Pays-Bas, *75*, 317, 405, 427, 925.

Vrij, A. (1968). J. Polymer Sci.-Part A-2 *6*, 1919.

Vroman, L., and Adams, A.L. (1987). ACS Symp. Series *343*, 154.

Vroman, L., Adams, A.L., Klings, M., and Fischer, G. (1975). Advan. Chem. Series *145*, 255.

Walport, M. (1989). In: *Immunology*, I.M. Roitt, J. Brostoff and D.K. Male, C.V. Mosby, St. Louis, MO, p. 113.

Walter, H., Brooks, D.E., and Fisher, D., Eds., (1985). *Partitioning in Aqueous Two-Phase Systems*, Academic Press, New York.

Ward, C.A. (1987). ACS Symp. Series *343*, 551.

Washburn, E.W., Ed. (1928). *International Critical Tables*. McGraw-Hill, New York, pp. 466, 467.

Weast, R.C., Ed. (1970/71). *Handbook of Chemistry and Physics*, *51st Edition*, Chemical Rubber Co., Cleveland, OH.

Weber, K., and Osborn, M. (1969). J. Biol. Chem. *244*, 4406.

Weiss, L. (1960). *Internatl. Rev. Cytology*, Vol. 9, G.H. Bourne and J.F. Danielli, Eds., Academic Press, New York, p. 187.

Weiss, L. (1961). Exptl. Cell Res. Suppl. *8*, 141.

Wenzel, R.N. (1949). J. Phys. Colloid Chem. *53*, 1466.

White, L.R. (1982). J. Colloid Interface Sci. *90*, 536.

White, L.R., Israelachvili, J.N., and Ninham, B.W. (1976). J. Chem. Soc. Faraday Trans. I *72*, 2526.

Wiersema, P.H. (1964). *On the Theory of Electrophoresis*, dissertation, University of Utrecht.

Wiersema, P.H., Loeb, A.L., and Overbeek, J.Th.G. (1966). J. Colloid Interface Sci. *22*, 78.

Wilson, M., Ed. (2002). *Bacterial Adhesion to Host Tissues*. Cambridge University Press, Cambridge, UK.

Wilson, M. and Devine, D., Eds. (2003). *Medical Implications of Biofilms*. Cambridge University Press, Camvridge, UK.

Winzler, R.J. (1972). In: *Glycoproteins*, Part B, Elsevier, Amsterdam, p. 1268.

Wood, J. and Sharma, R. (1995), Langmuir *11*, 4797.

Wu, W., Giese, R.F. and van Oss, C.J. (1994-a). Colloids Surfaces-A: Physicochemical and Engineering Aspects *89*, 241.

Wyman, J. (1931). J. Am. Chem. Soc. *53*, 3292.

Yalkowsky, S.H. and Banerjee, S. (1992). *Aqueous Solubility—Methods of Estimation for Organic Compounds*. Marcel Dekker, New York.

Yaminsky, V.V., Yushchenko, V.S., Amelina, E.A., and Shchukin, E.D. (1983). J. Colloid Interface Sci. *96*, 301.

Yang, D.S.C., Sax, M., Chakrabartty, A., and Hew, C.L. (1988). Nature *333*, 252.

Yang, X. (1990). *Middle Phase Microemulsion Study*, dissertation, SUNY, Buffalo.

Young, T. (1805). Phil. Trans. Roy. Soc. *95*, 65.

Yushchenko, V.S., Yaminsky, V.V., and Shchukin, E.D. (1983). J. Colloid Interface Sci. *96*, 307.

Zaslavsky, B.Y. (1995). *Aqueous Two-Phase Partitioning*. Marcel Dekker, New York.

Zeiller, K., Pascher, G., Wagner, G., Liebich, H.G., Holzberg, E., and Hannig, K. (1974). Immunology *26*, 995.

Zisman, W.A. (1964). Advan. Chem. Ser. *43*, 1.

Zucker-Franklin, D., Greaves, M.F., Grossi, C.E., and Marmont, A.M., Eds. (1988). *Atlas of Blood Cells–Function and Pathology*, Lea & Febiger, Philadelphia.

Zukoski, C.F., and Saville, D.A. (1986). J. Colloid Interface Sci. *114*, 32.

Index

Adhesion and adsorption in aqueous media, 382–383
  adsorption
    blotting, 347–349
    and denaturation/hysteresis, 356–357
    liquid chromatography, 357–359
    protein adsorption, 349–350, 351t, 352–354, 353f
    and protein configuration changes, 354–356
    size of solute as factor, 347
  cell adhesion, 360
    and cell fusion, 366–368
    and cell shape, 363
    influence of protein adsorption, 360–361
    microbial adhesions, 363–364
    negative/positive and microbial pathogenicity, 364–365
    opsonization and phagocytosis, 365–366
    prevention of, 361–362
    to high-/low-energy surfaces, 361, 363
  cell freezing/negative and positive adhesion to advancing ice fronts, 368–371, 369t
  interaction of two different materials, 345–346
  macro and microscopic-scale interactions, 236–347
  See also Ligand receptor interactions
Adhesion methods (particle adherence or adsorption of polymers), 176–177
Adsorption and adhesion in aqueous media. *See* Adhesion and adsorption in aqueous media

Adsorption (protein) onto metal oxide surfaces. *See* Protein adsorption onto metal oxide surfaces/kinetics and energetics of
Advancing solidification fronts, 174–175
  "critical velocity" concept, 175
Air-hysteresis, and contact angle measurement, 134–135
Amphiphilic compounds, and surface tension of water, 121–122
Apolar interactions, 45, 84
  apolar and polar surface tension component of liquids, 142–143
  apolar surface tension component and Hamaker constant, 19–24
  contributions to hydrophobic interactions, 231–234
  and Hildebrand's solubility parameter, 297–298
  mechanism of polymer phase separation in apolar media, 283
  solubility of apolar polymers, 300–301
Aqueous solubility approach (interfacial tension determination), 160, 161t, 162
  and surfactants, 162–164
Atomic force microscopy (AFM), 210
Average ℓ_0 value, 20–22
Bipolymer separation, use of aqueous two-phase systems, 285
Biopolymers, insolubilization of, 309–311
Blood, surface tension properties at 0°, 96–97
Blotting
  "Southern blotting," 347–349
  "Western blotting," 349
Brooks effect, 67–68

Printed in the United States
by Baker & Taylor Publisher Services